W0235176

LARGE SCALE STRUCTURE AND MOTIONS
IN THE UNIVERSE

ASTROPHYSICS AND
SPACE SCIENCE LIBRARY

A SERIES OF BOOKS ON THE RECENT DEVELOPMENTS
OF SPACE SCIENCE AND OF GENERAL GEOPHYSICS AND ASTROPHYSICS
PUBLISHED IN CONNECTION WITH THE JOURNAL
SPACE SCIENCE REVIEWS

VOLUME 151

PROCEEDINGS

LARGE SCALE STRUCTURE AND MOTIONS IN THE UNIVERSE

PROCEEDING OF AN INTERNATIONAL MEETING
HELD IN TRIESTE, ITALY, APRIL 6-9, 1988

Edited by

M. MEZZETTI

G. GIURICIN

F. MARDIROSSIAN

Dipartimento di Astronomia, Università degli Studi di Trieste, Italy

and

M. RAMELLA

Osservatorio Astronomico di Trieste, Italy

KLUWER ACADEMIC PUBLISHERS

DORDRECHT / BOSTON / LONDON

Library of Congress Cataloging in Publication Data

Large scale structure and motions in the universe : international
  meeting held in Trieste, Italy, April 6-9, 1988 / edited by M.
  Mezzetti ... [et al.].
        p.   cm. -- (Astrophysics and space science library)
  Sponsored by the European Physical Society, Astronomical Division.
  Includes indexes.

    1. Cosmology--Congresses. 2. Mechanics, Celestial--Congresses.
  I. Mezzetti, M. (Marino) II. European Physical Society. Astronomy
  and Astrophysics Division. III. Series.
  QB980.L36   1989
  523.1--dc19   88-39015

ISBN-13: 978-94-010-6895-6   e-ISBN-13: 978-94-009-0903-8
DOI: 10.1007/978-94-009-0903-8

Published by Kluwer Academic Publishers,
P.O. Box 17, 3300 AA Dordrecht, The Netherlands.

Kluwer Academic Publishers incorporates
the publishing programmes of
D. Reidel, Martinus Nijhoff, Dr W. Junk and MTP Press.

Sold and distributed in the U.S.A. and Canada
by Kluwer Academic Publishers,
101 Philip Drive, Norwell, MA 02061, U.S.A.

In all other countries, sold and distributed
by Kluwer Academic Publishers Group,
P.O. Box 322, 3300 AH Dordrecht, The Netherlands.

TABLE OF CONTENTS

POSTER PAPERS (in alphabetical order):

PREFACE

The 1980's have been times of great excitement in Astrophysics and Cosmology. Professors Dennis Sciama and Fabio Mardirossian and all the other Members of the Organizing Committees are to be congratulated for having given us a taste of this excitement in Trieste, by inviting the leaders of the subject to the meeting they have organized.

The excitement has come from the new observations of the three-dimensional structure of the universe through a large number of new measurements of redshifts. These have revealed that clusters of galaxies are distributed on the surface of big empty bubbles of diameters of the order of 20-50 Mpc. Additionally, there is some evidence for invisible dark matter (whose composition is not known) as well as evidence for the gravitational lens effect. To cap this has come the supernova of 1987, an event which last occurred 383 years ago. For the first time in history, the neutrino flux from the supernova was measured, giving limits to neutrino masses and numbers of neutrino types. (The dark matter problem is related to Particle Physics – beyond this standard model). It is good to be alive when all this happens and to try to comprehend this. Once again, our appreciation to the organisers and to those who presented their beautiful results.

Abdus Salam

Director
International Centre
for Theoretical Physics
Trieste

FOREWORD

The large-scale structure and motions in the Universe, and the presence of Active Galactic Nuclei inside it, have been rapidly developing subjects, specially in the eighties, and one of the most important research fields in astrophysics.

The meeting held in Trieste (Italy) from April 6th to 9th, 1988 is the third international meeting organized in Trieste devoted to these topics. In fact, it follows the meeting on "Clusters and Groups of Galaxies" held in September 1983 and that on the "Structure and Evolution of Active Galactic Nuclei" in April 1985.

The meeting was attended by about 140 participants from 19 nations.

The Scientific Organizing Committee consisted of Profs. J. Audouze, A. Cavaliere, G. Chincarini, G. Ellis, M. Geller, M. Hack, P. Osmer, M. Rees, D. Sciama (Chairman), G. Setti, F. Shandarin, G. Tammann. We acted as members of the Local Organizing Committee, together with Prof. M. Abramowicz.

The meeting was sponsored by the European Physical Society (Astronomical Division). It was organized and supported by the Dipartimento di Astronomia of the Universita' degli Studi di Trieste, the Consiglio Nazionale delle Ricerche - Gruppo nazionale di Astronomia, the International Centre for Theoretical Physics of Trieste, the International School for Advanced Studies of Trieste, the Osservatorio Astronomico di Trieste. It was also supported by the Commissariato del Governo nella Regione Friuli-Venezia Giulia, the Comune di Trieste, Informatica Friuli-Venezia Giulia SpA of Trieste, and Regione Autonoma a Statuto Speciale Friuli-Venezia Giulia.

We are very grateful for all the support we

received.   In   particular   we   wish   to   thank Prof.
P. Fusaroli, Rector of the University of   Trieste,   for   his
hearty   welcome,   and   Prof.   D. Amati (Director od the
International School for Advanced Studies of Trieste), Prof.
M. Hack (Director of the Department of Astronomy of the
University of Trieste), Prof.   A. Salam (Director of the
International Centre for Theoretical Physics of Trieste),
Prof.   G. Sedmak (Director of the Trieste Astronomical
Observatory), and the staffs of their institutions for their
help, which greatly contributed to the success of the
meeting.

Unforfunately the texts of the invited lectures by
Profs.   N. Bahcall   and   F. Melchiorri   have   not   arrived.
Moreover, we have collected   only   the   abstracts   of   those
posters   which   exceeded   the   length   requested by the
Publisher. All questions and answers have been typed from
the written version we have received from Questioners and
Authors. In this way we have been able to collect about 70%
of the discussion.

Editors

Marino Mezzetti
Giuliano Giuricin
Fabio Mardirossian
Massimo Ramella

WELCOME ADDRESSES

Gentili Signore, Signori, ill.mo professor Salam,

ho il piacere e l'onore di portare a questo autorevolissimo Convegno il saluto augurale del Consiglio della Regione Autonoma Friuli-Venezia Giulia e del suo Presidente e di esprimere il piu' cordiale benvenuto in questa meravigliosa citta' (credo di poterlo dire anche a nome del rappresentante del Sindaco), che della Regione e' il capoluogo.

E' significativo che incontri sempre piu' impegnati e di altissimo livello si svolgano a Trieste, che – oltre ad un parallelo potenziamento in chiave di innovazione tecnologica della sua struttura industriale pubblica – nel suo futuro vede a ragione un crescente sviluppo della dimensione scientifica e della ricerca.

Il tema dell'odierno Convegno e' particolarmente affascinante – e la prof.ssa Hack ce lo racconta spesso con le sue conferenze altrettanto affascinanti – perche' porta voi studiosi, voi scienziati ad indagare – attraverso esperienze le piu' diverse – il nostro mondo, questo universo ancora sconosciuto nei suoi possibili confini, per capire e per far conoscere a noi, profani, la vita nelle sue infinite sfaccettature.

Nell'augurarvi ancora un ottimo soggiorno a Trieste, sono certo che lavorerete con orizzonti sempre piu' aperti ai grandissimi ideali della pace e della cooperazione internazionale, di cui questa nostra umanita' ha tanto bisogno.

Claudio Tonel

Vice Presidente
Regione Autonoma Friuli-Venezia Giulia

Authorities, Rector, Ladies and Gentlemen, Dear Colleagues,

It is a great pleasure for me to welcome you. On this occasion I wish to thank all those institutions which have sponsored and given their contribution to the organization of this meeting. They are the Municipality of Trieste, the Government of the Regione Friuli-Venezia Giulia, the National Group of Astronomy of the National Council of Research, the European Society of Physics, the Informatica Friuli-Venezia Giulia, the International Centre of Theoretical Physics, the University of Trieste, the International School for Advanced Study of Trieste, and the Astronomical Observatory of Trieste.

As a stellar spectroscopist, used to studying very peculiar but relatively nearby objects, I am here for learning about the new results on the large scale structure of the Universe.

I think that, after the 60's, when a new era in cosmology started with the discovery of quasars in 1963 and of the background radiation at 3 K in 1965, these 80's have again been a period of exciting discoveries and intriguing problems, and an unexpected picture of the universe is now becoming more precise.

In the standard cosmological models the key events in the evolution of the Universe are supposed to have taken place within 10-25 sec of the Big Bang and thus are empirically unverifiable. However, many other problems can be attacked observationally. None of the standard cosmological models predicted the large-scale streaming of galaxies or the existence of structures on scales larger than 100 Mpc. The very existence of galaxies has never been adequately explained by any theory of cosmology, astrophysics or particle physics. It is unclear if the great attractor producing the stream of galaxies is simply an extraordinarily large supercluster or if it is a region of clumped dark matter.

Hence we face the problems of the distribution of galaxies in space, the large-scale motions and the nature of the great attractor, the puzzle of the dark matter, the origin of the X-ray background, the reality of the gravitational lenses and the origin of the recently discovered giant arcs, to mention just a few topics.

The marvellous technological progress of telescopes and, especially, the increased sensitivity of the detectors, the results from X-ray, UV and IR satellites have permitted us to give some answers to old problems, but have posed new questions and new problems. In the next ten years when the several new-technology-telescopes now planned or under construction and the Space Telescpe will, hopefully, start observing, we will certainly find some answers to our present problems, but also new surprises and new enigmas whose solution will have to wait for the astrophysicists of the next century.

I hope that this meeting, which is attended by some of the researchers who have given very large contributions to sketching the new picture of the Universe, will be useful to everybody, and I wish you a pleasant stay in Trieste.

Margherita Hack

Director
Dipartimento di Astronomia
Universita' degli Studi di Trieste

LIST OF PARTICIPANTS

ABRAMOWICZ, M., International School for Advanced Studies, Strada Costiera 11, 34014 Trieste, Italy
AMENDT, P., International School for Advanced Studies, Strada Costiera 11, 34014 Trieste, Italy
ANTONUCCIO-DELOGU, V.,International School for Advanced Studies, Strada Costiera 11, 34014 Trieste, Italy
BAHCALL, N., Space Telescope Science Institute Homewood Campus, Baltimore, U.S.A.
BALKOWSKI, C., Paris Observatory, DAEC, UA 173, Paris VII University F92195 Meudon Principal Cedex, France
BARCONS, X., Departamento de Fisica Moderna, Universidad de Cantabria, 39005 Santander, Spain
BATTANER, E., Universidad de Granada, Dpto Fisica Moderna, Granada, Spain
BERGVALL, N., Astronomiska Observatoriet, Box 515, S-751 20 Upsala, Sweden
BERTOLA. F., Dipartimento di Astronomia, Universita' di Padova, dell'Osservatorio 5, 35122 Padova, Italy
BINGGELI, B., Astronomiches Institut der Universitaet Basel, Binningen, Switzerland
BIVIANO, A., Dipartimento di Astronomia, Universita' di Trieste, Via G. B. Tiepolo 11, Trieste, Italy
BOERNER, G., MPI fuer Astrophysik, Karl-Schwarzschildstr. 1, D-8046 Garching bei Muenchen, Fed. Rep. of Germany
BOLDT, E., Laboratory for High Energy Astrophysics, Code 666, NASA Goddard Space Flight Center, Greenbelt, MD 20771, U.S.A.
BOSMA, A., Observatoire de Marseille, 2 Place Le Verrier, 13248 Marseille Cedex 04, France
BROSCH, N., Tel Aviv University, Ramat Aviv, Tel Aviv 69978, Israel
BURSTEIN, D., Arizona State University, Dept. of Physics, Tempe, Arizona, U.S.A.
BUSON, L., Osservatorio Astronomico, Vicolo dell'Osservatorio 5, 35122 Padova, Italy
BUZZONI, A., Osservatorio Astronomico di Brera, Via Brera 28, 20121 Milano, Italy
CALISSE, P., Dipartimento di Fisica, Universita' "La Sapienza", P.le A. Moro 2, 00185 Roma, Italy
CALVANI, M., International School for Advanced Studies Strada Costiera 11, 34014 Trieste, Italy
CAVALIERE, A., Dipartimento di Fisica, II Universita' degli

Studi di Roma, Via O. Raimondo, 00173 (La Romanina) Roma, Italy

CAYATTE, V., Observatoire de Paris, DAEC, 92195 Meudon Principal Cedex, France

CHU, Y., Center for Astrophysics, University of Science and Technology of China, Hefei, Hanhui, China

CLOWES, R., Royal Observatory, Blackford Hill, Edinburg EH9 3Hj, United Kingdom

COLAFRANCESCO, S., Dipartimento di Astronomia, Universita' di Padova, Vicolo dell'Osservatorio 5, 35122 Padova, Italy

COLE, S., Institute of Astronomy, Madinglay Road, Cambridge, CB3OHA, United Kingdom

COLEMAM, P., Kapteyn Laboratorium, Groningen, The Netherlands

COUCHMAN, H.M.P., Canadian Institute for Theoretical Astrophysics, University of Toronto, Toronto, ON M5S 1A1, Canada

DAVIES, R., Kitt Peak National Obs., P.O.Box 26732, Tucson, Arizona 85726, U.S.A.

de BERNARDIS, P., Dipartimento di Fisica, Universita' "La Sapienza" P.le A.Moro 2, 00185 Roma, Italy

DE LAPPARENT, V., Center for Astrophysics, 60 Garden Street Cambridge MA 02138, U.S.A.

DE ZOTTI, G., Padova Astronomical Observatory, Vicolo dell'Osservatorio 5, 35122 Padova, Italy

DOMINGUEZ-TENEIRO, R., Dpto di Fisica Teorica, Universidad Autonoma de Madrid, UAM Cantoblanco 28049 Madrid, Spain

DUBREUIL, D., C.E.A., CEN Saclay, 91191 Gif-Sur-Yvette Cedex, France

ELLINGSON, E., University of Arizona, Steward Observatory, Tucson, AZ 85721, U.S.A.

ELLIS, G., International school for Advanced Studies, Strada Costiera 11, 34014 Trieste, Italy

EVRARD, A.E., Institute of Astronomy, Madingley Road, Cambridge GB3 OHA, United Kingdom

FAIRALL,A. P., Department of Astronomy, University of Cape Town, Rondebosch, 7700 South Africa

FLORIDO, E., Universidad de Granada, Dpto. Fisica Moderna, Granada, Spain

FOCARDI, P., Dipartimento di Astronomia, Casella Postale 596, 40100 Bologna, Italy

FRANCESCHINI, A., Dipartimento di Astronomia, Universita' di Padova, Vicolo dell'Osservatorio 5, 35122 Padova, Italy

GELLER, M. J., Smithsonian Astrophysical Obs. Cambridge, Mass., U.S.A.

GERBAL, D., Department d'Astrophysique Extragalactique et de Cosmologie, Obsertatoire de Paris, 92195 - Meudon Principal Cedex, France

GIOIA, I.M., Harvard-Smithsonian Center for Astrophysics, 60

Garden st., Cambridge MA 02138, also from Istituto di Radioastronomia, CNR, Via Irnerio 46, 40126 Bologna, Italy

GIOVANNINI, G., Istituto di Radioastronomia CNR, via Irnerio 46, 40126 Bologna, Italy

GIURICIN, G., Dipartimento di Astronomia, Universita' di Trieste, Via G. B. Tiepolo 11, 34131 Trieste, Italy

GOICOECHEA, L.J., Departamento di Fisica Moderna Universidad de Cantabria, Avda. de los Castros, s/n 39005 Santander, Spain

GONDOLO, P., Dipartimento di Astronomia, Universita' di Trieste, Via G. B. Tiepolo 11, 34131 Trieste, Italy

GRABINSKA, T., Institute of Physics, Wroclaw Technical University, Wroclaw, 50-370 Poland

GUDEHUS, D.H., Randall Laboratory, University of Michigan, Ann Arbor, MI 48109, U.S.A.

HACK M., Dipartimento di Astronomia, Universita' di Trieste, Via G. B. Tiepolo 11, 34131 Trieste, Italy

HAMMER, F., Observatoire de Meudon-DAEC, 92195 Meudon Principal Cedex, France

HAUBOLD, H. J., Central Institute for Astrophysics, DDR 1591 Potsdam-Babelsberg, Fed. Rep. of Germany

HENNING, P.A., Astronomy Program, University of Maryland, College Park, MD 20742, U.S.A.

HICKSON, P., Dept. Geophysics and Astronomy, 2219 Main Mall, Vancover, BC, Canada

HOFFMAN, G.L., Department of Physics, Lafayette College, Easton, PA 18042, U.S.A.

HOWARD, K. C.,

HUCHTMEIER, W.K., Max-Planck-Institut fuer Radioastronomie, Auf dem Hugel 69, 5300 Bonn, Fed. Rep. of Germany

IKEUCHI, S., Tokyo Astronomical Observatory, Mitaka, Tokio 181, Japan

IOVINO, A., European Southern Observatory, Karl-Schwarzschild Str. 2, D-8046 Garching bei Muenchen, Fed. Rep. of Germany

KAISER, N., Institute of Astronomy, Madingley Rd., CB3 0HA, Cambridge, U.S.A.

KALINKOV, M., Dept.to of Astronomy, Bulg. Acad. Sci., blvd Lenin 72, Sofia 1784, Bulgaria

KALLOGLIAN, A.T., Byurakan Astrophysical Observatory, Armenia, U.S.S.R.

KERR, F.J., Astronomy Program, University of Maryland, College Park, MD 20742, U.S.A.

KOLLATSCHNY, W., Universitats-Sternwarte Gottingen, Geismar-landstr. 11, Gottingen, Fed. Rep. of Germany

KRAAN-KORTEWEG, R. C., Astronomisches Institut de Universitat Basel, Venusstr. 7, Ch-4102 Binningen, Switzerland

KRUSZEWSKI, A., Warsaw University Observatory, Al. Ujazdowskie 4, 00-478 Warsaw, Poland

KUNEVA, I., Dept.to of Astronomy, Bulg. Acad. Sc., blvd

Lenin 72, Sofia 1784, Bulgaria

LAHAV, O., Institute of Astronomy, Madingley Road, Cambridge CB3 0HA, United Kingdom

LANGSTON, G., Max Planck Institute fuer Radioastronomia, Auf dem Hugel 69, 5300 Bonn-1, Fed. Rep. of Germany

LE FEVRE, O., Canada-France-Hawaii Telescope Corporation, P.O. Box 1597, Kamuela, HI96743, USA and DAEC, Observatoire de Paris- Meudon, 92195 Meudon, France

LEORAT, J., DAEC, UA 173, Observatoire de Paris-Meudon, F-92195 Meudon Principal Cedex, France

LILIJE, B., Institute of Astronomy, Madingley Road, Cambridge CB3 OHA, United Kingdom

LINDER, E.V., Max-Planck-Institut fur Astrophysik, 8046 Garching bei Muenchen, Fed. Rep. of Germany

LUCCHIN, F., Dipartimento di Fisica "G. Galilei", Via Marzolo 8, 35100 Padova, Italy

LUKASH, V. N., Space Research Institute, Profsoyuznaja 84/32, Moscow 117810, U.S.S.R.

MACGILLIVRAY, H.T., Royal Observatory Edinburg, Blackford Hill, Edinburg EH9 3HJ, United Kingdom

MANDELBROT, B., IBM Yorktown Labs, P.O. Box 218, Yorktown Heights, NY 10598, U.S.A., and Mathematics Department, Yale University, New Haven CT 06520, U.S.A.

MARASCHI, L., Dipartimento di Fisica, Universita' di Milano, Via Celoria 16, I-20133 Milano, Italy

MARDIROSSIAN, F., Dipartimento di Astronomia, Universita' di Trieste, Via G.B. Tiepolo 11, 34131 Trieste, Italy

MARTIN-MIRONES, J.M., Departamento de Fisica Moderna, Universidad de Catabria, Avda. de los Catros, s/n, 39005 Santander, Spain

MARTINEZ, V. J., Departamento de Matematica Aplicada i Astronomia Universitat de Valencia, Burjassot, 46100 Valencia, Spain

MARTINEZ-GONZALES, E., Depto. de Fisica Moderna, Universidad de Cantabria, 39005 Santander, Spain

MASI, S., Dipartimento di Fisica, Universita' di Roma "La Sapienza", V.le A. Moro 2, 00185 Roma, Italy

MATARRESE, S., Dipartimento di Fisica "G. Galilei", Via Marzolo 8., 35131 Padova, Italy

MAYOR, M., Observatoire de Geneve, Ch. des Maillettes 51, CH-1290 Sauverny, Switzerland

MAZURE, A., CNRS, Universite de Montpellier, France

McGILL, C., Canadian institute for Theoretical Astrophysics, University of Toronto, Toronto, ON M5S 1A1, Canada

MELCHIORRI, F., Dipartimento di Fisica, Universita' "La Sapienza", V.le A. Moro 2, 00185 Roma, Italy

MENON, T. K., University of British Columbia, Department of Geophysics and Astronomy, 2219 Main Mall, Vancouver, B.C. V6T 1W5, Canada

MEZZETTI, M., Dipartimento di Astronomia, Universita'di Trieste, Via G.B. Tiepolo 11, 34131 Trieste, Italy

MICHALEC, A., Astronomical Observatory of the Jagellonian
    University, 30-244 Cracow, Poland
MISSANA, M., Osservatorio Astronomico di Brera, Via Brera
    28, 20121 Milano
MITRA, S., Astronomy Department, University of Texas, Austin
    TX 78712-1083, U.S.A.
NESCI, R., Istituto Astronomico dell'Universita' di Roma,
    Via Lancisi 29, 00161 Roma, Italy
NICHOLSON, D., Department of Astronomy, University of
    Edinburgh, Royal Observatory, Blackford Hill, EH9
    3HJ, Edinburg, United Kingdom
NOVIKOV, I.D., Space Research Institute, Academy of Sciences
    of the USSR, Moscow, U.S.S.R.
OCCHIONERO, F., Osservatorio Astronomico di Roma, Monte
    Mario, Via Parco Mellini 84, 00136 Roma, Italy
OSMER, P.S., National Optical Astron. Observatory, Box
    26732, Tucson, Arizona, U.S.A.
PALUMBO, G., Dipartimento di Astronomia, Via Zamboni 33,
    40125 Bologna, Italy
PANEK, M., Copernicus Astronomical Center, Ul. Barticka 18.
    00716 Warszawa, Poland
PARKER, Q. A., Royal Observatory, Blackford Hill, EH9 3HJ,
    Edinburg, United Kingdom
PARTRIDGE, R. B.,Haverford College, Astronomy 3340,
    Haverford PA 19041, U.S.A.
PEEBLES, P., Princeton University, Joseph Henry Labs.,
    Physics Department, Jadwin Hall, Princeton, New
    Jersey 08544, U.S.A.
PHILLIPPS, S., Department of Applied Mathematics and
    Astronomy, University College Cardiff, Wales, United
    Kingdom
PIERRE, P. A., European Southern Observatory, Karl-Schwarz-
    schild Str. 2, D-8046 Garching bei Muenchen,
    Fed. Rep. of Germany
PISANI, A., International School for Advanced Studies,
    Strada Costiera 11, Trieste, Italy
PORTILLA-MOLL,M., Departamento de Fisica Teorica,
    Universitat de Valencia, C/. Dr. Moliner, 50,
    46100 Burjassot, Valencia, Spain
PUCHE, D., Departement de Physique and Observatoire
    Astronomique du Mont Megantic, Universite de
    Montreal, Montreal Quebec, H3C, 3J7, Canada
RAYCHAUDHURY, S., Insitute of Astronomy, Madingley RD,
    Cambridge, CB3 OHA, United Kingdom
RHEE, G., Sterrewacht Leiden, P.O. Box 9513, 2300 RA
    Leiden, The Netherlands
RICHTER, O.-G., Space Telescope Science Institute, 3700
    Baltimore, MD 21218, U.S.A. also affiliated to the
    Astrophysics Division of the Space Science
    Department of E.S.A.
ROEDER, R. C., Department of Physics, Southwestern
    University, Georgetown, TX 78626, U.S.A.

ROOS, N., Astronomical Institute, Catholic University, Toernooiveld, 6525 ED Nijmegen, The Netherlands
ROWAN-ROBINSON, M., Queen Mary College, University of London, Mile End Road, London E1 4NS, United Kingdom
SAGLIA, R. P., Scuola Normale Superiore, 56100-I Pisa, Italy
SALAM, A., International Centre for Theoretical Physics, Strada Costiera 11, 34014 Trieste, Italy
SANCISI, R., Kapteyn Laboratory, University of Groningen, Post bus 800, 9700 Av Groningen, The Netherlands
SANZ, J. L., Departamento Fisica Moderna, Iniversidad de Cantabria, Avdas de los Castros, s/n 39005 Santander, Spain
SCARAMELLA, R., International School for Advance Studies , 34100 Trieste, Italy
SCIAMA, D. W., International School for Advanced Studies, Styrada Costiera 11, Trieste, Italy
SEDMAK, G., Osservatorio Astronomico di Trieste, Via G. B. Tiepolo 11, 34131 Trieste, Italy
SHAVER, P. A., European Southern Observatory, Karl-Schwarzschild Str. 2, D-8046 Garching bei Munchen, Fed. Rep. of Germany
STAVELEY-SMITH, L., Anglo-Australian Observatory, P.O. Box 296, Epping, NSW 2121, Australia
SULENTIC, J. W., Department of Physics and Astronomy, University of Alabama, Tuscaloosa, U.S.A.
SUTHERLAND, W.J., Institute of Astronomy, Madingley Rd, Cambridge CB3 0HA, United Kingdom
SZALAY, A., Eotvos University, Budapest, Hungary
TEAGUE, P., Dept. of Physics e Astronomy, Northwestern University, Evanston, 60208 Illinois, U.S.A.
TOFFOLATTI, L., Department of Astronomy, University of Padova, Vicolo dell'Osservatorio 5, 35122 Padova, Italy
TOSI, M., Osservatorio Astronomico di Bologna, Italy
TULLY, B., Institute for Astronomy, University of Hawaii, 2680 Woodlawn Drive, Honolulu Hawaii 96822, U.S.A.
van de WEYGAERT, R., Sterrewacht Leiden, Postbus 9513, 2300 RA Leiden, The Netherland
van KAMPEN, E.,Sterrewacht Leiden, Postbus 9513, 2300 RA Leiden, The Netherland
VENTURI, T., Dipartimento di Astronomia, Universita' di Bologna, via Irnerio 46, 40126 Bologna, Italy
VITTORIO, N., Dipartimento di Fisica, Universita' dell'Aquila, Italy
WAGNER, S., Landessternwarte Heidelberg, Koenigstuhl D-6900, Heidelberg, Fed. Rep. of Germany
WANAS, M. I., Department of Astronomy, Faculty of Science, Cairo University, Egypt
WATSON, R.,A., NRAL, Jodrel Bank, SKII 9DL, Chershire, U.K.
ZABIEROWSKI, M., Institute of Physics, Technical University, 50-370 Wroclaw, Poland
ZAMORANI, G., Osservatorio Astronomico di Trieste, Via G.B.

Tiepolo 11, 34131 Trieste, Italy

ZANINETTI, L., Istituto di Fisica Generale, Via Pietro Giuria 1, 10125 Torino, Italy

ZEILINGER, W. W., Dipartimento di Astronomia, Vicolo dell'Osservatorio 5, 35122 Padova, Italy

ZHU, X., Center of Astrophysics, University of Science and Technology of China, Hefei, Anhui, China

INVITED LECTURES

THE CENTER FOR ASTROPHYSICS REDSHIFT SURVEY: RECENT RESULTS

Margaret J. Geller and John P. Huchra
Harvard-Smithsonian Center for Astrophysics
60 Garden Street
Cambridge, MA 02138 USA

ABSTRACT: Six strips of the CfA redshift survey extension are now complete. The data continue to support a picture in which galaxies are on thin sheets which nearly surround vast low density voids. In this and similar surveys the largest structures are comparable with the extent of the survey. Voids like the one in Boötes are a common feature of the large-scale distribution of galaxies. The observed structure presents a serious challenge for models.

Within the context of the observed structure in redshift space, we discuss the issue of "fair samples" of the galaxy distribution. We examine statistical measures of the galaxy distribution including the two-point correlation functions. We comment on limits on large-scale flows and on the distribution of groups and clusters of galaxies in the survey.

I. INTRODUCTION

Mapping out the large-scale distribution of individual galaxies is a step toward understanding the origin and evolution of large-scale structure in the universe. Over the past few years, each new approach to this problem has uncovered unexpectedly large structures. The 21-cm surveys by Giovanelli and Haynes (1985) and Giovanelli et al. (1986) defined the Perseus-Pisces chain. Optical redshift surveys in deep probes uncovered the void in Boötes (Kirshner, Oemler, Schechter, and Shectman 1981, 1987; KOSS hereafter). The completion of the first slice in the Center for Astrophysics redshift survey extension (de Lapparent, Geller, and Huchra 1986; LGH) suggested that galaxies are on thin, sharply defined surfaces which surround (or nearly surround) vast voids.

This sequence of recent discoveries proves the power of redshift surveys. How-

3

M. Mezzetti et al. (eds.), Large Scale Structure and Motions in the Universe, 3–18.

ever, the volumes completely surveyed are at best comparable with the scale of the largest known inhomogeneities (Sections II and III). The Shane-Wirtanen map (Shane and Wirtanen 1967; Seldner, Siebers, Groth, and Peebles 1977) remains the only existing sample of the universe which is, in principle, large enough to examine the "typical" behavior of the galaxy distribution on large scales (see Geller, de Lapparent, and Kurtz 1984 and de Lapparent, Kurtz, and Geller 1986 for a discussion of the problems introduced by possible systematic errors in this catalog). The APM surveys at Cambridge (Efstathiou *et al.* 1988) should soon provide a much more reliable catalogue for the southern hemisphere.

II . THE DISTRIBUTION OF INDIVIDUAL GALAXIES

In 1978, Joêveer, Einasto, and Tago (1978) suggested that the large-scale distribution of galaxies has a "cellular" pattern in which rich clusters are connected by "filamentary" structures. The data at that time were incomplete and could only hint at such structure. The discovery of the void in Boötes (KOSS) and the 21-cm survey of the Pisces-Perseus chain (Haynes and Giovanelli 1986; Giovanelli *et al.* 1986) soon lent support to this picture. However, these first surveys gave no clear message about the frequency of the structures.

It is becoming increasingly clear that large-scale features in the galaxy distribution are ubiquitous. The deep surveys of Koo, Kron, and Szalay (1986) indicate that voids are common even at high redshift. Because these surveys are one-dimensional, the constraints on the sizes of the voids are poor. The AAT surveys also reveal voids along with thin structures perpendicular to the line-of-sight (Peterson *et al.* 1986). The continuing Arecibo survey delineates nearby voids and appears to support the interpretation of the Pisces-Perseus chain as a filamentary structure.

The extension of the CfA redshift survey indicates that bright galaxies are distributed on thin sheets — two-dimensional structures — which surround (or nearly surround) vast voids. Recent completion of a southern hemisphere survey (da Costa *et al.* 1988) gives some support to this picture although the survey is not sufficiently deep or dense to be directly comparable with the recent CfA results. The 21-cm data (Giovanelli and Haynes 1986) also reveal sheet-like structures in the Perseus-Pisces region. The message of all surveys is now clear: large structures are a *common* feature of all surveys big enough to contain them.

Surveys like the CfA redshift survey extension (Huchra *et al.* 1992) which are complete over a region of large angular scale are not the most efficient method for merely identifying the locations of large structures (see KOSS 1987 for a discussion of the sparse sampling approach which led to the discovery of the void in Boötes), but they are necessary for quantitative characterization of the distribution of galaxies over a range of scales. The goal of the Center for Astrophysics redshift survey extension is to measure redshifts for all galaxies in a merge of the Zwicky *et al.*

(1961-1968) and Nilson (1973) catalogs which have $m_{B(0)} \leq 15.5$ and $| b_{II} | \geq 40°$. There will be about 15,000 galaxies in the complete survey; more than 9,000 of these already have measured redshifts. About 2,500 of the galaxies with measured redshifts lie in three adjacent "slices" where the survey is now complete: (1) a slice with $8^h \leq \alpha \leq 17^h$ and $26.5° \leq \delta < 32.5°$ (LGH), (2) a slice with $8^h \leq \alpha \leq 17^h$ and $32.5° \leq \delta < 38.5°$, and (3) a slice with $8^h \leq \alpha \leq 17^h$ and $38.5° \leq \delta < 44.5°$. In the southern Galactic hemisphere 3 slices covering $6° \leq \delta > 12°$ and $18° \leq \delta < 30°$ with $0^h \leq \alpha \leq 4^h$ and $20^h \leq \alpha \leq 24^h$ are also complete. More than 60% of the redshift were measured with the 1.5-meter telescope at Mt. Hopkins. The mean external error in the redshift measurements is ~ 30 km s^{-1} .

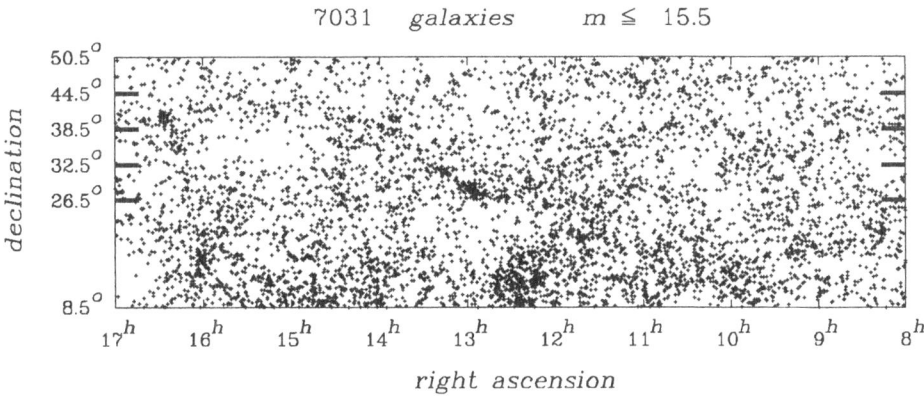

Figure 1: Positions of galaxies in the merged Zwicky-Nilson catalogues with $m_{B(0)} \leq 15.5$ in the northern galactic cap. The bold ticks indicate the limits of the complete strips.

Figure 1 shows the positions of 7031 galaxies in the Zwicky-Nilson merge which have $m_{B(0)} \leq 15.5$ and $8^h \leq \alpha \leq 17^h$ and $8.5° \leq \delta < 50.5°$. The grid is Cartesian in α and δ. The deficiency of galaxies west of 9^h and east of 16^h is caused by Galactic obscuration. The bold ticks indicate the limits of the three complete strips of the survey. The Coma cluster is the dense knot at $\alpha = 13^h$, $\delta = 30°$.

Figure 2 is a plot of the observed velocity versus right ascension for the strip centered at $29.5°$ (LGH): the strip is $6°$ wide in declination. The plot includes only the 1059 galaxies with redshifts $\leq 15,000$ km s^{-1}. A galaxy with the characteristic luminosity M* ($\simeq$ -19.4) is at 10,000 km s^{-1} in this survey. Nearly every

galaxy in this slice lies in an extended thin structure. The boundaries of the low
density regions are remarkably sharp. Several of these voids are surrounded by thin
structures in which the inter-galaxy separation ($\sim 3h^{-1}$ Mpc) is small compared
with the radius ($\sim 25h^{-1}$ Mpc) of the empty region. The edges of some of the
largest structures may be outside the right ascension limits of the survey. The only
pronounced velocity finger in this slice is the Coma cluster at $\sim 13^h$ — the torso of
the homunculus.

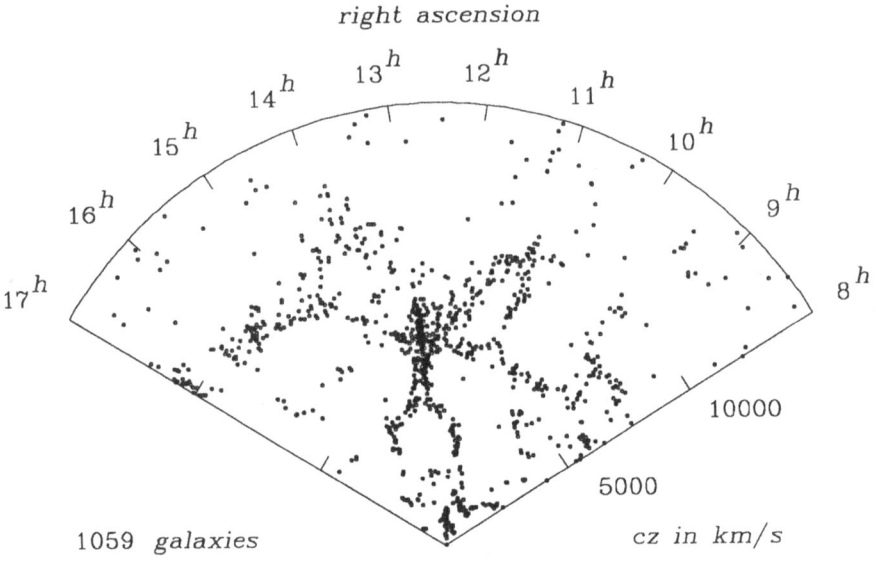

Figure 2: Observed velocity versus right ascension for the survey strip
centered at $\delta = 29.5°$. The strip is 6° in declination. Only the galaxies
with velocities $\leq$ 15,000 km s^{-1} are shown.

This first slice alone demonstrates that the thin structures in the distribution
of galaxies are cuts through two-dimensional sheets, *not* one-dimensional filaments.
If the $\sim$ 150 Mpc long structure which extends across the entire survey (from 9^h
to 16^h between 7,000 km s^{-1} and 10,000 km s^{-1}) is a filament, a thin linear
structure should be visible on the sky. This statement is particularly strong because the structure lies near the survey limit. The required filamentary structure
is absent from Figure 1. Because structure on the sky can be caused by patchy
obscuration and/or by inhomogeneities in the galaxy catalog, structure on the sky
cannot provide complete proof (or disproof) of the filamentary nature of a structure
in redshift space. A second argument against the filamentary nature of the structures in Figure 2 is that several thin, elongated structures lie in this single survey
slice: the intersection of a slice with a three-dimensional network of filaments is *a
priori* unlikely to be a two-dimensional network of filaments. Of course, we could

have been lucky (or unlucky).

A geometric structure in which thin sheets surround or nearly surround voids accounts for the data. Examples include "bubble-like" and "sponge-like" geometries (see Gott, Melott, and Dickinson 1986). I will use the word "bubble" to convey the image of a structure dominated by thin sheets and holes. Note that the "bubbles" are not necessarily round. In this picture the 150 Mpc "filament" is made up of portions of adjacent "bubbles" and the richest clusters like Coma lie in the interstitial regions (where several "bubbles" come together).

In this interpretation of the data, we assume that the maps are similar in redshift space and in real physical space. Both the cold dark matter (White *et al.* 1987) and the adiabatic models (Centrella *et al.* 1988) suggest that this assumption is reasonable (but see Kaiser 1987).

Maps of the adjacent slices support the *qualitative* picture suggested by the first slice. Figure 3 shows the slice centered at 35.5°, just to the north of the slice in Figure 2. Once again the galaxies are in thin structures. Furthermore these structures are natural extensions of the structures in Figure 2; the structures are highly correlated in the two slices. The two closed structures at $\sim 11^h$ (9000 km s^{-1} $\lesssim cz \lesssim$ 11,000 km s^{-1}) and at 14^h (7000 km s^{-1} $\lesssim cz \lesssim$ 11,000 km s^{-1}) are not so clearly delineated in Figure 3 as in Figure 2. Sampling of these structures is probably affected by variations in the limiting magnitude of the galaxy catalog.

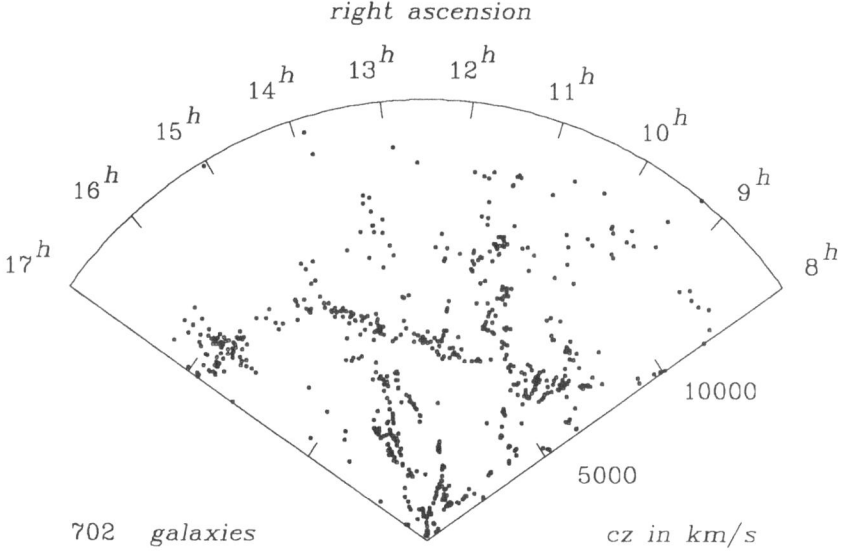

Figure 3: Same as Figure 2 for a slice centered at $\delta = 35.5$ °

Note that voids in the galaxy distribution are *not empty*: they are regions which are underdense relative to the global mean. The galaxies inside the large void have normal properties and their infra-red Tully-Fisher distances put them at the relative distances indicated by their redshifts (Geller *et al.* 1988). These galaxies may form a tenuous structure which would not be detected in a sparse survey like the KOSS survey of Boötes (1987).

Figure 4 shows the cone diagram for the third slice centered at $\delta = 41.5°$. At first glance this slice gives a somewhat different visual impression from the first two. The reason for the difference is that some of the surfaces lie in this slice; in particular, a portion of the structure surrounding the largest void is nearly in the plane of this slice and appears diffuse. Comparison of this slice with Figure 3 continues to support the large-scale coherence of the structures.

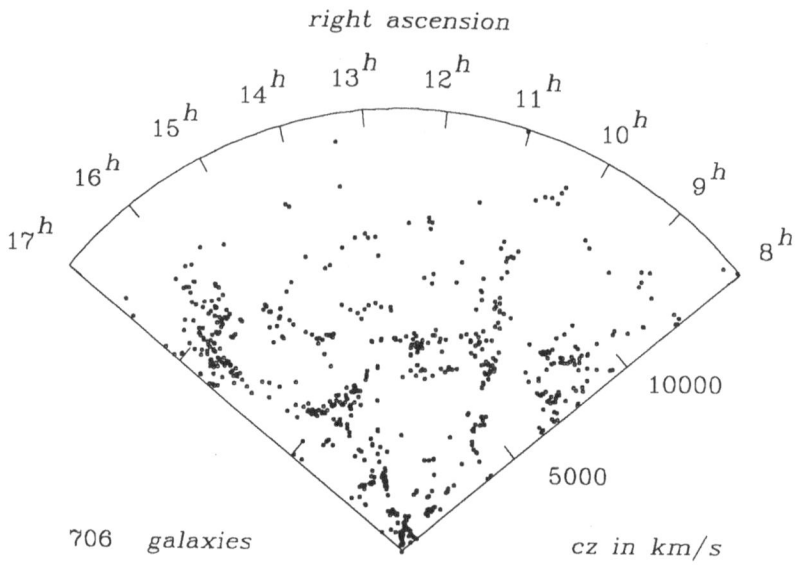

Figure 4: Same as Figures 2 and 3 for slice centered at $\delta = 41.5°$.

The study of structures in Figures 2–4 as a function of limiting apparent magnitude (see LGH 1986) along with deep probes through the 29.5° slice (Postman, Huchra, and Geller 1986) indicate that the large-scale galaxy distribution is insensitive to luminosity for $M_{B(0)} \lesssim -17.4$. More precisely, the density contrast between the high and low density regions (about a factor of 20) is insensitive to absolute magnitude over this range. Surveys to fainter limiting magnitudes *do* turn up galaxies in the voids, but they turn up proportionately more in the dense structures (see Binggeli's contribution to this meeting for further discussion of this issue and for discussion of the distribution of galaxies as a function of surface brightness).

The dependence of large-scale structure on the properties of individual galaxies can also be examined by comparing redshift surveys based on the IRAS catalog with those based on optical catalogs. At least two complete studies (Smith *et al.* 1987; Strauss and Huchra 1988) indicate that the IRAS galaxies trace the large-scale structures seen in optical catalogs except that IRAS galaxies are absent from the dense cores of rich clusters like Coma. Strauss and Huchra (1988) survey the void in Boötes and Smith *et al.* (1987) cover the redshift survey slice in Figure 2. Salzer *et al.* (1988) show that emission line galaxies also trace the structure in Figure 2 again with the caveat that they undersample the most dense regions.

III. IMPLICATIONS OF THE SURVEY

a). "Fair" Samples?

An important message of these surveys is that the largest inhomogeneities are comparable with the size of the sample. None of the existing redshift surveys are thus large enough to be "fair".

Perhaps even more sobering is that the largest inhomogeneities we detect are the largest we *could* detect within the limits set by the extent of the survey — we have few, if any, reliable direct limits on larger structures in the distribution of light-emitting matter. The size of the inhomogeneities relative to the volume of the surveys may underlie unexplained variations in traditional statistics of the galaxy distribution like the luminosity function (Schechter 1976; KOSS 1983; Bean *et al.* 1983; Davis and Huchra 1982) and the two-point correlation function at large scale (Groth and Peebles 1977; Davis and Peebles 1983; Kirshner, Oemler, and Schechter 1979; Shanks *et al.* 1983). When the inhomogeneities are large compared with the sample volume, mean quantities are not well-defined.

Two strip surveys to $z \simeq 0.1$ (Geller *et al.* 1988; KOSS 1988) could provide 10% measurements of the mean galaxy density. Of course, if they uncover yet larger structures with high contrast, even larger surveys will be necessary to meet the limit. We do not yet know how a large a survey is necessary for a "fair" sample.

b). Correlation Functions

The domination of the sample by large-scale coherent structures and the related 25% uncertainty in the mean density imply that the two-point correlation function is more poorly constrained than previously thought. Figure 5 shows the two-point correlation function $\xi(s)$ where

$$s = \frac{(V_i^2 + V_j^2 - 2V_iV_j\cos\theta_{ij})^{1/2}}{H_o} \tag{22}$$

for the 12° slice with declination between 26.5° and 38.5° (Figures 2 and 3). Here V_i and V_j are the velocities of two galaxies separated by θ_{ij} on the sky and H_o

is the Hubble constant. We make no correction for the r.m.s. pairwise peculiar velocities of $\lesssim 350$ km s^{-1} (Davis and Peebles 1983; de Lapparent, Geller, and Huchra 1988). The calculation of this correlation function is not seriously affected by the presence of the Coma cluster or of other more poorly sampled clusters in the sample.

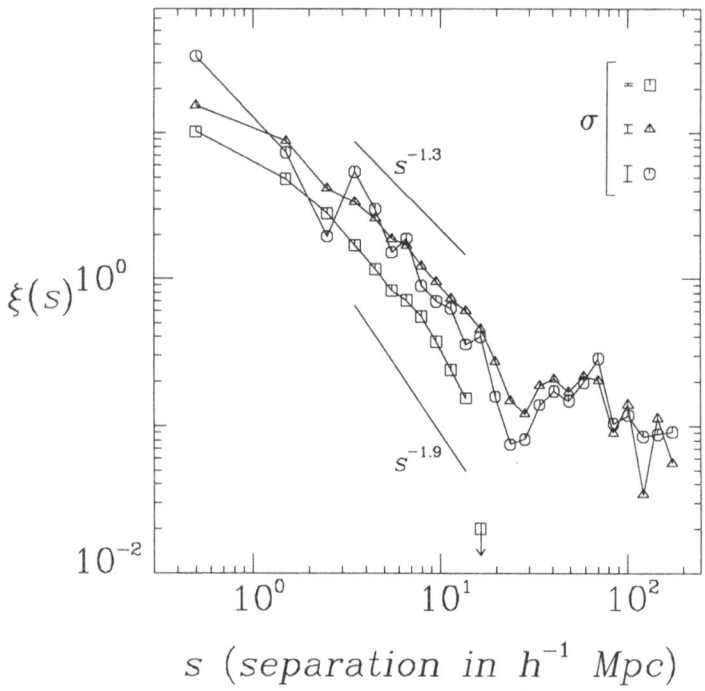

Figure 5: The two–point correlation function $\xi(s)$ for the sample in Figure 5. The symbols denote ξ_{11} ($\square$), $\xi_{\phi\phi}$ ($\triangle$), and $\xi_{1\phi}$ ($\bigcirc$). Note that the amplitude varies by a factor of two among these estimators.

Because of the large-scale coherent structures in the sample, the weighting scheme for the calculation of the correlation function *does* affects the result substantially. The calculation of the correlation function and the weighting schemes follow prescriptions in Davis and Peebles (1983). The subscript ϕ indicates that the galaxy is weighted by $\phi(V)^{-1}$, where ϕ (V) is the selection function for a magnitude limited sample. For these calculations the parameters of the luminosity function (equation (8)) are M* = -19.15, α = -1.2, and $\phi^{*} \simeq 0.025$ Mpc^{-3} mag^{-1} (de Lapparent, Geller, and Huchra 1988). The differences among the estimators in Figure

5 are symptoms of the lack of a fair sample.

We fit the data in the range $3.5 \leq s \leq 9.5h^{-1}$ Mpc to the standard power law form

$$\xi(s) = \left(\frac{s_o}{s}\right)^{\gamma} \qquad (23)$$

and find that $1.3 \leq \gamma \leq 1.9$ and $5 \leq s_o \leq 12$ h^{-1} Mpc. On scales larger than 20 h^{-1} Mpc the correlation function is indeterminate because its amplitude is comparable with the uncertainty in the mean density. For an average of the estimators in Figure 12, we obtain

$$\gamma = 1.5 \qquad (24)$$

and

$$s_o = 7.5h^{-1} \text{ Mpc.} \qquad (25)$$

These values agree with the one obtained by Davis and Peebles (1983) for the 14.5 CfA sample (see their Figure 1).

Calculation of the correlation function $\xi(r_p, \pi)$ (Davis and Peebles 1983) where

$$r_p = \frac{V_i + V_j}{H_o} \tan\frac{\theta_{ij}}{2} \qquad (26)$$

and

$$\pi = V_i - V_j \qquad (27)$$

is a method of examining the distortions in redshift space caused by peculiar velocities. The solid contours in Figure 6 show $\xi(r_p, \pi)$ for the data in Figures 2 and 3. All galaxies within 3° of the center of the Coma cluster have been removed from the sample. The weighting scheme is the same as for ξ_{11}. The dashed curves are the expectation for undistorted Hubble flow. The intercept of the $\xi = 1$ contour with the abscissa gives r_{po} and the mean radius of the $\xi = 1$ contour is s_o. The values of these quantities are underestimated relative to calculations weighted by the inverse selection function, ϕ^{-1}. Note that the elongation along the π axis occurs only for $r_p \lesssim 2h^{-1}$ Mpc. The elongation for the $\xi = 1$ contour implies an r.m.s. pairwise peculiar velocity of $\sim 450 \pm 150$ km s^{-1} in agreement with analyses of other catalogs (Davis and Peebles 1983; Bean et al. 1983). The r.m.s. distortions are also consistent with expectations based on an analysis of groups of galaxies in the 12° slice (Ramella, Geller, and Huchra 1988). In fact, removal of the groups from the survey removes the distortion of the $\xi(r_p, \pi)$ contours (Ramella, Geller, and Huchra 1988).

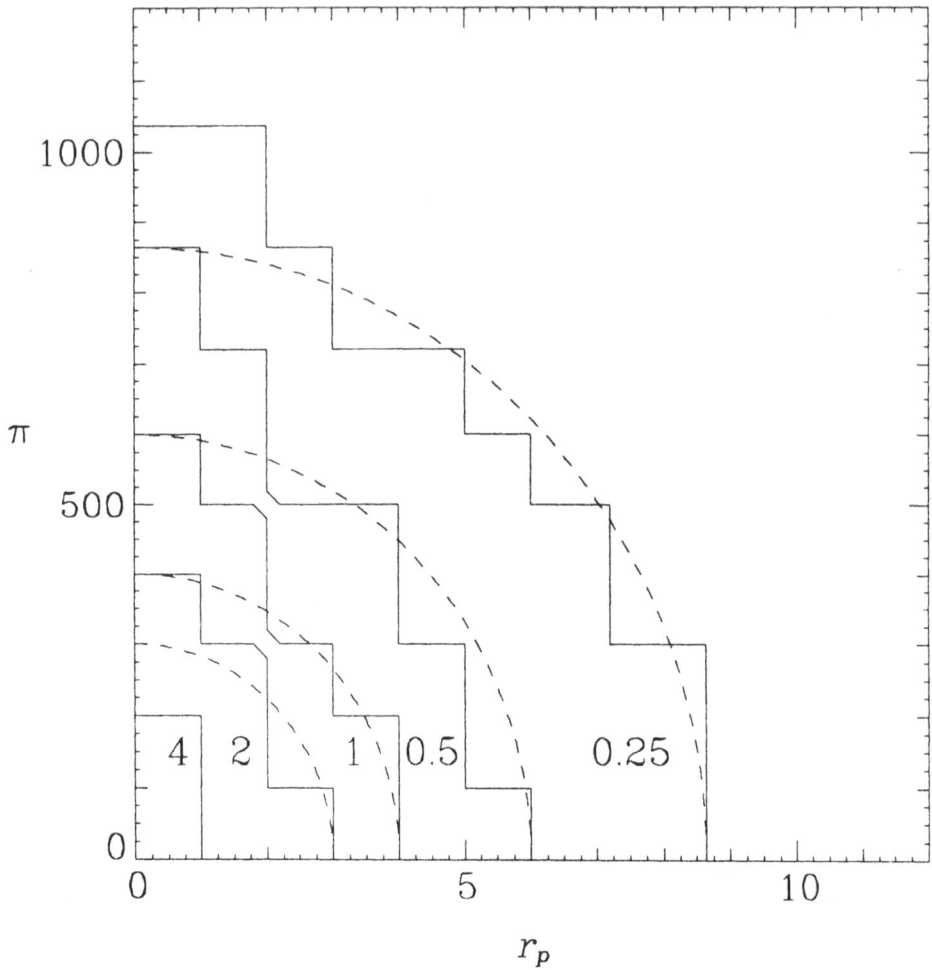

Figure 6: Contours of constant $\xi(r_p, \pi)$ (solid lines) in the plane defined by the projected separation perpendicular to the line-of-sight, r_p (in h^{-1} Mpc), and parallel to the line-of-sight, π (in km s^{-1}). The dashed lines are the expected pattern for pure Hubble flow.

Inclusion of the Coma cluster in the sample makes a substantial change in $\xi(r_p, \pi)$; the distortion along the π direction then extends to $\sim 10 h^{-1}$ Mpc (as in the analysis of the 14.5 sample which is influenced by both Virgo and Coma — see Davis and Peebles (1983)). These distortions are introduced by the virialized cores of the rich clusters, not by large-scale coherent departures from uniform Hubble flow.

c). Coherent Large-Scale Flows

Because of the limited amount of data and the difficulty of the measurements, the relationship between large-scale coherent flows and the structure in redshift surveys remains unclear. However, coherent flows should, for example, be associated with large low density regions. If the matter density inside a void is low compared with the average surroundings (i.e. if the galaxy density contrast is a measure of the matter density contrast), the voids expand relative to the average cosmological flow and the structures should appear elongated in redshift space. For an isolated adiabatic void in an $\Omega = 1$ universe, the outward peculiar velocity $v_{pec} \simeq 0.2 - 0.3$ v_H where v_H is the radius of the void in redshift space. For lower Ω the peculiar velocities are smaller (see Ostriker 1986 for a brief discussion and for references to other work).

The measurement of distances to galaxies in the structures offers a direct probe for large scale flows associated with voids. Because many spirals lie in the extended sheets, the infra-red Tully-Fisher technique (Aaronson, Huchra, and Mould 1979) can be used to obtain limits at the few hundred kilometers per second level on scales of fifty megaparsecs, within the theoretically predicted range (Geller et al. 1988). The current limit on peculiar velocities on the large shell in Figure 5 is $v_{pec}/v_H \lesssim 0.2$.

d). Comparison of the Data with Models

The frequent mention of the "size" of the Boötes void is a demonstration of the power of a measure of the scale of the "largest" observed structure. Sampling according to the procedure followed by KOSS, White et al. (1987) find that 3/25 simulations contain a void as large or larger. It is not clear whether the simulations can meet the challenge posed by the increasing number of surveys with more dense sampling than the Boötes survey. There are now five surveys large enough to contain ~ 5000 km s^{-1} voids — all of them do.

The spectrum of void sizes is an important test of models; the small-scale end is a constraint on hot dark matter models (Zel'dovich 1970; Doroshkevich et al. 1980; Centrella and Melott 1983; Centrella et al. 1988) and the large-scale end is most demanding for cold dark matter models (Davis et al. 1985; White et al. 1987) and for the explosive models (Ostriker and Cowie 1981; Ikeuchi 1981; Saarinen, Dekel, and Carr 1986). Determination of the distribution of sizes of voids requires samples much larger than those currently available.

The thickness, coherence, and filling factor of the sheets provide further constraints. The FWHM of the sheets is $\lesssim 500$ km s^{-1} (de Lapparent, Geller, and Huchra 1988a). The thickness as a function of orientation with respect to the line-of-sight restricts physical models. If the sheets were collapsing pancakes, we would expect them to the thinner when they are perpendicular to the line-of-sight than when they are parallel to it. On the other hand, any internal velocity dispersion will make the sheets appear thicker when they are perpendicular to the line-of-sight.

The fraction of the survey volume filled by the coherent structures in the distribution of individual galaxies can be calculated by appropriately binning and smoothing the data. The galaxies fill $\lesssim 20\%$ of the volume and the typical separation of galaxies in the sheets is $3h^{-1}$ Mpc at the survey depth D to which M^* galaxies are included. Remarkably this density is comparable with the surface density of the structures in the deep probes discussed of Koo, Kron, and Szalay (1986).

The "uniformity" of the sheets may provide a constraint on Ω (see Peebles 1986). If $\Omega = 1$ and the distribution of galaxies marks the distribution of matter, it is unlikely that a smooth shell can persist for a Hubble time; gravity causes the galaxies to clump up and "fingers" in redshift space should be apparent. If the actual matter density contrast in the sheets is small and the voids are full of nearly uniformly distributed dark matter (with Ω close to 1), the structures could still be in the linear regime. If, on the other hand, $\Omega = 0.2$ or less (as indicated by the dynamical estimates and by analysis of the abundance of the light elements), the structure could set in early on and then just stretch with the universal expansion.

IV. GROUPS, CLUSTERS AND REDSHIFT SURVEYS

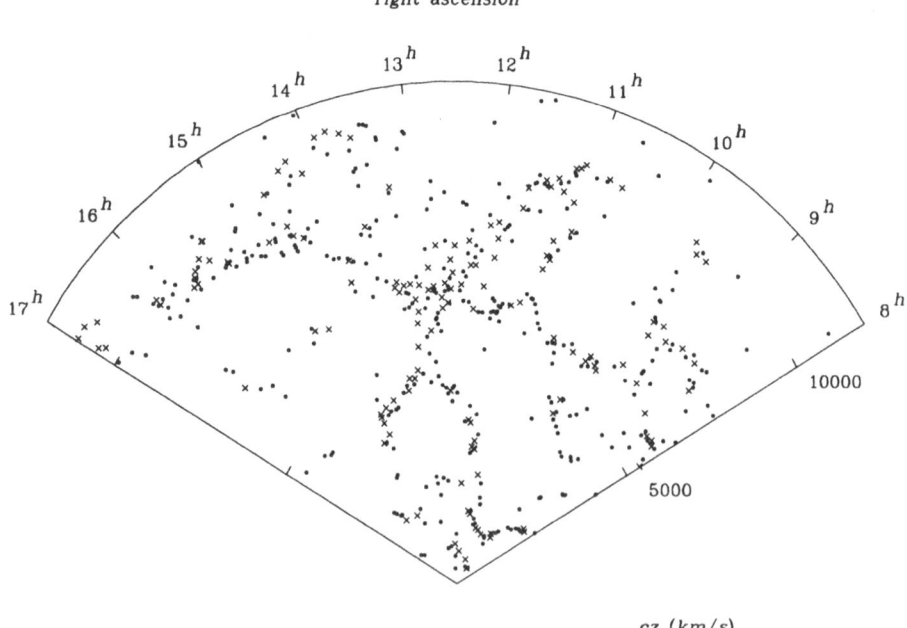

Figure 7: The slice of Figure 2 with "fingers" removed. The $\times$'s are binaries, groups, and clusters; the o's are individual galaxies not assigned to any system.

Small groups of galaxies are embedded in the structures in Figure 2–4 (Ramella, Geller, and Huchra 1988). The properties of the groups with 5 or more members are similar to the properties of such groups in the 14.5 survey (Geller and Huchra 1983); the median line-of-sight velocity dispersion $\sigma \simeq 230$ km s^{-1}, the typical scale of the systems is 500 kpc, and the median mass-to-light ratio is 200 $M_\odot/L_\odot$. The centers of these groups trace out the coherent structures in the survey. Figure 7 is the slice of Figure 2 after velocity fingers have been removed; all "systems" from binary galaxies to clusters have been replaced by a single point at the barycenter of the system ($\times$). The o's are individual galaxies not assigned to any condensation. There are 73 groups containing three or more members, 75 binaries, and 352 galaxies which were not assigned to a group. The groups were identified with the objective algorithm of Huchra and Geller (1982). Note that the plot in Figure 7 extends only to 12,000 km s^{-1}.

When the CfA survey is complete, it will include tens of Abell clusters and should provide a first *direct* measure of the relationship between individual galaxies and clusters of galaxies as tracers of the large-scale matter distribution.

V. DISCUSSION

Theories for the formation of large-scale structure must explain the observed structure in the distribution of galaxies. Is the observed structure consistent with models based on gravitational amplification of random-phase density fluctuations? White *et al.* (1987) and Centrella *et al.* (1988) make quantitative comparisons of the cold and hot dark matter models, respectively, with the available data. The models are $\Omega = 1$ universes. Both the cold and hot dark matter models can successfully reproduce some of the salient features of the data. Both produce large voids and some thin coherent structures. White *et al.* (1987) argue that a standard cold dark matter model with biased galaxy formation produces a distribution of galaxies which is hard to distinguish from the data in Figures 5–9 (see their Figure 10).

The sharpness of the structures in Figures 5–9 and the possibility that they imply non-random phase initial conditions motivate consideration of alternatives to the standard gravitational models for large-scale structure formation. One suggestion (Ostriker and Cowie 1981; Ikeuchi 1981) is that explosions promote or amplify galaxy formation. Although the origin of such explosions is pure speculation, they can produce voids with radii in the range 5 - 20 h^{-1} Mpc without violating the constraints on these models imposed by the smoothness of the microwave background (Vishniac and Ostriker 1986). Explosions or the decay of superconducting strings (Ostriker, Thompson, and Witten 1986) during the epoch of galaxy formation releasing 10^{61} - 10^{62} (or more) ergs would produce shells with radii of 5 - 6 h^{-1} Mpc or more. Coalescence of these structures is responsible for the production of the larger voids. Detailed investigation of these models would be useful for identifying statistical measures which can differentiate them from the more standard cold and hot dark matter models (see e.g. Ostriker and Strassler 1988). The hydrodynamic models have the appeal that they provide a natural explanation for sharp-edges structures. They may also provide a natural "biasing" mechanism.

Comparisons between models and the data are clearly limited by the absence of "fair" samples. Further progress in mapping out the large-scale structure of the universe requires deeper and more reliable photometric catalogs. Systematic variations in the magnitudes from one region to another in a single catalog (not to mention variations from one catalog to another) almost surely compromise detailed analyses of the properties of the structure. The advent of large format CCD's should go some way toward solving the problem by enabling digital sky surveys.

We thank Valérie de Lapparent and Massimo Ramella for discussions and for providing some of the figures. This research was supported in part by the Smithsonian Institution and by NASA grant NAGW-201.

REFERENCES

Aaronson, M., Huchra, J.P., and Mould, J. 1979, *Ap.J.*, **229**, 1.

Binggeli, B. 1988, this volume.

Centrella, J., Gallagher, J.S., Melott, A.S., and Bushouse, H.A. 1988, preprint.

Centrella, J., and Melott, A.S. 1983, *Nature*, **305**, 196.

da Costa, L.N., Pellegrini, P.S., Sargent, W.L.W., Tonry, J., Davis, M., Meiksin, A., and Latham, D.W. 1988, *Ap.J.*, **327**, 544.

Davis, M. Efstathiou, G., Frenk, C. and White, S.D.M. 1985, *Ap.J.*, **292**, 371.

Davis, M. and Huchra, J.P. 1982, *Ap.J.*, **254**, 437.

Davis, M. and Peebles, P.J.E. 1983, *Ap.J.*, **267**, 465.

Doroshkevich, A.G., Kotok, E.V., Novikov,I.D., Polyudiv, A.N., Shandarin, S.F., and Sigov, Yu.S. 1980, *M.N.R.A.S.*, **192**, 321.

Efstathiou et al. 1988, in preparation.

Geller, M.J. *et al.* 1988, in preparation.

Geller, M.J. and Huchra, J.P. 1983, *Ap.J. Suppl.*, **52**, 61.

Geller, M.J., and Huchra, J.P. 1988, in *Vatican Study Week on Large-Scale Motions in the Universe*, V.C. Rubin, ed. (Princeton University Press: Princeton).

Geller, M.J., de Lapparent, V., and Kurtz, M.J. 1984, *Ap.J. Letters*, **287**, L55.

Giovanelli, R. and Haynes, M.P. 1985, *A.J.*, **90**, 2445.

Giovanelli, R. and Haynes, M.P. 1986, *Ap. J.*, **292**, 404.

Giovanelli, R., Haynes, M.P., Myers, S.T., and Roth, J. 1986, *A.J.*, **92**, 250.

Gott, J.R., Melott, A., and Dickinson, M. 1986, *Ap.J.*, **306**, 341.

Groth, E.J. and Peebles, P.J.E. 1977, *Ap.J.*, **217**, 385.

Huchra *et al.* 1992, in preparation.

Huchra, J.P. and Geller, M.J. 1982, *Ap.J.*, **257**, 423.

Ikeuchi, S. 1981 *Publ. Astr. Soc. Japan*, **33**, 211.

Joeveer, M., Einasto, J., and Tago, E. 1978, *M.N.R.A.S.*, **185**, 357.

Kaiser, N. 1987, *M.N.R.A.S.*, **227**, 1.

Kirshner, R.P., Oemler, A., and Schechter, P. 1979, *A.J.*, **84**, 951.

Kirshner, R.P., Oemler, A. Jr., Schechter, P.L., and Shectman, S.A. 1981, *Ap.J. (Letters)*, **248**, L57.

Kirshner, R.P., Oemler, A., Schechter, P.L., and Shectman, S.A. 1983, *A.J.*, **88**, 1285

Kirshner, R.P., Oemler, A. Jr., Schechter, P.L., and Shectman, S.A. 198, *Ap.J.*, **314**, 493.

Kirshner, R.P., Oemler, A., Schechter, P, and Shectman, S. 1988, private communication.

Koo, D. Kron, R., and Szalay, A. 1986 in *13th Texas Symposium on Relativistic Astrophysics*, M. Ulmer, ed. (World Scientific: Singapore).

de Lapparent, V., Geller, M.J., and Huchra, J.P. 1986, *Ap.J. (Letters)*, **202**, L1.

de Lapparent, V., Geller, M.J., and Huchra, J.P. 1988, *Ap.J.*, in press.

de Lapparent, V., Kurtz, M.J., and Geller, M.J. 1986, *Ap.J.*, **304**, 585.

Nilson, P. 1973 *Uppsala General Catalog of Galaxies, Uppsala Astr. Obs. Ann.*, **6**.

Ostriker, J.P. 1986, in *Galaxy Distances and Deviations from the Hubble Flow*, B. Tully, ed.

Ostriker, J.P. and Cowie, L.L. 1981, *Ap.J. (Letters*, **243**, L127.

Ostriker, J.P. and Strassler, M. 1988, preprint.

Ostriker, J.P. ,Thompson, C., and Witten, E. 1986, *Phys. Rev. Lett. B*, **180**, 231.

Peebles, P.J.E. 1986, *Nature*, **321**, 27.

Peterson, B.A., Ellis, R.S., Efstathiou, G., Shanks, T., Bean, A.J., Fong, R. and Zen-Long, Z. 1986, *M.N.R.A.S.*, **221**, 233.

Postman, M., Geller, M.J., and Huchra, J.P. 1986 *A.J.*, **92**, 1238.

Ramella, M., Geller, M.J., and Huchra, J.P. 1988, in preparation.

Saarinen, S., Dekel, A., and Carr, B.J. 1987, *Nature*, **325**, 598.

Salzer, J., Aldering, G., Bothun, G.D., and Lonsdale, C. 1988, *A.J.*, in press.

Schechter, P.L. 1976, *Ap.J.*, **203**, 297

Seldner, M., Siebers, B., Groth, E.J., and Peebles, P.J.E. 1977, *A.J.*, **82**, 249.

Shane, C.D. and Wirtanen, C.A. 1967 *Publ. Lick Obs.*, **Vol. XXII**, Part 1

Shanks, T., Bean, A.J., Efstathiou, G., Ellis, R.S., Fong, R., and Peterson, B.A. 1983, *Ap.J.*, **274**, 529.

Smith, B., Kleinmann, S., Huchra, J.P., and Low, F. 1987, *Ap.J.*, **318**, 161.

Strauss, M., and Huchra, J.P. 1988, *A.J.*, **95**, 1602.

Vishniac, E.T. and Ostriker, J.P. 1985,*Societa Italiana di Fisica*, **1**, 157.

White, S.D.M., Frenk, C.S., Davis, M., and Efstathiou, G., 1987, *Ap. J.*, **313**, 505.

Zeldovich, Ya. B. 1970, *Astron. Astrophys.*, **5**, 84.

Zwicky, F.,Herzog, W., Wild, P, Karpowicz, M. and Kowal, C. 1961-1968, *Catalog of Galaxies and of Clusters of Galaxies*, (Pasadena: California Institute of Technology)

DISCUSSION

GUDEHUS: In the figure you showed of your Tully-Fisher calibration, there appeared to be about twice as many points below your mean line as above. Could you comment on this?

GELLER: It's carelessness in tracing the line onto the trasparency! The actual scatter is symmetric with respect to the fitted relation.

PEEBLES: I agree that larger scale redshift surveys, along the line of your very impressive work, are needed, to trace out the largest structure in the universe. But I think that for the purpose of estimating low order correlation function we already have a fair sample: the Lick catalog. You were right to challenge us to show that our old reduction of the Lick catalog is reliable. I think the recent papers in the Astronomical Journal by E. Groth, M. Brown and me pretty convincingly show it is.

GELLER: We certainly agree that larger redshift surveys are important. We still disagree about the reliability of the Shane-Wirtanen counts. Soon well-calibrated photometric surveys should be avaible to settle this issue.

RICHTER: Do the few galaxies within the detected voids show any peculiar or special spectral features?

GELLER: The galaxies in the survey which lie in the voids show no special features.

THE LARGE-SCALE DISTRIBUTION OF IRAS GALAXIES

Michael Rowan-Robinson
Astronomy Unit
School of Mathematical Science
Queen Mary College
Mile End Road
London E1 4NS.

ABSTRACT: Estimates of the IRAS surface brightness dipole are reviewed and results from a new complete and reliable 60μm galaxy catalogue compiled from the IRAS Point Source, Small Scale Structure and Large Galaxy Catalogs, are presented. The catalogue covers 82% of the sky and includes 17,710 sources. The dipole direction and amplitude agree with those of the earlier study by Yahil et al (1986). The interpretation of the IRAS dipole, including the results from recent redshift surveys, is reviewed.
The consistency of the IRAS data with the Great Attractor model is discussed and results on clustering of IRAS galaxies are briefly reviewed.

1. THE IRAS SURVEY

The IRAS Point Source Catalog (PSC) provides the first homogeneous uniformly-calibrated survey of 96% of the sky at 12, 25, 60 and 100 μm. Away from regions of high source density (which comprise some 14% of the sky - the region of the Milky Way at $|b| \leqslant$ 5°, $270^{\circ} \leqslant 1 \leqslant 90^{\circ}$, and regions like the LMC, SMC, Orion, Ophiucus etc) the survey is 99.9% reliable and 98% complete to 0.4, 0.4, 0.6 and 1.6 Jy at 12, 25, 60 and 100μm, respectively (Rowan-Robinson et al, 1984, IRAS Introductory Supplement, Chapter Vlll). The photometric accuracy is about 10%, corresponding to 0.1 mag,: for comparison the Zwicky Catalogue of Galaxies (and the Uppsala Catalogue) has a photometric accuracy of around 0.5 mag.

Away from the Galactic plane, the 12μm sources in the PSC are mainly stars, the 25μm sources are mainly stars with some galaxies, the 60μm sources are almost entirely galaxies (most stars being recognizable by their 25μm/60μm colour ratio or by identification

M. Mezzetti et al. (eds.), Large Scale Structure and Motions in the Universe, 19–39.

with optical catalogues) and the 100μm sources are galaxies or emission from interstellar dust (the "cirrus") (Rowan-Robinson et al 1986).

The IRAS 60μm survey is at present the deepest unbiased, homogeneous, uniformly calibrated galaxy sample and hence of the greatest importance for large-scale cosmological studies. Interstellar extinction is negligible at 60μm (the absence of any diminuition of source-density down to $|b| \sim 5^o$ implies that $\tau(60\mu m)/\mathrm{cosec} \ |b| <$ 0.01). Two effects that have to be carefully controlled are some residual effects of interstellar emission at 60μm and the fact that some nearby galaxies are extended with respect to the IRAS beam. These are also some residual effects of detector "hysteresis" at $|b| < 10^o$

2. THE IRAS DIPOLE

Yahil et al (1986) have shown how IRAS galaxies may be used to map the local gravitational field (see also Rowan-Robinson 1987a, 1988, Lahav et al 1988, Strauss and Davis 1988a,b). If we assume that the space-density of IRAS galaxies follows that of the total matter distribution and that there exists a universal luminosity function $\Phi(L)$, then the number of galaxies in luminosity range dL, volume element d^3r, can be written.

$$dN = D(\underline{r}) \ d^3r \ \Phi(L) \ dL \qquad (1)$$

where $D(\underline{r})$ is the local relative density function (so D=1 corresponds to the mean density of the universe).

Then the smoothed surface brightness due to sources as a function of flux-density S and direction $\hat{r}$ is

$$4 \ \pi \ \sigma(S,\hat{r}) \ = \ 4\pi \ S \ \frac{dN}{dsd\omega} = \int D(r)dr \int L \ \Phi(L) \ \delta(S - \frac{L}{4\pi r^2})dL. \qquad (2)$$

The dipole moment of this

$$4 \ \pi \ S \ \underline{\sigma} \ (S) \ = \ 3 \ S \int_0^{4\pi} \sigma \ (S,\hat{r}) \ \hat{\underline{r}} \ d\omega = \int_0^{\infty} L^2 \ \Phi(L) \ \frac{dG}{dr}dr \qquad (3)$$

$$\text{where } \underline{G}(r) \ = \ \frac{3}{4\pi} \int_0^r \int_0^{4\pi} D(\underline{r}') \ \frac{\underline{r}'}{r^3}d^3r' \ , \qquad (4)$$

The peculiar gravitational acceleration acting on the Local Group due to matter out to r.

Table 1 lists several determinations of the IRAS surface brightness dipole, with notes on how the sample is defined and how extended sources are treated. The most detailed study is that of Rowan-Robinson (1988). The basic mask consists of the coverage gaps together with lune bins in which the high-source density flag is set in any band (IRAS Introductory Supplement 1984), which comprises altogether 18% of the sky.

TABLE 1: MEASUREMENTS OF THE IRAS SURFACE-BRIGHTNESS DIPOLE

Authors	IRAS PSC version	Flux range	Cirrus rejection	correction for extended sources	% of sky	no. of sources	(l,b)
Yahil, Walker & Rowan-Robinson (1986)	1	0.5–31.6 Jy	CIRR 1 > 1	none	47	9903	(248,40)
Meiksin & Davis (1986)	1	S > 0.5 Jy	Worst areas excluded[1], unidentified sources excluded	none	76		(235,43.5)[2]
Harmon, Lahav & Meurs (1987)	2	S > 0.7 Jy	Colour selection	none	66	8985	(228,22)[3]
Rowan-Robinson (1988)	2	0.63–31.6 Jy[4]	Colour selection + HSD flag	LGC and SSS fluxes	79	12448	(248,40)
Strauss & Davis (1988a)	1	S > 1.936 Jy	Worst areas excluded	double PSC flux [5]	76	2176	(255,54)
Strauss & Davis (1988b)	2	S > 1.936 Jy	Worst areas excluded	LGC and ADDSCAN fluxes	76	2244	(231,48)

[1] Significant cirrus contamination at S < 2 Jy (Rowan-Robinson 1987a)

[2] Number-weighted dipole

[3] Out of line with other studies, but number-weighted dipole agrees with CBR

[4] But see Table 2

[5] Most IRAS sources flagged as extended in the PSC are not significantly extended and do not appear in the SSS catalog

```
PSC     = IRAS Point Source Catalog
SSS     = IRAS Small Scale Structure Catalog (Helou & Walker 1986)
LGC     = IRAS Large Galaxy Catalog (Rice et al 1988)
CIRR1   = Cirris 1 flag
HSD     = High Source Density flag
ADDSCAN = 1-D coaddition of IRAS raw data.
```

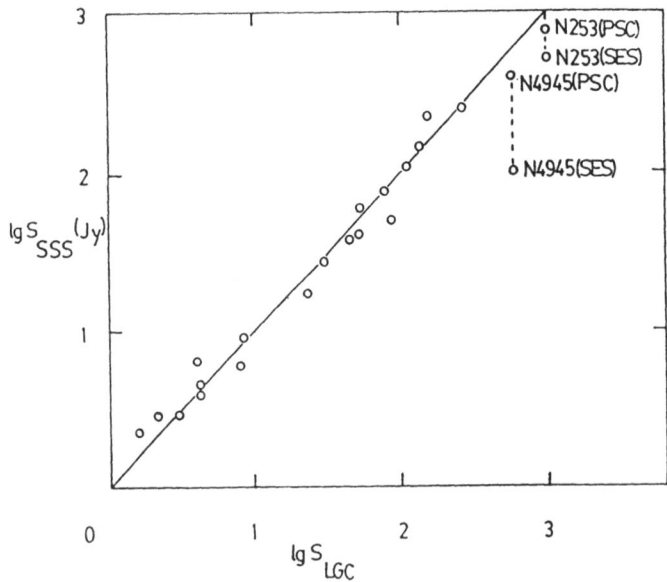

Fig. 1. Comparison of fluxes from IRAS Small Scale Structure and Large Galaxy Catalogs for 23 galaxies in common.

Outside this mask, colour constraints are used to eliminate stars, planetary nebulae and cirrus sources. Fluxes from the IRAS Large Galaxy Catalog (Rice et al 1988) and IRAS Small Scale Structure Catalogs (Helou and Walker 1985) are used for extended sources. Fluxes from the latter are remarkably good, in the mean, as can be seen from Fig 1 a comparison of LGC and SSS fluxes for sources in common. Fig 2 shows the distribution of the masked bins on the sky and Fig 3 shows the distribution of IRAS 60μm sources, satisfying the colour constraints, outside the mask.

Table 2 summarises the dependence of the IRAS dipole on the flux-range of the sample and choice of mask, for the sources of Fig 3. The remarkable stability of the IRAS dipole can be clearly seen.

Table 3 gives the quadrapole components of the surface brightness distribution determined for 2 flux ranges. The quadrapole is significantly detected and has amplitude comparable to the dipole, reflecting the fact that the galaxy distribution is far from dipolar. The two strongest components are those with axes towards b=90^{0} and (l,b) = (270^{0},0^{0}), close to the directions of the Virgo and Centaurus clusters.

3. REDSHIFT SURVEYS OF IRAS GALAXIES AND THE INFRARED LUMINOSITY FUNCTION.

The first complete redshift survey of IRAS galaxies was published by Lawrence et al (1986), the result of a collaboration between QMC and RGO. The sample consisted of IRAS 60μm sources with flux greater than 0.85 Jy in the North Galactic polar cap (b > 60^{0}, 0^{0} < l < 110^{0}). A third of the area was also surveyed in the flux range 0.5–0.85 Jy. Lawrence et al (1986) gave the first determination of the 60μm luminosity function for galaxies, which has been confirmed by subsequent studies. Rowan-Robinson et al (1987) studied IRAS galaxies with b > 60^{0}, m_{pg} < 14.5, which gave additional information at lower luminosities in the Lawrence et al luminosity function.

Rieke and Lebofsky (1986) compiled a bright IRAS galaxy sample from the literature and Soifer et al (1986) studied a sample consisting of 324 IRAS galaxies with S(60) > 5 Jy, b > 50^{0}. Both these studies gave luminosity functions consistent with that of Lawrence et al (1986) (see Rowan-Robinson 1987a). Smith et al (1987) studied a sample of 72 IRAS galaxies with S(60) > 2 Jy.

Davis, Strauss, Huchra and Yahil (Strauss and Davis 1988a, 1988b, Yahil 1988) have carried out a redshift survey of IRAS galaxies with S(60) > 2Jy, |b| 10^{0}, a total of 2176 galaxies, of which ~ 60% required new redshift measurements. Some results from this survey are discussed below. In a collaboration between QMC, Cambridge and Durham, Lawrence, Rowan-Robinson, Saunders, Crawford, Efstathiou, Kaiser, Ellis, Frank and Perry (1988, in preparation) are carrying out a survey of a random sample of 1 in 6

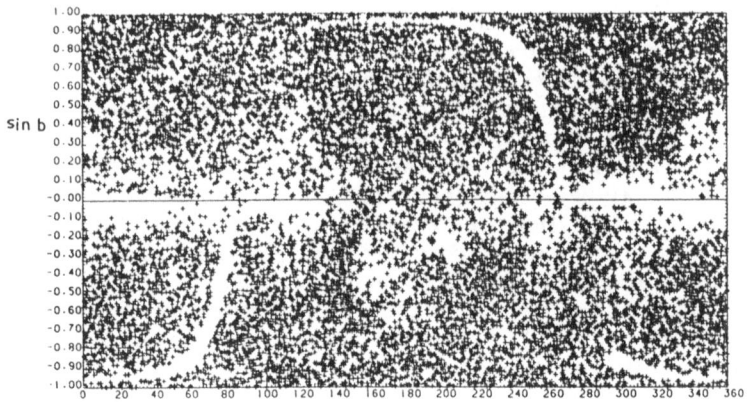

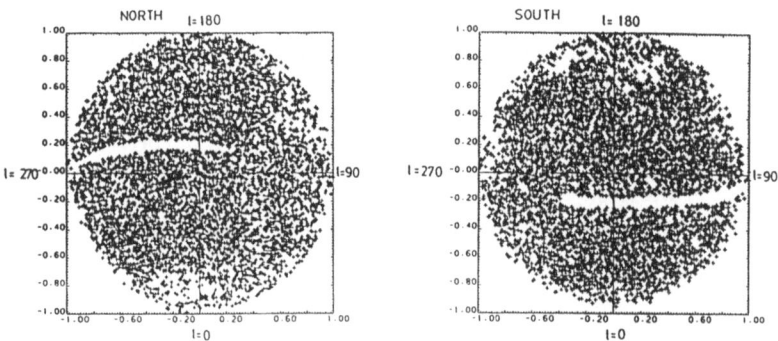

Fig. 2. Distribution of 17,710 IRAS 60 μm sources in the galaxy catalogue described in the text. (a) sin *b* vs *l*, (b) North Galactic Hemisphere, (c) South Galactic Hemisphere

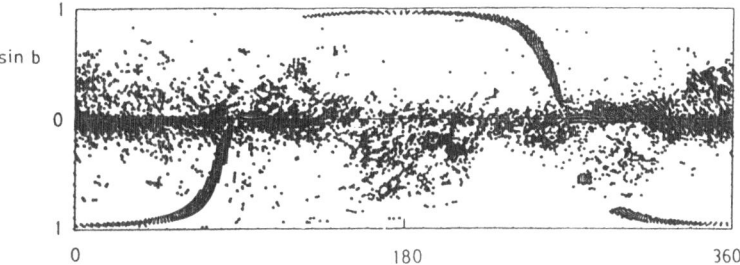

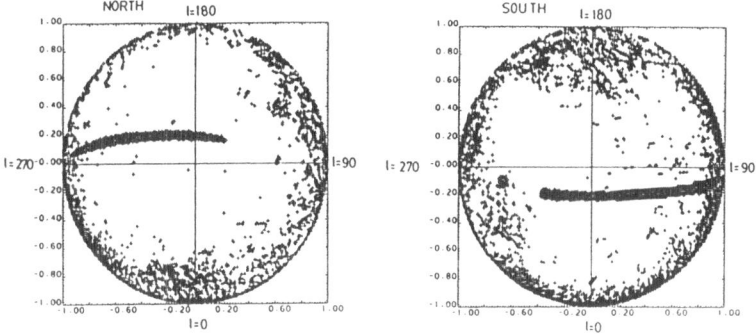

Fig. 3. Distribution of masked bins for same areas as in Fig. 2. The two tongues extending from the Galactic plane are the coverage gaps.

Table 2. The IRAS 60 μm dipole

sample	mask[a]	area (sr)	flux-range (Jy)	no. of sources	dipole amplitude $\|4\pi S\sigma\|$	$l°$	$b°$	θ^b_{CBR}
present	**basic** +$\|b\| < 5°$	9.94	0.63 - 31.6	12448	1342 ±148	248.2 ±9.6	39.5 ±9.6	20.7
"	"	"	0.63 - 31.6[c]	12448	1425	252.1	45.7	22.5
			0.63 - 200[c]	12495	1323	233.8	48.9	34.3
			0.63 - 63	12480	1347	246.3	41.2	22.8
			0.5 - 31.6	16694	1430	239.3	40.6	27.3
			1.0 - 31.6	6126	1468	247.9	44.1	23.6
			2.0 - 31.6	2285	1656	245.5	48.7	27.8
"	basic only	10.28	0.63 - 31.6	12950	1320	241.2	40.6	20.0
	basic +$\|b\| < 10°$	9.28	0.63 - 31.6	11715	1408	249.2	37.0	18.1
"	basic +$\|b\| < 20°$	7.63	0.63 - 31.6	9698	1375	256.9	37.7	14.2
	$CIRR1 > 1$	5.94	0.5 - 31.6	7089	1513	250.6	31.9	16.0
	$CIRR1 > 1$	5.94	0.5 - 31.6	9521	1518	248.8	38.1	19.6
YWR	$CIRR1 > 1$	5.94	0.5 - 31.6	9903	1550	248	40	21.0

[a]basic mask is coverage gaps + high source density bins
[b]CBR direction taken as $(l, b) = (268, 27)$
[c]straight average of flux bins (rest are error-weighted averages)

Table 3. Spherical Harmonic components

Component	magnitude (0.63 − 31.6 Jy)	(1 − 31.6 Jy)
l	8561 ± 92	6944
$\cos b \cos l$	−384 ± 157	−389
$\cos b \sin l$	−961 ± 153	−982
$\sin b$	854 ± 143	1044
$1.5\sin^2 b - 0.5$	557 ± 217	656
$\sin 2b \cos l$	615 ± 159	430
$\sin 2b \sin l$	−1092 ± 154	−864
$\cos^2 b \cos 2l$	632 ± 181	359
$\cos^2 b \sin 2l$	−1636 ± 177	−1832

IRAS 60μm sources brighter than 0.6Jy. The total sample is 2387 galaxies, of which 1600 new redshifts need to be measured. The survey is 85% complete to date.

Vader and Simon (1988a,b) have measured redshifts of IRAS galaxies in several selected areas. Strauss and Huchra (1988) have carried out a survey of IRAS galaxies in the direction of the Bootes void. Hacking et al (1988) have started a deep sample at the ecliptic pole and Lonsdale and Hacking (1988) have studied deep samples compiled from IRAS pointed observations. Wolstencroft et al (1988) have surveyed IRAS galaxies at the South Galactic pole (b > 80°).

Saunders et al (1988) have calculated the 60μm luminosity function for a combined sample including most of these surveys, by a new method which takes account of the local over-density of galaxies and of the tendency of galaxies to cluster. The method fits a maximum likelihood step-functional form to the data by iteration (Efstathiou et al 1988). The best-fitting simple function fitting this luminosity function is of the form

$$\Phi\,(L) = C(L/L_*)^{1-\alpha}\ \exp\ \{-\ (\log\ (1+L/L_*))^2\ /2\sigma^2\}\qquad(5)$$

with $C = 3.8\times10^{-2}\ h^3\ Mpc^{-3}$, $\alpha = 1.11$, $\sigma = 0.77$, $L_* = 10^{8.34}h^{-2}\ L_0$,

where Fig 4a, shows a comparison of this function with the combined data and Fig 4b, shows a comparison with the 2 power-law form which gave a good fit to the data of Lawrence et al (1986). The agreement for $\log L(60)\ /L_0 > 10$ is excellent. Below this luminosity the correction for the effect of clustering has significantly lowered $\Phi(L)$.

4. 60μm SOURCE COUNTS

Fig 5 shows the 60μm differential source-counts derived by Rowan-Robinson et al (1986) and Hacking and Houck (1987). The latter were derived by co-addition of the many survey scans over an area near the north ecliptic pole. Also shown in Fig 5 are predicted curves in an $\Omega_0 = 1$ universe with no evolution and with a simple luminosity evolution of the form $L(z) \propto \exp Q\ (1-1/(1+z))$, with $Q = 5,8$. Evolution of this form with $Q = 5-6$ gives an adequate fit to optical counts of quasars and to radio source-counts (Rowan-Robinson 1970,1987b). It can be seen that if the data of Hacking and Houck (1987) are representative, even steeper evolution is implied by the IRAS counts. Francheschini et al (1988) also obtain a good fit to 60μm counts with an evolutionary model for starburst and active galaxies which also fits the faint radio source-counts.

Lonsdale and Hacking (1988) obtained redshifts of 107 sources down to 150 mJy at 60μm, from pointed IRAS observations covering 18.4 sq deg. The luminosity function they obtain agrees in shape with that of Lawrence et al (1986), but the normalisation implies the

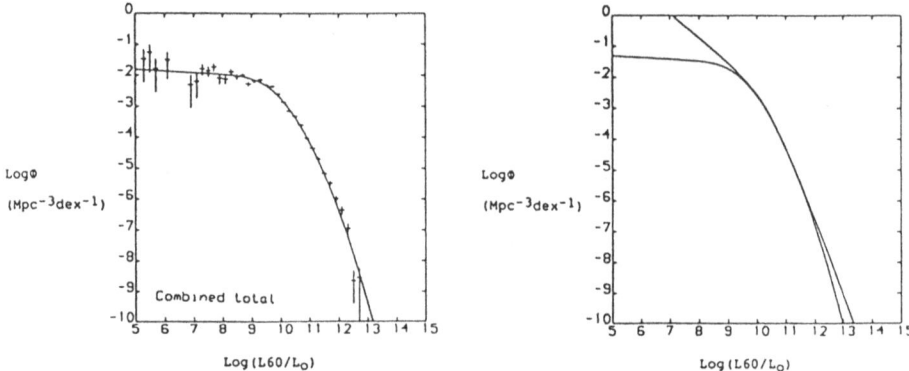

Fig 4: (a) Combined 60µm luminosity function for 8 samples comprising 1385 IRAS galaxies determined by iterative maximum likelihood method (Saunders et al 1988). Solid curve is eqn (5). (b) Comparison of eqn (5) with 2 power-law form of Lawrence et al (1986).

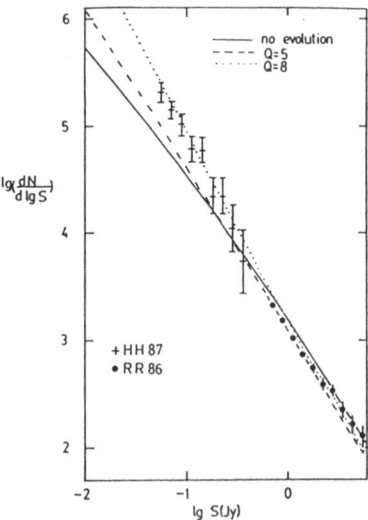

Fig 5: Differential 60µm source-counts due to Rowan-Robinson et al (1986) and Hacking and Houck (1987), compared with $\Omega_0 = 1$ models with no evolution and with luminosity evolution of the form $L(z) = L(0) \exp Q(1 - {}^1/(1+z))$, with Q=5,8.

existence either of evolution in the source population or of large-scale structure.

Hacking et al (1988) have obtained redshifts for galaxies in their very deep sample (50mJy) and conclude that the excess source-density is due to the presence of a cluster rather than to evolution.

5. INTERPRETATION OF THE IRAS DIPOLE

Returning to eqn (3), the analysis of Yahil et al (1986) can be summarised by noting that if $\Phi(L) = CL^{-2}$ for $L_1 \angle L \angle L_2$ and is negligible outside this range, and if $G(r)$ is generated over a range $rl \angle r \angle r_2$, where

$$\frac{L_1}{4\pi \ r_1^2} \ < \ S \ < \ \frac{L_2}{4\pi \ r_2^2} \ , \tag{6}$$

then $4 \ \pi \ S \ \underline{\sigma}(S) = C \ \underline{G}(r_2)$, independent of S.

The Pebbles (1980) formula calculated from linear theory

$$\underline{u} = \Omega_0^{0.6} \ \underline{G} \ / \ 3 \tag{7}$$

can then be used to derive Ω_0. Yahil et al (1986) found $\Omega_0 = 0.85$.

More generally (Lahav et al 1988), the density factor $D(\underline{r})$ may be expanded in spherical harmonics

$$D(\underline{r}) = 1 + \underline{a}(\underline{r}). \ \hat{\underline{r}} + \ .....$$

and then $\frac{dG}{dr} = \underline{a}(r)$, independently of the other harmonics.

The simplest possible model (Willemsen and Strauss 1987) is to take $\underline{a}(r) = a(r) \ \hat{\underline{n}}$

$$\begin{aligned} \text{with } a(r) &= 0, \ r < r_1, \\ &= a, \ r_1 \angle r \angle r_2, \\ &= 0 \end{aligned}$$

If we now use this model to solve eqn (3) for $\underline{a}(r) = \frac{dG}{dr}$
using the data of Rowan-Robinson (1988) and the luminosity function of Saunders et al (1988), we find

$$r_1 \ = 1600 \ \genfrac{}{}{0pt}{}{+}{-} \ \genfrac{}{}{0pt}{}{400}{1600} \ \text{km/s}, \qquad r_2 \ = 2500 \ \genfrac{}{}{0pt}{}{+}{-} \ \genfrac{}{}{0pt}{}{1500}{500} \ \text{km/s}$$

$$|\underline{G}| \ = \ 2130 \ \genfrac{}{}{0pt}{}{+}{-} \ \genfrac{}{}{0pt}{}{650}{180} \ \text{km/s}, \tag{8}$$

where distances have been multiplied by H_0, since H_0 does not enter into the solution. Fig 5 shows a comparison of theory and observation for this model.

$$\text{These values yield } \Omega_0 = 0.76 \begin{array}{c} + \\ - \end{array} \begin{array}{c} 0.12 \\ 0.28 \end{array}$$

This has to be corrected for non-linear effects. Yahil et al (1986) estimate that the magnitude of this correction is + 15%.

Lahav et al (1988) find that where this method is applied to Lahav's diameter-limited optical galaxy catalogue (Lahav 1987), a value for $\Omega_0 = 0.16 + 0.07$ is obtained. The contribution to the optical and infrared dipoles from different regions of the sky differ markedly. Fig 6 shows the cumulative amplitude as a function of angle from the microwave background dipole direction. Most of the optical dipole is generated within 45% of the apex, by the Virgo and Centaurus Superclusters, with an additional kick from the "Local Void" in the anti-apex direction. The IRAS dipole is generated more uniformly round the sky. The contribution to the total dipole amplitude from a 15° circle centred on $(l,b,) = (310^\circ, 29^\circ)$, i.e. Centaurus, is 41% for the optical dipole, but only 14% for the IRAS dipole. On the other hand both studies agree that Virgo makes only a small contribution to the dipole, with 18% of the optical amplitude, and 16% of the $60\mu m$ amplitude, coming from a 10° circle centred on Virgo. These figures translate to an infall velocity to Virgo of only ~ 100 km/s.

A possible interpretation of the different optical and IRAS results is that the universe is dominated by a dark matter constituent whose distribution is traced fairly faithfully by spiral galaxies. The rich, elliptical-dominated, clusters which dominate the optical picture and cause the higher dipole amplitude (and hence lower apperent value of Ω_0) would represent zones of higher elliptical galaxy density, but not of correspondingly high total matter density. Such a bias could arise if the galaxy density and primordial star formation rate were very strongly dependent on the mean density in density fluctuations.

An alternative, equally valid interpretation is that the optical picture is the true one and hence that $\Omega_0 \ll 1$. Because IRAS sees mainly spirals and these avoid the cores of rich clusters, IRAS underestimates the dipole amplitude and overestimates Ω_0.

Kaiser and Lahav (1988) have investigated cold dark matter models with biased galaxy formation and conclude that both optical and IRAS samples show evidence of biased galaxy formation. They infer an intermediate value of Ω_0, 0.4-0.6.

Strauss and Davis (1988a) have made a very significant contribution to the interpretation of the IRAS dipole by carrying out a redshift survey of IRAS galaxies brighter than 1.936 Jy which did not previously have known redshift.

The area of the sky at $|b| < 10^\circ$ and selected other regions of high latitude star formation are excluded. The fluxes of sources

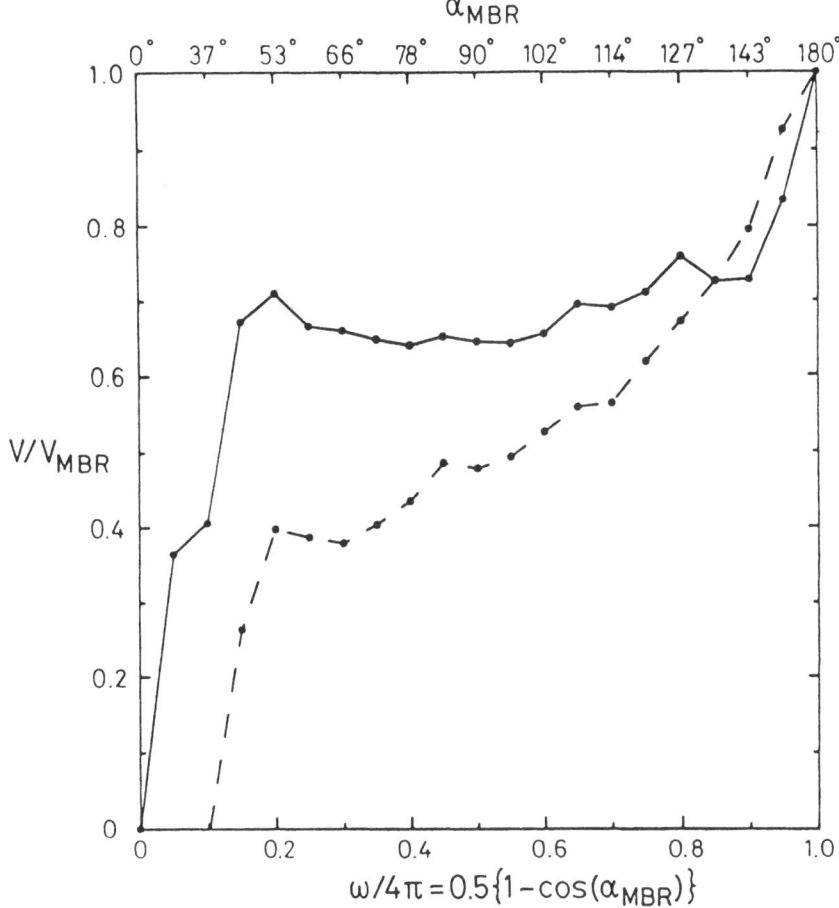

Fig 6: The cumulative contributions to the IRAS (broken curve) and optical (solid curve) dipoles from annular shells centred on the direction of the CBR dipole (Lahav et al 1988).

flagged as extended were doubled to correct for flux lying outside
the IRAS beam. This is likely to óvercorrect for extended emission,
since many PSC sources flagged as extended are bonafide point
sources (Rowan-Robinson 1988). Strauss and Davis (1988b) have use
LGC fluxes (Rice et al 1988) where available and used 1-dimensional
co-added fluxes (ADDSCAN) for extended sources otherwise. The
total sample consists of 2285 galaxies.

Fig 7 shows the cumulative gravitational acceleration acting on
the Local Group, calculated as (see eqn (6))

$$\underline{v} \ \Omega_0^{-0.6} = \frac{H_0}{4\pi n_1} \ \sum_i \ \frac{1}{\Phi(r_i)} \ \frac{r^i}{r_i^3} \tag{9}$$

where n_1 is the galaxy density, r_i is the vector to galaxy i, and
$\Phi(r_i)$ is the selection function at the distance of the galaxy i. From
this plot Strauss and Davis conclude that our peculiar acceleration
converges by $H_0 r = 4000$ km S^{-1}. The solution given by equ (8) is
consistent with this. The QCD survey, which is almost a factor of 2
deeper, will provide a useful test of this convergence.

Table 4 summarizes the values of Ω_0 found by Strauss and Davis
(1988a,b) and by Yahil (1988) under a variety of assumptions: these
range from 0.4 to 1.2.

6. THE "GREAT ATTRACTOR" MODEL

Lynden-Bell et al (1988) have revived an idea of Tammann (1984)
and Tammann and Sandage (1985), that our motion with respect to
the microwave background is the result of a simple vector sum of
attractions from Virgo and from the Hydra-Centaurus supercluster.
Lynden-Bell et al have in mind a rather larger mass concentration
than Tammann and Sandage, incorporating both the Hydra-Centaurus
and Pavo-Indus superclusters, the so-called "Great Attractor", and
with a mean redshift of 4350 km s^{-1} rather than the 3000 km s^{-1}
main peak of Hydra-Centaurus. A link between these two
superclusters had earlier been proposed by Fairall (1984, see also
Fairall 1988). Reviewing the motions of galaxies within 3000 km s^{-1},
Faber and Burstein (1988) also favour the "Great Attractor" model.
Dressler (1988) has carried out a redshift survey in the region
-35° < b < 45°, 290° < l < 350°, finding 2 velocity peaks at 3000 and
4500 km^{-1}. He argues that much of the higher velocity peak is
hidden by the Galactic plane and that the whole volume can be
thought of as a sphere of radius 3000 km s^{-1} centred at a distance
of 4000 km s^{-1}, with an average overdensity of $\Delta\rho / \rho = 2.1$. Such
an object would be an order of magnitude more massive than the
Local Supercluster, consistent with the "Great Attractor" concept of
Lynden-Bell et al (1988).

What do the IRAS data tell us about the "Great Attractor"?
First the IRAS galaxies in the area surveyed by Dressler do indeed
show the same double-velocity peak (Fig 8), though with the 3000 km

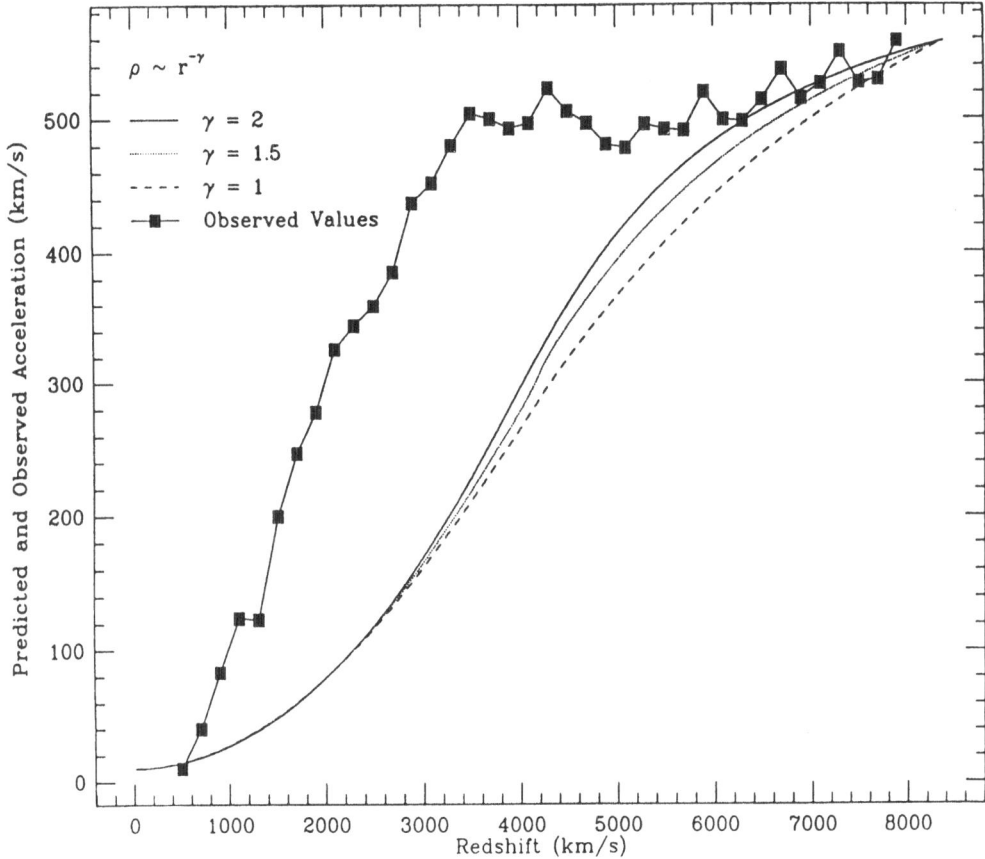

Fig 7: The cumulative amplitude of our peculiar acceleration as a function of redshift (Strauss and Davis 1988b). The smooth curves are the expected peculiar acceleration due to a Great Attractor placed at a redshift of 4200 km 5^{-1}. The density distribution around the GA was taken to be a power-law.

Table 4.   ESTIMATES OF Ω_0 DERIVED FROM 2 Jy REDSHIFT SURVEY

STRAUSS AND DAVIS (1988a⟩	Ω_0
using PSC fluxes	0.83
double fluxes for sources flagged as extended	0.67
also double contribution of regions of high density (to correct for ellipticals)	0.39
STRAUSS AND DAVIS (1988b)	
using LGC and ADDSCAN fluxes for extended sources	1.2
also double contribution of regions of high-density	1.0
YAHIL (1988)	
with correction for peculiar velocities of galaxies	0.5

S^{-1} peak much the stronger (Strauss and Davis 1988b). Strauss and Davis find an overdensity of 0.4 ± 0.2 in a sphere centred on the "Great Attractor" center and reaching out to our radius. The centre of mass is some 500 km s^{-1} nearer to us than the proposed Great Attractor centre.

In the distribution on the sky of IRAS galaxies in the catalogue of Rowan-Robinson (1988), shown in Fig 3, the Hydra-Centaurus and Pavo-Indus clusters can be clearly seen, centred at $(l,b) = (310,30)$ and $(343,-30)$. However the density of galaxies has fallen considerably, for both clusters, well before the masked area is reached. Only if a very considerable concentration of galaxies lies behind the masked area, can the concept of single large object incorporating both clusters be maintained.

Fig 7 shows Strauss and Davis (1988b)'s cumulative distribution of our peculiar acceleration as a function of redshift, compared with the simple "Great Attractor" model. There is a clear discrepancy. The IRAS dipole is generated at redshifts less than 4000 km s^{-1} and by several mass concentrations rather than just one or two. Of course, if a very large mass concentration lies behind the Galactic plane, the effect of this would not have been included in the IRAS dipole.

Strauss and Davis (1988b) have calculated the peculiar velocity of each IRAS galaxy, analogously to equ (9) and find good overall agreement with the peculiar velocity distributions derived by Faber and Burstein (1988) for elliptical and spiral galaxy samples.

Yahil (1988) has used the IRAS 2Jy sample to show projections of the 3-D distribution of the local gravitational field on various planes (eg Fig 9). These show the major centres of attraction locally to be Virgo, Hydra-Centaurus and Perseus-Pisces.

7. CLUSTERING OF IRAS GALAXIES

The covariance function for IRAS galaxies was estimated by Rowan-Robinson and Needham (1986). Using the redshift survey of Lawrence et al (1986) they used the 2-dimensional clustering function to derive

$$\zeta(r) = (r/r_0)^{-1.8}$$
$$\text{where} \quad r_0 = 2.6 \ h^{-1} \ \text{Mpc.}$$

This value of r_0 was considerably lower than the $\sim 5 \ h^{-1}$ Mpc found for optically selected galaxies, indicating IRAS galaxies are less strongly clustered than optically selected samples.

Using the more extensive QMC-Cambridge-Durham redshift survey, Efstathiou (1988) found the IRAS galaxy coverance function to be consistent with Davis and Geller's (1976) result for spiral galaxies

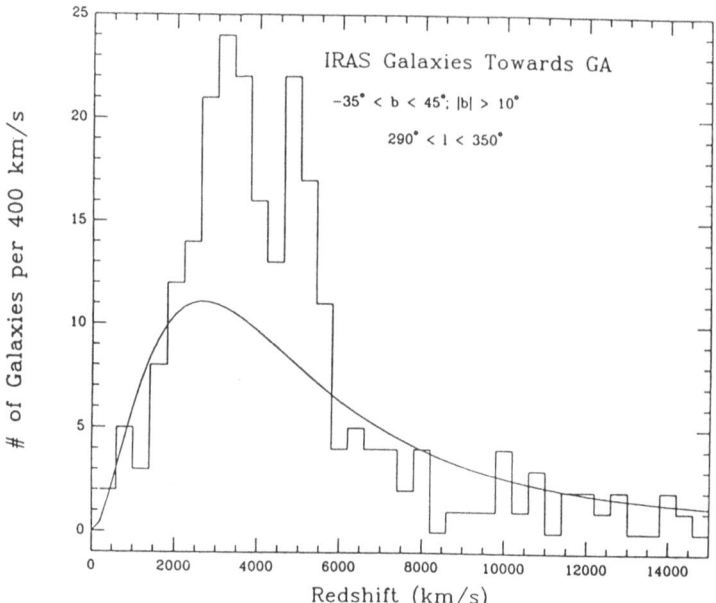

Fig 8: Velocity distribution of IRAS galaxies in the region
$290^0 < 1 < 350^0$, $\sim35^0 < b < 45^0$, $|b| < 10^0$ (Strauss and Davis 1988b).

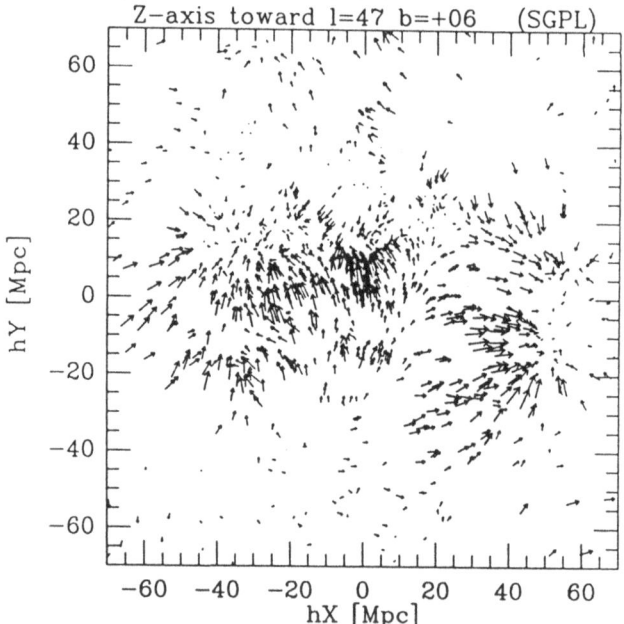

Fig 9: The gravitational field predicted by the IRAS galaxy
distribution (Yahil 1988). The effects of the major mass concentration
of the Virgo, Perseus-Pisces and Centaurus Superclusters can be
clearly seen.

$$\zeta(r) = (r/r_0)^{-1.69}$$
$$\text{with } r_0 = 3.6 \ h^{-1} \ \text{Mpc.}$$

Rowan-Robinson (1988, in preparation) has made a catalogue of IRAS clusters by 2-dimensional percolation, finding 174 clusters with > 7 members within a percolation radius of 1°. This represents the first complete and homogeneous cluster list covering a large fraction of the sky.

Strauss and Huchra (1988) have carried out a survey of IRAS galaxies towards the Bootes void. They do find some IRAS galaxies within the void but conclude that the void is indeed underdense by a factor of 4.

8. CONCLUSIONS

The tremendous potential of the IRAS survey for cosmological studies is still only just beginning to be explored. We can hope to gain a good picture of the local galaxy distribution out to a redshift of 10,000 km s^{-1}, to understand the origin of our peculiar motion, to determine he density parameter Ω_0 accurately, to study clustering on scales to 100 Mpc, and to define very interesting samples of moderate redshift (z =0.1-0.3) galaxies.

REFERENCES

Beichman,C.A., Neugebauer,G., Habing,H.J., Clegg,P.E., and
    Chester,T.J., 1984, IRAS Introductory Supplement (NASA,JPL
    D-1855)
Davis,M., and Geller,M.J., 1976, Astrophys.J. 208,13
Dressler,A., 1988, Astrophys.J. 329,519
Efstathiou,G., 1988, "Comets to Cosmology", ed. A. Lawrence
    (Springer), p. 312.
Efstathiou,G., Ellis, R.S., and Peterson,B.A., 1988, Mon. Not. R. Astr. soc.
    232,431.
Faber,S.M., and Burstein,D., 1988, Vatican Study Week, Large-Scale
    Motions in the Universe.
Fairall,A.P., 1984, Publ. Astr. Dept. Univ. Cape Town, No.6
Fairall,A.P., 1988, Mon. Not. R. astr. Soc. 230,69
Francheschini,A., Danese,L., de Zotti,G., and Xu,C., 1988, Mon. Not. R. astr.
    Soc. 233,175.
Hacking,P., and Houck,J.R., 1987, Astrophys. J. Supp.63
Hacking,P., et al, 1988, in preparation.
Harmon,R.T., Lahav,O., and Meurs,E.J.A., 1987, Mon. Not. R. astr. Soc.
    228,5p
Helou,G., and Walker,D.W., 1986, IRAS Small Scale Structure Catalog (NASA,

JPL D-2988)

Kaiser,N., and Lahav,O., 1988, Vatican Study Week, "Large Scale Motions in the Universe".

Lahav,O., 1987, Mon. Not. R. astr. Soc. 225,213

Lahav,O., Rowan-Robinson,M., and Lynden-Bell,D., 1988, Mon. Not. R. astr. Soc. (in press)

Lawrence,A., Walker,D., Rowan-Robinson,M., Leech,K.J., and Penston,M.V., 1986, Mon. Not. R. astr. Soc. 219,687

Lonsdale,C.J., and Hacking,P.B., 1988, Astrophys.J. (in press)

Lynden-Bell,D., Faber,S.M., Burstein,D., Davis,R.L., Dressler,A., Terlevich,R.J., and Wegner,G., 1988, Astrophys.J.

Meiksin,A., and Davis,M., 1986, Astron.J., 91,191.

Peebles,P.J.E., 1980, "Large Scale Structure of the Universe", (Princeton University Press)

Rice,W.L., Persson,C.J., Soifer,B.T., Neugebauer,G., and Kopan,E.L., 1988, Astrophys.J. Supp.

Rieke,G.H., and Lebofsky,M.J., 1986, Astrophys.J. 304,326

Rowan-Robinson,M., 1970, Mon. Not. R. Astr. soc. 149,365.

Rowan-Robinson,M., 1987a, IAU Symposium No. 124, "Observational Cosmology", ed. A. Hewitt et al (Reidel) P.229

Rowan-Robinson,M., 1987b, Vatican Conference, "Cosmology", ed. W. Stoeger, p.401

Rowan-Robinson,M., 1988, "Comets to Cosmology", ed. A. Lawrence (Springer) p.348

Rowan-Robinson,M., and Needham,G., 1986, Mon. Not. R. astr. Soc. 222,611

Rowan-Robinson,M., Walker,D., Chester,I., Soifer,T., and Fairclough,J., 1986, Mon. Not. R. astr. Soc. 219,273.

Rowan-Robinson,M., Walker,D., and Helou,G., 1987, Mon. Not. R. astr. Soc. 227,589

Saunders,W., Rowan-Robinson,M., Lawrence,A., Efstathiou,G., Kaiser,N., Ellis,R.S., and Freck, C.S., 1988, in preparation.

Smith,B.J., Kleinman,G.G., Huchra,J.P., and Low,F.J., 1987, Astrophys.J.

Soifer,B.T., Sanders,D.B., Neugebauer,G., Danielson,G.E., Lonsdale,C.J., Madore,B.F., and Persson,S.E., 1986, Astrophys.J. 303,L41

Strauss,M.A., and Davis,M., 1988a, "Comets to Cosmology", ed. A. Lawrence (Springer) p. 361.

Strauss,M.A., and Davis,M., 1988b, Vatican Study Week, "Large-Scale Motions in the Universe.

Strauss,M.A., and Huchra,J., 1988, Astrophys.J. (in press)

Tammann,G.A., 1984, "Clusters and Groups of Galaxies", ed. F. Mardirossian et al (Reidel) p.529.

Tammann,G.A., and Sandage,A., 1985, Astrophys.J. 294, 81.

Vader,J.P., and Simon,M., 1988a, Astron.J. 94,636

Vader,J.P., and Simon,M., 1988b, Astron.J. 96,854

Wolstencroft,R.D., Savage,A., Clowes,R.G., MacGillivray,H.T., Leggett,S.K., and Kalafi,M., 1986, Mon. Not. R. astr. Soc. 223,279

Yahil,A., 1988, Vatican Study Week, "Large-Scale Motions in the Universe",

DISCUSSION

GUDEHUS: The fact that the microwave background dipole and your IRAS data dipole point in the same direction implies that there should be very little infall of the Local Group towards Virgo. What limits do you find for the infall to Virgo?

ROWAN-ROBINSON: About Vinfall < 100 km/s!

GUDEHUS:That value is consistent with my m* data which gives a neglible infall of the Local Group to Virgo.

STRUCTURE ON A SCALE OF A TENTH THE EVENT HORIZON

R. Brent Tully
Institute for Astronomy, University of Hawaii
2680 Woodlawn Drive
Honolulu, Hawaii 96822 USA

In two recent articles (Tully 1986, 1987), I have argued for the existence of clustering of Abell clusters on a scale of 0.1 c. The purported structures would involve of order 60 Abell clusters and 10^{18} $M_\odot$ each. Further, I have claimed that we must reside in something I have called the Pisces-Cetus Supercluster Complex, which has the extraordinary property that it is concentrated to a plane coincident with the much more nearby plane of the Local Supercluster.

At the conference, I will show some computer-generated renderings of the distribution of galaxies and rich clusters. Two are included with this manuscript. Fig. 1 illustrates the extreme concentration of nearby galaxies to the equator of the Supergalactic coordinate system defined by de Vaucouleurs (1964), and Fig. 2 shows that Abell clusters in the south galactic hemisphere are also strongly concentrated toward this same equator. The scale of Fig. 2 is twenty times the scale of Fig. 1.

The problem is to evaluate the reality of the apparent large-scale structures. Here I will dwell on only one argument—the strongest one in favor of the physical nature of structure on a scale of 20,000 km s^{-1}. It involves a one-dimensional, two-point correlation function analysis. The test evaluates the statistical significance of the apparent coincidence between the planar distributions of nearby galaxies and distant clusters in the Pisces-Cetus feature.

Imagine that objects are confined to a plane of finite thickness. Consider two objects drawn randomly from the sample, as illustrated in Fig. 3. The projected separation of these two objects onto an arbitrary vector is Δr. The relative numbers of pairs as a function of projected separation can be described by the relation $N(\Delta r) \propto \Delta r^\gamma$. When the slope γ is calculated with real data for randomly selected position angles of the arbitrary vector, one finds $\gamma \simeq 0$, which is the anticipated result based on what is found in the cases of the three- and two-dimensional correlation functions. By contrast, if the vector is aligned normal to the plane, then there will be a large number of pairs with small Δr separations and relatively few with large Δr separations, whence $|\gamma| \gg 0$. The actual value of γ will not have a specific physical meaning but depends on the thickness of the plane and the fraction of the sample that is concentrated to the plane.

Of course, biases in the sample could create features in a display of the correlation slope vs. position angle of the arbitrary vector. Fortunately, it is possible to neutralize the two most obvious biases caused by galactic obscuration and incompletion with distance. We take recourse to the Cartesian version of the Supergalactic coordinate system described in Tully (1986, 1987). The SGY axis is directed within 6° of the galactic poles so, to

41

M. Mezzetti et al. (eds.), Large Scale Structure and Motions in the Universe, 41–46.
© 1989 by Kluwer Academic Publishers.

42

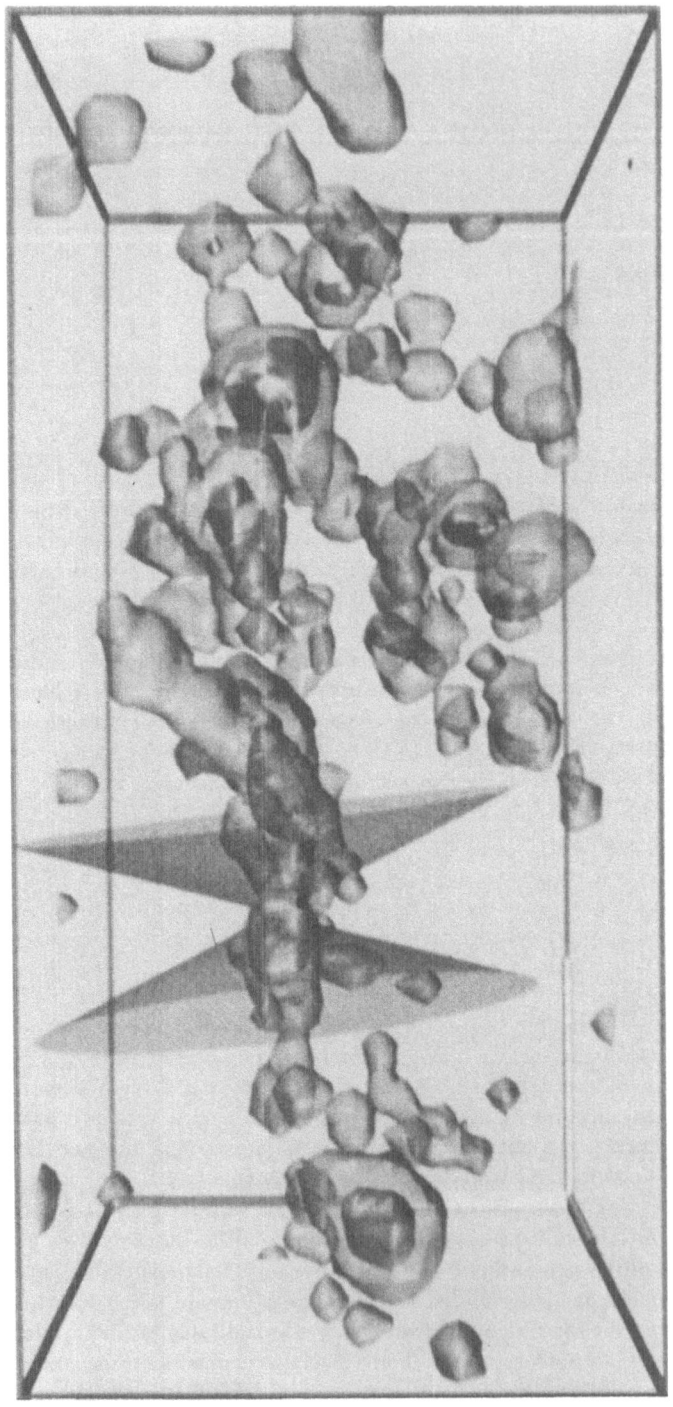

Fig. 1. Density contour map of nearby galaxies. The longest axis corresponds to a distance of 1500 km s^{-1} or 20 Mpc if $H_0 = 75$ km s^{-1} Mpc^{-1}. With this value of H_0, the outer surface represents a density of 1 galaxy Mpc^{-3}, and the inner surface represents a density of 4 galaxies Mpc^{-3}. Our Galaxy is located at the apex of the two cones, which delineate $b = \pm 20°$. The region between the cones is obscured. The supergalactic equator lies horizontally and passes through the position of the Galaxy.

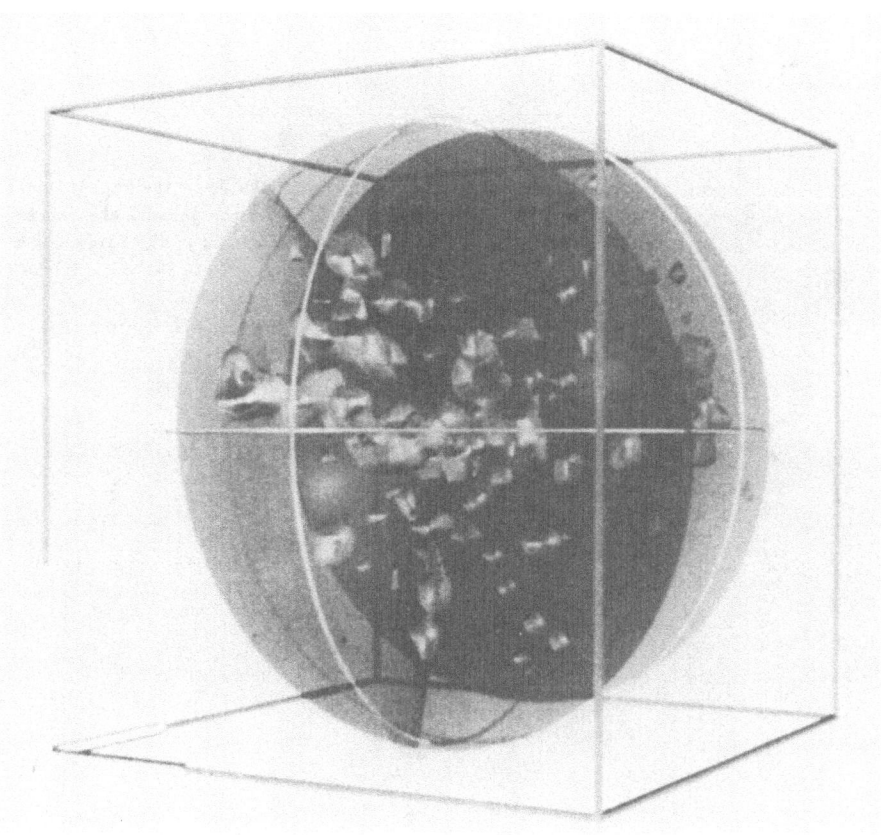

Fig. 2. Density contour map of rich clusters of galaxies. To the foreground in this view are clusters south of the galactic equator. The radius of the sphere containing data is 0.1 c. The horizontal circumferential line lies on the supergalactic equator. With $H_0 = 75$ km s^{-1} Mpc^{-1}, the outer surface represents a density of 1.7×10^{-5} clusters Mpc^{-3}, and the inner surface represents a density of 5×10^{-5} clusters Mpc^{-3}. The view is looking into the cone that delineates $b = -20°$. Another cone is seen edge-on that delineates $\delta = -27°$, the southern limit of the Abell survey. Non-Abell clusters with known velocities that are south of this limit are mapped, but there is considerable incompletion below $\delta = -27°$.

within this accuracy, vectors drawn in the SGX-SGZ plane are symmetric with respect to obscuration and distance incompletion.

The results of the one-dimensional correlation analysis are shown in Fig. 4. In the case of the top panel, the analysis was performed on a sample of 2367 galaxies with velocities <3000 km s^{-1} (the *Nearby Galaxies Catalog* sample), which is drawn essentially from the domain that was involved in the delineation of the Supergalactic coordinate system. There

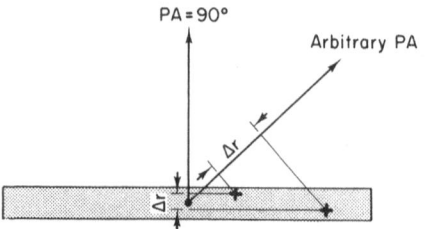

Fig. 3. One-dimensional two-point correlation. Two galaxies represented by the crosses would have the projection Δr onto a vector of arbitrary position angle (PA). If the galaxies are confined to a plane of finite thickness, the projected separations onto a vector normal to the plane (PA = 90°) are small.

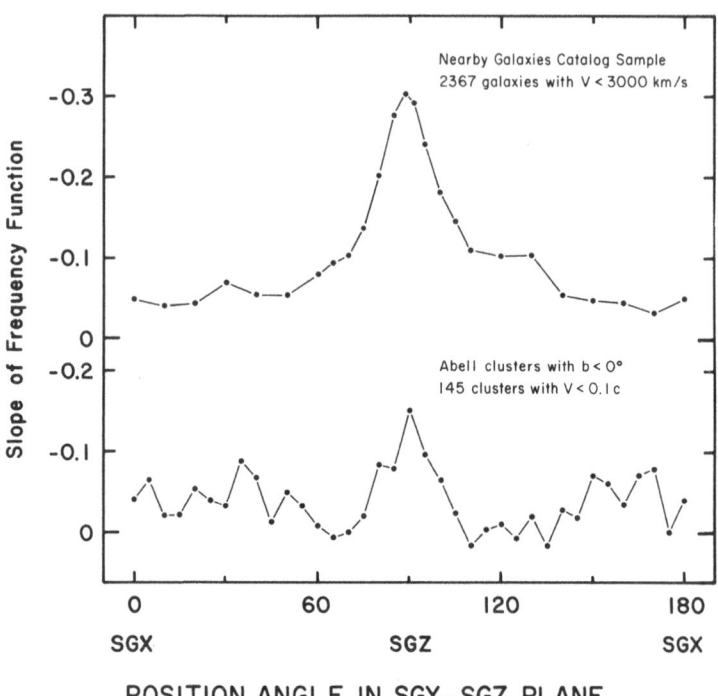

POSITION ANGLE IN SGX-SGZ PLANE

Fig. 4. Slope of the frequency function(unnormalized correlation function) versus position angle in the SGX-SGZ plane. PA = 90° corresponds to a vector along the SGZ axis, normal to the supergalactic equator. Top panel: nearby galaxies. The signal at PA = 90° is expected because it was the perception of a plane in the distribution of nearby galaxies that lead to the definition of supergalactic coordinates. Bottom panel: Abell clusters. The signal at PA = 90° has 3 σ significance.

is a well-known tendency for nearby galaxies to concentrate to a plane in this region, so it is no surprise that there is a highly significant signal at the position angle of the SGZ axis where projections are onto a vector normal to the known plane.

In the case of the bottom panel of Fig. 4, the analysis was repeated on a sample of 145 Abell clusters with $b < 0$ and $V < 0.1$ c. There is again a signal at the position angle of the SGZ axis that is evaluated to have a significance of 3.0 σ. This signal is seen in spite of a negative bias. The Abell sample is restricted to $\delta > -27°$ which, together with the restriction $b < 0$, defines a half-moon region in projection with the short cross-section of the moon roughly aligned with the SGX axis and the long cross-section roughly aligned with the SGZ axis. This geometry allows larger values of Δr to be found in projection onto the SGZ axis than onto the SGX axis and accounts for the broad minimum underlying the signal in the case of the clusters. If this effect is taken into consideration, the significance of the signal is even greater.

The interpretation of this test is that a substantial fraction of Abell clusters with $b < 0$ and $V < 0.1$ c are in strata parallel to the plane of the Local Supercluster. In itself, the test does not require that these clusters be in the *same* plane as the Local Supercluster nor that there be only one plane. However, other ways of looking at the data reveal that most of the contribution to the positive signal is coming from the concentration of clusters in Pisces-Cetus at typical redshifts of 15,000–20,000 km s^{-1} and, indeed, that these clusters live in a stratum that is remarkably coincident with the plane of the Local Supercluster. The vertical displacement of the planes is $0 \pm 10\%$ of the distance between Pisces-Cetus and ourselves. There is a tantalizing hint that some of the power in the signal seen in the bottom panel of Fig. 4 comes from other strata, but the statistics are still too poor to explore this possibility.

SUMMARY

The number of clusters with known redshifts is still so small that it is difficult to provide convincing statistical evidence for the reality of structure on scales up to 0.1 c, especially in the face of potential biases in the cluster samples. However, the one-dimensional two-point correlation analysis does give a 3 σ confirmation of the reality of the nearest structure. This analysis neutralizes the two obvious biases that result from galactic obscuration and incompletion with distance. Nearby galaxies and distant clusters in Pisces-Cetus are concentrated to the same great circle on the sky but *not* to the same lines of sight, since most of the nearby galaxies are in the north galactic hemisphere, while most of the clusters in the Pisces-Cetus Supercluster Complex are in the south.

The evidence from the test that has been described is for a phenomenon considerably more bizarre than simply structure on a scale of 0.1 c. The evidence is that *the structure adheres to a flattened plane* across such a great distance.

REFERENCES

de Vaucouleurs, G., and de Vaucouleurs, A. 1964, *Reference Catalogue of Bright Galaxies* (Austin: University of Texas).

Tully, R. B. 1986, *Astrophys. J.* **303**, 25.

Tully, R. B. 1987, *Astrophys. J.* **323**, 1.

DISCUSSION

RICHTER: Can you determine our distance from the plane defined by Abell clusters?

TULLY: We lie within about 1 Mpc of the best-fit plane of the Local Supercluster. The planes of the Local Supercluster and the Abell clusters in Pisces-Cetus are coincident to within +/-10% of the distance between us and the Pisces-Cetus feature, i.e. 0+/-20 Mpc.

DE ZOTTI: We (Persic et al., in preparation) have analyzed the surface brightness of the hard X-ray background and do not find any structure. This result is very hard to reconcile with the existence of very large-scale structure involving rich clusters of galaxies, given that clusters are powerful hard X-ray sources.

TULLY: You speak of an attractive approach, but remember that no significant structure is seen in the surface distribution of Abell clusters either. And all the filamentary structure seen in the distribution of nearby galaxies is not obvious in two dimensions. Your conclusion might be different if you had velocity information and could resolve in distance.

RHEE: Structure on 0.1 c scales is visible in smoothed density maps of the Abell catalogue; why are these structures not visible when one looks at stereogram of the unsmoothed distribution of points?

TULLY: I have noticed myself that it is much more difficult to see structure in a field of points than to see structure in smoothed iso-density surface. Maybe the smoothing simplifies the presentation enough that the mind is able to digest the image.

PEEBLES: To help us judge the significance of your contour maps of the Abell cluster distribution, it would be most helpful to have contour maps constructed the same way for model distributions of points with the same low-order correlation functions as the Abell clusters but no filamentary structure.

TULLY: Sounds reasonable. It is the real data that have the fiercest "boundary conditions". Simulated data should be similarly constrained. I would be happy to work with anyone with simulated data in an effort to display it and the real data in a common way.

DWARF GALAXIES AND LARGE-SCALE STRUCTURE

Bruno Binggeli
Osservatorio Astrofisico di Arcetri
Largo E. Fermi 5
I-50125 Firenze, Italy

ABSTRACT. Recent models based on the concept of "biased galaxy formation" predict that dwarf galaxies should be more uniformly distributed
than luminous galaxies on large scales. In particular, dwarf galaxies
are expected to fill the voids. This paper reviews all available
information on the large-scale distribution and clustering properties
of different types of dwarfs, including objects of extremely low
surface brightness. It is shown that the observational evidence is
AGAINST a large-scale segregation between giants and dwarfs. Residual
differences on smaller scales can be attributed to the morphology-
density relation. Consequences for "biased galaxy formation" are briefly
discussed. A future merit of mapping dwarf galaxies is that they may
reveal the FINE-STRUCTURE of the large-scale structure of the Universe.

1. INTRODUCTION

The grand picture of the large-scale structure of the Universe, as it
emerged over the last 15 years - with its surprising non-uniformities,
filaments, sheets, and voids (see,e.g., Chincarini and Vettolani 1987) -
is based entirely on high-luminosity, giant galaxies. These objects are
known to constitute merely the tip of an iceberg of mostly low-luminosity,
dwarf galaxies. How, then, can we be sure that giant galaxies are fair
tracers of the large-scale structure? - Dwarf galaxies, with their
characteristically low surface brightness, are hard to detect even
nearby; they could easily fill the prominent voids that appear in the
distribution of bright galaxies, while going unnoticed. The principal
possibility has always been clear. What is new, however, is the suspicion, and even prediction that giants and dwarfs are spatially seg-
regated in this way - based on a concept called "biased galaxy formation".
The central idea of this concept (Kaiser 1986, Bardeen 1986) is that
luminous galaxies were formed preferentially in high-density regions
of the Universe, thus giving us a biased, i.e. exaggerated, contrast-
enhanced view of the large-scale distribution of matter. The main
motivation is to reconcile $\Omega \approx 0.1$, as traced by the luminous matter,
with the theoretically preferred $\Omega = 1$ (inflation); there are other

47

M. Mezzetti et al. (eds.), Large Scale Structure and Motions in the Universe, 47–63.
© *1989 by Kluwer Academic Publishers.*

merits of biased galaxy formation (see Dekel and Rees 1987). A convenient
formal description of biasing – in lack of a physical basis – is to set a
density threshold of 2.5 or 3 times the mean (1σ) density fluctuation for
the formation, or identification of normal, giant galaxies; these are then
automatically stronger clustered than the underlying dark matter, provided
the fluctuation spectrum is Gaussian (Kaiser 1984, Politzer and Wise
1984).- But all of this tells us nothing about dwarf galaxies. It should
be emphasized that "biased galaxy formation" does not necessarily imply
a large-scale segregation between giants and dwarfs; only between
luminosity (bright galaxies) and mass (dark matter); dwarf galaxies are
negligible in terms of luminosity as well as mass. To bring in the dwarfs,
additionally requires a model of the formation of dwarf galaxies. Such a
model has been worked out by Dekel and Silk (1986) within the framework
of the "cold dark matter" scenario, where the low-surface brightness
dwarfs originate from 1σ fluctuations and thus are expected to be much
more uniformly distributed than the rare high-density peaks, i.e.the
giants. Dwarf galaxies should essentially trace the dark mass on large
scales – right into the voids.

To test this hypothesis is obviousely not an easy task. Dwarf galaxies do give the observer a foretaste of dark matter. Nevertheless, the
last few years have seen fast progress in the detection and mapping of
low-surface brightness objects; decisive surveys are now underway.-
It will be argued below that enough is known about the abundance and
distribution of dwarf galaxies already to strongly dismiss the above
hypothesis. Dwarf galaxies closely follow the distribution of giants on
large scales. Residual differences on smaller scales are part of the
general morphology-density relation (Dressler 1980). The consequences of
this negative result for "biased galaxy formation" will be sorted out in
the concluding section.

In the following we will review all available information on the
large-scale distribution of dwarf galaxies, beginning with the direct
approach of dwarf surveys in low-density regions, turning to indirect
evidence by means of clustering properties and luminosity functions, and
finally discussing the different types of dwarfs case by case.

2. SEARCHES FOR DWARFS IN THE VOIDS

It is an obvious task to look for dwarf galaxies in the voids, since they
are predicted to be there. But before initiating new surveys one can use
existing galaxy catalogs to check for systematic trends in the distribution
of galaxies with respect to luminosity or surface brightness. The only
catalog to date that contains a reasonably large number of intrinsically
faint galaxies to serve this purpose is the diameter-limited Uppsala
General Catalog of Galaxies (UGC, Nilson 1973). Bothun, Beers, Mould, and
Huchra (1986), by measuring the redshifts of many low-surface brightness
galaxies, arrived at a total of ~4600 UGC's with known redshifts; this
sample they divided into three regimes of surface brightness to test for
variations in the redshift distribution (of galaxies in a given area) as
a function of surface brightness. The result turned out to be negative;
low-surface brightness galaxies seem to obey the voids as well. There are

some shortcomings in Bothun et al.'s (1986) analysis, however. The red-shift incompleteness of their sample of low-surface brightness galaxies is unknown; low-surface brightness galaxies need not be dwarf galaxies; and the surface brightness values themselves are highly uncertain because they are defined by eye-estimeted magnitudes and diameters.

These difficulties have been avoided by Thuan, Gott, and Schneider (1987), who collected a small but complete redshift sample of ~ 60 UGC galaxies that lie in the area of the Center for Astrophysics (CfA) "slice of the Universe" through the Coma Cluster (de Lapparant et al. 1986) and are classified as dwarf galaxies. A plot of these objects in the CfA wedge diagram is quite revealing: The dwarfs are well confined to the web-like structure delineated by the bright CfA galaxies; they do not fill the voids. - Four systems do appear somewhat isolated, but here Thuan et al. (1987) argue that these may indicate possible bridges of faint (yet unmapped) galaxies between the bright structures. We will support this view in Sec.5. A possible caveat of Thuan et al.'s analysis (although not a critical one for their conclusions) is that all UGC "dwarfs" beyond a velocity of 3000 km/s - where the big voids in the CfA slice show up - are brighter than absolute blue magnitude $M_B \approx -17.5$, i.e. are barely dwarf galaxies. Hence Thuan et al. could not really test for typical (faint) dwarfs in the voids. The important fact, however, is that at low redshifts the clumpy and cloudy distribution of giants on the scale of superclusters is well followed by dwarfs as faint as $M_B \approx -13$. This is clearly not in favor of Dekel and Silk (1986).

A different area of sky, containing a smaller void in the CfA universe, was probed for dwarf galaxies by Eder, Schombert, Dekel, and Oemler (see Oemler 1987) - again with the result that the dwarfs did not fill in.

A basic deficiency of all studies mentioned so far, however, is that they are restricted to gas-rich dwarfs, i.e. dwarf irregulars. This is mainly because dwarf elliptical galaxies (the gas-less types) are fainter and of lower surface brightness on average and thus figure rarely, e.g., in the UGC catalog. Also,for irregulars it is easy (in principle) to get a redshift from the 21cm line. - Yet clusters of galaxies are known to be swamped with dE systems; 800 out of 1300 members of the Virgo Cluster are dwarf ellipticals (see Binggeli, Sandage, and Tammann 1985). If this high abundance of dE's persisted in the field (outside of clusters), the voids might be swamped as well. Motivated by this perspective, Binggeli, Tarenghi, and Sandage (1988) conducted a survey of low-surface brightness dwarfs on deep (2-3 hour exposure) IIIaJ-Schmidt plates covering two northern strips of 900 square degrees of sky. A depth corresponding to $v \approx 3000$ km/s was reached for systems of $M_B \approx -16$. Surprisingly few dwarfs were found on the whole and, judged from the surface distribution of galaxies, almost all of them (including the dwarf ellipticals) are con-fined to groups and clouds of bright galaxies (see also Fig.1), in good accord with Thuan et al. (1987). The relative abundance of dE systems in the field is definitively much lower than in clusters. This fundamental fact will be adressed in Sec.4.

The new Palomar Sky Survey on fine-grain emulsion will bring low-surface brightness galaxies out of the shadow on a large scale. Schombert and Bothun (1988) have begun to catalog these objects while the survey is underway; this will lead to a data base far superior of the UGC. But the ultimate direct check for faint,low-surface brightness dwarfs in the prominent voids (- not achieved so far!) will come from David Malin's technique of photographic amplification; his method has recently been used by Impey, Bothun, and Malin (1988) to uncover a special brand of ultra-low-surface brightness galaxies in the Virgo Cluster; to be dis-cussed in Sec.6. (Note, however, that in this case one will be forced to argue with sky-projected densities of faint objects, because redshifts will not be measurable for most of them.)

3. CORRELATION FUNCTIONS

Why endeaver to reach distant voids while the truth lies in our backyard? According to "biased galaxy formation", luminous galaxies should be more strongly clustered than the dark matter on all scales. The numerical calculations for a "biased cold dark matter" scheme of Davis, Efstathiou, Frenk, and White (1985), and White, Frenk, Davis, and Efstathiou (1987), setting a density threshold of 2.5σ for luminous matter ("galaxies"), - show the amplitude of the (spatial two-point) "galaxy" correlation function to be almost a factor of 10 above that of the mass correlation function on small scales (separations $\ll 1$ Mpc) - with a gradual decline of this difference to a factor of ~ 4 at separations of ~ 30 Mpc (H_o=50), i.e. the "galaxy" correlation function is also somewhat steeper than the mass correlation function. If, therefore, dwarf galaxies would trace the dark mass (as Dekel and Silk (1986) suggest), we should be able to see this already in the correlation function of nearby dwarf galaxies.

The first (apparent!) observational support for "biased galaxy formation" did indeed come from correlation functions. Davis and Djor-govski (1985), working with UGC objects, found a strong dependence of the galaxy correlation function on surface brightness, in the sense that correlations are weaker and shallower for low-surface brightness (dwarfish) galaxies than for high-surface brightness (normal) galaxies. However, as pointed out by Bothun et al.(1986), and Thuan et al.(1987), there is a fundamental flaw in this analysis regarding the redshift distribution of low-surface brightness galaxies. On the other hand, it has been shown long ago by Davis and Geller (1976) that galaxy correlat-ions do depend on morphological type; elliptical galaxies are more strongly clustered than S0 galaxies, S0's more than spirals. This is equivalent to Dressler's (1980) well-known morphology-density relation, where the frequency of a type appears as a function of local density. Giovanelli, Haynes, and Chincarini (1986), by calculating angular correlation functions, as well as Dressler diagrams (type frequency versus local density) for galaxies in the Perseus-Pisces Supercluster, - have demonstrated that the morphological segregation holds even for the spiral subtypes; the "later" the spirals the weaker the clustering - all the way down to systems later than Sc, where we enter the domain of dwarf irregular galaxies. Here again some saw the first indications of a luminosity segregation on large scales. What is overlooked, however, is that irregulars are not representative of all dwarfs. For dwarf ellipticals, in fact, the trend is just reversed: they are strongly clustered (see below).

The $\underline{\text{spatial}}$ correlation function of all types of dwarf galaxies is still a desired quantity. It is indeed difficult to get because redshifts are required for the dwarfs - either for all dwarfs (including the diffuse dE's!) to construct the spatial correlation function directly; or for a $\underline{\text{representative}}$ sample of dwarfs to transform an $\underline{\text{angular}}$ into a $\underline{\text{spatial}}$ correlation function. The only knowledge to date on the variation of $\underline{\text{spatial}}$ galaxy correlations with luminosity is restricted to normal (bright) galaxies. Phillips and Shanks (1987), who invoked an ingenious method to estimate the spatial correlation function on the basis of the excess density of galaxies around a given center galaxy whose distance is known, - have tested for the magnitude range ($-17 \leq M \leq -20$) and found no variation of the correlation function (see, however, Börner, Deng, and Xia in this volume).

In summary: the route via correlation functions is not practicable at present. We have to rely on more qualitative knowledge of the clustering properties of dwarf galaxies.

4. A SIMPLE ARGUMENT BASED ON LUMINOSITY FUNCTIONS

A very simple yet quantitative argument can be produced to dismiss "biasing" in the mode of Dekel and Silk (1986) at once. It is based on the luminosity function of galaxies. Suppose that faint galaxies indeed are more uniformly distributed than bright galaxies: this should reflect in the luminosity function, in the sense that the number ratio of faint-to-bright galaxies is higher in low-density regions than in high-density regions. Applied to typical environments this means that, relative to giants, dwarf galaxies are expected to be more abundant in the field than in clusters - or technically speaking: the faint-end slope of the luminosity function should be steeper in the field than in clusters. What is observed, however, is just the $\underline{\text{opposite}}$ trend! -

The case is not trivial: until recently any variations of the luminosity function with environment have not been acknowledged. The reason why is that the faint end of the luminosity function was essentially unknown: - one simply used to model it by fitting a Schechter (1976) function to the $\underline{\text{bright}}$ end; this yielded a faint-end slope parameter $\alpha \approx -1.25$ for $\underline{\text{both}}$ field and clusters. It is almost surprising that the luminosity function of Virgo Cluster galaxies has recently been found to be in good accord with this prediction down to $M_B \sim -12$ (Sandage, Binggeli, and Tammann 1985); so - for clusters of galaxies the exponential faint end is a valid description. However, as the magnitude limits of $\underline{\text{field}}$ studies are pushed faintward, it is becoming clear that the luminosity function of field galaxies is essentially $\underline{\text{flat}}$ at the faint end ($M_B > -17$), i.e., has rather $\alpha \approx -1$ instead of $\alpha \approx -1.25$ (cf. Binggeli, Sandage, and Tammann 1988). The most recent evidence comes from the field dwarf survey of Binggeli, Tarenghi, and Sandage (1988), whose integrated dwarf counts are down by a factor of three compared to what would be expected based on a cluster luminosity function; being consistent with a flat faint end.

All of this means that dwarf galaxies are more strongly clustered than giant galaxies on the scale of clusters and superclusters of galaxies.

Again: this is opposite to the prediction of Dekel and Silk (1986).
Of course one might object that we do not know the luminosity function
specifically for regions of very low density (the voids, that is).
However, the general luminosity function can be written as the sum over
all type-specific luminosity functions (which approximately have universal
shape) weighted by the environment-specific type fractions (Binggeli,
Sandage, and Tammann 1988). The type mixture, on the other hand, has been
shown by Postman and Geller (1984) to become independent of the local
density at densities below ~1/10 galaxy/Mpc^3 (galaxies brighter than
$M_B=-17.5$; $H_0=50$). Hence we do not expect the general luminosity function
to change any further by going from the low-density "field" into the
very low-density "voids".

It is indeed the density-dependent mixture of galaxy types (with a ~
constant shape of the luminosity function for any given type) which
explains the difference of the faint-end slope of the total luminosity
function for clusters and the field. The total faint-end slope is
essentially determined by the relative abundance of dwarf elliptical
galaxies which, taken alone, show a steeply rising luminosity function:
in the field, however, their number is so small that the dwarf irregulars
with a broad Gaussian luminosity function take over - and one ends up
with a flat faint end (cf. Fig.1 in Binggeli, Sandage, and Tammann 1988).

Here, at the latest, we have to differentiate our analysis with
respect to morphological type. The simple argument presented above holds
only for dwarf elliptical galaxies; for irregulars it is incorrect.
Although this does not weaken the evidence against Dekel and Silk -
(because in their model the difference between dE and Irr is only of
secular origin, i.e. one can put them together here) - we now want to
push further by asking - Is there any single type of dwarf galaxy that is
much more uniformly distributed than the rest? -

5. EARLY-TYPE VERSUS LATE-TYPE DWARFS

The diffuse dwarf ellipticals, which escape our attention more easily
than the irregulars, are regarded by theoreticians as the prime candidate
for an "unbiased" (less clustered) population of faint galaxies (Dekel
and Silk 1986; Dekel and Rees 1987; Silk, Wyse, and Shields 1987). But,
in fact, they are the worst possible candidate! This is clear after what
has been said above. Dwarf galaxies are much rarer in the field than in
clusters - and those that do appear outside of clusters typically are
satellites of massive giants; i.e. there is strong clustering also on
small scales. Fig.1 serves as an illustration. Very few truly isolated
dE systems were found in the survey of Binggeli et al.(1988). On the whole,
dwarf ellipticals seem to be clustered like giant ellipticals (see also
Binggeli, Tammann, and Sandage 1987).

Remain the dwarf irregulars - which indeed are the least clustered
type of galaxies in the Universe (cf. Sec.3, and the contributions to this
volume by Shaver; and Iovino, Melnick, and Shaver). But whether or not
irregulars may constitute the sought-for species of galaxies cannot be
judged from correlation analysis alone; one has to look at their distribution directly. By doing so, one finds them closely confined to groups

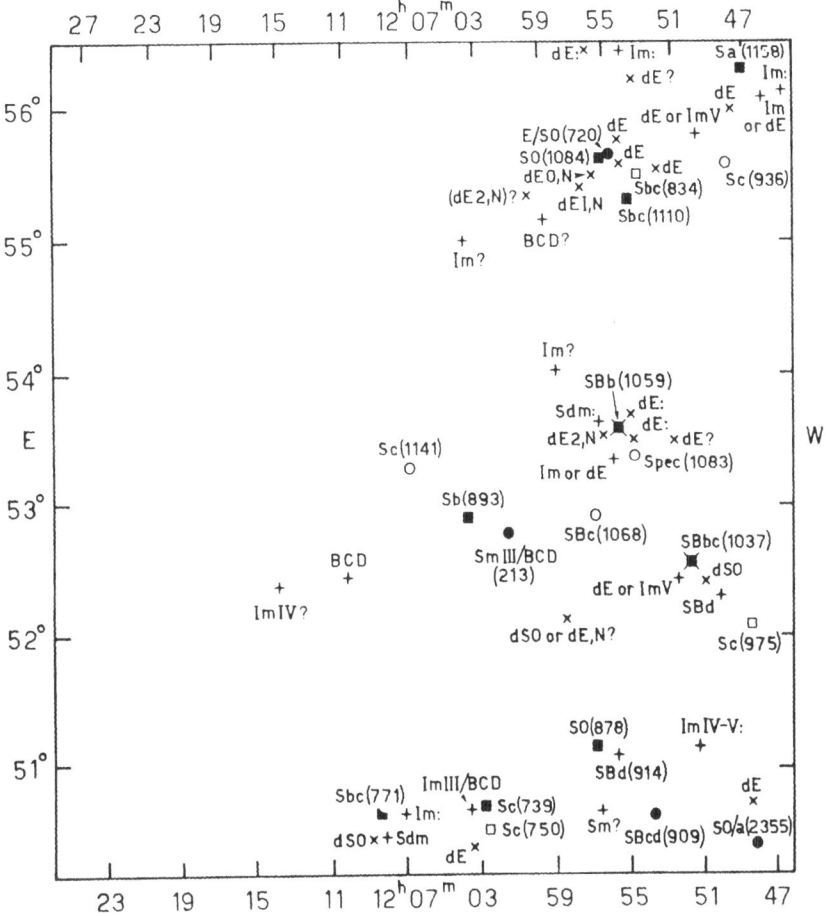

Figure 1. Field No.269 of the Zwicky Catalog (Zwicky et al.1961-68). Shown with the original magnitude coding (squares and circles, open and filled) are all Zwicky galaxies that have heliocentric v≤2500 km/s and $m_z \leq 14.5$. Additionally shown are many dwarf galaxies (non-Zwicky's), most of which have recently been discovered in the deep Schmidt survey of Binggeli, Tarenghi, and Sandage (1988). Early-type dwarfs (dE,dS0) are indicated by crosses, late-type dwarfs (Sm,Im,BCD) by pluses; uncertain types have question marks. The morphological types are from the RSA Catalog (Sandage and Tammann 1981), or from the above-mentioned survey work of Binggeli et al. The numbers in parentheses give heliocentric velocities (from Huchra et al.1983). The field contains a small portion of the "Ursa Major Cloud", which covers the right half of the figure. Notice the absence of dwarfs ($M_B \leq -14$) outside the cloud (on the left). ⁻

and clouds of bright galaxies. Clouds are large, unbound concentrations
of galaxies that comprise almost all of the "field" galaxy population
(Tully 1982; Tully and Fisher 1987). One such cloud is "Ursa Major" (at
roughly the distance of the Virgo Cluster), part of which is shown in
Fig.1. It is clearly seen that late-type dwarfs are the most scattered
objects - but also that they, too, respect the boundary of the cloud.
Irregulars are far from being uniformly distributed! The confinement of
dwarfs to the structures formed by bright galaxies has been shown by
Thuan et al.(1987) to hold also in 3D space. All of this has been
discussed in Sec.2. The situation can thus be summarized as follows.
For dwarf galaxis, the great voids are still out of reach; but there are
many smaller voids - between the clouds - and those are void of all types
of dwarfs! -

Notice (in Fig.1) how groups seem to be embedded in the cloud, made
up of one or few massive galaxies that are surrounded by (predominantly)
dE satellites. The groups, if bound, constitute "hot spots" in a quietly
expanding stratum of spirals and irregulars. (In fact, the densest core
of the U Ma Cloud is called the U Ma Cluster; see Tully and Fisher 1987.)

The same features can be found in our local neighbourhood. Consider
Fig.2. The only known dwarf ellipticals in that region (excluding the
M81/N2403 Group) are satellites of the two most massive galaxies: the
Milky Way (Fornax, Sculptor, LeoI, LeoII, Carina, U Mi, and Draco), and
the Andromeda galaxy (N205,N147, N185, AndI-III). The completeness of this
sample is of course a big question; extreme systems like Draco could hide
almost everywhere (in fact, this is the topic of the following section).
However, dwarfs like N147 or Fornax would easily be detected in the Virgo
Cluster. It seems unlikely that the concentration of dE's around the two
giants is simply an observational bias.

The late-type dwarfs, on the other hand, are organized in clouds
together with the spirals. Again there are regions void of galaxies.
One well-defined cloud is Canes Venatici I (CnV I)"; another one may be
comprised of what is separately described as the "Local Group", and the
"South Polar Group".

This point deserves special attention. The classical boundary of the
LG has been put at a radius of 1.5 Mpc, centered on the Galaxy. A glance
at Fig.2 makes clear, however, that this definition is artificial and
physically meaningless. There is a high abundance of "LG members" in the
south galactic polar cap out to the magic boundary, behind of which
immediately follows the "South Polar Group". It seems more likely that one
deals here with one structure: an unbound cloud of late-type galaxies,
containing (at least) two hot spots - the Galaxy and M31 (with their
satellites) which, together with a few other "LG members", constitute a
larger hot spot, i.e. bound group: the LG proper. Of course, we argue
here only from a morphological point of view; the dynamics have to be
considered as well. (Note that in the nomenclature of Tully and Fisher
(1987) our small clouds are still called groups, even if they are not
bound, e.g. CnV I Group instead of CnV I Cloud.)

The most remarkable feature in Fig.2, however, is the concentration
of late-type dwarfs between the LG and CnV I (Leo A, GR 8, Sex A, Sex B,
D187, N4163). (The LG membership of Leo A and GR 8 has sometimes been
debated, which now appears as a meaningless task.) The point is that

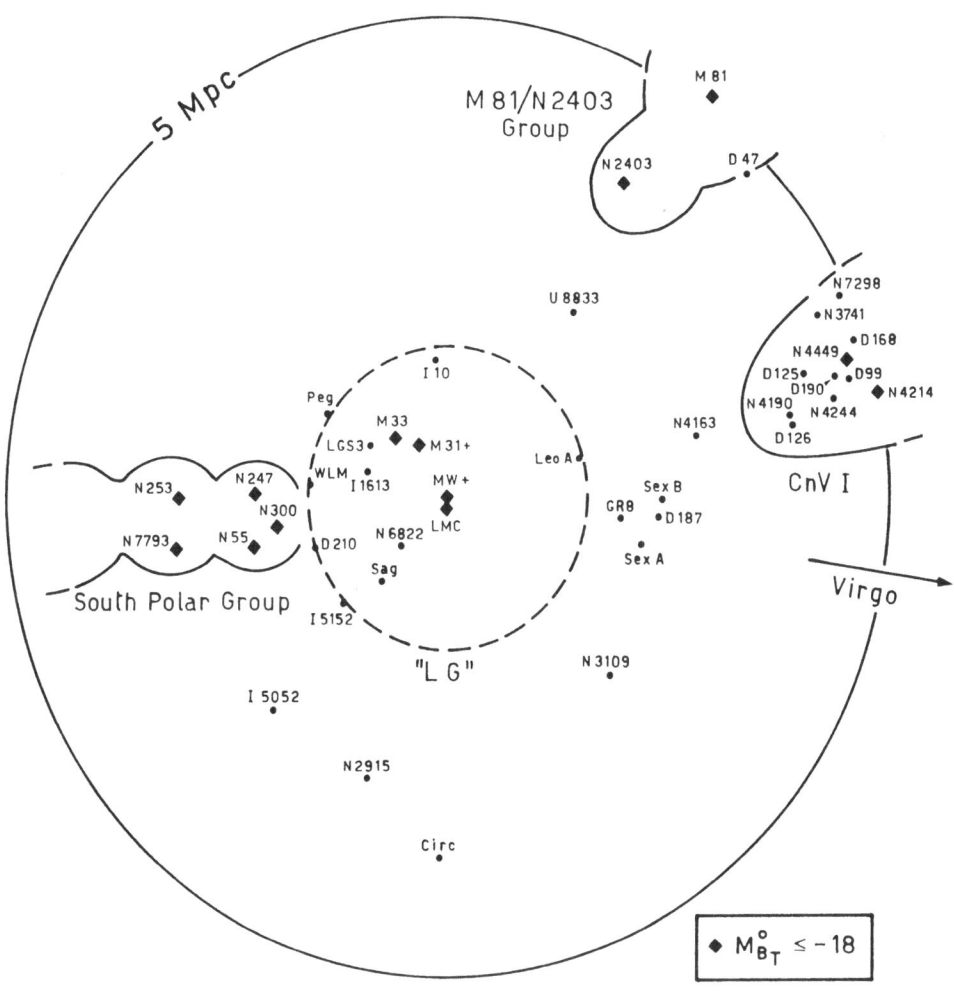

Figure 2. The neighbours of the Milky Way Galaxy out to a distance of 5 Mpc, projected onto the supergalactic plane. The very close (and faint) satellites of our Galaxy and M31 are not shown. A rough distinction between bright (giant) and faint (dwarf) galaxies is made with a division line at $M_B=-18$. The faintest dwarfs shown have $M_B \approx -11$. Various groupings are indicated: The Local Group (LG), with its classically adopted boundary radius of 1.5 Mpc (broken circle); the South Polar (or Sculptor) Group, and the M81/N2403 Group, both shown without dwarfs in lack of individual distances; and the Canes Venatici (CnV I) Cloud, which is assumed to be freely expanding. The direction to the Virgo Cluster (at a distance of ≈ 22 Mpc) is also indicated. With few exceptions the data are from Kraan-Korteweg (1986) and rely on $H_0=50$, and a local infall velocity towards Virgo of 220 km/s. Notice - (1), the absence of dwarfs (and giants) in large regions of space (-not sufficiently explained by the zone of avoidance); (2), the concentration (bridge?) of dwarfs between CnV I and the Local Group; (3), the ill-fated definition of the Local Group. -

dwarfs can be clustered without being associated directly with bright
galaxies; - yet these dwarf groupings are not isolated in space but seem
to form faint bridges between the prominent structures!- The overall
impression (in Fig.2) is indeed that of a filamentary (prolate) structure
that runs from the (SPG+LG) cloud over to CnV I. The feature extends in
fact well beyond CnV I and is known as the "Coma-Sculptor Cloud" (Tully
and Fisher 1987). It is interesting to note that Thuan et al.(1987) also
found some apparently isolated dwarfs (in the CfA slice) that could be
interpreted as traces of faint connecting filaments (cf. Sec.2). Thus it
appears that dwarf galaxies do, after all, play an important rôle in
the mapping of the large-scale structure of the Universe. They may reveal
the fine-structure of the large-scale structure.

6. GALAXIES OF VERY LOW SURFACE BRIGHTNESS

As the predicted population of non-clustered ("unbiased") faint galaxies
cannot be identified with any of the classical dwarf types, these galaxies
simply do not exist -- or they are invisible. We still have to check on
this latter possibility.
    That the realm of galaxies could be much more diverse than what is
contained in our catalogs has been realized a long time ago (Arp 1965,
Disney 1976). Disney and Phillips (1983) have shown that galaxies of
"normal" characteristics are preferentially selected in a photographic
survey with limits on magnitude and angular size. Essentially, it is the
central surface brightness that decides whether or not a galaxy is detectable at all. Large numbers of galaxies of very low surface brightness
could simply be outshined by the night sky and therefore have been missed.
Some galaxies of this type are known to exist: the local dwarf spheroidals
(cf. the preceding section) that were discovered accidentally because
being nearby they are resolved into stars. Thus it is meaningful (and for
theoreticians natural) to pursue the possibility that the "unbiased"
dwarf population consists of galaxies of (nearly) undetectably low
surface brightness. These systems, if containing HI, might even be responsible for the Lyman alpha absorption systems observed in the spectra
of distant quasars (Sargent, Young, Boksenberg, and Tytler 1980).
    Recent progress in photographic techniques of enhancing faint
features due to D.Malin (for refs. see Impey et al.1988) allows now the
routine detection of systems of central surface brightness as low as
26 B mag/deg^2, which is at least one magnitude fainter than what could be
achieved before. Applying Malin's technique to the Virgo Cluster (using
Schmidt plates), Impey et al.(1988) promptly uncovered a large number
of extremely diffuse objects in that cluster. This caused quite some
euphoria, since the predicted, hidden sea of diffuse dwarfs may just have
been found! -
    However, it can be argued that this is not likely the case. First
of all, 110 of the 137 "Malin objects" in Virgo studied by Impey et al.
are already listed in the Virgo Cluster Catalog. The newly discovered
systems, on the other hand, appear to constitute the faint extension of
a known, special class of dwarfs that are unusually large (i.e. have
unusually low surface brightness) at a given total magnitude, described

by Sandage and Binggeli (1984). The distribution of these dwarfs in the Virgo Cluster is shown in Fig.3. Notice the strong confinement to the distribution of luminous cluster galaxies, - demonstrating that these objects are indeed cluster members, and hence that the abundance of "Malin dwarfs" in the field (foreground and background of the cluster) must be very low.

This is supported by the field survey of Binggeli, Tarenghi, and Sandage (1988), in which a single one objects with "Malin" characteristics showed up - in the densest part of Ursa Major. Malin himself has noted the sparseness of low-surface brightness objects on high-contrast prints of non-cluster areas (cf. Phillips, Disney, Kibblewhite, and Cawson 1987, p.510); but a systematic field survey with Malin's technique has yet to be carried out. It seems not unreasonable to speculate that the large, low-surface brightness types are tidally distorted dwarfs (or tidal debris), which naturally would explain their occurrence in clusters.

One "Malin object" in the Virgo Cluster did, in fact, turn out to lie in the background: "Malin 1" - a huge spiral galaxy at a velocity of ≈ 25000 km/s! (Bothun, Impey, Malin, and Mould 1987). This object is as fascinating as unique. Given its distance, and taking into account the area covered by Impey et al.(1988), one can derive a rough upper limit for the spatial density of "Malin 1" systems of one per 10^5 Mpc^3. Thus, even if distributed uniformly in the Universe, these objects have no bearing on our subject. Other "Malin 1's" may be discovered, but they are expected to be rare (cf. also Bosma, this volume).

By now the predicted sea of non-clustered "galaxies" has vaporized almost completely. If such a population exists, its members have to be truly invisible - below a surface brightness level of ~ 27 B/deg^2, even at the center! Systems of such low stellar densities - if there are stars at all - should essentially be gas clouds. The ultimate possibility we are left to consider is therefore a Universe filled with optically invisible gas clouds ("protogalaxies").

A handle on the abundance of intergalactic HI clouds is provided by direct searches for 21cm radio emission, and by studies of absorption features in the optical spectra of quasars. A number of radio surveys have given upper limits on the frequency of such clouds, the most strin- gent one by Hoffman, Helou, Salpeter, and Lewis (1988; cf. also for earlier references), who did not find any significant number of HI clouds without accompanying optical emission in the Virgo Cluster. A similar, negative result, based on the quasar absorption line method, is reported by Brosch (this volume). The Lyα systems in the spectra of quasars (which show that there is much intervening hydrogen, if not nearby) have been used to derive the clustering properties of the Lyα absorbers, in particular, whether or not they are consistent with the presence of voids in the large-scale structure. The latest trend is more or less to deny the existance of voids in the Lyα forest. Pierre, Shaver, and Iovino (1988, and this volume; cf. also for earlier refs.) have found that the voids are less pronounced in the distribution of Lyα absorbers at high redshifts than they are in the distribution of galaxies today. It is impossible, however, to conclude that this means that there are, after all, invisible clouds also in the nearby voids; the voids could simply have become bigger in the course of time (evolution of the large-

58

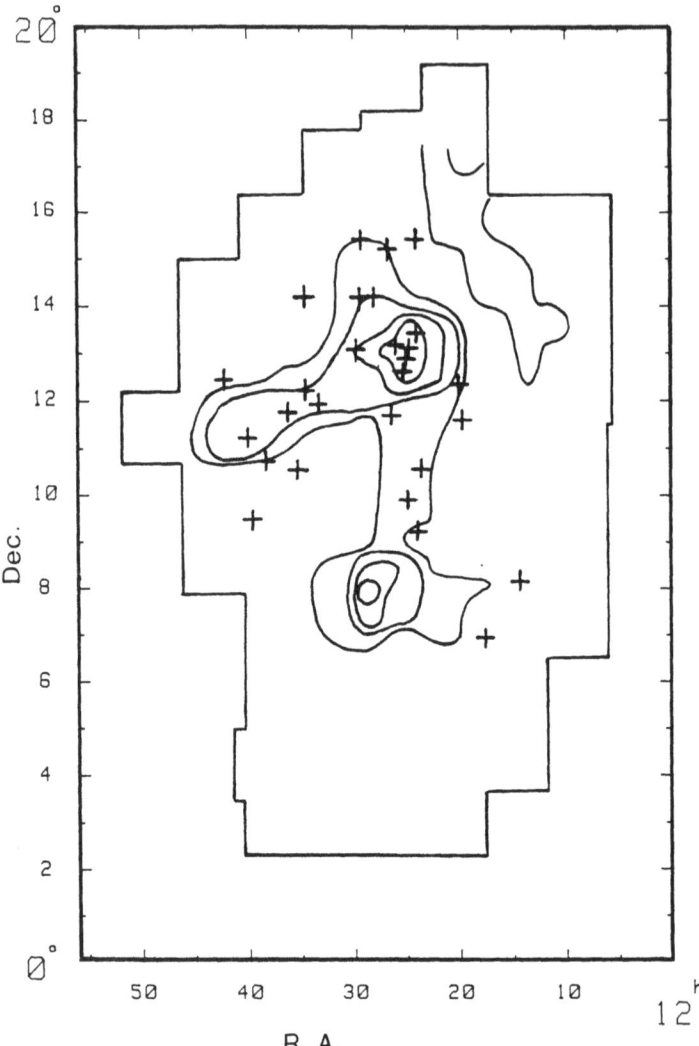

Figure 3. Distribution of dwarfs of very low surface brightness in the Virgo Cluster. This class of galaxies has been identified, and is illustrated in, Sandage and Binggeli (1984). The objects plotted (+) are listed in Table XIV of Binggeli et al.(1985). The contour lines, taken from Binggeli et al.(1987), are lines of constant <u>luminosity</u> density of the Virgo Cluster. The irregular frame gives the area covered by the Virgo Cluster survey (Binggeli et al.1985). The class of huge, diffuse dwarf galaxies has been generalized, and extended to fainter magnitudes by Impey et al.(1988), giving weight to the suspicion that the Universe is filled with elusive "Malin objects". Notice, however, that our Virgo "Malin's" are strongly confined to the cluster itself (in contrast to irregulars, which are much more dispersed). -

scale structure). - Yet another possibility is to have ionized gas....;
but fortunately we are already far away from our topic which is the
large-scale distribution of dwarf galaxies. The large-scale distribution
of baryonic matter in general will briefly be alluded to in the last
section.

7. SUMMARY AND CONCLUSIONS

In summary, there is no indication of a large-scale segregation between
giant and dwarf galaxies as predicted by Dekel and Silk (1986). On the
contrary, much evidence could be produced to show that dwarf galaxies
closely follow the large-scale structures delineated by luminous galaxies.
    There is segregation not with respect to luminosity but morphological
type. Dwarf ellipticals are strongly clustered like giant E's and S0's;
they occur in dense clusters and groups, or else they are satellites of
massive giants in the field. Dwarf irregulars, on the other hand, are
scattered in space much like the spirals; in fact, there is a progression
of decreasing clustering from early-type spirals to the latest irregulars.
But the dispersion of late-type systems is not complete. Spirals and
irregulars constitute large clouds of galaxies - with regions void of all
types between. A general description of the morphological segregation is
provided by the morphology-density relation (Dressler 1980).
    The present subject matter has often been reduced to the question -
Do· dwarfs fill the voids?, by "voids" always meaning the large, Boötes-
like voids. Although the big voids have not yet directly been probed for
faint dwarf galaxies, the answer is almost certainly no. First of all,
there are many smaller, nearby voids which are known to be void of giants
and dwarfs (Figs.1 and 2); - why should the large voids be different? -
The greatest unknown at present is the rôle played by the elusive "Malin
objects", because these have not yet been mapped outside of clusters.
First indications, however, strongly suggest that their abundance in the
field is low.
    What are the consequences for "biased galaxy formation"? - It has
been made clear in Sec.1 that the concept as such was not at stake. The
model of Dekel and Silk (1986) is only one version of "biased galaxy
formation", which, however, seemed to offer the least severe biasing
(because only luminous galaxies would have been affected) and one that
could be understood as a natural consequence of the formation history of
dwarf galaxies. Insisting in Ω=1 (i.e. having dark matter in the voids),
the biasing has simply to be stronger: - ALL types of galaxies (giants
and dwarfs alike) have to be prevented from forming in large regions of
space. - Two cases can further be distinguished. (1) - Baryonic matter
as a whole does trace the dark (non-baryonic) mass on large scales. Then -
since galaxies do not - there has to be a lot of gas in the voids. This
gas must be hot, because otherwise it would have formed galaxies, or
at least fragmented into HI clouds (protogalaxies), contrary to observation (Sec.6). (2) - Galaxies are representitive of baryonic matter on
large scales, i.e. there is not even gas (but only dark matter) in the
voids. This would be the strongest possible biasing. - Either way, the
primordial gas in the voids had to be reheated or removed altogether.

Whether the physical processes for biasing in the "cold dark matter" scenario suggested by Dekel and Rees (1987) are sufficient in terms of energy and range to do this job, remains to be shown. Both the voidness of the voids, i.e. the strong spatial segregation of dark matter and galaxies (if Ω=1), and the sharp, round boundaries of the voids (see Geller, this volume) seem to call for non-gravitational, dissipative processes that work on large scales. Models of this type, invoking "large-scale explosions", have been developed by Ostriker and Cowie (1981), and Ikeuchi (1981).

Alternatively, instead of embarking on ever stronger biasing, it seems possible to dismiss "biased galaxy formation" altogether and to believe in what one sees: luminous galaxies give us a correct idea of the distribution of mass on large scales, there is no dark matter in the voids, and hence $\Omega\sim0.1$ as observed. A scenario in accord with these premises is Peebles' "baryonic isocurvature model" (Peebles, this volume).

Finally, we emphasize that dwarf galaxies are potentially important for the mapping and understanding of the large-scale structure also in a positive sense. Although nothing new is apparently gained by adding the dwarfs, since by and large they follow the distribution of giant galaxies, there are indications of fine-structure features - faint bridges or arms connected to the bright structures - that are delineated only by dwarf galaxies. Thus going fainter, ever finer details of the large-scale structure of the Universe may be uncovered -- to the amazement of observers, and the embarrassment of theorists.

Partial financial support by the Swiss National Science Foundation is gratefully acknowledged.

References

Arp, H.C. (1965). Astrophys. J. 142, 402.

Bardeen, J.M. (1986). Inner Space/Outer Space, ed. E.W.Kolb, M.S.Turner, D.Lindley, K.Olive, and D.Seckel (University of Chicago Press, Chicago), p.212.

Binggeli, B., Sandage, A., and Tammann, G.A. (1985). Astron. J. 90, 1681.

Binggeli, B., Sandage, A., and Tammann, G.A. (1988). Ann. Rev. Astron. Astrophys. 26, 509.

Binggeli, B., Tammann, G.A., and Sandage, A. (1987). Astron. J. 94, 251.

Binggeli, B., Tarenghi, M., and Sandage, A. (1988). In preparation.

Bothun, G.D., Beers, T.C., Mould, J.R., and Huchra, J.P. (1986). Astrophys. J. 308, 510.

Bothun, G.D., Impey, C.D., Malin, D.F., and Mould, J.R. (1987). Astron. J. 94, 23.

Chincarini, G. and Vettolani, G. (1987). Observational Cosmology (IAU Symp. 124), ed. A.Hewitt, G.Burbidge, and L.Z.Fang (Reidel, Dordrecht), p.275.

Davis, M. and Djorgovski, S. (1985). Astrophys. J. 299, 15.

Davis, M., Efstathiou, G., Frenk, G.S., and White, S.D.M. (1985). Astrophys. J. 292, 371.

Davis, M. and Geller, M.J. (1976). Astrophys. J. 208, 15.

Dekel, A. and Rees, M. (1987). Nature 326, 455.

Dekel, A. and Silk, J. (1986). Astrophys. J. 303, 39.

de Lapparent, V., Geller, M.J., and Huchra, J.P. (1986). Astrophys. J.
Lett. 302, L1.

Disney, M.J. (1976). Nature 263, 573.

Disney, M.J. and Phillips, S. (1983). Mon. Not. R. Astron. Soc. 205, 1253.

Dressler, A. (1980). Astrophys. J. 236, 351.

Giovanelli, R., Haynes, M.P.,and Chincarini, G.L. (1986). Astrophys. J.
300, 77.

Hoffman, G.L., Helou, G., Salpeter, E.E., and Lewis, B.M. (1988). Preprint.

Huchra, J., Davis, M., Latham, D., and Tonry, J. (1983). Astrophys. J.
Suppl. 52, 89.

Ikeuchi, S. (1981). Pub. Astr. Soc. Japan 33, 221.

Impey, C., Bothun, G., and Malin, D. (1988). Astrophys. J., in press.

Kaiser, N. (1984). Astrophys. J. Lett. 284, L9.

Kaiser, N. (1986). Inner Space/Outer Space, see ref. Bardeen (1986),p.258.

Kraan-Korteweg, R. (1986). Astron. Astrophys. Suppl. 66, 255.

Nilson, P. (1973). Uppsala General Catalog of Galaxies (Uppsala Astr. Obs.
Ann., Vol. 6) (UGC).

Oemler, A. (1987). Nearly Normal Galaxies, ed. S.M.Faber (Springer, New
York), p.213.

Ostriker, J.P. and Cowie, L.L. (1981). Astrophys. J. Lett. 243, L127.

Phillips, S., Disney, M.J., Kibblewhite, E.J., and Cawson, M.G.M. (1987).
Mon. Not. R. Astron. Soc. 229, 505.

Phillips, S. and Shanks, T. (1987). Mon. Not. R. Astron. Soc. 229, 621.

Pierre, M., Shaver, P.A., and Iovino, A. (1988). Astron. Astrophys. Lett.,
in press.

Politzer, D.H. and Wise, M.B. (1984). Astrophys. J. Lett. 285, L1.

Postman, M. and Geller, M.J. (1984). Astrophys. J. 281, 95.

Sandage, A. and Binggeli, B. (1984). Astron. J. 89, 919.

Sandage, A., Binggeli, B., and Tammann, G.A. (1985). Astron. J. 90, 1759.

Sandage, A. and Tammann, G.A. (1981). A Revised Shapley-Ames Catalog of
Bright Galxies (Carnegie Institution of Washington, Washington D.C.).

Sargent, W.L.W., Young, P.J., Boksenberg, A., and Tytler, D. (1980).
Astrophys. J. Suppl. 42, 41.

Schechter, P. (1976). Astrophys. J. 203, 297.

Schombert, J.M. and Bothun, G.D. (1988). Astron. J. 95, 1389.

Silk, J., Wyse, R.F.G., and Shields, G.A. (1987). Astrophys. J. Lett.
322, L59.

Thuan, T.X., Gott, R., and Schneider, S.F. (1987). Astrophys. J. Lett.
315, L93.

Tully, R.B. (1982). Astrophys. J. 257, 389.

Tully, R.B. and Fisher, J.R. (1987). Nearby Galaxies Atlas (Cambridge
University Press, Cambridge).

White, S.D.M., Frenk, C.S., Davis, M., and Efstathiou, G. (1987). Astrophys. J. 313, 505.

Zwicky, F., Herzog, E., Karpowicz, M., Kowal, C.T., and Wild, P. (1961-68).
Catalog of Galaxies and Clusters of Galaxies, 6 Vols. (California
Institute of Technology Press, Pasadena).

DISCUSSION

PARKER: Your Virgo cluster luminosity contour plot (Fig. 3) showed 2 main peaks one of which was overlaid with many dwarf galaxies whilst the other had no overlaid dwarfs - in fact the dwarfs seemed to avoid this second peak - do you think this is significant?

BINGGELI: Yes, I think the difference is significant. In our Virgo (page VI, Binggeli et al. 1987) we have argued that the southern peak is probably a small, less dense cluster of its own, being $\sim$5 Mpc behind the peak which constitutes the Virgo cluster proper. The asymmetry in the distribution of Malin dwarfs you noticed is therefore not surprising.

TULLY: On the matter of the morphology-density relation, recently Simon White and I compared the results of cold dark matter n-body simulations, with data from my Nearby Galaxies Catalog and found good compatibility. We actually compared relation velocities versus local density. In both the simulations and the data, there are subtle dependencies in the sense that more dwarf-like systems tend to be found in lower density regims.

BURSTEIN: Two parts to the question: 1) What is the sample used to derive the dwarf LF for the field? 2) How accurate is that LF? In particular what possible systematic problems could affect its derivation?

BINGGELI: The LF of field dE's is largely unknown. That shown in the figure was simply made to have the same shape as the LF of VIRGO dE's, scaled down to reproduce the number of known, local dE's at the bright end. We known that the number of dE's in the field is small, but just because of this we cannot model the LF shape; as a first approximation we assume it to be universally constant.

WAGNER: As far as the distribution of matter is concerned it is not only important to study the abundance of dwarf galaxies but rather their fraction of the mass density. What do we know about the fraction of total mass in dwarf galaxies in regions where they are well studied-e.g. in the Virgo cluster?

BINGGELI: The fraction of the total luminosity tied to dwarf galaxies is of course exactly known in the case of the Virgo cluster. It is extremely low. Galaxies with $M_B>=-16$, despite the exponential luminosity function, contribute less than 5% of the total cluster light (cfr. Fig. 23 and 24 of Sandage et al. 1985). Their mass function is less secure

becouse we don't know the M/L ratios. But even with
(M/L)dwarf $\sim$ 10 (M/L)giant, which is very unlikely the case,
the dwarfs would contribute less than 50% of the total mass.
In the field, where the number ratio of dwarf-to-giant
galaxies is smaller than in clusters, the dwarfs "weight"
less still.

PEEBLES: Another constraint on the abundance of objects
with low luminosity per object or low surface brightness is
that these objects would increase the mean surface
brightness of the extragalactic sky; the expected surface
brightness due to ordinary galaxies already is close to the
observational bounds.

DISTRIBUTION OF DARK MATTER IN GALAXIES

A. Bosma
Observatoire de Marseille
2 Place Le Verrier
13248 Marseille Cedex 4
FRANCE

ABSTRACT. The current evidence for dark matter in galaxies is briefly reviewed. While for ellipticals the evidence for dark halos is not yet overwhelming, for disk galaxies we can already start to constrain the distribution of the dark matter. The global M/L-ratios out to the largest radii where the tracers of the mass are still reliable are below the values of M/L currently found in groups. Attention is drawn to galaxies with low surface brightness disks, which may have peculiar halos.

1. INTRODUCTION

Dark matter is now widely thought to be a pervasive constituent of the Universe. The evidence for it is based on Newton's laws; alternative theories modifying Newtonian dynamics on a large scale (e.g. Milgrom 1983, Sanders 1986) have not yet gained much acceptance. The most convincing evidence for the existence of dark matter comes from data on individual galaxies, and it is here where the dynamical arguments can be developed most easily to constrain its distribution. However, for larger scales the situation is far from satisfactory : there is no convincing evidence for $\Omega = 1$, and most dynamical arguments on the scales of groups and clusters leads to values for Ω of order 0.1 to 0.2 (e.g. Yahil 1987). Yet the arguments for the existence of dark matter do not tell us much about the distribution : the question is still open whether mass follows light.

On the scale of clusters this question is unclear (e.g. Merritt 1987), while the fair degree of subclustering now found to be present (e.g. Dressler and Shectman 1988) might complicate the analysis. From a study of groups of galaxies Tully (1987) concludes that M/L = 120 h at radial scales of 100 - 800 h^{-1} kpc, and that the M/L-ratio does not increase in this radius range. Thus the mass could be in the galaxies, provided these reach global M/L values of 120 h. However, for binary galaxies Schweizer (1987) finds M/L = 40 h for spirals and 80 h for ellipticals at radial scales of 30 - 50 h^{-1} kpc, and no further increase in the mass beyond these radii.

In this review I will concentrate on the individual galaxies, and discuss 1) the evidence for dark matter in individual galaxies, and 2) for spiral galaxies, where more information is available, constraints on the distribution of dark matter.

65

M. Mezzetti et al. (eds.), Large Scale Structure and Motions in the Universe, 65–72.
© *1989 by Kluwer Academic Publishers.*

2. ELLIPTICAL GALAXIES.

M/L-ratios in the central parts of elliptical galaxies, derived from velocity dispersions, are generally of order 10 h (Schechter 1980). The evidence for dark matter beyond the effective radius is still rather uncertain. X-ray data indicate M/L's of order 10 h - > 200 h (Forman et al. 1985), but except for M87, the data on the temperature gradient of the X-ray emitting gas are too insecure to put stringent limits on the M/L-increase (Trinchieri et al. 1986). For M87 the velocity dispersion of the globular cluster system remains high at large radii (Mould et al. 1987, Huchra and Brodie 1987), and thus strengthens the case for a dark halo around this galaxy situated in the centre of the Virgo cluster.
       For other ellipticals the situation is less clear. Hernquist and Quinn (1987) have argued for an external origin of the shells found around many elliptical galaxies by Malin and Carter (1982), and for the multi-shell system NGC 3923 propose a model which needs a very heavy halo around this galaxy. However, even though the external origin is very likely, the actual situation is confused by the discovery of additional shells very deep in the potential well (Prieur 1988), so that the modelling should take dynamical friction into account. A simplified treatment of this by Dupraz and Combes (1987) indicates that dark matter may not be necessary, but fully self-consistent models have yet to be constructed.
       HI- (and extended HII-) emission provides another indicator. Data for NGC 4278 (Raimond et al. 1981), NGC 1052 (Van Gorkom et al. 1987), NGC 5666 (Lake et al. 1987) and NGC 2974 (Kim et al. 1988) indicate relatively flat rotation curves (if the gas is indeed in normal rotation), and therefore local M/L-ratios which are increasing with radius. However, the data do not go very far out, and indicate global M/L-values of order 10 - 25 h. Finally several other indicators are being sharpened : gravitational lenses, planetary nebulae (Ford et al., 1988), and the dynamics of satellites (Dressler et al. 1986), all of these pointing to higher M/L-values in the outer parts of ellipticals, and thus towards the existence of dark halos.

3. DWARF GALAXIES

Several categories are comprised under this heading. For dwarf ellipticals no information is available yet. For dwarf spheroidals the situation has been reviewed by Aaronson (1987) : most dwarf spheroidals have M/L-ratios below 10; only for the faintest ones, Draco and Ursa Minor, the measurements of the radial velocities of a handful of stars indicates a high velocity dispersion and M/L-ratios of order 60 - 80. Godwin and Lynden-Bell (1987) argue for caution in deriving velocity dispersions for such galaxies. For dwarf spirals and irregulars the situation is similar (Freeman 1987) : again the smallest ones indicate high M/L-ratios above 20.
       An interesting observation has been reported recently by Brinks and Klein (1988). They observed II Zw 40, normally known as a blue compact dwarf, and find a very large HI disk (or perhaps two disks in collision,

as they prefer) of diameter 17 x 2 kpc, much larger than the optical image. Moreover, the centre indicated by the kinematics is completely offset from the optical image. There are two directions along which the velocity gradients reach full amplitude (about 150 km/s end-to-end), one along the long axis of the HI disk, and one at ~ 60° angle. Such a situation suggests 1) the existence of dark matter not well centered around the optical image, and 2) the luminous part of the galaxy might have formed relatively recently. Huge HI envelopes with large velocity gradients are known around several other late type irregulars, e.g. NGC 4449 (Van Woerden et al. 1974).

4. S0 - GALAXIES

For normal S0-galaxies not much information is available, but some peculiar S0 galaxies have shed light on the dark matter question.

S0-galaxies with polar rings provide information about the shape of the dark matter distribution, as follows. Schweizer et al. (1983) and Whitmore et al. (1987) measured position - velocity curves along the major axis of the S0 galaxy (which is seen edge-on) and along the polar ring (annulus in some cases), and find in all (three) cases studied that the velocities almost match when plotted as function of radius. The rotation curves inferred are flat, and this is further confirmed by the HI studies of Schechter et al. (1985) and Van Gorkom et al. (1987). The near equality of the equatorial and polar velocities suggests that there is a halo component, with a shape which is not far from round.

Certain S0-galaxies have detectable HI gas, and recent mapping of several of these (e.g. Van Driel 1987) shows that some of them have outer rings, which may well be in a polar ring configuration. One of the most striking cases is NGC 2787, reported by Shostak (1987). Again the rotation curve is flat, and the global M/L-ratio is of order 67 h.

5. SPIRAL GALAXIES

5.1 Composite bulge/disk/halo models

The most popular way to model the mass distribution in spiral galaxies is to decompose them into two or three components : a bulge, a disk, and a dark halo. For each of the luminous components the mass-to-light ratio is assumed to be constant as function of radius, so that available surface photometry data can be used to evaluate its contribution to the rotation curve. Since bulges are following roughly a $r^{\frac{1}{4}}$-law and disks are nearly exponential, it follows that for a rotation curve which is flat out to sufficiently large distances (3 - 5 disk scalelengths at least) a dark halo component is needed. Constant mass-to-light ratios are justified by noting that colour gradients appear to be minor (Wevers 1984, Kent 1986), so that each component may be treated as a mixture of stellar populations which globally have a single value of M/L. In fact, the minor colour changes which do exist show that the disk becomes bluer outwards, so that a constant M/L overestimates the disk surface density.

5.2 Bulge constraints

The M/L-ratio of the bulge is usually maximized, i.e. the highest M/L-value compatible with the observed rotation curve is taken. As a check Kent (1987) devised several ways to compute the central velocity dispersion, which for a number of galaxies can be compared with the observations. For most galaxies there is reasonable agreement, but for some Sa-galaxies there is a large discrepancy in the sense that the M/L-values are too low to be compatible with the observed velocity dispersion (Kent 1988), suggesting that the observed radial velocities in large bulges may not indicate properly the true rotation velocity.

An independent check on the 'maximum bulge' hypothesis has recently come from the work on the gravitational lens system 2237+05, where the lensing is thought to be due to the bulge. Schneider et al. (1988) estimate M/L = 9.5 h, with an error of 20%. Such high M/L-values occur only if the bulge is taken maximum.

5.3 Disk constraints

The simplest disk constraint is to require 'maximum disk', i.e. the highest M/L-value compatible with the observed rotation curve. If done so, in many cases the rotation data do not extend far enough in radius to require the presence of a dark halo (Kalnajs 1983, Kent 1986). If a halo has to be present to resolve the discrepancy between the observed and calculated rotation curves, then it seems natural to require that it does not have a hollow core (e.g. Van Albada et al. 1985). However, models with lower M/L-values for the disk cannot be excluded.

Several dynamical constraints can be applied to reduce the range of possible M/L-values. Swing amplifier spiral structure theory (Toomre 1981) puts strong constraints on the ratio of epicyclic frequency to the product of the number of arms and the local disk surface density. Thus for a galaxy which is predominantly two-armed, the disk has to be strong enough to allow amplification of $m = 2$ structures, but not so strong as to allow amplification of $m = 1$ structures. These considerations have led Athanassoula et al. (1987) to propose two limiting series of models : 'no $m = 1$'- models, for which the disk M/L is taken as high as possible without allowing amplification of $m = 1$ structures, and 'no $m = 2$'- models, where the amplification of $m = 2$ structures is just suppressed. In most cases the requirement of 'maximum disk without hollow core' satisfies the 'no $m = 1$' criterion as well. The range in mass-to-light ratios so constrained is about 0.3 dex.

Disk velocity dispersions can also be used, at least in principle. If the data are good enough to allow it, and using reasonable assumptions about the z-component of the observed velocity dispersion, the run with radius of Toomre's Q-parameter (Toomre 1964) can be calculated. For the galaxy NGC 488 the 'no $m = 1$' model gives Q of order 1.4, and the 'no $m = 2$' model gives Q of order 2.8. If all the mass is assumed to be in a disk, it would actually be unstable to axisymmetric instabilities ($Q < 1$). Yet the uncertainties in the actual data values is very large, and the data themselves are still scarce.

A further constraint on the entire modelling, which gives also a clue to the shape of the halo potential, comes from the consideration of the vertical equilibrium of the gaseous and stellar disk in the combined disk-halo potential (cf. Bergeron and Gunn 1977 for an early attempt). For several galaxies with complete decomposition models (gas, stars and halo components) one can predict the run of the gas layer thickness with radius, assuming a contant scale height for the stellar disk (cf. van der Kruit and Searle 1982) and either a spherical or a flat shape of the halo. The gas layer is predicted to flare, and for the spherical halo case significantly so at radii where the stellar disk ceases to dominate. Such a behaviour has been observed for our Galaxy (Knapp 1987), and inferred for M31 (Brinks and Burton 1986). More work on this for several edge-on galaxies is in progress.

5.4 Is the 'maximum disk with no m = 1' the best model ?

The models with maximum disk which still suppress m = 1 instabilities seem at present the 'best' models available. They have reasonable gas fractions, mass-to-light ratios of the visible components in rough agreement with current thoughts about galaxy evolution, reasonable Q-values, and are compatible with the number and the extent of the spiral arms (cf. Athanassoula et al. 1987). However, this does not mean that they are the only models allowed : any model having a M/L-ratio of the disk in between the 'no m = 2' model and the 'no m = 1' model are within the dynamical constraints discussed above. For Sc galaxies 'no m = 2' models are dominated by the dark matter, and the gas constitutes an important component of the visible disk.

      It is possible that, at least for some galaxies, the maximum disk situation might not be attained. For our Galaxy recent work indicates a surface density in the solar neighbourhood of 45 M_0 pc^{-2} (Kuyken and Gilmore 1987), corresponding to a disk rotation velocity of 110 km/sec, which excludes the maximum disk solution. Van der Kruit (1988) argues for a similar situation in the edge-on galaxy NGC 891.

      On a slightly different line, some results concerning the total mass of our Galaxy, and of the Local Group, point to low masses. Little and Tremaine (1987) argue for a total mass of the Galaxy of only 3 - 5 10^{11} M_0, or alternatively a low rotation velocity at 50 kpc, and Sandage (1986) finds a Local Group mass much lower than the classical few times 10^{12} M_0. If this is correct, 'maximum disk' models may not be compatible with these numbers.

      Theoretical studies also are not always in agreement with 'maximum disk' models. Thus the studies of Blumenthal et al. (1986) and Ryden and Gunn (1987) concerning the compression of the halo by the infalling baryons during galaxy formation produce models which are closer to 'no m = 2' models. A detailed discussion of these models, as well as new numerical simulations of this infall scenario (Athanassoula 1988) reveal, however, that 'no m = 1' models are possible.

5.5 Spiral galaxies with low surface brightness disks

Recent discoveries of giant galaxies with low surface brightness disks, e.g. Malin 1 (Bothun et al. 1987) and GP 1444 (Davies et al. 1988) rekindle interest in the old question whether the central surface brightness of spiral galaxies is really nearly constant (e.g. Freeman 1970, Disney 1976, Van der Kruit 1987). While Malin 1 and GP 1444 are as yet difficult to study in detail, some milder versions of this phenomenon are accessible.

One such example is the Sc galaxy NGC 5963, studied by Romanishin et al. (1982) and recently by Bosma et al. (1988). They find for this galaxy that i) the HI surface density as function of radius is 'normal' compared to such standard Sc galaxies as NGC's 2403, 3198, 4395 and 6503, ii) the outer disk is of lower surface density, and iii) the halo has a smaller core radius, both in absolute value (factor 2) and even more so when expressed in disk scalelengths.

An evaluation of the minimum velocity dispersion necessary to prevent axisymmetric instabilities shows that in the outer disk the values are below 8 km/sec, i.e. smaller than the velocity dispersion of the gas (cf. Van der Kruit and Shostak 1984). Thus the gas may be Jeans stable, and not form stars. A similar situation exists at the edge of every disk galaxy (see Athanassoula and Bosma 1988, and Fall and Efstathiou 1980). What remains to be clarified is why NGC 5963 's disk lacks the factor of 2 or so of baryons compared to normal disk galaxies.

A further question is how many of such galaxies there are. A partial answer to this has come from work by Bosma and Freeman (see Freeman 1979, Bosma 1983). They compared diameters of galaxies on various sky surveys. Low surface brightness galaxies are much larger on deeper surveys, a fact recently confirmed for the new Palomar survey (Schombert and Bothun 1988). Only 15 % of the spirals measured by Bosma and Freeman show low surface brightness extensions, and the Malin 1's, where all of the disk is below the detection limit of the old Palomar survey, are rare. Of course (cf discussion) these galaxies are still recognized because of the presence of a high surface brightness core (bulge or lens). Very faint galaxies of large angular extent have recently been found in the Virgo cluster and several other clusters (see Binggeli, this volume), but no information is available about the amount of dark matter around them (if any).

6. Concluding remarks.

Present evidence for dark matter around spirals is relatively strong, and we can already address questions about its distribution. The extent of the dark halo is quite unclear, however. Global M/L-ratios do not exceed 20-50, still far from the values of about 100 found for groups. For ellipticals, only the X-ray data indicate values around 50 - 100. Further work is needed to try to get limits on the extent of dark halos, and to address the question whether on larger scales mass follows light.

REFERENCES :

Aaronson, M.A.: 1987, in "Nearly Normal Galaxies", ed. S.M. Faber, Springer (Berlin), p. 57

Athanassoula, E., Bosma, A., Papaioannou, S.: 1987, Astr. Ap. 179, 23.
Athanassoula, E., Bosma, A.: 1988, in "The Structure of the Unviverse", I.A.U. Symp. 130, ed. J.Audouze. (in press)
Athanassoula, E.: 1988, in "Dark Matter", proc. XX Rencontre de Moriond, eds. J. Audouze and J. Tranh Vanh Van. (in press)
Bergeron, J., Gunn, J.E.: 1977, Ap. J. 217, 892
Blumenthal, G., Faber, S.M., Flores, R., Primack, J.: 1986, Ap. J. 310, 27
Bosma, A.: 1983, in "Internal Kinematics and Dynamics of Galaxies", I.A.U. Symp. 100, ed. E. Athanassoula, Reidel (Dordrecht), p. 253
Bosma, A., Van der Hulst, J.M., Athanassoula, E.: 1988, Astr. Ap. 198, 100
Bothun, G., Impey, C., Malin, D., Mould, J.: 1987, A. J. 94, 23
Brinks, E., Burton, W.B.: 1986, Astr. Ap. 141, 195
Brinks, E., Klein, U.: 1988, M.N.R.A.S. 231, 63P
Davies, J.I., Phillips, S., Disney, M.J.: 1988, M.N.R.A.S. 231, 69P
Disney, M.J.: 1976, Nature 263, 573
Dressler, A., Shectman, S.E.: 1988, Ap. J. 95, 985
Dressler, A., Schechter, P.L., Rose, J.A.: 1986, A. J. 91, 1059
Dupraz, C., Combes, F.: 1987, Astr. Ap. 185, L1
Fall, S.M., Efstathiou, G.: 1980, M.N.R.A.S. 193, 189
Ford, H.C., Ciardullo, R., Jacoby, G.H., Hui, X.: 1988, in "Planetary Nebulae, I.A.U. Symp. 131, ed. S. Torres-Peimbert (in press)
Forman W., Jones, C., Tucker, W.: 1985, Ap. J. 193, 102
Freeman, K.C.: 1970, Ap. J. 160, 811
Freeman, K.C.: 1979, in "Photometry, Kinematics and Dynamics of Galaxies", ed. D.S. Evans, Univ. of Texas (Austin), p. 85
Freeman, K.C.: 1987, in "Nearly Normal Galaxies", ed. S.M. Faber, Springer (Berlin), p. 317
Godwin, P.J., Lynden-Bell, D.: 1987, M.N.R.A.S. 229, 7P
Hernquist, L., Quinn, P.J.: 1987, Ap. J. 312, 1
Huchra, J., Brodie, J.: 1987, A. J. 93, 779
Kalnajs, A.J.: 1983, in "Internal Kinematics and Dynmics of Galaxies", I.A.U. Symp. 100, ed. E. Athanassoula, Reidel (Dordrecht), p. 87
Kent, S.M.: 1986, A. J. 91, 1301
Kent, S.M.: 1987, A. J. 93, 816
Kent, S.M.: 1988, A. J. (in press)
Kim, D.W., Guhathakurta, P., Van Gorkom, J.H., Jura, M., Knapp, G.R.: 1988, Ap. J. 330, 684
Knapp, G.: 1987, P.A.S.P. 99, 1134
Kuijken, K.H., Gilmore, G.: 1987, Bull. A. A. S. 19, 1108
Lake, G., Schommer, R.A., Van Gorkom, J.H.: 1987, Ap. J. 314, 57
Little, B., Tremaine, S.: 1987, Ap. J. 320, 493
Malin, D., Carter, D.: 1983, Ap. J. 274, 534
Merritt, D.: 1987, Ap. J. 313, 121
Milgrom, M.: 1983, Ap. J. 270, 365 & 371
Mould, J.R., Oke, J.B., Nemec, J.M.: 1987, A. J. 92, 53
Prieur, J.P.: 1988, Ap. J. 326, 596
Raimond, E., Faber, S.M., Gallagher, J.S., Knapp, G.R.: 1981, Ap. J. 246, 708
Romanishin, W., Strom, S.E., Strom, K.M.: 1982, Ap. J. 258, 77

Ryden, B., Gunn, J.E.: 1987, Ap. J. 318, 15
Sandage, A.: 1986, Astrophys. J. 307, 1
Sanders, R.H.: 1986, Astr. Ap. 154, 315
Schechter, P.L.: 1980, A. J. 95, 801
Schechter, P.L., Sancisi, R., Van Woerden, H., Lynds, C.R.: 1984,
    M.N.R.A.S. 208, 111
Schneider, D.P., Turner, E.L., Gunn, J.E., Hewitt, J.N., Schmidt, M.,
    Lawrence, C.R.: 1988, A. J. 95, 1619
Schombert, J.M., Bothun, G.D.: 1988, A. J. 95, 1389
Schweizer, F., Whitmore, B.M., Rubin, V.C.: 1983, A. J. 88, 909
Schweizer, L.Y.: 1987, Ap. J. Suppl. 64, 427
Shostak, G.S.: 1987, Astr. Ap. 175, 4
Toomre, A.: 1964, Ap. J. 139, 1217
Toomre, A.: 1981, in "Structure and Evolution of Normal Galaxies", eds.
    S.M. Fall and D. Lynden-Bell, Cambridge, p. 111
Trinchieri, G., Fabbiano, G., Canizares, C.R.: 1986, Ap. J. 310, 637
Tully, R.B.: 1987, Astrophys. J. 321, 280
Van Albada, T.S., Bahcall, J.N., Begeman, K., Sancisi, R.: 1985,
    Ap. J. 295, 305
Van der Kruit, P.C.: 1987, Astr. Ap. 173, 59
Van der Kruit, P.C.: 1988, Astr. Ap. 192, 117
Van der Kruit, P.C., Searle, L.T., 1982, Astr. Ap. 110, 95
Van der Kruit, P.C., Shostak, G.S.: 1984, Astr. Ap. 134, 258
Van Driel, W.: 1987, Ph.D. Thesis, University of Groningen
Van Gorkom, J.H., Schechter, P.L., Kristian, J.: 1987, Ap. J. 314, 258
Van Gorkom, J.H., Knapp, G.R., Raimond, E., Faber, S.M., Gallagher,
    J.S.: 1986, A. J. 91, 791
Van Woerden, H., Bosma, A., Mebold, U.: 1974, in "La Dynamique des
    Galaxies Spirales", ed. L. Weliachew, C.N.R.S. coll. 241, p. 483
Wevers, B.H.M.R.,: 1984, Ph.D. thesis, University of Groningen
Whitmore, B.M., McElroy, D.B., Schweizer, F.: 1987, Ap. J. 314, 439
Yahil, A.: 1987, in "Nearly Normal Galaxies", ed. S.M. Faber, Springer
    (Berlin), p. 332

DISCUSSION

BAHCALL: For NGC 3198, the galaxy with the longest rotation curve
measured, is the global M/L-ratio up to the last point measured also
consistent with ~ 10 - 20 ?

BOSMA : The global measured M/L is ~ 25

PHILLIPS : A comment on the Malin 1 type galaxies. Both Malin 1 and GP
1444 were first noticed because of the prominent bulge components. We
still do not know if there are many Malin 1 type disks without bulges.
Moreover, very low surface brightness disks would not appear on your
comparison of diameters on Palomar and UK-Schmidt plates since they
would be too faint to show up on the Palomar plates at all.

BOSMA : OK, but we selected only galaxies larger than 2' on the UK-
Schmidt plates/films, thus excluding most of these systems implicitly.

STRUCTURE AT INTERMEDIATE SCALES: CLUSTERS OF GALAXIES

A. Cavaliere
Astrofisica, Dip. di Fisica, II Università di Roma, Italy
S. Colafrancesco
Dip. di Astronomia, Università di Padova, Italy.

ABSTRACT: We discuss how diverse, ongoing theoretical efforts (N-body simulations, scaling laws and numerical computations for the luminosity functions) compare with the growing data base especially in X-rays in describing the morphologies and in predicting the development of Clusters and Groups of galaxies.

1. INTRODUCTION

Groups and Clusters of galaxies contribute conspicuously to local structure at intermediate scales, in the mass range $M \sim 10^{13} - 10^{15} \, M_{\odot}$. They fill the wide gap beyond the galaxies (dissipative, settled configurations of high density contrast) up to large scale structures (still of small amplitude).

In the picture envisaging structure formation by gravitational instability, they fit in as relatively young objects: perturbations gone non-linear and recollapsed but only partly virialized, especially near the upper end of the mass range (cf. Forman and Jones 1982, Geller 1988). Here cosmogony should comprise ongoing non-linear dynamics and initial conditions still remembered. If so, one may look for relics of the initial conditions in the local objects, and for non-linear clustering in the making at modest redshifts.

X-rays constitute a good band to probe both these aspects. Groups and Clusters are known since the UHURU era (Cavaliere, Gursky and Tucker 1971, Sarazin 1988) to be strong X-ray sources with luminosities in the range $L \sim 10^{43} - 10^{45}$ erg/s. Power

$$L \propto \hat{\rho} \hat{M} T^{1/2} \tag{1.1}$$

and spectral shape (continuum approximated by $E^{-\alpha} \, e^{-E/kT}$ with $\alpha \sim 0.4$, plus high-excitation emission lines) are accounted for in terms of optically thin, thermal bremsstrahlung emission from hot Intra-Cluster Plasma (ICP) at temperatures $T \lesssim 10^{8}$ K and densities $\hat{\rho}$ corresponding to $\lesssim 10^{-3}$ particles/cm^3, with masses $\hat{M}$ up to $\sim 10^{14} M_{\odot}$ filling the global gravitational potential wells and their localized minima associated with clumps of members or even with single large galaxies.

Specifically, X-rays constitute an efficient means for mapping the detailed morphology of nearby Clusters, because the bremsstrahlung emission/unit volume $\propto \hat{\rho}^2 T^{1/2}$ stresses physical condensations of the total mass (including any dark component). X-rays are also potentially apt to survey and count distant Clusters as resolved sources distinct from,

M. Mezzetti et al. (eds.), Large Scale Structure and Motions in the Universe, 73–92.

though more difficult to measure than the other extragalactic class of comparable power, the compact AGNs; for example, if the emissivity $\propto \hat{\rho} T^{1/2}$ were to increase outwards following closely the increase of the cosmic density, it should be feasible to detect such Olbers-like behaviour out to redshifts where the hot ICP forms.

2. MORPHOLOGICAL COSMOGONY

The ICP has fairly fast microscopic relaxations, sound crossing time $\lesssim$ dynamical times and cooling time generally long. So it may be assumed in fair equilibrium with the gravitational potential over Mpcs (cf. Cavaliere 1980, Sarazin 1988) and may be described by a hydrostatic disposition, implying at lowest order a scale height matching that of the galaxies (except for cooling flows localized in the densest $\sim 10^2$ kpcs, cf. Fabian 1988).

Whence one may proceed, in cases such as A2255 or Coma where spherical symmetry applies to first order, with analytical models (using, e.g., a polytropic equation of state) that provide parametric patterns for the run of surface brightness and of temperature out to several Mpcs, and yield bounds to the ICP mass, see Cavaliere 1980, Sarazin 1988.

On the other hand, optical studies (like those of Geller and Beers 1982, Binggeli et al. 1987) and extended X-ray mappings by EINSTEIN (cf. Forman and Jones 1982) stressed the *variance* present even within the set of the rich Clusters, with clumpy and irregular morphologies prevailing in many objects: examples are provided by A194 or A2151 up to such extreme cases as A1367, but even Coma shows evidence of some inner asymmetry (see, e.g., Fitchett and Webster 1987). Such observational resolution can be matched only by computations resolved in space-time, like the N-body simulations of the dynamical structure pioneered by Peebles 1970 and White 1976, then taken up by Cavaliere et al. 1986 using extended sets of initial conditions and a hydro-module to produce also simulated X-ray maps from the ICP resolved down to nearly galactic scales.

Such codes follow in time and in 3-D detail the non-linear action of the gravitational instability on a field of small density enhancements of the total density as emerged from equidensity and recombination. In particular, they can implement rather directly the guidelines of the Hierarchical Clustering Scenario (HCS), the most definite and least unsatisfactory cosmogony to now (cf. Peebles 1980, White 1982, Dekel 1987), where the instability proceeds bottom up dissipationlessly from galaxies clustering into Groups, these into poor and eventually into rich Clusters.

In the simplest version of the HCS, the field of the initial overdensities $\delta \equiv \delta\rho/\rho$ on a mass scale M follows a Gaussian distribution $p(\delta|M)$ with variance $\sigma \propto M^{-(n+3)/6}$, and may be Fourier represented by random-phased components with power spectrum $|\delta_k|^2 \propto k^n$, extending longward from galactic or even subgalactic masses.

Beyond the linear regime $\delta_k(t) \propto t^{2/3}$ ($\Omega_o \to 1$), the non-linear development is the domain proper to N-body simulations, whose results may be described in terms of two coupled sequences: I) overall condensation: the expansion detaches from the Hubble flow, and halts, then recollapse follows; II) internal subclustering: clumps of a few galaxies single out, merge into larger and larger units, a few large clumps fall together. Eventually, the systems violently relax into roughly spherical configurations (with some residual asymmetry often persisting); similar events occur on smaller and faster scales in the preceding subclusters; mass from weaker, slower substructures infalls onto the prompter condensations, and so does (depending on Ω_o) some bound matter surrounding the Cluster.

The essentials of sequence I) may be recovered from an analytical model, spherical and uniform (Gunn and Gott 1972) collapsing on the time scale $\sim R/\langle v^2 \rangle^{1/2} \propto |E|^{-3/2} \propto \delta^{-3/2}$, that unifies the non-linear with the linear regime in terms of the conserved total energy E. For *typical* units along the hierarchy corresponding to $\delta \sim \sigma$ and virializing (density contrast $\gtrsim 170$) at a redshift z density, mass, size and specific energy scale like

$$\rho_c \propto (1+z)^3, \quad M_c \propto (1+z)^{-6/(n+3)}, \quad R_c \propto (M_c/\rho_c)^{1/3}, \quad v_c^2 \propto (1+z)^{(n-1)/(n+3)} \quad (2.1)$$

(Peebles 1974, White 1982). Values of the index n in the range $-3 < n < 1$ (with preferred values $n \sim -1 \pm 0.5$, as we shall see) preserve with increasing cosmic time or increasing mass a neat sequential increase of the average collapse epochs $(1+z_c)^{-3/2} \propto M^{(n+3)/4}$ (so formation is diachronic) and of the average specific energies $v^2 \propto M^{(1-n)/6}$ (so mergings into larger units erase previous structure).

But we know – particularly from observed spreads in galaxy and in Cluster formation epochs – that such a sequence is much too rigid, so a crucial point concerns *variances and dispersions*. What the N-body simulations really add is small scale inhomogeneity and its effects, that is, they add the sequence II) and its coupling with sequence I): the gravitational instability amplifies statistical inhomogeneities or fine-grained correlations in the initial galactic bindings to form macroscopic substructures; in turn, extensive subclustering slows down the overall collapse, because much of the initial binding energy is drained and frozen – as it were – into the smaller scales. Considerable consequences follow:

a) The effective collapse times are distributed in a wide range (differing by factors up to $2.5 \cdot 2.5$, Cavaliere et al. 1986) with the more clumpy objects being more delayed, which translates into a broad morphological range at the present epoch. Equivalently, the variance of the mass distribution at given z is enhanced beyond the contribution from the linear $p(\delta|M)$. To wit, some runs are found to hang on in unevolved configurations similar to A1367 after many canonical crossing times when others are already well relaxed, yet having same mass and close total energy, but differing mainly in the fine-grained statistical distribution of the initial bindings. Such sparse, unevolved configurations where the definition of a single "core" loses meaning, see fig. 1, are nearly as frequent as the relaxed ones at the present epoch; at moderate z they are to become rapidly more frequent, and may require aimed searches.

b) A conspicuous class of inhomogeneity is constituted by the bimodal configurations; these developed in a fair percentage (another $\lesssim 1/3$) of the runs by Cavaliere et al. 1986 and persisted for long times, providing models for such objects as A548 traced by Geller and Beers 1982, and A754 mapped and analyzed by Fabricant et al. 1986, see fig. 2.

c) The X-ray mapping of subclusters (both real and simulated) is favoured by two circumstances: the ICP is a continuous medium in local equilibrium, as opposed to discrete galaxies with their small numbers and possibly anisotropic velocities; in addition, the emission density $\propto \bar{\rho}^2 T^{1/2}$ stresses and selects physical ICP density bumps out of random galaxy superpositions. Cavaliere et al. 1986 illustrated how these circumstances help confirming substructure signaled by optical tracings, concurrently with spectroscopic evidence, cf. Geller 1988, Dressler and Schectman 1988. The other side of this coin is that strong clumpiness may further enhance the variance of Cluster luminosity in X-rays over that in the optical band.

d) The above simulations started from white-noise initial distributions corresponding in a Fourier representation to $n = 0$. Smaller values of n with equal normalization at $\sim 10^{15} M_\odot$ imply less inhomogeneity (less power initially on small scales), more easily erased in the center of mergers (the specific energies increase more steeply with M). West et al. 1988 have evolved initial conditions with values of n down to -2. At the present epoch (in principle defined by the absolute clock provided by the evolution of galaxy correlation as suggested by Davis and Peebles 1983, in practice contending with a correlation length uncertain at best by a factor of 2, cf. Geller 1988) they find a larger proportion of centrally smooth systems, but also rich subclustering outwards of a few Mpcs consistent with the trend expected. They also sample the alternative top-down scenario with only perturbations on large scales initially given, and find (with no dissipative fragmentation included) virtually no substructure.

e) We are testing the following hypothesis: the outline of the cluster structure is imprinted by the granularity of the total mass (including any dark collisionless component) which for $z \gtrsim 3$ is bound to galaxies already settling rather than to the shallow potential well of the hardly looming Cluster, even though in the next development much of the matter is stripped away and diffused throughout the system. The final distribution of the optical L/M may settle as governed by added dissipations (Evrard 1987, West and Richstone 1988).

To sum up, the existing simulations qualitatively validate the predictions of the HCS but still differ in important structural detail, with those given more initial inhomogeneity better reproducing morphologies and variances as observed. The ongoing work concentrates on two lines: i) Minimal amount of initial fine-grained inhomogeneity needed to generate the structure observed locally; this implies calibration of the initial conditions as to clock, and as to perturbation profile, power spectrum and content in DM (the Cold Dark Matter curved spectrum tending to relatively underplay power on small scales).
ii) Assessment of shapes, structures and luminosities expected for distant Clusters, in preparation to forthcoming surveys from ROSAT (cf. Trümper 1988) and AXAF (cf. Giacconi 1980, Tananbaum 1988) that ought to catch most those Clusters in the making.

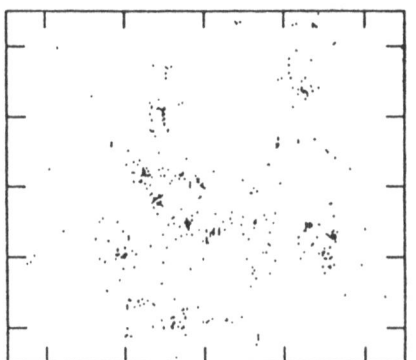 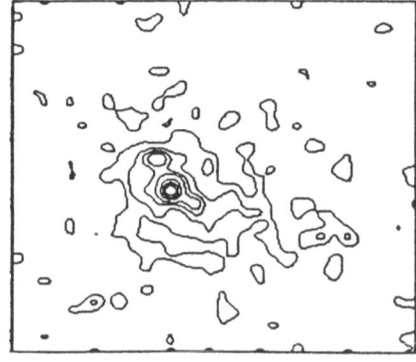

fig. 1. An unevolved configuration of intermediate age from the simulations of Cavaliere et al. 1986. Size = 8 Mpc. Notice how some outlying emission of low surface brightness is drowned into the background with its noise.

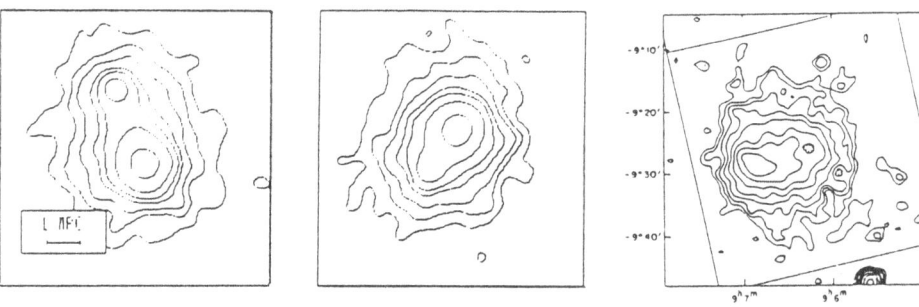

fig. 2. Two orthogonal projections of the same simulated Cluster in a persisting bimodal configuration, are compared with the map of A754 that indicates a similar projection effect. (Fabricant et al. 1986).

fig. 3. Simulations of West et al. 1988, corresponding to HCS with $n = 0, -2$, and to a top-down ("pancake") scenario. Present epoch, size $= 20$ Mpc.

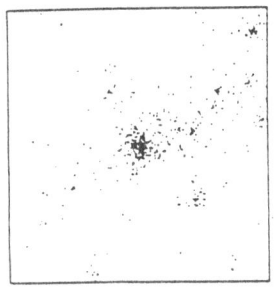
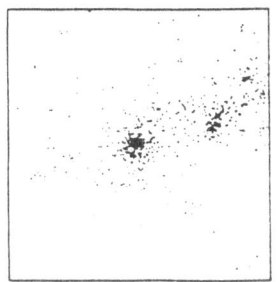
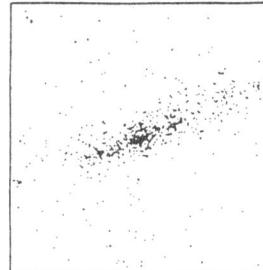

3. CLUSTERS IN THE MAKING: SCALINGS

Relatively few distant Clusters have been detected so far. The existing data base outlines the (comoving) local luminosity function $N(L, z \lesssim 0.1)$ of Clusters and Groups in the several keV range, mainly from HEAO 1 observations (Piccinotti et al. 1982, Henry et al. 1982, Johnson et al. 1983, Kowalski et al. 1984). A power law $N(L) \propto L^{-\gamma}$ is indicated with $\gamma \simeq 1.7$, gradually cut off above $\sim 5 \; 10^{44}$ erg/s, see fig. 4.

For $z > 0.15$, the existing information comes instead from EINSTEIN in the softer range $0.3 - 3.5$ keV, and concerns the number counts (see fig. 6) and the redshift distribution (see fig. 7), two integrals of $N(L, z)$ out to $z \simeq 0.5$ visualized by fig. 5. As for the redshift distribution, the preliminary detections from the ongoing Medium Sensitivity Survey out of archived material by Gioia et al. 1987, show a *dearth* of objects at $z \gtrsim 0.3$. Concurrently, the counts-flux relationship $N(> F)$ derived so far by Gioia et al. 1987 is apparently *flat* as shown by fig. 6: considerably flatter than the "Euclidean" $\beta = 1.5$, certainly quite flatter than found in the same survey for the evolutionary AGNs.

Now, cosmogonies like the simplest HCS imply a considerable degree of *evolution* also for Groups and Clusters. In general, as one looks back into the past light cone, one expects (Kaiser 1986) the luminosity functions to scale with z in amplitude $\mathcal{N}(z)$ and in shape, governed by object number and by typical luminosity L_c after

$$N(L, z) \propto \mathcal{N}(z) \, L_c^{\gamma - 1}(z) \,, \tag{3.1}$$

where we assumed a shape $L^{-\gamma}$. As for the simplest HCS in particular, looking back one expects from the scalings 2.1 more, smaller, denser and also cooler units *assuming* $T \propto v_c^2$ for the ICP in equilibrium (Kaiser 1986). So, *assuming* constant comoving mass the number increases ("density evolution") like

$$\mathcal{N}(z) \propto M_c^{-1}(z) \propto (1 + z)^{6/(n+3)} \tag{3.2}$$

while, *assuming* constant $\hat{M}/M$, the luminosities

$$L_c(z) \propto (1 + z)^{(5+7n)/2(n+3)} \tag{3.3}$$

decrease ("anti-evolve") only moderately and even less does $L_c^{\gamma - 1}$ in the scaling 3.1. The result for $N(L, z)$ is an evolution strong and close to the "density" kind, at least out to such z where the typical virializing mass goes below $\sim 10^{13} M_\odot$. By itself, this circumstance would tend to *steepen* the counts and to enrich the redshift distribution at its *upper* end.

But other effects intervene. Cosmology and spectral decline combine to weaken the flux after $F = L(z)(1+z)^{1-\alpha}/4\pi D^2(z)$ and so flatten the counts, to an eventual saturation when the spectral cut off at $E \sim kT(z)/(1 + z)$ of the bremsstrahlung emission is shifted into the observational window.

To gauge the balance between these competing trends CC 1988a use as a guide the first order approximation to the counts

$$N(> F) \sim F^{-3/2}[1 - B(F_o/F)^{1/2}] \,,$$

$$B = [2(1 + \alpha) - (dN/Ndz)_o] \frac{3D_o}{4R_H} \frac{\langle \ell^2 \rangle}{\langle \ell^{3/2} \rangle} \,, \tag{3.4}$$

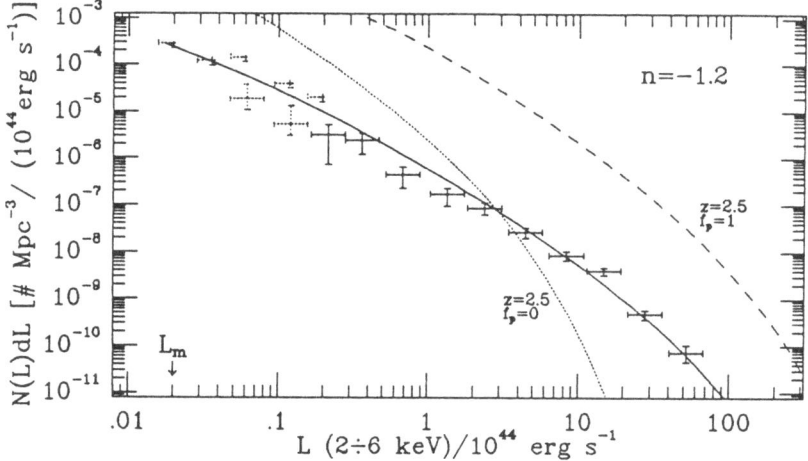

fig. 4. The local X-ray luminosity function for our standard model is compared with the data from Kowalski et al. 1984; normalization at $L \sim 10^{45}$ erg/s. The luminosity functions out to $z = 2.5$ are shown for $f_p = 0$, $\xi = 1$ (no early ICP and constant build up, dotted curve) and $f_p = 1$ (all early ICP, dashed curve). The faint end asymptotes to $L^{-(11-n)/8}$. $H_o = 50$ km/s Mpc, $\Omega_o = 1$ unless otherwise specified.

fig. 5. To visualize the build up of counts and of z-distribution, this overview of the $L - z$ plane shows schematic plots of $N(L, z)$ and the isoflux lines corresponding to the windows of HEAO 1, EINSTEIN and ROSAT.

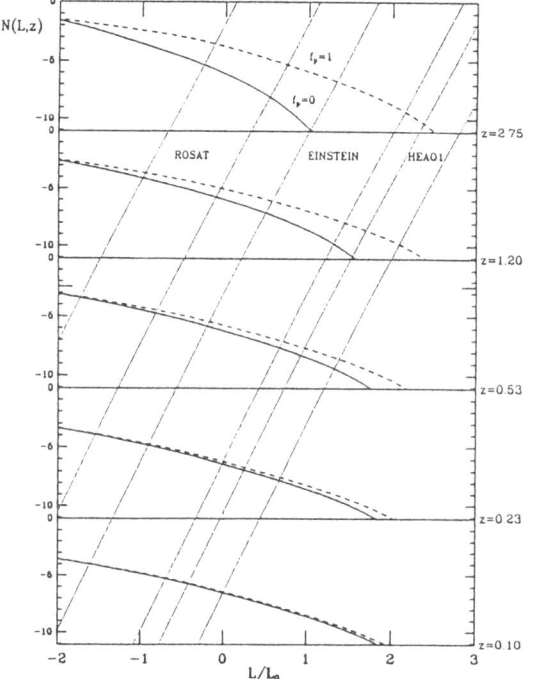

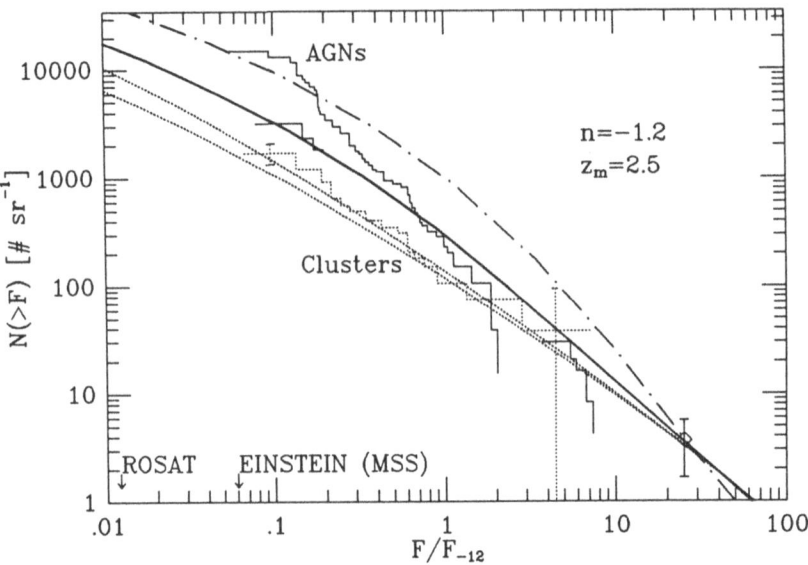

fig. 6. Number counts of X-ray Clusters for various fractions f_p of early ICP. Dotted lines: $f_p = 0$, constant ICP build up rate ($\xi = 1$), with $\Omega_o = 0.2, 1$ in ascending order. Continuous line: $f_p = 0.3$, with constant rate for the late component. Dot-dashed line: $f_p = 1$. The histograms represent the counts by Gioia et. al. 1984 (dotted line) and by Gioia et al. 1987 (continuous line).

fig. 7. Redshift distributions in the flux ranges of EINSTEIN and ROSAT, normalized to their area. Thick lines correspond to the luminosity functions for $f_p = 0$ ($\xi = 1$) and thin lines correspond to those for $f_p = 1$, in fig. 4. The preliminary results from the MSS (Gioia et al. 1987) are close to the former distribution.

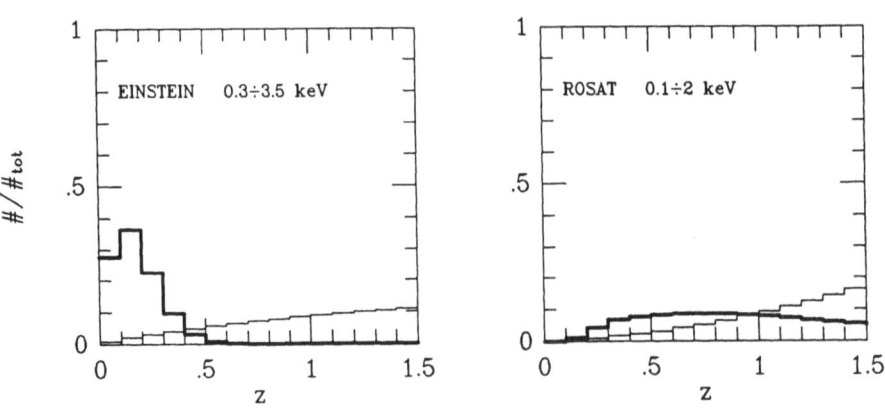

valid for $z \ll 1$, that is, for distances $\ll R_H$, the Hubble radius, or close to $D_o = (L_o/4\pi F_o)^{1/2}$ ($L_o \sim 10^{45}$ erg/s, $F_o \sim 10^{-11}$ erg/cm^2 s). The previous scalings together yield $N \propto (1+z)^\kappa$ with $\kappa = [7 + 5\gamma + 7n(\gamma - 1)]/2(n+3)$. The adimensional moments $\langle \ell^2 \rangle / \langle \ell^{3/2} \rangle$ involve to some extent the local shape of the luminosity function.

Whence it is seen that the combined implications of the simplest HCS: fast number increase and slow luminosity decrease, together nearly cancel those of cosmology and spectrum ($\kappa \gtrsim 2.7$ vs. $2 + 2\alpha \simeq 2.8$), conserving a slope ~ 1.5 well outside the local domain. The result persists for $n \sim -1 \pm 0.5$. Second order effects, discussed below, do not change materially the result.

So flatness of the counts, if confirmed, will call for a flatter increase of $\mathcal{N}(z)$ or for a steeper decrease of $L_c(z)$ than provided by the scalings 3.2 or 3.3. For example, a natural complement to the latter within the framework of the HCS, envisages a gradual *build up* in cosmic time of the ICP mass $\hat{M}$. This may be parameterized in the form $\hat{M}/M \propto g(z)$, with $g(t) \propto (t - t_m)^\xi$ ($\xi > 0$) increasing after a threshold because of ICP being produced by stars or infalling from outside over several Gyr: a constant build up rate corresponds to $\xi = 1$. The build up translates into an extra factor $g^2(z) \to (1+z)^{-3\xi}$ to eq. 3.3 and adds in eq. 3.4 a considerable anti-evolutionary term $\Delta\kappa = -3\xi(\gamma - 1)$. Similarly, if the number effectively observed scaled $\propto t^x$, then $\Delta\kappa = -3x/2$ would obtain.

The scalings thus completed are yet limited to FRW cosmologies with $\Omega_o = 1$ and neglected several refinements, including the important effects of the shape of the luminosity functions at $z > 0$, and the spread $T(L)$ at any given z discussed below. Moreover, two components to the ICP may be present, namely the early one already present in the condensations at the Groups era, and a slowly growing component. These additions are included in numerical computations as outlined below.

4. VARIANCES AND DISPERSIONS

Beyond the heuristics, the evolution of the luminosity function is given by the formalism of the continuity equation in the form

$$\Delta N(L,z) = \int_z^{z_m} dz_i \; s(L_i, z_i) \frac{dL_i(z_i)}{dL} \tag{4.1}$$

(see CC 1988a,b) that includes changes of the object number s either by destruction or by formation, and a subsequent increase in luminosity from the initial value $L_i(z_i)$ since the Group formation epoch z_m.

As for the average luminosities, the ICP build up is included with

$$L \propto g^2(z) M \rho T^{1/2} . \tag{4.2}$$

A late and an early (fraction f_p) component of the ICP are accounted for by using instead of $g(z)$ the compound mass fraction $G(z) = [f_p + (1 - f_p)g(z)]$.

As for the rates of object destruction or production

$$s(L,z) = \frac{dN(M,z)}{dz} \frac{dM}{dL} \tag{4.3}$$

to be inserted in the solution 4.1 with $L \propto M^{4/3}$ from eqs. 4.2 and 2.1, they are to quantify the reshuffling of smaller condensations into larger ones typical of the HCS. Press and Schechter 1974 compute $N(M, z)$ in a cutely simple way from the initial Gaussian field $p(\delta/M)$ assuming a schematic collapse prescription and dynamics independently completed at each step of the hierarchy, to yield

$$N(M, z) \propto \frac{1}{M^2} \frac{d\ln\nu}{d\ln M} \; \nu \; exp[-\nu^2/2]$$

$$\nu \equiv \frac{\delta_c}{\sigma(M, z)} = \delta_c(1+z)\left(\frac{M}{M_o}\right)^{(3+n)/6} , \qquad (4.4)$$

where $M_o \propto M_c(0)$ is essentially the mass going non-linear at the present epoch; we adopt the minimal (conservative, as we shall see) value of the collapse factor $\delta_c \sim 1.3$ insuring central virialization or a realistic density contrast for bona fide Clusters (cf. Bardeen et al. 1987).

The point to stress is that the above *standard model* introduces a crucial *variance* of the non-linear variables around the typical values given by the rigid scalings eqs. 2.1, inasmuch generated – at objects' turnaround – from the unbiased statistics of the initial perturbations $p(\delta|M)$: additional variances are discussed in the next §. Meanwhile, notice that the scalings 2.1 themselves imply not only an average spectral evolution due to the temperature changes $T_c(z)$ but also at any given z a correlation $T \propto G_i^{-1} \rho_i^{-1/4} \; L^{1/2}$ (corresponding to objects just virialized), that will be dispersed by the Gaussian field toward the flatter and higher line $T \propto L^{(1-n)/8}$ (corresponding to forerunners) as found by Cavaliere, Danese and De Zotti 1978; observations have given an average correlation $T \propto L^{2/5}$, cf. Mushotzky 1988.

Counts and redshift distributions are derived using standard relations (see Weinberg 1982), in the appropriate instrumental windows. The results are illustrated in figs. 6 and 7. The bottom line reads: counts always flat *and* redshift distributions skewed toward low z obtain when the fraction of early or promptly produced ICP is well under 50%.

Most sensible variations of parameters from the conservative setting: $L_c(0) \sim 10^{45}$ erg/s, $M_c(0) \sim 10^{15} \, M_\odot$, $L_m \sim 5 \; 10^{42}$ erg/s, $z_m \simeq 2.5$, confirm or enhance this result, see the figures with their captions, and the discussion by CC 1988a,b. A possible exception is related to the shape of the luminosity function.

5. LUMINOSITY FUNCTION AND COSMOGONY

Such shape is more relevant to the counts at intermediate fluxes than might be expected from the high-flux approximation 3.4. This is understood on viewing at the (differential) counts as the envelope of the z-dependent luminosity functions weighted at each z with the appropriate cosmological volume, cf. CC 1988b. As such, the counts can be only as steep as permitted by the slope of the dominant luminosity functions, namely, the nearly local ones for $f_p = 0$ and those at large z for $f_p \to 1$; in the latter case, the counts-flux relationship tends to follow the shape of the distant luminosity functions. Correspondingly, low or high z will dominate the redshift distribution.

Here we outline the dependence on cosmogony of the shape of the luminosity function. Our standard model was based on the simple *Press-Schechter theory*, but in our numerical code we take into account the efforts being devoted to inserting additional parameters into that rather inflexible formalism. In brief, we find the faint end steepened in the *CDM model*, and flattened in models with collapses localized at the fluctuation *peaks*. At the

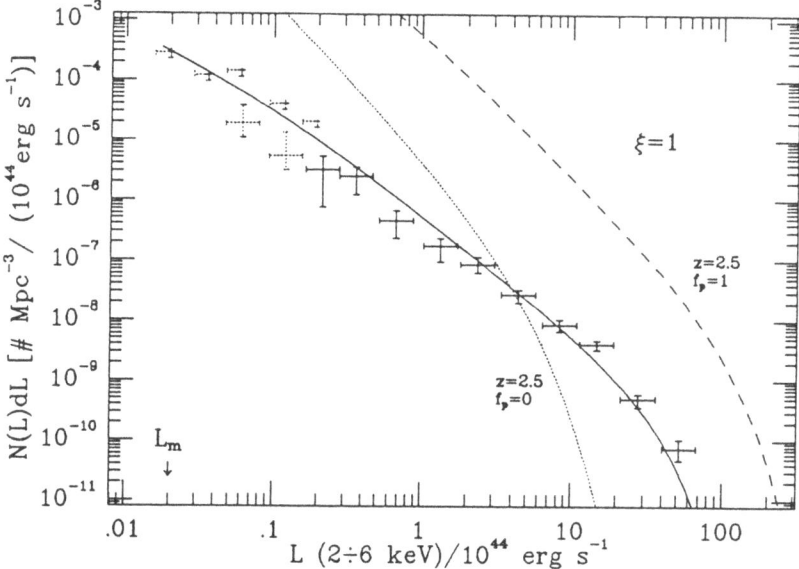

fig. 8. The luminosity functions for a CDM scenario (with overall $b = 1$, see text), to be compared with fig. 6. Note how slope of the luminosity function and all evolutionary rates increase as n decreases with M.

fig. 9. Local luminosity function with variance modified relative to our standard case ($b = 1$, continuous curve): the dotted curve is for $b = 1.5$ while the dashed curve is for $b = 2/3$ to simulate the effects of non-linear dynamics or those of a non-Gaussian statistics.

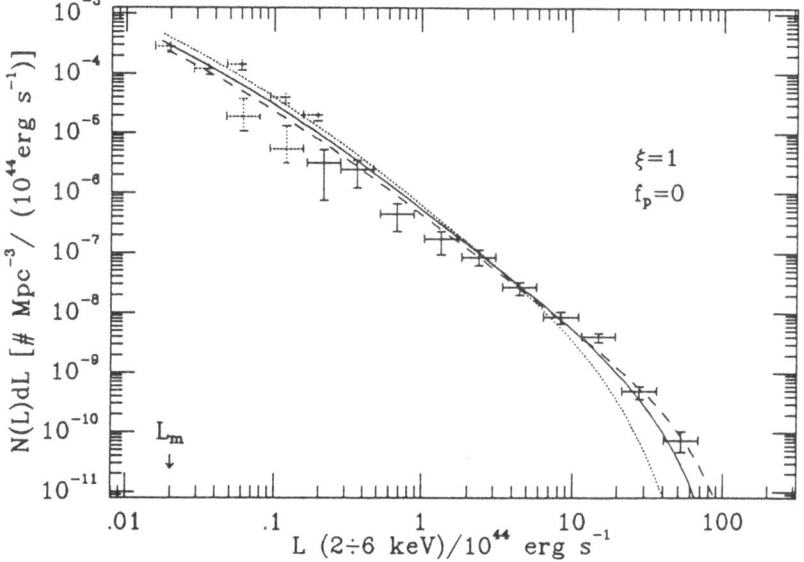

bright end the variance is decreased by biasing, but increased by non-linear dynamics and by non-Gaussian statistics. Details follow.

The CDM model (cf. Blumenthal et al. 1984) contains two main additional degres of freedom. First, the limited growth of the lighter perturbations entering the horizon during the radiation era produces curvature of the variance $\sigma(M)$ as described by the transfer function given by Davis and Peebles 1983. The luminosity functions may be still represented in the form 4.4, where however $d\ln\nu/d\ln M = [n(M)+3]/6$ defines an *effective* $n(M)$ decreasing from $\simeq -1$ to $\simeq -1.7$ when M decreases from $\sim 10^{15}$ to $\sim 10^{13} M_\odot$. Heuristically, the previous scalings still may be used with an effective $n(M)$. It follows that $N \to L^{-(11-n(M))/8}$ is to steepen at its faint end. The counts steepen somewhat at intermediate fluxes because of this, and also because they are affected by the scaling $N \propto L_c(z)^{\gamma-1}/M_c(z)$ with a luminosity decline somewhat steeper but the number increase steeper yet; a sharper terminal saturation is caused by the faster spectral cut off. All that is consistent with the numerical findings shown in figs. 8 and 10.

Second, because CDM naturally associates with $\Omega_o = 1$ but galaxies apparently do not trace such a large mass, a biasing factor $b > 1$ is required to renormalize the variance $\sigma_o(\sim 8$ Mpc$) = 1/b$ (cf. Bardeen et al. 1988). The effect on the luminosity function is twofold: $N \propto b\delta_c exp[-(b\delta_c)^2]$. While normalizations depend also on the uncertain total infall, the dependence of the luminosity dispersion is obviously sensitive (see fig. 9), and tends in turn to steepen the counts correspondingly (see fig. 10).

In terms of the X-ray data, the local luminosity function apparently favours a dispersion corresponding to an effective $b \sim 1$. However, additional contributions to the effective variance $B\sigma$ are envisaged from non-linear dynamics described in § 2, so that the dependence $N \propto exp[-(b\delta_c/B)^2]$ is expected; the cut off is further smoothed by the non-linear X-ray L/M after eq. 4.3. In fact, the optical luminosity function (less vulnerable to emission non-linearities, though statistically more uncertain) requires a larger bias up to $b \sim 2$, see fig. 11. In the balance, with our conservative δ_c that allows a maximal b and a minimal B, the present context is not inconsistent with a linear bias $b \sim 1.5$, cf. also Evrard in these Proceedings.

Another cause for increased variance may be constituted by a non-Gaussian hierarchical statistics of the perturbation field (as discussed by Lucchin and Matarrese 1988), in which case a higher linear bias $b > 2$ would be required by the optical and allowed by the X-ray data.

Finally, the peaks saga. Collapses are expected to prefer peaks, particularly the higher peaks, of the initial perturbation field (Peebles 1980). Whence one envisages a flattened luminosity functions at the faint end corresponding to a deficit of failed perturbations, and a relative excess of massive systems. In fact, a number of Authors (including Bardeen et al. 1986, Bond 1988, Lucchin and Matarrese 1988, Colafrancesco et al. 1988) rederive the mass distribution from this approach, to find a class of modified Press-Schechter results of the form $N \propto \nu^\epsilon N_{PS} \propto (M/M_c)^{\epsilon(n+3)/6} N_{PS}(M,z)$ with $0 \leq \epsilon < 1$, the lower limit corresponding to the minimal and the upper bound to a nearly maximal collapsed mass (no subsequent infall). We note the limiting form represented in fig. 12, $N \propto L^{1/5} N(L,0)$ in terms of our standard model, set by the existing X-ray *lower* bounds; the corresponding limiting counts are flattened at their faint end even for $f_p = 1$ but lie still above the data existing at the bright end, see fig. 13. The limit corresponds to $\epsilon \lesssim 1$ when the collapsed mass associated with a peak is taken always close to the filtering mass; but a general,

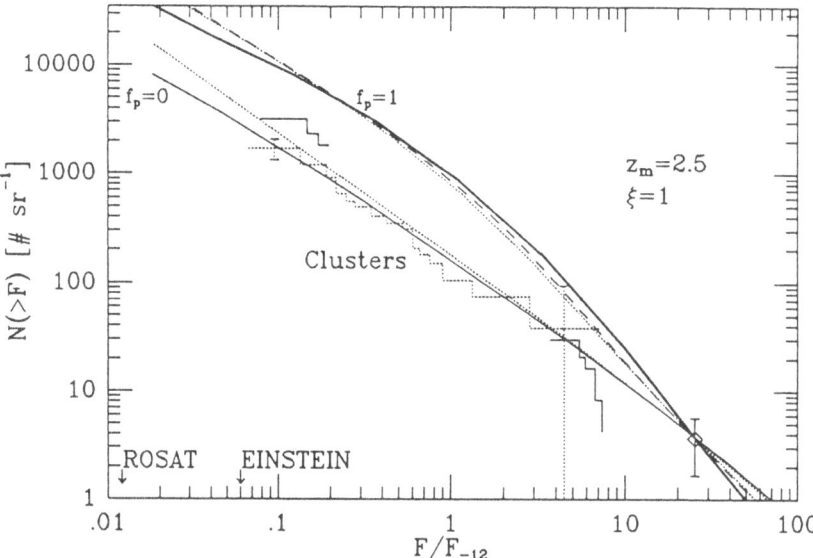

fig. 10. Counts for CDM models compared to those for the simple HCS (continuous curves). CDM models with $b = 1$ (dotted curve) and with $b = 1.5$ (dashed curve) provide lines nearly superposing.

fig. 11. The optical luminosity function from the CDM model for two values of the linear bias: $b = 2$ (continuous line) and $b = 1.5$ (dotted line). Data from Bahcall 1979. A comparison with the X-ray data of fig. 9 suggests that there non-linear effects enhance the variance of the luminosity function (equivalent to decrease the effective b).

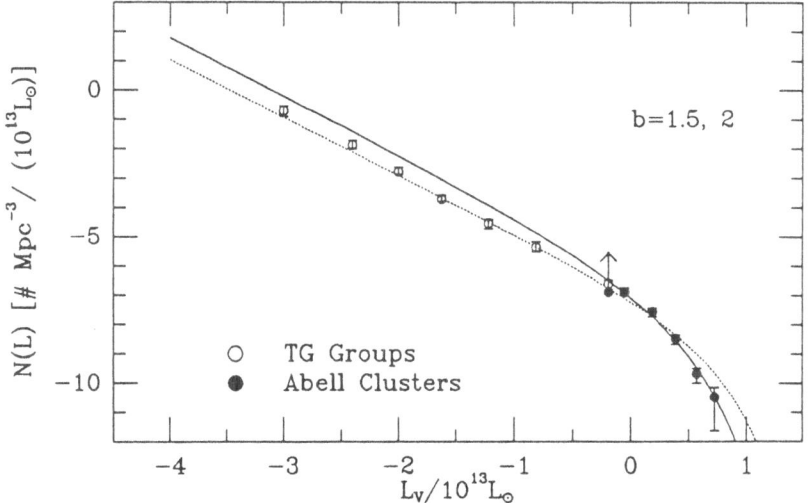

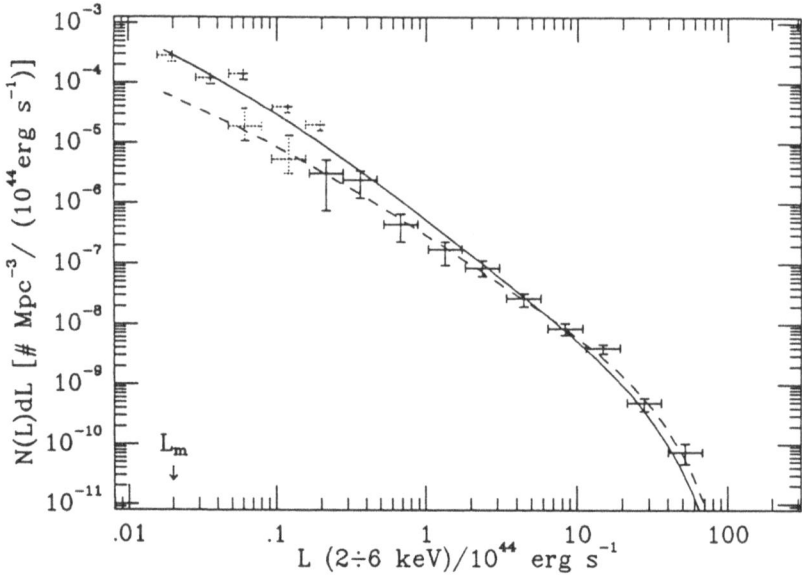

fig. 12. The local luminosity function for a model of collapse around density peaks with $N \propto \nu N_{PS}$ is compared with the CDM result (solid curve). Note that the former model is the flattest allowed by the data of Kowalski et al. 1984.

fig. 13. Counts based on the previous local luminosity functions and all early ICP (solid line, dashed line as in fig. 12). Also the counts for $\mathcal{N} \propto t/M_c(z)$ are represented (dotted line).

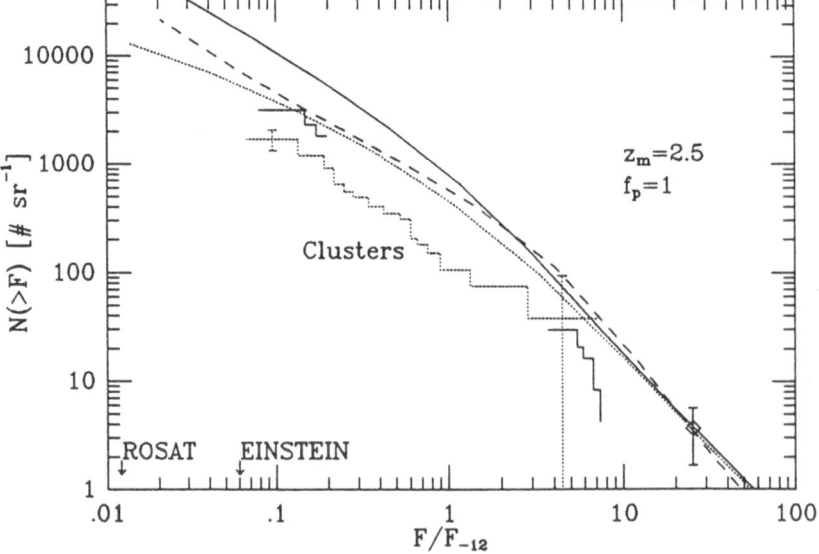

non-trivial uncertainty persists concerning the masses actually collapsing around peaks especially of low amplitudes and high z, cf. discussion by Bond 1988.

The above is to stress the rich cosmogonic information that is potentially contained in the faint and in the bright ends of the luminosity function.

6. DISCUSSION AND CONCLUSIONS

Rich local morphology and a dearth of distant Clusters are the observational indications. The former obtains within the Hierarchical Clustering Scenario once a fair amount of variance is granted. The latter feature (based on counts and redshift distribution as currently known) is apparently mismatched to the evolutionary behaviour envisaged by the simplest HCS: number strongly increasing in look back time with luminosity only slowly declining. The CDM model by itself is no solution. Nor directly is the amount of variance allowed by the dispersion of the local luminosity function as known from Kowalski et al. 1984, that yet includes such extremes as A1367 and A426.

The discrepancy constitutes a problem either for the data or for the theory: here we discuss briefly three lines of explanation ranging from undersampling to intrinsic effects.

i) Extended faint sources might be missed altogether by finding procedures optimized for the steep gradients permitted by point sources, an issue to be left for the observers to control. In a similar, subtler vein the *high* X-ray background with its fluctuations coupled with the intrinsic *substructures* of the sources hinder total flux calibration out to a common isophotal diameter (as fig. 1 indicates). In fact, after the HCS Clusters and Groups are far from being standard candles, in contrasting ways: with z increasing, their overall emission decreases and their overall shape grows larger and sparser, yet the condensed fragments' intrinsic size $R_c(z)$ decreases; the intrinsic surface brightness makes up for the cosmological dimming and helps detection of *condensed* subclusters. Problems of identity may arise, however, analogous to that pointed out by Geller 1988 for bimodal or multiple configurations: are they counted as one cluster with summed flux, or as separate objects to increase the number in their flux ranges? For earlier objects the superpositions may be less severe, but in gauging the balance the N-body simulations may be really instrumental.

ii) Poor Clusters and Groups may be *fewer* or *dimmer* or *cooler* than predicted by the simplest scalings.

Fewer Groups: The number of observed objects may scale up with z less than expected for spheres at constant comoving mass: subclusters with marginal contrast for collapse (size larger than average, irregular shape and slow condensation) easily yield to tidal effects and to cannibalism, or even as they survive are liable to sink into background because of lower surface brightness, cf. fig. 1. Fig. 13 shows one example of the corresponding count loss affecting mainly low fluxes, preliminary to aimed simulations.

Cooler Groups: with increasing z the photon flux becomes and appears indeed cooler, tending to leave unfilled the upper end of the instrumental window. But out to $z \sim 0.5$ the effect is small; even a lower effective efficiency of EINSTEIN below 0.5 keV would not be crucial. More of a problem might be constituted by a lower normalization of the correlation $T \propto T_i(z)L^{1/2}$, corresponding to incomplete virializations: however, to flatten the counts even only below $\sim 10^{-12}$ erg/cm^{-2}s it takes a value $T_o \lesssim 1.3$ keV for richness 1 Clusters, low compared with the data in Jones and Forman 1984, Mushotzki 1988.

Dimmer Groups: counts always flat and z-distribution skewed to low z obtain when $L_c(z)$ is dimmer than expected at constant $\hat{M}/M$. This may be the effect of the ICP own history, if most ICP is *younger* than the first structures (themselves younger than $z \sim 2.5$ in the CDM model): meaning only a small amount already present at $z \sim 2$, with an even build up subsequently. Still the two options remain open for varying the ratio $\hat{M}/M$: bulk ICP production by the stars or infall of $\gtrsim 50\%$ H from outside, with the proviso that Fe and other "metals" betrayed by the emission lines must in any case result from astration in the member galaxies (cf. Matteucci and Vettolani 1988). If production by stars is dominant, then star formation or starburst activity must be decaying very slowly on gross average. If it is infall, this must be continuous and favouring gaseous barionic matter, e.g., debris of failed galaxies (Dekel and Silk 1986); this would produce the strongest dilution of the ratio Z/H from $z \sim 0.5$.

iii) A luminosity function *flattened* at its faint end as suggested by the peak approach will generate counts correspondingly flattened toward their faint end, yet steep at bright fluxes as prescribed by the simplest HCS with all early ICP, see fig. 13. At such high fluxes statistics may be a problem, but for consistency a redshift distribution rich at $z \gtrsim 0.5$ should still obtain. To be sure, a lower limit to the faint slope of $N(L, z \sim 0)$ has been provided by the HEAO 1 data of Kowalski et al. 1984 but these are most sensitive at 2 - 6 keV and a softer data base homogeneous with the counts is obviously needed (see also Boldt, these Proceedings).

A final point concerns an inference from the intermediate structures here considered to the structure of the Universe as a whole, besides the obvious but delicate gauging of Ω_o from the weak dependence of the run of the luminosity functions (see CC 1988a). While the Cluster counts may well be affected by some incompleteness, it looks hardly possible to make up in these terms the deficit by a factor $\lesssim 1/3$ relative to the AGN counts (cf. fig. 6), which in turn may be rather lower bounds (cf. Boldt, these Proceedings). The discrepancy spells difficulties not only for cosmologies like the Steady State that leave no room for different intrinsic evolutions of the diverse source classes, but also for those like the chronometric cosmology (see Segal 1987) that attempt to ascribe a common slope steeper than Euclidean entirely to the effects of the overall space-time structure.

To conclude, our cross section into the matter intended to probe how much variations must be added to the rigid frame of the gravitational instability to account for the rich morphological zoo and for the local and distant statistics. In the present context the HCS, with $n \sim -1$ at intermediate scales and with $b \sim 1.5$, to stand the body of the current evidence needs help by a good pinch of small scale inhomogeneity, and from some differential effect breaking approximate scale invariance at least as for observability. Another area of intense development is constituted by the shape of the luminosity function. For all these developments the EINSTEIN archives and the forthcoming deeper, softer survey from ROSAT will provide a crucial testing ground. It is just by the paradigm that relevant planning seems in turn to require preliminary theoretical explorations like here outlined.

Aknowledgements. We are grateful to G. De Zotti, M. Geller, R. Giacconi, I. Gioia, F. Lucchin and A. Renzini for many helpful discussions, and to N. Vittorio for critical reading.

REFERENCES

Bahcall, N.A. 1979, *Ap. J.*, **232**, 689.

Bardeen, J.M., Bond, J.R., Kaiser, N., and Szalay, A.S. 1986, *Ap. J.*, **304**, 15.

Bardeen, J.M., Bond, J.R., and Efstathiou, G. 1987, Ap. J. **321**, 28.

Binggeli, B., Tammann, G.A., and Sandage, A. 1987, preprint.

Bond, J.R. 1987, in Proc. of the Pontifical Academy of Sciences Study Week 27 *Large Scale Motions in the Universe*.

Blumenthal, G.R., Faber, S.M., Primack, J.R., and Rees, M.J. 1984, *Nature*, **311**, 517.

Cavaliere, A., Santangelo, P., Tarquini, G., and Vittorio, N. 1986, *Ap. J.*, **305**, 651.

Cavaliere, A., Danese, L., and De Zotti, G. 1978, *Ap. J.*, **221**, 399.

Cavaliere, A. 1980 in *X-ray Astronomy*, R. Giacconi and G. Setti eds., (Dordrecht, Reidel).

Cavaliere, A., Gursky, H., Tucker, W.H. 1971, *Nature*, **231**, 437.

Cavaliere, A., and Colafrancesco, S. 1988 a, *Ap. J.*, in press; 1988 b, in Proc. of the NASI *Hot Thin Plasmas in Astrophysics*, in press; (CC 1988a,b).

Colafrancesco, S., Lucchin, F., and Matarrese, S. 1988, preprint.

Davis, M., and Peebles, P.J.E. 1983, *Ap. J.*,**267**, 465.

Dekel, A. 1987, in Proc. IAU Symposium 124 *Observational Cosmology*, in press.

Dekel, A., and Silk, J. 1986, *Ap. J.*, **303**, 39.

Dressler, A., and Shectman, S.A. 1988, Ap. J., in press.

Evrard, A.E. 1987, *Ap. J.* **316**, 36.

Fabian, A.C., in Proc. of the NASI *Hot Thin Plasmas in Astrophysics*, in press.

Fabricant, D., Beers, T.C., Geller, M.J., Gorenstein, P., Huchra, J.P., and Kurtz, M.J. 1986, *Ap. J.*, **308**, 530.

Fitchett, M., and Webster, R. 1987, *Ap. J.*, **317**, 653.

Forman, W., and Jones, C. 1982, *Ann. Rev. Astr. Ap.*, **20**, 547.

Geller, M.J. 1988, to appear in Saas Fee Lectures, *Large-Scale Structure in the Universe*.

Geller, M.J., and Beers, T.C. 1982, *P.A.S.P.*, **94**, 421.

Giacconi, R., et al. 1980, *The Advanced X-ray Astrophysics Facility*, Science Working Group Report, NASA Report No. TM-78285.

Gioia, I.M., Maccacaro, T., Schild, R.E., Stocke, J.T., Liebert, J.W., Danziger, I.J., Kunth, D., and Lub, J. 1984, *Ap. J.*, **283**, 495.

Gioia, I.M., Maccacaro, T., Morris, S.L., Schild, R.E., Stocke, J.T., and Wolter, A. 1987, preprint.

Gunn, J.E., and Gott, J.R. 1972, *Ap. J.* **176**, 1.

Johnson, M.W., Cruddace, R.G., Ulmer, M.P., Kowalski, M.P., and Wood, K.S. 1983, *Ap. J.*, **266**, 425.

Jones, C., and Forman, W. 1984, *Ap. J.* **276**, 38.

Kaiser, N. 1986, *M.N.R.A.S.* **222**, 323.

Kowalski, M.P., Ulmer, M.P., Cruddace, R.G., and Wood, K.S. 1984, *Ap. J. Suppl.*, **56**, 403.

Lucchin, F., and Matarrese, S. 1988, *Ap. J.*, in press.

Matteucci, F., and Vettolani P. 1988, preprint.

Mushotzky, R.F. 1984, *Physica Scripta*, **T7**, 157.

Mushotzky, R.F. 1988, in Proc. of the NASI *Hot Thin Plasmas in Astrophysics*, in press.

Peebles, P.J. 1970, *A. J.* **75**, 13.

Peebles, P.J. 1974, *Ap. J. Lett.* **189**, L51.

Peebles, P.J. 1980, *The Large Scale Structure of the Universe* (Princeton Univ. Press).

Piccinotti, G., Mushotzky, R.F., Boldt, E.A., Holt, S.S., Marshall, F.E., Serlemitsos, P.J., and Shafer, R.A. 1982, *Ap. J.*, **253**, 485.

Press, W.H., and Schechter, P. 1974, *Ap. J.*, **187**, 425.

Sarazin, C.L. 1988, *X-ray Emission from Clusters of Galaxies*, (Cambridge, Cambridge University Press).

Segal, I.E. 1987, *Ap. J.*, **320**, 135.

Trümper, J. 1988, in Proc. of the NASI *Hot Thin Plasmas in Astrophysics*, in press.

Weinberg, S. 1972, *Gravitation and Cosmology* (New York, Wiley).

West, M.J., Oemler, A., and Dekel, A. 1988, *Ap. J.*, **327**, 1.

West, M.J., and Richstone, D.O. 1988, preprint.

White, S.D.M. 1982, in *Morphology and Dynamics of Galaxies* ed. by Martinet, L. and Mayor, M. (Geneva Observatory, Geneva), p. 289.

DISCUSSION

Hammer: I am surprised by the way you reject the Wampler study on bright distant QSO and radiogalaxies: in fact, he deduced a high value of q_o by considering overluminosities due to space curvature in a homogeneous Universe. However, you may consider that, in a inhomogeneous universe, the q_o value depends on the line of sight (by the mean of gravitational lensing) and reconcile the Wampler study with yours; except if you assume that there is no selection effects in the observations of bright distant QSO and galaxies.

Cavaliere: My point concerns the apparent discrepancy between AGN and Cluster counts in the EMSS. To me, this indicates different intrinsic astrophysics affecting the space distribution of either class of sources, rather than the common action of the overall structure of the Universe. Gravitational lensing would flatten, if anything, the AGN counts and little affect the Clusters' in view of their large size and modest redshift; the luminosity distribution of the latters bears even less signs of lensing than the formers', cf. Turner 1988, in Tucson Workshop on Optical Surveys for Quasars. Unless you assume selections to play real havoc with the relative counts of the two classes.

Zaninetti: How much sure is the assumption that the X-ray emission from clusters is of thermal origin? I remember that you previously mentioned evolution of quasars and radiosources where the radiation is surely nonthermal.

Cavaliere: Thermal bremsstrahlung origin for the *bulk* of Cluster emission in X-rays is consistent with the shape of the continuum below ~ 10 keV. Whence one expects also very high excitation emission lines, like the Fe complex at ~ 7 keV, that in fact are observed in many objects. Sunyaev-Zel'dovich effects and head-tail radiogalaxies provide confirming evidence. This is not to deny, however, that in some objects minor components of different origin exist, like the hard, point source associated with the active galaxy NGC1275 in Perseus Cluster.

As for evolution, it may be of some interest to stress that the deep gravitational potential wells at the compact cores of QSOs and Radiosources, emitting non thermal components detectable or even dominant, apparently processed more power on average in the past to dim or quench afterwards, running down as it were. On the other hand, from the relatively shallow Group or Cluster potentials that accumulate waste mass and thermalized energy, the emission increases toward us.

Rhee: You assume the existence of substructure on 1 Mpc scales but there is considerable controversy on this issue in literature? (e.g., Mazure et. al. find substructure in Coma, Dressler does not). Could you comment on this issue?

Cavaliere: Substructure in X-rays in many, perhaps in most Clusters is beyond reasonable doubt, cf. reviews by Forman Jones 1982, Sarazin 1988. Typical sparse, very clumpy configurations are A1367 (where however plumes ans halos around single galaxies may be confusing), A194, A566, A2151, A2069, also A496 (Mazure et al. 1986, A & A 157, 159). In the optical, the situation is more complex, as expected: the candidate clumps are comprised of a relatively small number of galaxies, and Poisson superpositions may contaminate the samples. However, spectroscopy by Dressler and Shectman 1988 confirmed the percentage ($\sim 40\%$) of substructured objects found by Geller and Berrs 1982 from isodensity maps. Even West et al. 1988 question frequency and mass of the clumps, but do not deny substructure within some Clusters. As to Coma, the discrepancy you mention is to be set in the canonical perspective holding that Cluster to be prototypical of the smooth, spherical, relaxed class. Several works stressed inner $\sim$ bimodal structure (Fitchett and Webster 1987; Mellier, Mazure, Mathez, Chauvineau and Proust, 1987, preprint), consistent with the slight elongation of the X-ray maps first stressed by Johnson, Cruddace, Fritz, Shulman and Friedman 1979, Ap. J. 231, L45. Dressler and Schectman simply do not find evidence of strong structure in v space.

Peebles: The Press-Schechter approximation you use is quite crude. Have you or someone else cheked this approximation against the result of N-body model simulations?

Cavaliere: As to the PS *expression* for the mass distribution, several recent works including Efstathiou and Rees 1988, MNRAS 230, 5p; Cole and Kaiser 1988, preprint; Davies, Efstathiou, Frenk and White 1988, preprint; Carlsberg and Couchman, 1988, preprint, find it remarkably effective in describing N-body results. ER, for example, find the abundance of clumps of high contrast in the simulations to agree with the PS expression modified after the CDM prescriptions to include the effective variance $\sigma(M)$, or equivalently the effective index $n(M)$, with a threshold $\delta_c = 1.33$. Davies et al. extend the comparison to a range of power law power spectra.

As to *derivation*, you pointed out (Peebles 1980) that an analysis based upon collapses at the peaks of the linear density field would be much more satisfactory. In fact, such analyses are being produced by a number of workers (including Bond 1988, Colafrancesco, Lucchin, Matarrese 1988) to find modified PS distributions with the expected relative deficit of small objects and excess of large masses.

I stress we use such variants in our numerical code to compute t-dependent luminosity function and related counts-flux relation, but find the latter to be as flat as preliminarily indicated by the EMSS only when the former is considerably flatter than indicated by the local data. These, however, come mainly from HEAO1 and though Soltan and Henry 1983, Ap. J. 271, 442, have attempted to combine them with EINSTEINs' at lower energies, calibration match often constitutes a problem: direct, complete local statistics from EINSTEIN archives is badly wanted.

De Zotti: The cooling time of rich clusters today is close to the Hubble time. If early Clusters were denser, they were cooling in a time that may be much shorter that the Hubble time. What may be the effect of that on predicted counts?

Cavaliere: Prevalence of cooling flows is inevitable if the ICP is present within the first structures forming at $z > 2$ with density scaling $\propto (1 + z)^3$, as stressed by Fabian and Rees 1978, MNRAS 185, 109, and White 1982. If so, single objects will be conspicuous in soft X-rays and enhanced soft counts are expected. On the other hand, if the ICP mass decreases relative to the total toward the past, the density at $z \sim 2$ may be well be too low for that. In fact, this constitutes a test for young ICP: no soft excess; in addition, $Fe/H >$ solar at $z \gtrsim 0.5$ if most H comes from infall.

QUASARS AND LARGE SCALE STRUCTURE

PATRICK S. OSMER
National Optical Astronomy Observatories
Kitt Peak National Observatory
P. O. Box 26732, Tucson, Arizona 85726, U.S.A.

PAUL C. HEWETT
Institute of Astronomy, Madingley Road,
Cambridge CB3 0HA, England

ABSTRACT. Several large scale quasar surveys are now being completed and analyzed for their spatial distribution. They provide the best available information on the structure of the universe for $0.5 < z < 3$. Despite the continued, occasional detection of localized groups of quasars, there is no evidence for departures from uniformity in the population as a whole for scales of $50 - 500 \ h^{-1}$ Mpc.

INTRODUCTION

The microwave background radiation in the universe is observed to be remarkably smooth, while nearby galaxies are observed to have a very clustered and non-uniform distribution. The questions of how the clustering structure originated in the universe and how it subsequently evolved constitute major topics in astronomical research today.

At present quasars are the best probes of structure in the universe for redshifts between one half and three. Galaxies at such redshifts are still too faint to study in large numbers with currently existing telescopes and instrumentation. The space distribution of quasars themselves yields important information on possible structure at early epochs; in addition, quasars reveal gas clouds and galaxies along the line of sight, which can be studied by the absorption lines they produce in quasar spectra. These data together can give a useful picture of relations among the gas, the galaxies, and the quasars.

Admittedly the objection can be raised that quasars may not reflect either the distribution of galaxies or of matter in the universe. However, there is strong evidence for quasars being associated with galaxies and galaxy groups for redshifts up to about 0.6 (e.g., the work of Green and Yee (Yee 1987)). To the extent that quasar evolution means that activity was much more frequent in the past in galaxies, then the quasars would indeed reflect the galaxy distribution and clustering would be expected among faint quasars

93

M. Mezzetti et al. (eds.), Large Scale Structure and Motions in the Universe, 93–100.
© 1989 by Kluwer Academic Publishers.

(Setti and Woltjer 1977). In any event, the quasar distribution and its evolution are fundamental data for observational cosmology.

The observational evidence for quasar clustering has had a curious development. On the one hand, a number of intriguing pairs and groups have been found (e.g., Burbidge et al 1980, Oort, Arp, and de Ruiter 1981, Osmer 1981, Webster 1982, Crampton, Cowley, and Hartwick 1987). On the other hand, quantitative analyses of quasar surveys have not shown any significant indication that the population as a whole is clustered on scales greater than $10 \ h^{-1}$ Mpc. There is recent evidence now in favor of clustering being detected on scales less than 10 Mpc; Peter Shaver discusses these fascinating results in his talk. It is safe to say that much of the uncertainty about quasar clustering and the significance of the groups that have been found results from the lack of large, well-calibrated surveys.

RECENT DEVELOPMENTS

In the last two years there have been substantial advances in both the size and nature of quasar surveys. The use of plate scanning machines and automatic detection software enables substantially larger and deeper surveys to be done. The advent of multi-object spectrometers permits much larger samples of quasars with accurate redshifts to be compiled. A partial list of surveys described at the NOAO workshop on quasar surveys held in January, 1988 (Osmer et al, 1988) shows that over 800 quasars are becoming available, as indicated below in Table 1.

TABLE I

NEW SURVEYS WITH SPECTROSCOPICALLY CONFIRMED REDSHIFTS

GROUP	SAMPLE SIZE	COMMENTS
Shanks, Boyle	423	UVX technique, separated fields
Peterson		
Crampton, Cowley Hartwick	>200	Grens technique, separated fields
Barbieri, Cristiani Iovino, Nota	80	3 degree square area
Osmer, Hewett	127	Grism technique, connected strip of sky

For this talk I will discuss the quasar distribution for large scales, 50 - 500 h^{-1} Mpc, for which the evidence continues to support the concept of an unclustered population. For example, Shanks et al (1988) have now observed 34 fields, each 40' in diameter, with a total of 423 quasars. They obtained a surface density of 38 ± 9.8 quasars per sq. deg, with no evidence for field to field variations. This not only refutes the qualitative impression that many of us observers have had that some fields are poorer in quasars than others, but it provides strong quantitative support for quasars being distributed uniformly on large scales. Indeed, the application by Shanks et al of the correlation function showed no evidence for structure over the range in scales of 10 to 2000 Mpc.

Crampton et al (1987, 1988) also found no evidence for clustering in their sample, although they do have a very interesting group of 16 quasars with an average redshift of 1.11.

Osmer and Hewett (1988) have been carrying out a survey specifically designed to look for clustering in the quasar population; they, too, find no positive detection of clustering. Their work, as described below, illustrates some of the issues involved in the search.

They chose a new field on the sky at 12^H 10^M, -11° in the northern galactic hemisphere to obtain results completely independent of previous surveys. This was done because there has been a tendency to repeat known fields or quasars in some surveys, which makes it difficult to evaluate the statistical significance of the results found. Osmer and Hewett also covered a long, thin, connected strip on the sky, 0.5 x 12 degrees, as numerical simulations (Osmer 1981) had shown this configuration to be sensitive to any clustering that might exist on a wide range of scales. In fact, it is a first step toward assembling a slice of the universe diagram that we have seen used so successfully in galaxy surveys (de Lapparent, Geller, and Huchra 1986).

It should be noted that with the availability of redshifts and the positions on the sky, it is now common in quasar surveys to carry out the different tests, such as the correlation function or nearest neighbor, in three dimensions (cf Osmer 1981). It is also important to note that the different selection and edge or shape effects must be taken into account in any quantitative analysis of quasar surveys. Because of the relatively small areas of sky covered, the often disconnected positioning of the fields, and the uneven sampling in redshift, these effects will dominate the analysis. Today, analyses usually correct for these problems by using simulated catalogs of data in which data points are drawn at random from distributions that follow the large scale trends in the redshift and positional data. Or, random mixing of the redshifts and positions is used to generate catalogs that also follow the general trends in the data but for which small scale structure should no longer exist.

In the case of the Osmer Hewett survey, the data can also be displayed very instructively in a so-called pie diagram of comoving r vs right ascension, because the strip is thin in declination and little information is lost in projecting onto the r - theta plane.

Figure 1a shows a plot of comoving r (for a Friedmann cosmology with $H_o = 100$ and $q_o = 0.5$) vs right ascension. This is the best format for visual inspection of the data because it allows the eye maximum opportunity to look for structure while at the same time allowing it to judge the noise. It is intriguing to see that one can pick out small groups, arcs, and even apparent voids in the data. Given that statistical tests can be somewhat coarse tools in the sense that they may not be very sensitive or may be sensitive only to certain kinds of structures, it is important just to look at the data.

However, before we get too interested in the structures in Figure 1a, consider Figure 1b , which is a plot in the same format of a simulated catalog. The simulation assumes equal probability in right ascension but follows the smoothed redshift distribution. Even though we know there are some plate to plate effects that should be allowed for and make Figure 1b smoother than it should be for comparison with Figure 1a, we can see immediately that there are interesting groups, arcs, and voids in Figure 1b. Evidently the structures in the real data cannot be very significant.

Similar results obtain from application of the nearest neighbor test, the correlation function analysis, and a test for voids in the real data. Numerical values and graphs are shown by Osmer and Hewett (1988). They, in agreement with Shanks et al (1988) and Crampton et al (1988), conclude that there is no evidence for structure on the 50 – 500 h^{-1} Mpc scales, with a 1 σ limit of about 0.1 in the correlation function.

It should be pointed out that these results taken together provide perhaps the best evidence to date for the uniformity of the universe on such scales for the redshift range $0.5 < z < 2.5$. Given that ever larger structures are being found in galaxy and galaxy cluster surveys, this takes on importance for our understanding of cosmology and for the evolution of structure in the universe (see discussion by Shanks et al 1987). After all, the standard cosmological models assume a uniform distribution of matter in the universe.

FUTURE WORK

At present, the subject of the three dimensional distribution of quasars is in its infancy when compared with the state of galaxy surveys. As we look ahead, we see every reason to expect continued rapid development of the subject as interest in the topic grows and technology permits order of magnitude increases in survey sizes. When it becomes possible to study galaxies and their associations with quasars at $z > 0.6$, the results will undoubtedly teach us more about the nature of the quasar phenomenon itself as well as the large scale distribution of matter at high redshift.

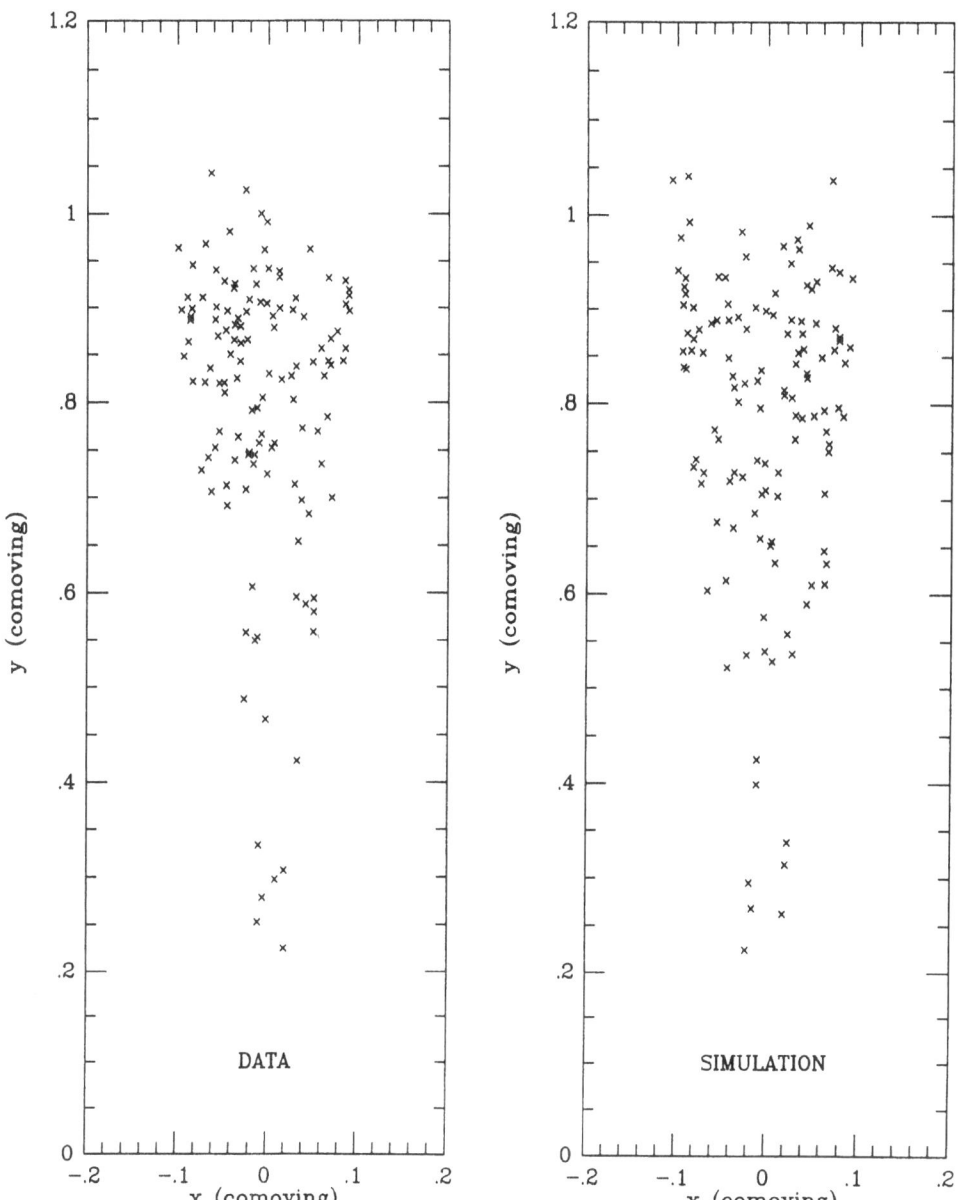

Fig. 1a (left). A pie diagram (comoving r vs alpha) for the 127 quasars in the Osmer Hewett sample. Unit distance corresponds to z = 3 or 3000 Mpc for H_o = 100 and q_o = 0.5. The earth is at the origin of the coordinate system.

Fig. 1b (right). The same diagram for a simulated catalog of 127 quasars that follow the smoothed redshift distribution of the real objects and are uniformly distributed in alpha.

REFERENCES

Burbidge, E. M., Junkkarinen, V. T., Koski, A. T., Smith, H. E., and
     Hoag, A. A. 1980, Ap. J. (Letters), **242**, L55.
Crampton, D., Cowley, A. P., and Hartwick, F. D. A. 1987, Ap. J., **314**,
     129.
Crampton, D., Cowley, A. P., and Hartwick, F. D. A. 1988, Proc. of a
     Workshop on Optical Surveys for Quasars, ASP Conference Series,
     in press.
de Lapparent, V., Geller, M. J., and Huchra, J. P. 1986, Ap. J.
     (Letters), **302**, L1.
Oort, J. H., Arp, H., and de Ruiter, H. 1981, Astron. Astrophys., **95**,
     7.
Osmer, P. S. 1981, Ap. J., **247**, 762.
Osmer, P. S., and Hewett, P. C. 1988, in Proc. of a Workshop on
     Optical Surveys for Quasars, ASP Conference Series, in press.
Osmer, P. S., Porter, A., Foltz, C. and Green, R. F. editors 1988,
     Proc. of a Workshop on Optical Surveys for Quasars, ASP
     Conference Series, in press.
Setti, G., and Woltjer, L. 1977, Ap. J. (Letters), **218**, L33.
Shanks, T., Boyle, B. J., and Peterson, B. A. 1988, in Proc. of a
     Workshop on Optical Surveys for Quasars, ASP Conference Series,
     in press.
Shanks, T., Fong, R., Boyle, B. J., and Peterson, B. A. 1987,
     M.N.R.A.S., **227**, 739.
Webster, A. 1982, M.N.R.A.S., **199**, 683.
Yee, H.K.C. 1987, A. J., **94**, 1461.

DISCUSSION

BURSTEIN: Pat — If the quasars you are observing preferentially reside in spiral galaxies, both the evidence of clustering on $\lesssim$10 Mpc scales, and the lack on >20 Mpc scales might be consistent with what we know about the clustering properties of nearby galaxies.

OSMER: That is a good point. It would be very interesting if we could determine the type of galaxies in which quasars at Z>0.5 reside. Additional factors are the associations of quasars with galaxy groups at Z>0.5 and the apparent triggering of AGN activity by galaxy interactions.

KAISER: Reagarding future extensions, one can imagine two approaches; either extending the angular extent or trying to increase the space density in a fixed solid angle. Could you comment on the pros and cons of these approaches?

OSMER: A circular or square field with the highest possible surface density would be very good for investigating clustering on small scales. At larger scales, I think that extending the angular length of a strip of sky, still at high surface density would be best.

BAHCALL: Kruszewski as well as Shaver have previously shown QSO correlations to scales of $100h^{-1}$ Mpc (for $Z \lesssim 1.5$). What happened to these correlations at $\gtrsim 10h^{-1}$ Mpc — did the data or analysis change the results, or do you simply consider these correlations too weak?

OSMER: The data and analysis have not changed. The weak correlations are still there at about a 3-sigma level. Much more work is need in order to improve this detection.

KRUSZEWSKI: I would like to make a distinction between the short range quasar correlations on separations up to 10 or 16 Mpc where the presence of quasar clustering seems to be well established, and the large scale correlations on separations from 16 to 60 or even 100 Mpc where there is still only 1-sigma detection.

OSMER: I agree that all the evidence in favor of whatever clustering there is occurs for the small scales. At scales greater than 10 Mpc I think there is no convincing evidence for clustering.

EVRARD: In your simulated catalogues you used the redshift distribution of the data which appeared to have some (albeit small) features. Can you comment on the sensitivity of your results if you used a different, perhaps much smoother, redshift weigthing distribution in setting up the simulations?

OSMER: You raise a good point. I should try different forms of the redshift distribution to investigate the sensitivity of the results. However, I have used a scrambled version of the redshift distribution, which preserves the selection effects exactly, and obtained virtually identical results.

THE EVOLUTION OF STRUCTURE

P.A. SHAVER, A. IOVINO, M. PIERRE
European Southern Observatory
Karl-Schwarzschild Str. 2
D-8046 Garching bei München, W. Germany

ABSTRACT. A summary is given of recent observational work pertaining to the evolution of large scale structure, with particular reference to quasar clustering and the Lyα forest.

1. INTRODUCTION

It is well known that the microwave background, thought to represent conditions in the early ($z \approx 1000$) universe, is extremely smooth, with fluctuations less than 10^{-4}. Today's universe, by contrast, is very inhomogeneous, with density fluctuations exceeding unity on scales of the order of $10h^{-1}$ Mpc. Theories and models abound which attempt to account for the transition from the very early universe to that of today. Numerical simulations have been extensively used in attempts to reproduce the observed correlation functions of galaxies and clusters, the filamentary structure of superclusters and the associated presence of large voids, without violating the observational constraints imposed by the microwave background.

Many free parameters are involved in these models - the value of the cosmological constants Ω and Λ, the nature of the possible dark matter, the nature of the initial density fluctuations, the possible presence of bias in galaxy formation - and there are more exotic possibilities involving cosmic strings or explosive galaxy formation. These different scenarios imply different evolutionary paths from the early universe to today, and observations pertaining to the intervening 99 percent of the history of the universe would provide important constraints.

Until recently, observational work on objects in the high redshift universe on the one hand, and the structure of the low redshift universe on the other, have been regarded as relatively distinct disciplines, with little contact between them. New observational developments are rapidly moving to close this gap, however, and a comprehensive picture of the detailed evolution of structure is now becoming thinkable. In the paper we review some of these developments, starting with a discussion of possible tracers of large scale structure, followed by reviews of recent work on the clustering of quasars and void structure

M. Mezzetti et al. (eds.), Large Scale Structure and Motions in the Universe, 101–118.
© *1989 by Kluwer Academic Publishers.*

in the Lyα forest.

2. TRACERS OF STRUCTURE

Is there an ideal tracer of the evolution of large scale structure? One which is observable over a large range in redshift, unaffected by astrophysical processes, and which faithfully follows the structure without bias on all scales as it evolves? Or is it necessary to combine information gleaned from several different types of objects? What kinds of tracers *can* we use at low and high redshifts? We first consider the hierarchy of structure that exists at low redshift.

2.1 The Hierarchy of Structure

It has been argued that virtually all luminous matter observable at low redshifts respects the voids defined by the distribution of normal galaxies (*cf.* Binggeli, these Proceedings, and references therein). If that is the case, then the concept of a "minimal structure" naturally arises, the structure determined only by the disposition of these voids. Objects placed at random in the regions between the voids would have the smallest spatial correlation possible.

We can simulate one such universe through the process of "Voronoi tessellation" (Voronoi, 1908). Details are given in Pierre, Shaver and Iovino (1988). We divide the universe into three-dimensional cells of a range of sizes, separated from each other by walls of constant thickness. A cut through one such simulation is shown in fig. 1.

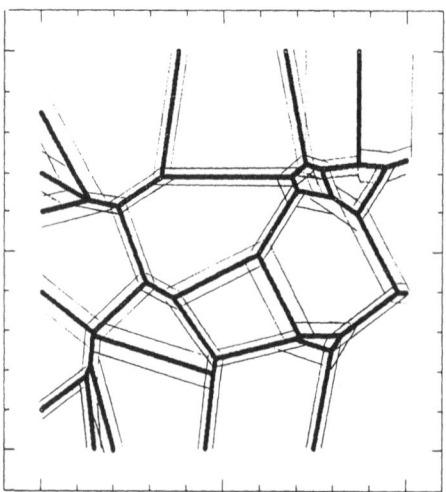

Fig. 1. A cut through the three-dimensional cellular structure of the simulated universe. The thick lines are the cell boundaries, and the thin lines the void walls, which have uniform thickness.

These walls are then populated by objects distributed randomly within them. No other objects which respect the voids can be less clustered. Any breaks in the walls, or clustering within the walls, will result in stronger spatial correlations.

The real universe is then represented by assigning a value of $25h^{-1}$ Mpc to the characteristic void size, and $4h^{-1}$ Mpc to the wall thickness ($H_o = 100\ h\ km\ s^{-1} Mpc^{-1}$ and $q_o = 0.5$ used throughout this paper), values obtained from the CfA redshift survey by de Lapparent (1986). The distribution of cell sizes is Gaussian, centered around that corresponding to $25h^{-1}$ Mpc voids, as shown in fig. 2. A slice through this universe is compared with the CfA slice through the real universe in fig. 3. The simulated universe appears less clustered as expected, because no clustering is permitted within the walls.

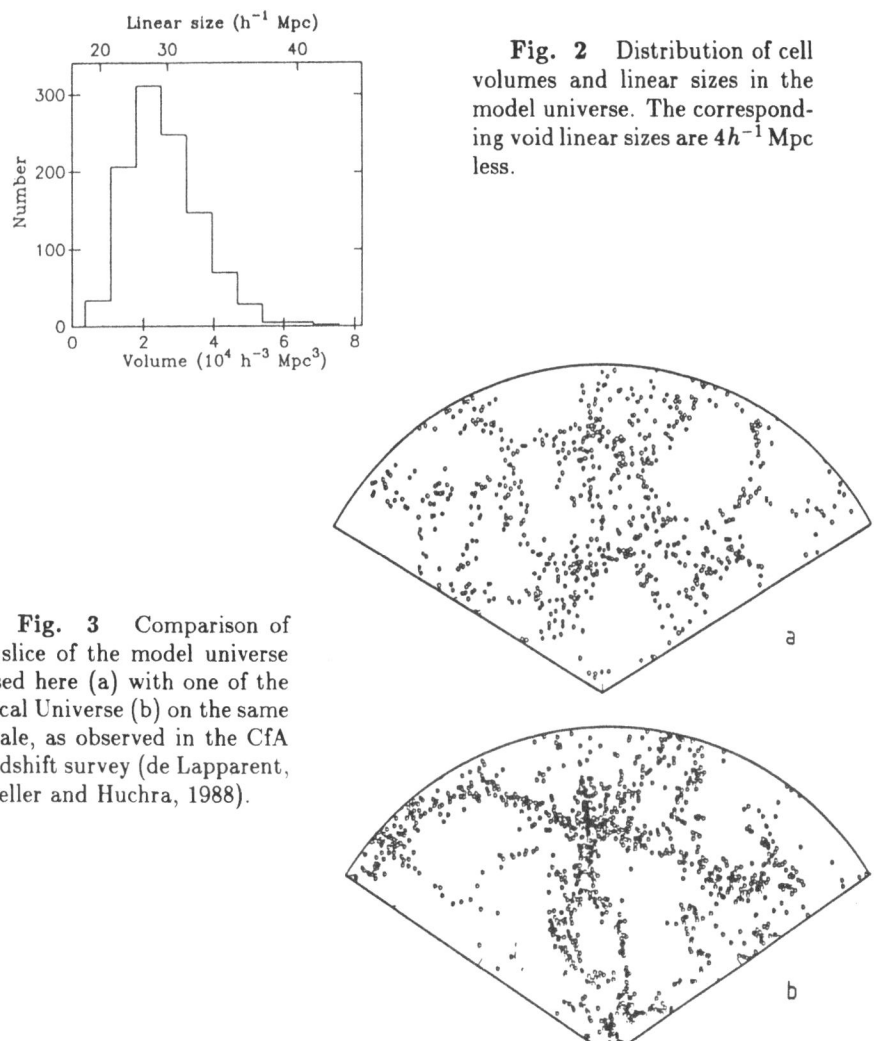

Fig. 2 Distribution of cell volumes and linear sizes in the model universe. The corresponding void linear sizes are $4h^{-1}$ Mpc less.

Fig. 3 Comparison of a slice of the model universe used here (a) with one of the local Universe (b) on the same scale, as observed in the CfA redshift survey (de Lapparent, Geller and Huchra, 1988).

The three-dimensional two-point correlation function for the objects in this simulated universe (dotted curve in fig. 4) has the lowest amplitude possible for any objects which do not occupy the voids. For points randomly distributed on a flat two-dimensional surface the three-dimensional two-point correlation function is a power law of slope -1 (which is approached asymptotically at small separations by the dashed curve in fig. 4); it will be shallower if there is a three-dimensional cellular structure (dashed curve at larger separations), and shallower still if the walls are thick (dotted curve), and can only be steeper and have a greater amplitude if there is structure within the walls.

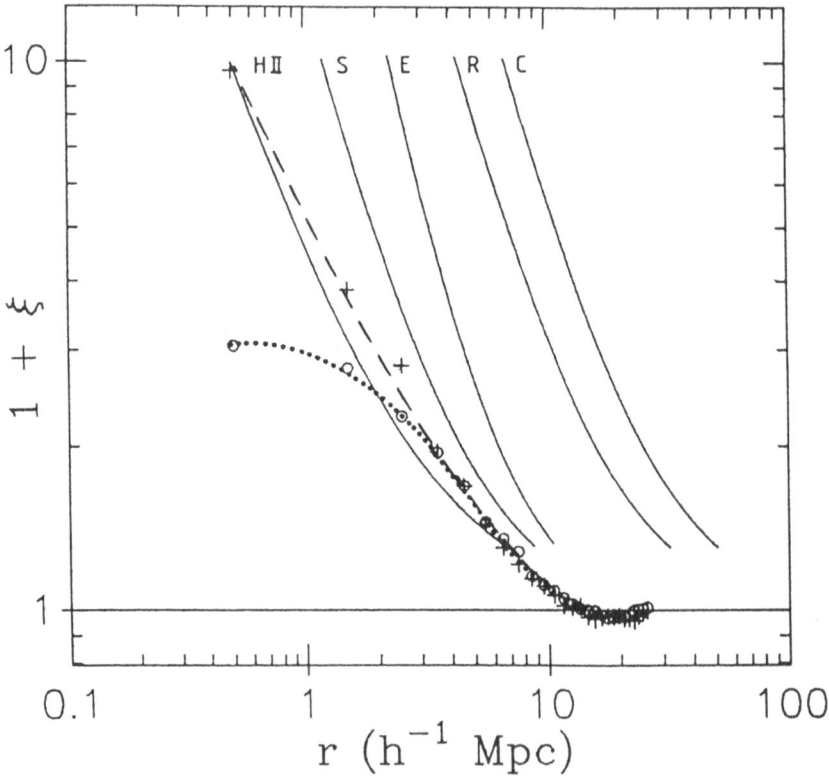

Fig. 4 Three-dimensional two-point correlation functions for the model universe with cell sizes as in fig. 2 and wall thicknesses of 0 and $4h^{-1}$ Mpc (dashed and dotted lines respectively), HII galaxies (HII - Iovino, Melnick and Shaver, 1988), spiral and elliptical galaxies (S,E - Davis and Geller, 1976), radio galaxies of Fanaroff- Riley class I (R - Peacock et al., 1988), and clusters of galaxies (C - Bahcall and Soneira, 1983).

Dwarf irregulars are thought to be more dispersed than any other type of galaxy, but even they still appear to be concentrated to some extent in groups (*cf.* Bingelli, this volume). Thuan, Gott and Schneider (1987) have plotted the positions of several such systems in the CfA survey region, showing that they also seem to respect the voids, and follow the structure defined by more massive galaxies. Thus it may be expected that their correlation function would come closest to the minimal correlation function determined above, and that indeed appears to be the case. Iovino, Melnick and Shaver (1988, and this volume) have for the first time measured the correlation function of HII galaxies, a class of gas-rich dwarf galaxies. Fig. 4 shows that it closely follows the minimal correlation function at separations above a few Mpc; it deviates at small separations, presumably reflecting small scale clustering within the walls.

Normal galaxies (predominantly spirals) are more strongly correlated, as shown in fig. 4, indicating greater clustering within the walls. Spirals are known to be concentrated around groups and rich clusters, although they avoid the higher density inner regions. Ellipticals are still more strongly correlated on scales of a few Mpc, and their correlation function is steeper (Davis and Geller, 1976), reflecting their concentration towards the cores of clusters. This "morphological segregation" has been known for some time (Dressler, 1980; Postman and Geller, 1984; Giovanelli, Haynes and Chincarini, 1986), and is illustrated schematically in fig. 5. Even stronger correlations appear amongst radio galaxies of Fanaroff-Riley class I, which are known to reside in clusters of Abell richness class 0 (Peacock et al., 1988).

Finally, clusters of galaxies themselves appear to be very strongly correlated, with a correlation length of $23h^{-1}$ Mpc for richness classes of 1 or greater, and superclusters still more so (Bahcall, this volume, and references therein). There appears to be a relationship between richness and clustering amplitude, which is still not understood.

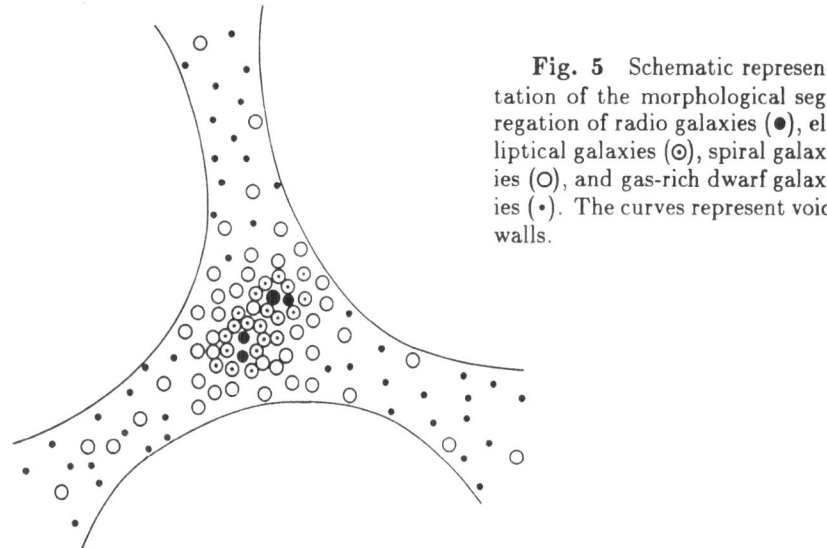

Fig. 5 Schematic representation of the morphological segregation of radio galaxies (●), elliptical galaxies (⊙), spiral galaxies (O), and gas-rich dwarf galaxies (·). The curves represent void walls.

Thus, there is a hierarchy of structure, ranging from the minimal correlations implied by the presence of voids to the strong correlations of clusters and superclusters. It is of course not very accurately defined - the void structure may be more complex than a simple cellular structure, the voids may not be totally empty, and there are significant uncertainties in measuring precise correlation functions - but at least the presence of a hierarchy seems to be well established.

There are simple astrophysical reasons why some such hierarchy should exist. Radio galaxies (and quasars) are commonly thought to be triggered or fueled by interactions, which occur most frequently in the dense cores of rich clusters. Massive elliptical galaxies may develop in clusters through cannibalism. Spirals are gravitationally attracted to clusters, but can be tidally stripped of their gas if they venture within. Gas-rich dwarf irregulars are more easily stripped, and are even less concentrated around clusters. It has also been argued that the morphological segregation was in place almost from the beginning, perhaps due to astrophysical processes which were important during the epoch of galaxy formation. In that case objects of a given type would always have inhabited the same type of environment.

It therefore appears that no single type of object is likely to give a comprehensive picture of the evolution of structure. Different objects sample different environments, and the typical environment of a given type of object may conceivably change with time due to continuing astrophysical processes. A complete understanding of the evolution of large scale structure will probably require complementary information from as many different types of objects as possible.

2.2 Links between the Low and High Redshift Universe

The counterparts of all the objects that make up the low redshift hierarchy must also exist at high redshifts, but observing them, identifying them, and relating them uniquely to objects at low redshifts will not be an easy task. It will, however, be a rewarding one, as it should provide information not only on the evolution of structure, but also on the evolution of individual types of objects and the astrophysical processes involved.

The minimal structure defined by the voids at low redshift may also be present in the distribution of high redshift objects, and would probably best be studied using the Lyα forest absorption lines, in view of their relatively high number density. These may conceivably be the counterparts of dwarf irregular galaxies. Their clustering properties can be studied at least over the redshift range $\sim 1.5 - 4.4$, and Space Telescope may extend this to lower redshifts and identify their low redshift counterparts.

Normal galaxies can be studied directly to redshifts of 0.5-1.0, and some work has already been done on the evolution of galaxy clustering over this range. It appears that the shape of the correlation function remains constant with redshift, and the amplitude decreases at least as fast as expected for stable clustering (Koo and Szalay, 1984; Loh, 1988; Jones et al., 1988). Void structure can possibly also be traced using deep galaxy surveys out to these redshifts (*e.g.* Szalay, this volume), and this may eventually be an important probe of the evolution of large scale structure over half the age of the universe. It appears that some if not all metal absorption lines in quasar spectra originate in the halos of normal galaxies (Bergeron, 1988); their clustering properties may then give information on galaxy clustering up to the highest redshifts, although there is the possible problem of contamination by multiple absorptions in the halos of individual galaxies (e.g. Sargent et al., 1980). Correlated absorption in the spectra of pairs or groups of quasars can provide information on structure on very large scales.

Galaxies are known to be concentrated to some extent around quasars of modest redshift ($\sim 0.5 - 1.0$). The companions of such quasars are sometimes individual distorted or compact galaxies, often compact groups of galaxies, or, particularly for radio-loud quasars, rich clusters of galaxies (*cf.* Yee and Green, 1987, and references therein). Plausible counterparts at higher redshifts are absorption line systems which are found in excess near the redshifts of radio-loud quasars (Foltz, 1988) but not radio-quiet quasars (Sargent, 1988). The concentration of absorbers near quasars can also be studied using associated absorption in quasar pairs (Shaver and Robertson, 1984). The clustering of objects around quasars can therefore in principle be studied up to the highest redshifts.

Radio galaxies *per se* can only be observed directly at high redshift in rare cases (Spinrad, 1988), and clusters of galaxies not at all above $z \sim 1$. But quasars, which often appear to be associated with groups or clusters, can be observed at all redshifts up to 4.4, though their space density is so low that clustering studies are difficult.

Other constraints on high-redshift structure may include the X-ray background (Barcons and Fabian, 1988; Meszaros, 1988; de Zotti, this volume), the infrared background (Boughn et al., 1986), and the microwave background (Toffolatti et al., this volume). Further constraints may become available, and the observational situation with regard to the possibilities discussed above should certainly improve dramatically with the advent of Space Telescope and the new generation of large ground-based telescopes, but at the moment the best observational handles on the evolution of structure are galaxies at modest redshifts, quasars, and quasar absorption lines. Below we discuss recent work on the clustering of quasars and void structure in the Lyα forest.

3. THE CLUSTERING OF QUASARS

As extremely luminous objects visible to very high redshifts, quasars have long offered the promise of directly exploring the structure of the distant universe. Quasar samples of sufficient size and depth for such studies have not been available until recently, however, and so it is only now that this promise is beginning to be realized. On the very largest scales quasars may be the best probes for structure, because of the huge volumes they sample; Osmer discusses this aspect in an accompanying paper in this volume. Structure on virtually any scale will, however, ultimately leave its imprint in the form of an excess of pairs at small separations, so it is natural to search for clustering on small scales.

Two approaches have been taken to obtain samples with sufficiently large numbers of close quasar pairs. One has been to resort to large quasar catalogues; these contain several thousand quasars, but they are heterogeneous, and special techniques are required to analyze them for clustering. The other has been to obtain homogeneous samples which are both deep and large. Deep samples containing a few hundred quasars are becoming available, and it is now possible to obtain a strong clustering signal by combining several of them.

Table 1 lists measurements of the quasar correlation function which have been made to date, and fig. 6 shows the results as a function of redshift. All of these results have been obtained in just the last four years, a measure of the pace of development in this field. Quasar clustering has now been confirmed beyond doubt. If one combines the results from the deep homogeneous samples studied by Iovino et al. (this volume) and Boyle et al. (1988), one finds, in a total sample of 806 quasars, 40 pairs of separation $\leq 10h^{-1}$ Mpc, where 19 would be expected if quasar were distributed at random, a clear 5σ detection of clustering for quasars regardless of redshift. Most of these studies also find the clustering to be strongest at low redshifts, although with a relatively low statistical significance at present. In summary then, quasar clustering has been confirmed, but the question of evolution still remains open.

TABLE I   Measurements of the Quasar Correlation Function

Sample	Method	Reference
Véron Catalogue *	2a	Shaver (1984)
UVX QSOs (170)	1a	Boyle (1986); Shanks et al. (1987)
Véron Catalogue *	2b	Kruszewski (1988a); Shaver (1988a)
3 Deep Homogeneous Samples (376)	1a	Iovino & Shaver (1988)
9 Homogeneous Samples (629)	1c	Kruszewski (1988b)
UVX QSOs (398)	1a	Boyle et al. (1988)
Hewitt-Burbidge Catalogue **	2a	Zhu & Chu (1988); Chu & Zhu (1988)
Véron Catalogue *	1b 2a	Anderson et al. (1988) Shaver (1988b)
Grens QSOs (150)	1a	Crampton et al. (1988)
17 Homogeneous Samples (1254)	1c	Kruszewski (1988c,d)
Augmented Véron Catalogue*	2b	Kruszewski (1988d)
4 Deep Homogeneous Samples (579)	1a	Iovino et al. (1988)

* Véron-Cetty & Véron (1987)
** Hewitt & Burbidge (1987)
(Numbers in parentheses indicate sample size)

Methods

1. Random Comparison Sample
    (a) 3D randomization
    (b) z randomization
    (c) z scrambling

2. Normalization to Large Scales
    (a) radial and tangential coordinates separately
    (b) radial and tangential coordinates combined

Several related questions are now being addressed. Is there a dependence on radio properties? Chu and Zhu (1988) and Shaver (1988c) find a $2 - 3\sigma$ tendency for radio-loud quasars to be more strongly clustered. Is there serious contamination from gravitational lensing? There certainly can be on very small scales ($\leq 10''$, corresponding to $\sim 0.1h^{-1}$ Mpc comoving) where gravitational lensing due to galaxies and clusters is expected to occur, but much of the clustering is detected on considerably larger (Mpc) scales; the reverse problem of identifying *bona fide* gravitational lenses is probably far more severe.

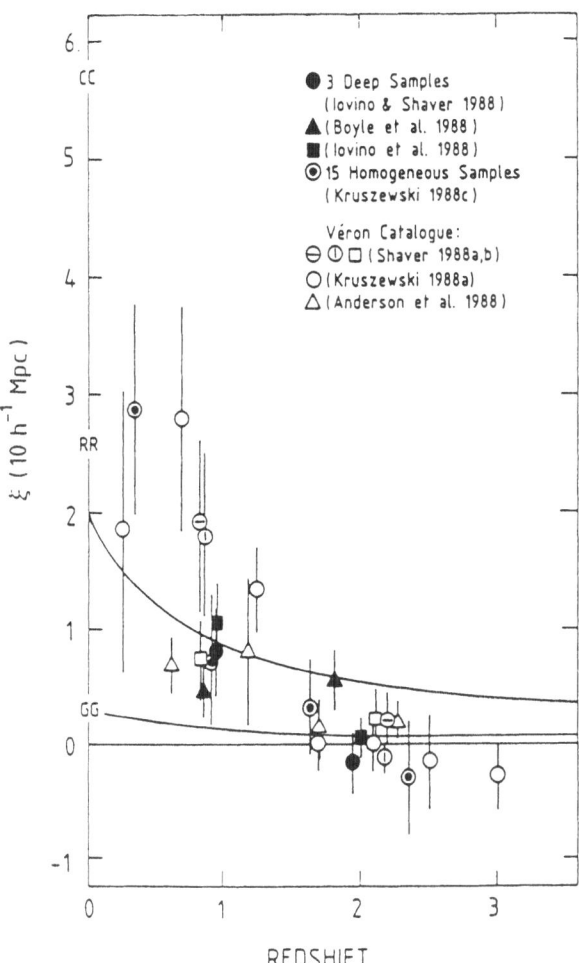

Fig. 6   Amplitude of the quasar correlation function at $10h^{-1}$ Mpc as a function of redshift, from several different studies as indicated. On the left axis are marked the amplitudes of the correlation function at low redshift for galaxies (GG), clusters (CC - Bahcall and Soneira, 1983), and radio galaxies (RR - Peacock et al., 1988). The two curves show the expected evolution for stable clustering.

If the clustering of quasars on all scales increases towards low redshifts, then at least a statistical distinction can be made (Shaver, 1988b), as gravitational lensing should be more prevalent amongst high redshift quasars.

The expected evolution of quasar clustering is not yet clear. On the one hand, if quasars reside in groups or clusters of galaxies and are triggered/fed by interactions, then their spatial correlations may lie in the non-linear regime, and rapid evolution may be expected. There may be complications due to astrophysical processes; for example, a selective depletion of fuel in the densest clusters where quasars may have formed first could cause a *decrease* in the overall quasar correlation function after some time. On the other hand, if quasars are regarded simply as "markers" of (primordial) regions of high density, then they may be comoving, and no evolution of the correlation function might be expected.

In any case, still larger deep samples will be necessary to map out the evolution of quasar clustering and investigate further issues such as the dependence on radio and optical luminosity. Substantial amounts of large telescope time are required to obtain the thousands of redshifts needed. Individual samples will be increasingly inadequate on their own, and the answers will ultimately come from the combination and analysis of many such samples.

4. VOID STRUCTURE IN THE LYMAN ALPHA FOREST

The Lyα absorbers may in fact come close to the ideal tracer of large scale structure. They are numerous, easily observable up to very high redshifts, and presumably follow the fundamental structure on the largest scales where the linear approximation should remain valid. Even for these objects, however, irrelevant astrophysical processes may produce complications - the number density is known to increase strongly with redshift (Hunstead, 1988). It is possible at present to study the evolution of their clustering properties from $z = 4.4$, the redshift of the most distant known quasar, to $z = 1.5$, corresponding to the atmospheric cutoff, and tentative evidence for evolution has already been found by Webb (1987). With Space Telescope it should be possible to extend this study to lower redshifts, if the number density at these low redshifts is not too small.

A related question is whether the void structure as defined by galaxies today is also present in the distribution of Lyα absorbers at high redshift. This has a bearing on the nature of the Lyα absorbers themselves, and on the nature and evolution of voids. A few recent studies (Carswell and Rees, 1987; Crotts, 1987; Ostriker, Bajtlik and Duncan, 1988) conclude that the filling factor of large $(40 - 50h^{-1}$ Mpc comoving) voids in the Lyα forest must be small, $\leq 5 - 15\%$.

A different approach to this problem has been used by Pierre, Shaver and Iovino (1988), taking into account a range of smaller, more typical void sizes. A model of the present universe was made containing the least structure consistent with the presence of voids, as described in Section 2 above. It contains voids of characteristic size $25h^{-1}$ Mpc, separated by walls of constant thickness $4h^{-1}$ Mpc which contain absorbers distributed at random within them.

This model universe, fixed in comoving coordinates, was then transposed to high redshift, to permit a direct comparison with the Lyα forest. Lines of sight generated at random through it encounter the absorbers which have given cross-sections, column densities, and velocity dispersions. The cross-sections were chosen to reproduce the observed number density in the Lyα forest; in practical terms they also had to be considerably smaller than the wall thickness but large enough that the model universe did not contain too many objects to handle. The column densities were fixed by randomly assigning widths to the 'spectral lines' in accordance with the observed distribution, thereby taking into account both the intrinsic widths and instrumental resolution. And the velocity dispersions were taken into account by modifying the absorption line redshifts in accordance with a Gaussian distribution of velocities with $\sigma(z) = \sigma_o(1 + z)^{-1/2}$ (Davis et al., 1982), where $\sigma_o \sim 250 \; kms^{-1}$ for galaxies at low redshifts (Davis and Peebles, 1983; de Lapparent, 1986).

These simulations were then compared with the correlation analysis made by Webb (1987) on the highest-resolution observations available of the Lyα forest. Webb's analysis provides the first probable (4σ) detection of clustering in the Lyα forest; the clustering is weak, and the correlation function appears to be relatively flat ($\gamma \sim 0.8$). Fig. 7 shows that the correlation resulting from the model universe, using the observed number density and line width distribution from Webb's data and an identical analysis, is considerably stronger than the observed Lyα correlation.

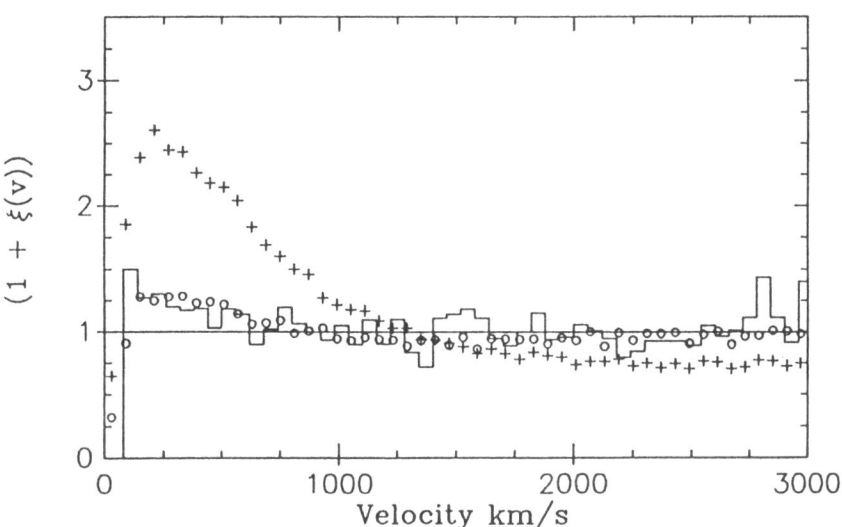

Fig. 7 One-dimensional two-point correlation function for the model universe compared with that of Lyα absorbers measured by Webb (1987). The crosses and circles refer to models with $4h^{-1}$ Mpc thick cell walls and mean void sizes of 25 and $7.5h^{-1}$ Mpc respectively.

The low correlation amplitude observed in the Lyα forest therefore appears to be incompatible with the void structure observed at low redshift. It is difficult to significantly reduce the correlation from the model without flagrantly violating the low-redshift observational constraints. If the cell walls are thickened or the void sizes reduced sufficiently to match the observed Lyα correlation, the resulting model looks very unlike the CfA slice

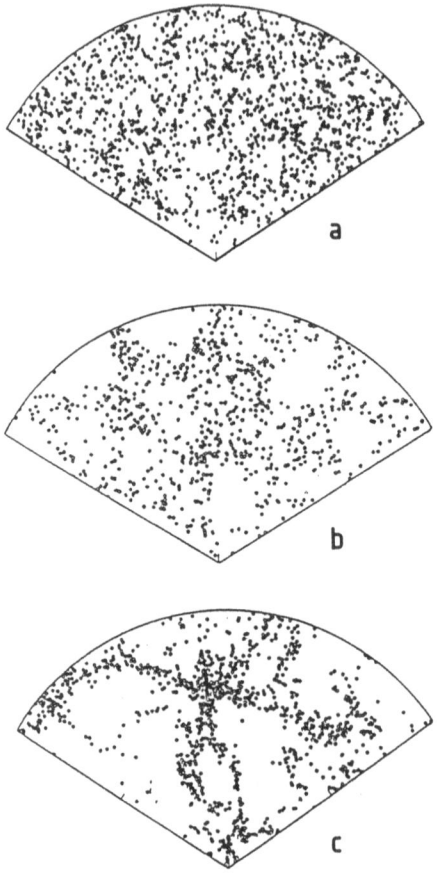

Fig. 8 Comparison of slices of two model universes (a,b) which fit the Lyα correlation in fig. 7, with one of the local universe (c) as observed in the CfA redshift survey (de Lapparent, Geller and Huchra, 1988), on the same scale. The two models have the same void filling factor of 28%; the void size and wall thickness is 7.5 and $4h^{-1}$ Mpc for model (a) and 19 and $10h^{-1}$ Mpc for model (b).

(fig. 8). The actual void filling factor in the low redshift universe may if anything be larger than the 64% in the model (de Lapparent, 1986), resulting in still larger correlations. Nor can a different spectrum of void sizes be easily concocted which could reconcile the low and high redshift data; a broader distribution would imply more voids of large size, which are limited both by the direct searches for large gaps in the Lyα forest mentioned above and the similarity in the Lyα statistics towards different quasars (Sargent et al., 1980). Line blending already seems excessive in the simulations compared with Webb's data (the downturn in the first few bins in fig. 7), and random motions would have to be increased to well over $1000 km s^{-1}$ to reduce the correlation sufficiently. Finally, if there were any intrinsic velocity structure in the absorbers, Webb's result would overestimate the spatial correlation, and the difference with the simulations would be even greater.

We therefore conclude that any structure amongst the Lyα absorbers is less pronounced than the void structure defined by the distribution of galaxies today. Perhaps the Lyα absorbers occupy the voids; Space Telescope should determine whether this is the case at low redshift. The other possibility is that the Lyα absorbers respect the voids, but the voids at high redshift were smaller in comoving coordinates than they are today, with thicker walls, or simply more numerous. This would be consistent with the tentative direct evidence (Webb, 1987) for evolution in the Lyα correlation at high redshift.

If indeed the Lyα absorbers are contained within the cell walls, then this provides a direct measure of the evolution of large scale structure over the history of the universe. The evolution could have taken the form of collapsing walls, merging voids, or both. Agreement with the observed Lyα correlation can be achieved with a smaller filling factor ($\approx 28\%$) corresponding to smaller void sizes of $19 h^{-1}$ Mpc and thicker $(10 h^{-1}$ Mpc) walls, or still smaller voids of $7 h^{-1}$ Mpc and the original $4 h^{-1}$ Mpc walls. The latter would be consistent with the evolution expected (*cf.* Shandarin, 1988) from linear theory with a primordial density perturbation spectrum of index -1.

It should be possible to make further progress on the large scale structure at high redshift using Lyα absorbers, for example by measuring more accurately the evolution of the Lyα clustering, or distinguishing between different geometries (cellular structures with different characteristic scales, clusters superimposed on a random background, etc.). There are various statistical tests, particularly percolation analysis, which can be applied. Absorption line observations of quasar pairs add an extra dimension to this study. Cross-correlation of two close lines of sight may be more sensitive to spatial clustering than autocorrelation in one line of sight with the same number density and total number of absorption lines. More important is the potential use of quasar pairs in determining the scale and nature of any spatial structure which may be present amongst the Lyα absorbers. Fig. 9 shows, for example, the cross-correlation function for two lines of sight as a function of angular separation, for two cases similar to those mentioned above which match Webb's Lyα correlation. It appears that a clear distinction may eventually be possible.

5. CONCLUSIONS

A diverse range of observational data is now becoming available which can provide information on large scale structure at high redshift, and its evolution. While no single class of object is likely to tell the entire story on its own, it is hoped that the evolution of structure can be pieced together by combining complementary evidence from different objects which reflect structure on different scales. Such information will place increasingly strong constraints on models of the evolution of structure and galaxy formation.

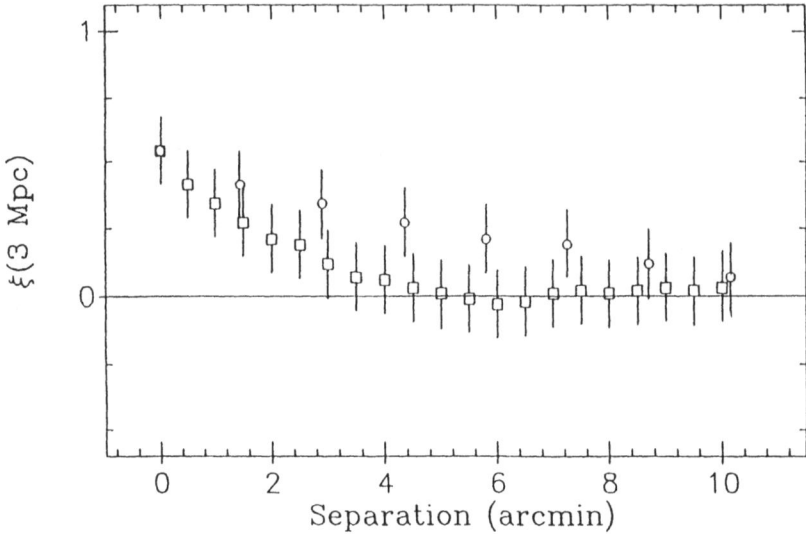

Fig. 9 Cross-correlation amplitude (averaged up to $3h^{-1}$ Mpc) as a function of angular separation of two lines of sight for two model universes similar to those which fit the Lyα correlation in fig. 7. The two models have the same void filling factor of 22%; the void size and wall thickness is 6 and $4h^{-1}$ Mpc ($\square$) and 17.5 and $11.5h^{-1}$ Mpc ($\circ$). The error bars represent the case in which just one quasar pair is used, with the same Lyα number density as in Webb's (1987) data, restricted to the spectral range between the Lyα and Lyβ emission lines. Line blending and random motions, which can decrease the correlation amplitude, are not included here.

In three cases where structural information can presently be obtained over some range in redshift - galaxies, quasars, and Lyα absorbers - tentative evidence is found for evolution. In the case of quasars the statistical significance of the reported evolution is not high, and there are conflicting reports, so the situation is unclear; the interpretation of any such evolution may also not be straightforward. In the case of the Lyα absorbers, the direct evidence for evolution is weak, and the stronger indirect evidence relies on the assumption that the absorbers do not occupy the voids.

Thus, while a convincing measurement of the evolution of structure on some scale has perhaps yet to be made, the tools are now at hand, and rapid progress in determining the evolution of large scale structure over most of the history of the universe can be anticipated.

REFERENCES

Anderson, N., Kunth, D.,Sargent, W.L.W., 1988, preprint

Bahcall, N.A., Soneira, R.M., 1983, *Ap. J.*, **270**, 20.

Barcons, X., Fabian, A.C., 1988, *M.N.R.A.S.*, in press.

Bergeron, J., 1988, in *The Post-Recombination Universe (NATO ASI)*, in press.

Boughn, S.P., Saulson, P.R., Uson, J.M., 1986, *Ap. J.*, **301**, 17.

Boyle, B.J., 1986, Ph. D. thesis, University of Durham.

Boyle, B.J., et al., 1988 in *The Post-Recombination Universe (NATO ASI)*, in press.

Carswell, R.F., Rees, M.J., 1987, *M.N.R.A.S.*, **224**, 13p.

Chu, Y., Zhu, X., 1988, preprint.

Crampton, D., Cowley, A.P., Hartwick, F.D.A., 1988, in *Optical Survey for Quasars,* (e.d. P. Osmer), P.A.S.P, in press

Crotts, A.P.S., 1987, *M.N.R.A.S.*, **228**, 41p.

Davis, M., Geller, M.J., 1976, *Ap. J.*, **208**, 13.

Davis, M., Peebles, P.J.E., 1983, *Ap. J.*, **267**, 465.

Davis, M., Huchra, J., Latham, D.W., Tonry, J., 1982, *Ap. J.*, **306**, 341.

Dressler, A., 1980, *Ap. J.*, **236**, 351.

Foltz, C.B., Chaffee, F.H., Weymann, R.J., Anderson, S.F. 1988, in *QSO Absorption Lines: Probing the Universe*, (eds. Blades, J.C., Turnshek, D.A., Norman, C.A.), in press.

Giovanelli, R., Haynes, M., Chincarini, G., 1986, *Ap. J.*, **300**, 77.

Hunstead, R.W., 1988, in *QSO Absorption Lines: Probing the Universe*, (eds. Blades, J.C., Turnshek, D.A., Norman, C.A.), in press.

Hewitt, A., Burbidge, G., 1987, *Ap. J. Suppl.*, **63**, 1

Iovino, A., Melnick, J., Shaver, P.A., 1988, *Ap. J. Letts.*, in press.

Iovino, A., Shaver, P.A., 1988, *Ap. J. Letts.*, in press.

Iovino, A., Shaver, P.A., Crampton, D., Cowley, A.P., Hartwick, F.D.A., Osmer, P., Hewitt, P., Barbieri, C., Christiani, S., 1988, this volume

Jones, L.R., Shanks, T., Fong, R., 1988, in *Evolution of Large Scale Structures in the Universe*, (eds. J. Audouze & A. Szalay; Reidel, Dordrecht), in press.

Koo, D.C., Szalay, A.S., 1984, *Ap. J.*, **282**, 390.

Kruszewski, A. 1988a, preprint.

Kruszewski, A., 1988b, in *Proceedings of 2nd Ringberg Workshop on High Energy Astrophysics*, in press.

Kruszewski, A. 1988c, *Acta Astr.*, in press.

Kruszewski, A. 1988d, this volume.

de Lapparent, V., 1986, Ph.D. Thesis, Universite de Paris 7.

de Lapparent, V., Geller, M.J., Huchra, J.P., 1988, *Ap. J.*, in press.

Loh, E.D., 1988, in *Evolution of Large Scale Structures in the Universe*, (eds. J. Audouze & A. Szalay; Reidel, Dordrecht), in press.

Meszaros, P., 1988, in *Evolution of Large Scale Structures in the Universe*, (eds. J. Audouze & A. Szalay; Reidel, Dordrecht), in press.

Ostriker, J.P., Bajtlik, S., Duncan, R.C. 1988, *Ap. J. Letts.*, **327**, L35.

Peacock, J.A., Miller, L., Collins, C.A., Nicholson, D., Lilly, S.J., 1988, in *Evolution of Large Scale Structures in the Universe*, (eds. J. Audouze & A. Szalay; Reidel, Dordrecht), in press.

Pierre, M., Shaver, P.A., Iovino, A., 1988, *Astr. Ap. Letts*, in press.

Postman, M., Geller, M.J., 1984, *Ap. J.*, **281**, 95.

Sargent, W.L.W., 1988, in *QSO Absorption Lines: Probing the Universe*, (eds. Blades, J.C., Turnshek, D.A., Norman, C.A.), in press.

Sargent, W.L.W., Young, P.J., Boksenberg, A., Tytler, D., 1980, *Ap. J. Suppl.*, **42**, 41.

Shandarin, S.F., 1988, in *Evolution of Large Scale Structures in the Universe*, (eds. J. Audouze & A. Szalay; Reidel, Dordrecht), in press.

Shanks, T., Fong, R., Boyle, B.J., Peterson, B.A., 1987, *M.N.R.A.S.*, **227**, 739.

Shaver, P.A., 1984, *Astr. Ap.*, **136**, L9.

Shaver, P.A., 1988a, in *Evolution of Large Scale Structures in the Universe*, (eds. J. Audouze & A. Szalay; Reidel, Dordrecht), in press.

Shaver, P.A., 1988b, in *The Post-Recombination Universe (NATO ASI)*, in press.

Shaver, P.A., 1988c, in *Optical Surveys for Quasars*, (ed. P. Osmer), P.A.S.P., in press.

Shaver, P.A., Robertson, J.G., 1984, in *Frontiers of Astronomy and Astrophysics*, (ed. R. Pallavicini), p. 201.

Spinrad, H., 1988, in *High Redshift and Primeval Galaxies*, (eds. J. Bergeron et al.), p. 59

Thuan, T.X., Gott, J.R. III, Schneider, S.E., 1987, *Ap. J. Letts.*, **315**, L93.

Véron-Cetty, M.-P., Véron, P., 1987, ESO Scientific Report No. 5

Voronoi, G., 1908, *J. Reine Angew. Math.*, **134**, 198.

Webb, J.K., 1987, in *Observational Cosmology*, (ed. A. Hewitt, G. Burbidge, L.-Z. Fang; Reidel, Dordrecht), p. 803.

Yee, H.K.C., Green, R.F., 1987, *Ap. J.*, **319**, 28.

Zhu, X., Chu, Y., 1988, in *Proceedings of 2nd Ringberg Workshop on High Energy Astrophysics*, in press.

DISCUSSION

KAISER: Regarding the low clustering strength of the HII galaxies, I wonder if much of the difference might not be due to differences in the volume sampled − from the wedge diagram you showed they seemed to be concentrated in our vicinity. If the clustering strength in our local neighbourhood was anomalously low then this would also account for the low amplitude found by Rowan-Robinson and Needham for the IRAS galaxies. While the IRAS sample goes deep, the steep luminosity function results in a distribution of pairs which is strongly concentrated towards the observer.

SHAVER: We imposed a low redshift cutoff of 2000 km/s on our sample, so this effect should not be very significant. Furthermore, we also analyzed a galaxy sample covering the same redshift range using exactly the same method, and as our conclusions are based on a direct comparison of the two, any effect of the type you mention should largely cancel out.

SZALAY: One should be aware that fluctuations in the photoionizing flux due to e.g. a foreground QSO can cause large modulations in the density of the $Ly\alpha$ forest, which could be perceived as a void.

SHAVER: Yes, so the upper limit on real void structure in the distribution of $Ly\alpha$ absorbers would be even stronger.

BAHCALL: Radio QSOs appear to differ somewhat from radio-quiet QSOs in their associations with galaxies and clusters. It may therefore suggest that the clustering of radio QSOs may be stronger than that of optical QSOs. Have you checked that effect?

SHAVER: Chu and Zhu have found weak evidence that radio-loud quasars may indeed be more strongly clustered than radio-quiet quasars, using quasars of all redshifts. I have tried to check whether this could just be due to the fact that the catalogued radio-loud quasars are predominantly at low redshifts, where clustering in general may be stronger, and also conclude tentatively that radio-loud quasars in a given redshift range appear to be more strongly clustered. The statistical significance of this results is low (2-3 sigma), and it will be some time before it can be appreciably strenghthened, because of the scarcity of radio-loud quasars.

OSMER: With regard to the point about the detectability of voids, I can add that White has described a nice test for voids in the Monthly Notices. It complements the correlation function approach.

KERR: The work that you and Pat Osmer have been describing makes shorter and shorter the time available for the formation of voids. Have you or anyone else considered the implications of this?

SHAVER: Our results are not inconsistent with current models for the evolution of voids, for example that described by Shandarin.

RHEE: The two point correlation function of Ly α absorbers does not produce evidence for voids. Have you tried using other statistics specifically tuned to detecting voids?

SHAVER: We are exploring other possible tests to look for whatever void structure there may be in the Ly-α forest, including the use of cross-correlations or direct comparisons of the absorption spectra of close quasar pairs.

RHEE: Icke and v.d. Weygaert (A and A 1987) have provided a physical basis for the Voronoi foam description of the large structure. Rather than take a random distribution of expansion centers would it not be physical to use the fluctuation spectrum to produce an initial distribution of expansion centers?

SHAVER: One could certainly try other patterns, but I don't think they will change the basic conclusion that the distribution of Ly-α absorbers is inconsistent with their respecting the void structure as defined by galaxies.

ZANINETTI: Did you have some particular reason to simulate the distribution of galaxies using the tesselation model rather than other existing approaches like percolation theory?

SHAVER: We wanted a very simple and straightforward model to describe a three-dimensional universe whith voids and uniform walls, so we were led naturally to the Voronoi tessellation - before we even knew it had a name!

INFLATION AND THE BARYON ISOCURVATURE MODEL

P. J. E. Peebles
Joseph Henry Laboratories
Princeton University
Princeton N. J. 08544 U. S. A.

ABSTRACT. I describe a model for the origin of the initial conditions of the baryon isocurvature model within the inflation picture. The isocurvature fluctuations are assumed to have come from dilution of the primeval baryon to entropy ratio by decay of a massive field produced as quantum fluctuations leaving the horizon during the inflation era. This scenario requires an expansion time history that would be produced by a model inflaton field that has negative kinetic energy, a case that is strange but perhaps not impossible.

1. INTRODUCTION

In the baryon isocurvature model the mass density in the early universe is accurately homogeneous, so space curvature is not perturbed, while the baryon to photon number density ratio is a random function of position. (For discussions of the physics of the isocurvature model see Doroshkevich, Zel'dovich and Sunyaev 1978; Hogan and Kaiser 1983; Peebles 1980, 1983, 1987a,b) Under the assumption that the primeval fluctuations in the baryon to entropy ratio had an approximately flat power spectrum longward of a cutoff ~ 1 Mpc, with density contrast on the order of unity, this model fits two sets of observations that seem puzzling within the cold dark matter (CDM) picture (Frenk et al. 1988 and references therein). The first is the evidence of large-scale correlations in the distributions of galaxies (Haynes and Giovanelli 1986; Tully 1988) and mass (as inferred fron the peculiar velocity field: Dressler et al. 1987; Burstein 1988). As discussed by Bertschinger and Juszkiewicz (1988) this is unexpected in the CDM picture because the primeval scale-invariant spectrum makes the mass distribution anticorrelated on large scales. In the isocurvature model large-scale density fluctuations are produced at the matter-radiation Jeans length ~ 100 Mpc during decoupling of matter and radiation (Hogan and Kaiser 1983; Peebles 1987a). The second set of observations is the evidence that galaxies at redshift $z \sim 1$ had old star populations, suggesting bright galaxies formed at high redshifts (Lilly and Longair 1984, Hamilton 1985). This may be

M. Mezzetti et al. (eds.), Large Scale Structure and Motions in the Universe, 119–132.

a problem in CDM because the scale-invariant spectrum limits the amplitude of small-scale fluctuations (Frenk *et al.* 1988). It may be a virtue of the isocurvature picture as well as the explosion (Ostriker Thompson and Witten 1986) and cosmic string (Brandenberger, Kaiser and Turok 1987) pictures that galaxies can form at high redshift without violating the limits on the anisotropy of the microwave background radiation (Peebles 1987b; Efstathiou and Bond 1987).

To explore the observational consequences of the isocurvature picture in more detail we need a reasonably motivated prescription for the primeval distribution of baryons relative to entropy. A particular problem is that a phenomenologically reasonable power spectrum makes the rms density contrast nonlinear on small scales, which means we cannot try the usual random phase assumption. I propose here a prescription based on a model for the origin of isocurvature perturbations during an inflationary phase. A cautionary note is in order: as in any discussion of the physics of the universe at epochs approaching the planck time the details necessessarily are speculative and incomplete. A useful model, which the one presented here may prove to be, will at least tend to focus the discussion, as indeed has happened in the cold dark matter picture.

2. THE DECAYING MASSIVE FIELD MODEL

2.1. Assumptions

The model follows the remark of Kofman and Linde (1987), that isocurvature perturbations would be produced by dilution of an initially uniform baryon to entropy ratio by decay of an inhomogeneously distributed field. The field could be generated as quantum fluctuations leaving the horizon during inflation.

This idea will be developed under the following assumptions. Space is taken to be cosmologically flat, as suggested by inflation. The important dynamical actors are baryons and radiation, with negligible exotic dark matter. The density parameter Ω may be less than unity, the remainder being provided by a cosmological constant-like term. Consistent with this, galaxies are taken to trace the large-scale mass distribution. Galaxies and clusters of galaxies are assumed to have grown out of isocurvature fluctuations produced by decay of a massive nearly free field $\psi(\mathbf{x}_i)$ that was initially a random gaussian process.

The power spectrum $|\psi_k|^2$ of ψ is taken to be

$$k^3|\psi_k|^2 \sim \chi^2, \quad x_l^{-1} < k < x_{min}^{-1}; \tag{1a}$$

$$\sim \chi^2(kx_l)^\nu, \quad k < x_l^{-1}, \tag{1b}$$

where χ and ν are constants and $0 < \nu < 3$. The length x_l will be adjusted to fit the observed fluctuations in galaxy counts. The length $x_{min} \ll x_l$ is some very short wavelength cutoff. The autocorrelation function belonging to this spectrum is a power law at separations larger than x_l:

$$C(x) = \frac{\langle \psi(0)\psi(\mathbf{x})\rangle}{\langle \psi^2\rangle} \sim \frac{(x_l/x)^\nu}{\log \Lambda}, \qquad \Lambda = x_l/x_{min}. \tag{1c}$$

As outlined below, if $\nu > 1.5$ and the coherence length is adjusted to $x_l \sim 1$ Mpc at the present epoch then this picture would imply the formation of protogalaxies at high redshift, and it would imply large-scale mass fluctuations roughly consistent with the Great Attractor velocity field. A scenario for inflation that leads to equation (1) with $\nu = 2$ is presented in the next section.

The final assumptions concern the mass m and lifetime t_d of the field ψ. The total mass density at epoch t is

$$\rho \sim (m_{pl}/t)^2, \qquad m_{pl} = G^{-1/2},$$

where G is Newton's constant. Here and below I drop numerical factors of order unity because the estimates in any case are very uncertain. The field ψ is assumed to have lifetime large compared to its Compton wavelength, $t_d m \gg 1$, but small enough that when nonrelativistic ψ particles decay at $t \sim t_d$ lengths of astronomical interest are much larger than t_d. Fourier components of ψ that enter the horizon at $t < m^{-1}$ act like relativistic particles with amplitudes that decay as $a(t)^{-1}$, where $a(t)$ is the cosmological expansion parameter. It follows that at $t \sim m^{-1}$ the power spectrum of ψ is given by equation (1) at long wavelengths, $k/a < m$, and is suppressed by the factor $\propto k^{-2}$ at shorter wavelengths. This means that most of the ψ particles are nonrelativistic at $t \sim m^{-1}$. The mass density in the field is $\rho_\psi \sim m^2 \psi(\mathbf{x}_i)^2$, and the ratio of the mean value to the total mass density is $\rho_\psi/\rho \sim (\chi/m_{pl})^2 \log(\Lambda)$. Thus under the natural condition $\chi \ll m_{pl}$ the field ψ makes only a small contribution to the mass density at $t \sim m^{-1}$. At $m^{-1} < t < t_d$ the field oscillates with time and ρ_ψ decreases as $a(t)^{-3}$, which means ρ_ψ grows relative to radiation as $a(t)$. To get the wanted large fluctuations in the final baryon to photon ratio we must assume that at $t \sim t_d$ the mass density in ψ has grown to dominate. This happens if, as will be assumed,

$$m\, t_d \gg (m_{pl}/\chi)^4. \tag{2}$$

2.2. Statistics of the Baryon Distribution at Decoupling

The initial ratio of baryon number to entropy is assumed to be a constant, η_i. When the ψ field decays it adds to the entropy density, making the ratio of baryon number to entropy to a function of position, as follows. At time $t \sim m^{-1}$ the number density of ψ particles is $\sim m\psi^2$ and the entropy density (with temperature unit chosen so Boltzmann's constant is unity) is

$$s_m \sim \rho_r^{3/4} \sim (m\, m_{pl})^{3/2},$$

where ρ_r is the mass density in radiation, so the entropy per ψ particle is

$$s_\psi \sim \frac{s_m}{n_\psi} \sim \frac{m^{1/2} m_{pl}^{3/2}}{\psi(\mathbf{x}_i)^2}.$$

Since density fluctuations of interest have scales large compared to the horizon we can take it that s_ψ is conserved until ψ decay at $t \sim t_d$. It clarifies the discussion if we imagine that the decay is instantaneous. Just before decay the mass density is the sum of radiation, $\rho_r \sim s^{4/3}$, where s is the entropy density, and $\rho_\psi \sim m\, s/s_\psi$:

$$\rho_d \sim \left(\frac{m_{pl}}{t_d}\right)^2 \sim s^{4/3} + \frac{s\, m^{1/2} \psi(\mathbf{x}_i)^2}{m_{pl}^{3/2}}.$$

Under the assumption of equation (2) the second term dominates so the entropy density just before it is appreciably increased by ψ decay is

$$s(\mathbf{x}_i) \sim \frac{m_{pl}^{7/2}}{m^{1/2} t_d^2 \psi(\mathbf{x}_i)^2}.$$

This is a function of the original comoving position $\mathbf{x}_i$.

The baryon density at $t \sim t_d$ is $n_B = \eta_i s(\mathbf{x}_i)$. Just after ψ decay the universe is again radiation dominated with entropy density $s_f \sim (m_{pl}/t_d)^{3/2}$, so the local entropy per baryon is now

$$\frac{1}{\eta(\mathbf{x}_i)} = \frac{s_f}{n_B} \sim \frac{(t_d m)^{1/2} \psi(\mathbf{x}_i)^2}{\eta_i m_{pl}^2}, \tag{3}$$

and the ratio of entropy to baryon number in the volume V_i placed at the initial position of the baryons is

$$\frac{S_f}{N_B} \sim \frac{(t_d m)^{1/2}}{\eta_i m_{pl}^2} \int \frac{dV_i}{V_i} \psi(\mathbf{x}_i)^2.$$

Equation (3) gives the entropy per baryon as a function of comoving position of the baryons before the energy density in ψ became appreciable. Relative comoving positions at $t \sim t_d$ are perturbed by the different expansion factors in regions of different initial values of ψ. We can avoid dealing with this complication when V_i is large because the estimate of S_f/N_B for a region defined by a fixed value of N_B yields a reasonable approximation to an estimate of N_B for a region defined by a given S_f.

At redshift $z \sim 1000$ radiation pressure with damping by photon diffusion will have smoothed the radiation on scales less than the matter-radiation Jeans length $\sim 100\,\Omega^{-1}h^{-2}$ Mpc (Peebles 1987a). (Here h is Hubble's constant in units of 100 km sec^{-1}Mpc^{-1}.) Thus the baryon mass density contrast, δ, at decoupling, on the scale of protogalaxies and larger is given by the equation

$$\frac{S_f}{N_B} \sim \frac{(1-\delta)}{\eta},$$

where η is the present mean value of the baryon number per unit of entropy.

It is assumed that ψ is a random gaussian process so the moments of δ are determined by the autocorrelation function $C(x)$ of ψ (eq. [1c]). For example, the second moment of δ depends on the fourth moment of ψ through the equation

$$\langle \psi(\mathbf{x}_1)^2 \psi(\mathbf{x}_2)^2 \rangle = \langle \psi^2 \rangle^2 + 2\langle \psi(\mathbf{x}_1)\psi(\mathbf{x}_2) \rangle^2$$
$$= \langle \psi^2 \rangle^2 [1 + 2C(x_{12})^2],$$

and the third moment depends on the sixth moment of ψ,

$$\langle \psi(\mathbf{x}_1)^2 \psi(\mathbf{x}_2)^2 \psi(\mathbf{x}_3)^2 \rangle = \langle \psi^2 \rangle^3$$
$$+ 2\langle \psi^2 \rangle \langle \psi(\mathbf{x}_1)\psi(\mathbf{x}_2) \rangle^2 + \text{cycl.}$$
$$+ 8\langle \psi(\mathbf{x}_1)\psi(\mathbf{x}_2) \rangle \langle \psi(\mathbf{x}_2)\psi(\mathbf{x}_3) \rangle \langle \psi(\mathbf{x}_3)\psi(\mathbf{x}_1) \rangle.$$

The second and third moments of δ are

$$\langle \delta^2 \rangle = 2 \int \frac{d^2 V}{V^2} C(x_{12})^2,$$

$$\langle \delta^3 \rangle = -8 \int \frac{d^3 V}{V^3} C(x_{12})C(x_{23})C(x_{31}). \tag{4}$$

It was noted above that to get a reasonable model we had to assume that δ roughly approximates white noise, $\langle \delta^2 \rangle \sim V^{-1}$, on large scales, with $\langle \delta^2 \rangle$ flat and on the order of unity on small scales. We see that this follows in the ψ model if $\nu \geq 3/2$ so $\int dV\, C(x)^2$ converges at large x.

If $\nu \leq 3/2$ then

$$\langle \delta^3 \rangle^{1/3} \propto \langle \delta^2 \rangle^{1/2},$$

and the central limit theorem does not apply. If $\nu > 3/2$ then the distribution of δ approaches a gaussian at large V. In the inflation model to be outlined below $\nu = 2$. This value will be adopted in the remainder of the paper.

2.3. Protogalaxy Formation

We can estimate the time evolution of the variance of δ and so say something about galaxy formation in the model as follows. After radiation drag becomes unimportant fluctuations in the baryon distribution grow by gravitational instability as $\langle \delta^2 \rangle \propto \Delta(t)^2$ (Peebles and Groth 1976), where $\Delta(t)$ is the growing solution to to the equation describing linear fluctuations in the baryon distribution interacting with a smooth radiation field in an expanding universe (eq. [50] in Peebles 1971). Since density fluctuations are on the order of unity on small scales it is reasonable to suppose that stars form rapidly enough at redshift $z < 1000$ to keep the bulk of the matter ionized, which suggests that the full radiation drag term should be included in this equation.

I will quote numerical results for one set of parameters, density parameter $\Omega = 0.1$, which is in rough agreement with the dynamical estimates, and Hubble constant $H = 90$ km sec^{-1} Mpc^{-1}, which makes the age of the universe 14 billion years, as is indicated by some recent discussions (Hesser et al. 1987). The redshift at which the radiation drag term for the baryon peculiar velocity is equal to the expansion rate is $z_1 \sim 150$. This is the epoch at which fluctuations in the baryon distribution with $\langle \delta^2 \rangle \sim 1$ would start to break away from the radiation drag. The densities in in the first generation of bound objects would be comparable to the background density at epoch z_1, $n_1 \sim 3$ protons cm^{-3}.

We can scale the typical size of the first generation objects that form at $z \sim z_1$ from the observed fluctuations in galaxy counts under the assumption that galaxies trace mass. The present mean square fluctuation in mass found within volume V with radius ~ 10 Mpc is, with equation (4),

$$\left(\frac{\delta M}{M} \right)^2 = \int \frac{d^2 V}{V^2} \xi(x_{12}) \sim \frac{4\pi J_3}{a_o^3 V} \sim \frac{2}{V} \left(\frac{\Delta_o}{\Delta_i} \right)^2 \int d^3x \, C(x)^2,$$

$$J_3 = a_o^3 \int \xi(x) x^2 \, dx \sim 600 \, h^{-3} \; \text{Mpc}^3,$$

where a_o is the present value of the expansion parameter. The estimate of the integral of the galaxy two-point correlation function, ξ, is taken from the Lick (Clutton-Brock and Peebles 1981) and CfA (Davis and Peebles 1983) catalogs. The linear growth factor,

$$\Delta_o / \Delta_i \sim 67,$$

is the ratio of the present value of the linear density contrast $\Delta(t)$ to the value at $z \gg 100$. This fixes the normalization of ψ through the equation

$$I = a_o^3 \int d^3x \, C(x)^2 = 2\pi J_3 (\Delta_i / \Delta_o)^2. \tag{5}$$

Now let us consider how galaxy formation might be understood in this model. At redshifts in the range $z_1 < z < 1000$ mass fluctuations on small scales, where $\langle \delta^2 \rangle > 1$, would collapse and presumably form stars. However the star formation rate appears to be self limiting because where massive stars form they ionize the surroundings and so recouple the matter to the radiation. This means that matter in the neighborhood is required to expand with the radiation, suppressing further star formation. The situation changes at redshift $z \sim z_1$ when radiation drag has weakened to the point that ionized clouds can collapse faster than the universe expands. At this point the bulk of the matter in regions where the density contrast is greater than unity must collapse to a dissipationless state such as a star cluster. We can estimate the typical mass in this first generation from equation (4). When the averaging volume V_1 is chosen so the rms value of the density contrast is $\langle \delta^2 \rangle^{1/2} = 1/2$ the typical mass peak has $\delta \sim 1$ and so will gravitationally collapse once radiation drag allows it. Thus equation (4) says

$$V_1 \sim 8I.$$

The mass in the first generation is on the order of the mass in V_1,

$$M_1 \sim 16\pi J_3 \rho_o (\Delta_i/\Delta_o)^2 \sim 2 \times 10^{11} \mathrm{M}_\odot,$$

where ρ_o is the present mean mass density. (An earlier estimate in Peebles 1987b yielded smaller Δ_o/Δ_i and hence somewhat larger M_1 because it was assumed that the universe is open with $\Lambda = 0$.)

It seems an encouraging coincidence that M_1 is comparable to what is observed in the bright parts of large galaxies. Because the timescale for fragmentation and star formation is comparable to the time for collapse on the scale M_1 the collapse factor need not be large so the final density in this generation would be on the order of the mean density at the epoch z_1, which we have seen also is about right for the bright parts of galaxies. Discs of spiral and S0 galaxies would have to form well after z_1 out of debris from formation of the spheroid components, to allow for the large collapse factor needed so the discs end up rotationally supported.

These ideas could be pursued in greater detail through numerical simulations, which would help check whether the self-limiting process really suppresses large-scale star formation at $z < z_1$ and would give a better estimate of n_1; and might indicate also whether the model could account for systematics such as the Faber-Jackson Tully-Fisher correlation of luminosity and velocity dispersion.

2.4. Primordial Nucleosynthesis

It seems reasonable to assume that t_d is much less than the epoch of light element nucleosynthesis that commences at redshift $z \sim 10^{10}$ so the mean entropy per baryon number at nucleosynthesis is set by present observations in the usual way.

The local ratio of baryon to entropy density, η, is a function of position (eq. [3]). Using the assumptions that the baryons were uniformly distributed in initial position x_i and that $\psi(x_i)$ has a gaussian distribution one finds that the distribution in η weighted by baryon number is

$$\frac{dP}{d\eta} \propto e^{-\eta_o/\eta} \frac{d\eta}{\eta^{3/2}},$$

where η_o is a constant. The distribution in η weighted by volume is then

$$\frac{dP}{d\eta} \propto e^{-\eta_o/\eta} \frac{d\eta}{\eta^{5/2}}.$$

The distribution is cut off at some maximum value, η_i. If $\eta_i \gg \eta_o$ then the upper and lower quartiles of this distribution are 2.0 and 0.6 times the median. Yang et al. (1984), using a distribution of η that has a second moment, placed a bound on the standard deviation, $\delta\eta/\eta$ less than about 1.5. It appears from this that the model is not in serious danger of violating nucleosynthesis bounds.

3. INFLATION MODEL

Now let us consider a model for the origin of the wanted power spectrum of $\psi(x)$ in an inflation model. In the power spectrum in equation (1) the variance per octave increases with decreasing scale. It will be recalled that during inflation the rms value of ψ leaving the horizon is on the order of the expansion rate, and that smaller scale fluctuations leave the horizon later. Thus we have to assume that the expansion rate during inflation increases with time, in the 'super-inflation' case of Lucchin and Matarrese (1985). Since the expansion rate is $\sim \rho^{1/2}/m_{pl}$ the mass density ρ would have to increase with time.

Let us note furthermore that the small-scale part of the spectrum in equation (1) is the usual scale-invariant spectrum produced by a nearly constant expansion rate (eg. Bardeen, Steinhardt and Turner 1983). The large-scale part with $\nu = 2$ is the familiar form of zero point fluctuations in a free field in Minkowski space. Thus we get the wanted spectrum if the early universe passed from a quasistatic epoch to an exponential expansion phase that ended with the usual epoch of reheating. We can arrange this as follows.

In the conventional inflaton model with lagrangian $\dot{\phi}^2/2 - V(\phi)$ the mass density varies with time as $\dot{\rho} = -3\dot{a}\dot{\phi}^2/a$, which is opposite to what is wanted. We reverse the sign of $\dot{\rho}$ by adopting a lagrangian with negative kinetic energy, to get

$$L = -\frac{\dot{\phi}^2}{2} - V(\phi), \qquad \frac{d\rho}{dt} = 3\frac{\dot{a}}{a}\dot{\phi}^2. \tag{6}$$

This form is not popular, for good reason, though it has been discussed (Glauber 1986). In the present context the problem with it is that if we made the lagrangian covariant by replacing the kinetic energy with the square of the gradient of ϕ it would introduce an unbounded spectrum of negative energy exitations that couple to gravity, so the vacuum could decay into a pair of exitations plus a graviton. To avoid this we must imagine that equation (6) is only a model for some deeper and we trust consistent theory.

A quasistatic phase is obtained if initially $\dot{\phi}$ and $V(\phi)$ are small. This would evolve into exponential expansion if the initial value of $V(\phi)$ were a minimum in an otherwise nearly level and large value of $V(\phi)$. An example is the potential

$$V = \frac{(\chi m_{pl})^2}{8\pi} [\sin^2(\pi^{1/2}\phi/m_{pl})(2\sin^2(\pi^{1/2}\phi/m_{pl}) + 1)], \qquad (7)$$

where χ is a constant. The differential equations for ϕ and the expansion rate are

$$\ddot{\phi} + 3\frac{\dot{a}}{a}\dot{\phi} = \frac{dV}{d\phi}, \qquad \frac{\dot{a}^2}{a^2} = \frac{8\pi}{3m_{pl}^2}(V - \dot{\phi}^2/2). \qquad (8)$$

If ϕ and $\dot{\phi}$ initially are small the solution is

$$\phi(t) = \pi^{-1/2}m_{pl}\tan^{-1}e^{\chi t/2}, \qquad a(t) = a_l(1 + e^{\chi t}). \qquad (9)$$

As check that this solution is stable I numerically integrated equations (7) and (8), and found that a factor of two change in $\dot{\phi}$ at $\chi t = -20$ has little effect on the subsequent time evolution of $a(t)$, indicating that a substantial range of initial conditions would lead to the expansion time history of equation (9).

With the expansion rate of equation (9) one finds by the usual methods the power spectrum in equation (1), where the scale-free part (eq. [1a]) leaves the horizon at $\chi t > 1$ and the large-scale part at $\chi t \sim 1$, with the break at scale

$$a_l x_l \sim \chi^{-1}.$$

There are two parameters in the model, χ and the the expansion factor a_R/a_l during the exponential expansion phase, from a_l to the epoch of reheating at a_R. These must be adjusted to get the wanted value of the coherence length x_l. Let us verify that these parameters can be chosen to fall in what generally are considered to be reasonable ranges.

The break in the power spectrum (eq. [1]) is chosen to fit the large-scale fluctuations in galaxy counts. We have from equation (5)

$$I \sim (a_o x_l)^3/\log^2\Lambda \sim (1\ \text{Mpc})^3.$$

The characteristic scale at the break in the spectrum, evaluated at the present epoch, is then

$$a_o x_l \sim \chi^{-1} \frac{a_R}{a_l} \frac{a_o}{a_R} \sim 1 \text{ Mpc } \log^{2/3} \Lambda.$$

The first factor in the middle expression is the proper value of the characteristic scale at the start of the exponential expansion phase, the second is the expansion factor in the exponential phase, and the third is the redshift at reheating,

$$z_R = \frac{a_o}{a_R} \sim \frac{T_R}{T_o} \sim \frac{(\chi m_{pl})^{1/2}}{T_o},$$

where the present background temperature is $T_o \sim 1 \text{ mm}^{-1}$. Thus to make the coherence length agree with the observations we need

$$\frac{a_R}{a_l} \sim 10^{25} \log^{2/3} \Lambda \left(\frac{\chi}{m_{pl}} \right)^{1/2}.$$

The parameter $t_d m$ must be adjusted to fit the present value of the baryon number per unit of entropy, η. Equations (2) and (3) indicate

$$\left(\frac{m_{pl}}{\chi} \right)^4 \ll t_d m \sim \left(\frac{\eta_i}{\eta} \right)^2 \left(\frac{m_{pl}}{\chi} \right)^4.$$

The graviton amplitude is (Veryaskin, Rubakov and Sazhin 1983)

$$h \sim \chi/m_{pl},$$

per octave of wavelength from x_l to $x_l a_l / a_R$. If for example we chose $h \sim 10^{-6}$ we would have

$$t_d m \gg 10^{24}, \quad a_R / a_l \sim 10^{22}, \quad z_R \sim 10^{-3} m_{pl} / T_o.$$

Since $\eta \sim 10^{-9}$ there is considerable room for the inequalities

$$\eta \ll \eta_i \ll 1.$$

The adiabatic density fluctuations associated with the nearly free fluctuating part of the inflaton field would be

$$\frac{\delta \rho}{\rho} \sim \left(\frac{\chi}{m_{pl}} \right)^2,$$

which can be arranged to be negligibly small.

4. SUMMARY

The baryon isocurvature model presented here is phenomenologically interesting because as we have discussed the first generation of rapid star formation would be expected to occur in star clusters with characteristic masses and radii comparable to those of galaxies. One also finds that the large-scale mass fluctuations in the model are at least roughly comparable to what is indicated by the recent observations discussed at this meeting (Peebles 1987a; Bertschinger and Juszkiewicz 1988). On the theoretical side, it might be significant that the spectrum of the field ψ that is supposed to produce the isocurvature fluctuations has two familiar parts: the small-scale end is the usual scale-invariant spectrum of zero-point fluctuations produced in exponential inflation, and an acceptable form for the large-scale part is the spectrum of zero-point fluctuations of a nearly free field in Minkowski space. I have not presented a rational inflation model that could produce both effects, but it seems not unreasonable to hope that it could be done, as has been illustrated by the lagrangian in equation (6).

On the cosmological side, the most immediate challenge will be to decide whether the baryon isocurvature picture really could produce acceptable proto-galaxies. The model presented here provides a prescription for the baryon distribution at $z \sim 1000$ that may be a useful basis for more detailed studies.

I am grateful to Bharat Ratra for stimulating discussions. This research was supported in part by the U. S. National Science Foundation.

REFERENCES

Bardeen, J. M., Steinhardt, P. J., and Turner, M. S. 1983, *Phys. Rev.*, **D28**, 679.

Bertschinger, E., and Juszkiewicz, R. 1988, *Ap. J.*, in the press.

Brandenberger, R., Kaiser, N., and Turok, N. 1987, *Phys Rev.*, **D36**, 2242.

Burstein, D. 1988, this conference.

Clutton-Brock M., and Peebles, P.J. E. 1981, *A. J.*., **86**, 1115.

Davis, M., and Peebles, P. J. E. 1983, *Ap. J.*, **267**, 465.

Doroshkevich, A. G., Zel'dovich, Ya. B., and Sunyaev, R. A. 1978, *Soviet Astron—A. J.*, **22**, 523.

Dressler, A. *et al.* 1987, *Ap. J. Letters*, **313**, L37.

Efstathiou, G., and Bond, J. R. 1987, *M.N.R.A.S.*, **227**, 33P.

Frenk, C. S., White, S. D. M., Davis, M., and Efstathiou, G. 1988, *Ap. J.*, **327**, 507.

Glauber, R. J. 1986, *Ann. N. Y. Acad. Sci.*, **480**, 336.

Hamilton, D. 1985, *Ap. J.*, **297**, 371.

Haynes, M. P., and Giovanelli, R. 1986, *Ap. J. Letters*, **306**, L55.

Hesser, J. E., Harris, W. E., Vandenbergh, D. A., Allwright, J. W. B., Schott, P., and Stetson, P. B. 1987, *Publ. A.S.P.*, **99**, 739.

Hogan, C. J., and Kaiser, N. 1983, *Ap. J.*, **274**, 7.

Kofman, L. A., and Linde, A. D. 1987, *Nuclear Physics*, **B282**, 555.

Lilly, S. J., and Longair, M. S. 1984, *M.N.R.A.S.*, **211**, 833.

Lucchin, F., and Matarrese, S. 1985, *Physics Letters*, **164B**, 282.

Ostriker, J. P., Thompson, C., and Witten, E. 1986, *Phys Lett*, **B280**, 231.

Peebles, P. J. E. 1971, *Physical Cosmology* (Princeton: Princeton Univ. Press).

Peebles, P. J. E. 1980, *Physica Scripta*, **21**, 720.

Peebles, P. J. E. 1983, in *The Origin and Evolution of Galaxies*, eds. B. J. T. Jones and J. E. Jones (Dordrecht: Reidel) p. 143.

Peebles, P. J. E. 1987a, *Nature*, **327**, 210.

Peebles, P. J. E. 1987b, *Ap. J. Letters*, **315**, L73.

Peebles, P. J. E., and Groth, E. J. 1976, *Astr. Ap.* **53**, 131.

Tully, R. B. 1988, this conference.

Veryaskin, A. V., Rubakov, V. A. and Sazhin, M. V. 1983, *Soviet. Astron.*, **27**, 16.

Yang, J., Turner, M. S., Steigman, G., Schramm, D. N., and Olive, K. A. 1984, *Ap. J.*, **281**, 493.

DISCUSSION

LUKASH: I would expect in your baryonic model large perturbations in $\Delta T/T$ on scales $\theta \sim 1° - 10°$ due to the Sachs-Wolfe effect. As I tried to show in my talk, isocurvature perturbations in baryons set at the radiation dominated epoch will induce the 'growing' adiabatic mode after the equality epoch of about the same amplitude. This is important for scales which were larger or about the horizon at the equality epoch.

PEEBLES: I find that on scales large compared to the matter-radiation Jeans lenght an isocurvature perturbation remains isocurvature in the matter-dominated epoch. That is because the expanding universe conserves space curvature when pressure gradients can be neglected, so no adiabatic component develops. Thus I think there is no problem with the Sachs-Wolfe effect.

SCIAMA: What effect does the spatial fluctuation in n_b/n_γ have on primordial nucleosynthesis?

PEEBLES: In the published version of the paper I work out the effect; it seems to be just within the range allowed by Yang et al.

KAISER: Your calculations are very similar to those made by Craig Hogan and myself. We noted the pleasant consequences for the large-scale fluctuations and the acceptable $\Delta T/T$. The main difference is that we envisaged the entropy fluctuations to be generated by pre-galactic object around z_{EQ} or later – you have reproduced our results for $z \simeq z_{EQ}$ since the final result is indipendent of z for higher generation redshift. What benefits accrue from your assumption of a much earlier generation epoch?

PEEBLES: Thanks for pointing out the paper by you and Hogan. As far as I have been able to discover your paper is the first in which one clearly sees the growth at decoupling of the peak in the power spectrum at the matter-radiation Jeans length λ_J ; you are to be congratulated for discovering an important effect. I am a little less enthusiastic about your discussion of large-scale structure, because your proposal that pancake collapse at $\lambda \sim \lambda_J$ produced the observed sheet-like galaxy distribution would put the sheets kind of far apart. I think a better application of the effect is to the great attractor. Your idea for producing isocurvature perturbations is interesting and I certainly would encourage you to follow it up. I would worry about producing and thermalizing the energy needed at $z \sim z_{EQ}$.

ELLIS: The inflationary model proposed is consistent with $\Omega = 0.1$ today, even if $\Lambda = 0$ today. Could it not be preferable to $\Lambda \neq 0$?

PEEBLES: I am influenced by the simple argument that if the expansion factor during inflation were large enough to stretch out density gradients to unobservably small values then it would seem to be reasonable to expect that the expansion would do the same thing to space curvature. But in fact the baryon isocurvature model does about equally well in an open model with $\Lambda = 0$ as in the flat model with Λ , the main difference that I have found being that the lower growth factor in an open model raises the anisotropy of the cosmic background radiation to a level about comparable to the Davies et al. result.

SUTHERLAND: How do you account for the break in the galaxy correlation function, which now seems well established?

PEEBLES: I don't have a natural explanation of the break; however, the CDM model also has problems with this.

SUTHERLAND: According to George Efstathiou CDM fits the break quite well, though it may not be quite sharp enough in the model.

MILLER: A crucial question for your picture is how the initial conditions were arrived at. In some sense you are pushing the problem back into the initial conditions. Can you say anything more about how they may have been produced?

PEEBLES: Perhaps when we understand the physics well enough we will be able to dispense with initial condition in cosmology; but we are not at that point even if the inflation concept is correct. I introduced initial conditions "ad hoc" to fit my purpose. The test will be the observational succes (if any) of the model.

COUCHMAN: I should take to draw your attention to some work done by D. Bond and myself. We show that the break in the angular correlation of galaxies in scales of $\sim 3°$ in the Lick data may impose important constraints on cosmic fluctuation spectra with large-scale power. In particular, whilst the standard CDM model is generally consistent with this scale, break isocurvature baryon models seem not to be.

PEEBLES: It's a good point. I note, however that the inflation-inspired model providentially yields a flat spectrum of fluctuations of the baryon distribution, so in linear approximation the mass autocorrelation function vanishes. The serious challenge for the model will be to account for the shape of the two-point correlation function that develops through non-linear growth of clustering. I presume a full non-linear computation will be needed to test for this.

Large Scale Structure: Its possible imprint on the CMB?

R.A.Watson[1,2], R.Rebolo[1], J.E.Beckman[1],R.D.Davies[2] and A.N.Lasenby[3]
1. Instituto de Astrofisica de Canarias, Tenerife, Spain.
2. University of Manchester, NRAL, Jodrell Bank, UK
3. MRAO, Cavendish Laboratories, Cambridge, UK.

ABSTRACT: We present the new results of the Jodrell-IAC Cosmic Microwave Background (CMB) anisotropy experiment for declination $+40°$. A similar level of sensitivity has been achieved with respect to the reported previous experiment at an angular scale of 8.8° in Davies *et al.* 1987. A positive temperature deviation of $\approx$ 0.3mK at 10.46GHz is identified at RA 14^h55^m; known point sources with flux densities greater than 6Jy are seen in other sections of the scan, but none are known at the position of the 14^h55^m feature. This feature appears, albeit weakly, in both the 5.6° and the 8.8° data. We note that this position coincides with that of the large void in Boötes (Kirshner *et al.* 1981) where it crosses $\delta = +40°$. Also at this same position there appears an unusual string of galaxies, tentatively attributed by Tago *et al.* 1986, to the relic of a density singularity. It seems unlikely that the apparent level of anisotropy can be caused by these structures, at least by gravitational means alone, but there remains the possibility that the origin, apart from it being intrinsic to the CMB, could be galactic.

1. Introduction

New preliminary results from the Jodrell-IAC CMB anisotropy experiment at declination $+40°$ are presented here. The original results, also at the same declination, were reported in Davies *et al.* 1987, and consisted of a tentative detection of fluctuations in the sky signal for measurements with a triple beam configuration, of an amplitude of $\delta T/T = 3.7 \times 10^{-5}$ on a scale of 8.8°. This is greater than the possible contribution from known discrete sources, which are mainly quasars. A similar scan at the same centre frequency of 10.46GHz, has now been completed with the same triple beam configuration, but with beam sizes of 5.6° FWHM (Full Width Half Maximum), instead of 8.8°. Both sets of data are compared and particular attention is paid to possible correlations with large scale structure in the Boötes region, which corresponds to the most sensitive section of the data (RA 12^h to 18^h). By coincidence, this is also a portion of the sky covered by the Centre for Astrophysics (CfA) redshift survey (Geller *et al.* 1987), in which strong indications of large scale structure have been found, especially the Boötes void (Kirshner *et al.* 1981, 1987) and a prominent string of galaxies ,which Tago *et al.* 1986 attribute to the relic of an initial density singularity.

M. Mezzetti et al. (eds.), Large Scale Structure and Motions in the Universe, 133–137.
© 1989 by Kluwer Academic Publishers.

2. Observations

The experiment, located in Izaña (altitude 2400m), Tenerife, is a dual-beam, dual-channel, switching radiometer with cryogenically cooled receivers. A slower second stage of switching involving a rocking mirror produces the triple beam pattern of beams of FWHM 5.6°, separated by 8.1°, of amplitudes −0.5, +1.0 and −0.5 from east to west on the sky, as shown by the galactic plane crossings in figure 1 for both the old and the new experiments.

The new data presented here, were obtained over March and April 1986 and July 1987. From this period 35 days of data, unaffected by weather, were included in the final scan shown in figure 1a. The equivalent data for the 8.8° experiment are shown in figure 1b.

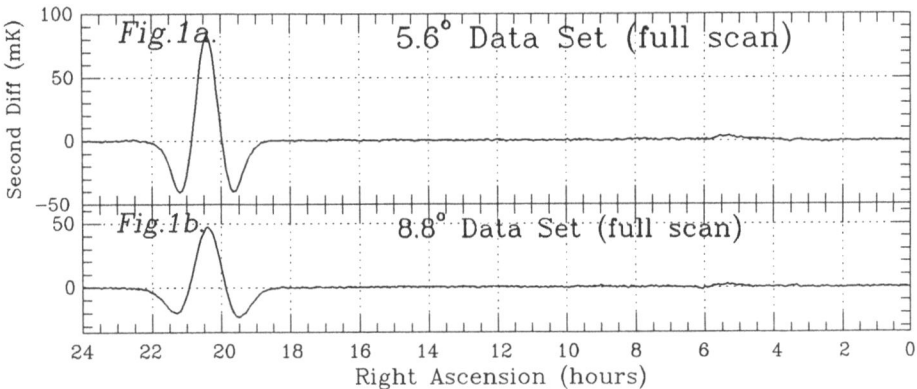

Figure 1
a) Full RA scan of beam temperature difference with new 5.6° beamwidth system.
b) Same as (a) but of the original 8.8° system.

3. Preliminary Analysis.

The main galactic crossing at RA 20^h30^m, which includes the strong thermal emission from the Cygnus-X region, was used to align and calibrate the scans as shown in figure 1. Polynomials were fitted to the data in order to remove the slight baseline drifts, the order used was dependent upon the length of the scan; the highest allowed being a 5th order for the $2\frac{1}{2}$ days of data that could be held on a single disk. This is sufficiently low enough so as not affect the beam response, which covers about 3 hours of data.

The weighted summation was carried out on the fitted scans, the weighting used being the inverse of the variance evaluated over each 80 second integration convolved with a gaussian of FWHM 1 hour. Therefore this weighting can be viewed as a quality parameter, which we attribute to slowly varying atmospheric conditions. The added 5.6° data between RA 12^h to 18^h are shown in figure 2a. The 8.8° were reduced in the same manner and are shown in figure 2b.

Due to the higher resolution of the new experiment, it is 2.5 times more sensitive to strong discrete sources. The point source contribution was evaluated by convolving the beam profile with strong sources ($\geq$ 1Jy at 10.6GHz) drawn from the 1Jy catalogue (Kühr et al. 1981). In the most sensitive part of the data a blend of two extra-galactic sources (1633+38 [6Jy] and 3C345=1641+39 [10Jy]) are clearly present and at the expected amplitude. In this range there are no other sources above

2Jy, which makes the excess emission found at 14^h55^m in both the 8.8° and 5.6° data at least suggestive of an intrinsic 'Hot-Spot'.

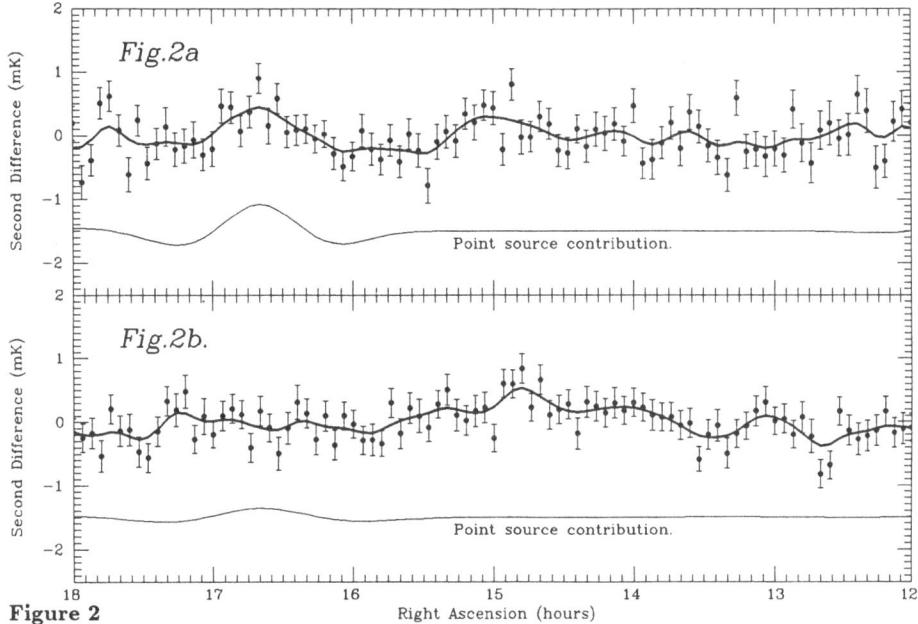

Figure 2

a) The most sensitive part of the scan of 5.6° data, between RA 12^h and 18^h
showing 1 sigma error-bars and a line of a running gaussian mean (of width 2.8°)
and, below, the predicted point source contribution.
b) Same as (a) but of the original 8.8° system.

With the sources removed the mean standard deviation per 4^mRA point (σ_i) is .27mK and the standard deviation calculated directly from the scatter in the points in the final scan (σ_d) is .31mK. In the 8.8° scan σ_i is .23mK and σ_d is .27mK, which is compatible with the longer period of observation and better quality data. The line shown passing through the data in figures 2a and 2b are the data convolved with a gaussian with half the width of the beam.

In order to determine if the 5.6° data contain excess variance on the scale of the beam size the data was broken down into bins separated by 40 minutes RA weighted with a gaussian the same width of the beam. The analysis used was that of Lasenby and Davies 1983, where the excess variance of the scatter in the final scan is sought over that pure attributable to instrumental noise. We find that although the 95% confidence limit allowed by the system noise is $\sigma_i = 0.082$mK, the observed standard deviation of the data is $\sigma_d = 0.088$mK, just barely a detection. The major part of this excess seems to be due to this feature at 14^h55^m , as repeated analysis in other parts of the scan, excluding this region, finds similar instrumental limits, but lower σ_d estimates.

Finally we should point out that it is only valid to sum the 8.8° and 5.6° data sets, as in figure 3., if the structure is the same size as the beam, so that its amplitude in both experiments is comparable. In order to do the summation correctly one must reduce the two data sets in parallel, coupled through a common model of the sky brightness distribution pattern (Lasenby, *in preparation*).

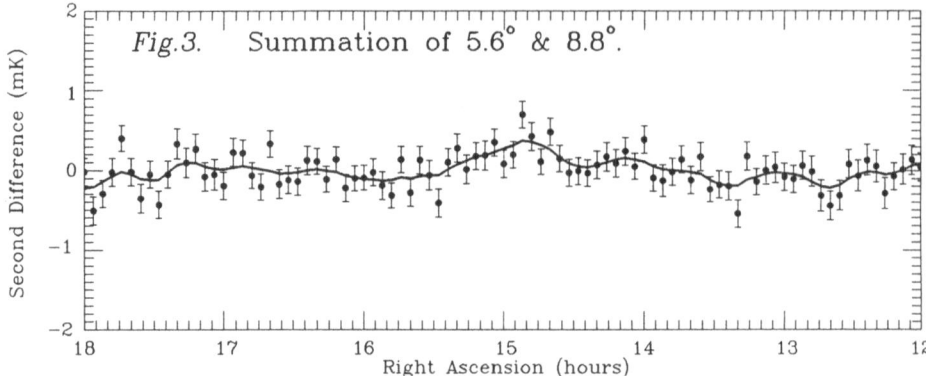

Figure 3. Summation of the 5.6° and 8.8° data sets, with point sources removed.

4. Discussion.

The results from the 5.6° experiment appear to support the previous reported indication of aniso-tropy, although as yet there may systematic problems due to incomplete baseline and/or point source subtraction. A feature at RA 14^h55^m repeats in the two scans and is a little larger in the 5.6° scan suggesting, if it is real, that it is slightly smaller than the 8.8° beam. It is interesting to note that it appears to coincide with the Boötes void (Kirshner *et al.* 1981, 1987), which covers a velocity range of 6000km s^{-1} centred on 15500km s^{-1}. The effect of such voids on the CMB as expressed in Thompson and Vishniac 1987, is $\delta T/T \approx 10^{-6}$, an order of magnitude less than detectable in our current measurements. Also the void covers 30° and this cannot be reconciled with the apparent feature size of 5.6°, unless it was associated with a dense portion of the void wall. A possible alternative source is an unusual string of galaxies that lie on the boundary of the void that Tago *et al.* suggest is the relic of a collapsed density singularity. The mass and velocities involved seem to be not enough to perturb the CMB by gravitational interaction alone, but this effect could arise from large scale structure at higher redshift. Finally, we can also appeal to the intrinsic fluctuations in the CMB at recombination, which allows 'hot spots' arising from peaks in the perturbation spectrum at long wavelengths ($> 50h^{-1}$ Mpc), which can also drive large scale structure and streaming motions (Juszkiewicz *et al.* 1987).

References

Davies, R.D., Lasenby, A.N., Watson, R.A., Daintree E.J., Hopkins, J., Beckman, J.E., Sanchez-Almeida, J. and Rebolo., R. 1987, Nature,**326**, 611

Geller, M.J., Huchra, J.P. and Lapparent. V. in *Observational Cosmology,* IAU Symposium 124, (Dordrecht: D. Reidel)

Juszkiewicz, R., Gorski, K. and Silk,J. 1987, Ap.J.,**323**, L1

Kirshner, R.P., Oemler, A., Schechter, P.L. and Shectman,S.A. 1981, Ap.J., **248**, L57

Kirshner, R.P., Oemler, A., Schechter, P.L. and Shectman,S.A. 1987, Ap.J., **314**, 493

Kühr, H., Witzel, A., Pauliny-Toth, I.I.K. and Nauber, U. 1981, A & A supp. ser., **45**, 367

Lasenby, A.N. and Davies, R.D. 1983, MNRAS, **203**, 1137

Tago, E., Einasto, J. and Saar, E. 1986, MNRAS, **218**, 177

Thompson, K.L. and Vishniac, E.T. 1987, Ap J, **313**, 517

DISCUSSION

PARTRIDGE: Two questions: 1) How certain are you that differential ground pick-up was absent (as the mirror switched positions)? 2) Suppose you consider your data with the single large "lump" at 15 hours removed. Are the point-by-point measurements then consistent with a gaussian distribution? To me, there appear to be too many "outliers", suggestives that your individual bars may be too small.

WATSON: 1) We have done tests by placing absorber and aluminum plate around the mirror and have seen no such effect. 2) I must admit my eye seems to see again too many "outliers"as well. It may be that this preliminary analysis has not entirely solved the problem of the baseline drift and the resulting distribution is not completely gaussian.

LUKASH: One cannot directly connect a possibility of spot $\Delta T/T \sim 0.3$ mK on scales $\theta \sim 5^\circ$ with the RELIC experiment since the latter observed larger scales. However, it is possible to do if one assumes that we know a spectrum of the primordial cosmological perturbations.

LARGE STRUCTURE PARAMETERS IN THE MULTICOMPONENT UNIVERSE

V.N.Lukash
Space Research Institute
Academy of Sciences of the USSR

Light weakly interacting particles which are ilusive now
both dynamically and experimentally, strongly influenced the cosmological evolution in the past. We found that the main parameters
of the large scale structure of the Universe depend sensitively
on the total amount of the cosmic light particles. We also discuss
the plausibility for the non-dominating dark matter components
to be highly perturbed nowadays.

Recent developmentsof the fundamental physics and observational astronomy evidence for the non-baryonic nature of the dark
matter in the Universe. Cosmological models most elaborated by
now assume the dynamically dominating dark matter component in
the form of collisionless particles which gravitationally manifest themselves now as a pressureless medium. In such an approach
a question arises on the other components of dark matter which
are now non-dominating and cannot be disclosed but played a crucial role in the early history of the Universe thus having left
prints in the cosmic large structures - primordial cosmological
perturbation spectrum , cosmic void and supercluster scales, angular anisotropy of the relic radiation etc. Among such "concealed
parameters are the two very important for cosmology:

M. Mezzetti et al. (eds.), Large Scale Structure and Motions in the Universe, 139–158.
© *1989 by Kluwer Academic Publishers.*

(•) a total amount of light non-interacting particles which
   were relativistic at the time of equality of matter densities
   of the non-relativistic and relativistic components, and
(••) an amount of baryons in cosmic voids and islands as compared
   to the mean baryon density.

Below we briefly discuss physical processes which are substantially influenced by the missing mass parameters mentioned above,
for the Universe with a "cold" variant of the leading dark matter component.

I. A new test for the number of species of light
particles in the Universe

First, let us show how light weakly interacting particles
change the spectrum of the primordial cosmological perturbations
in the cold dark matter Universe with critical density.

Besides usual baryons and photons (γ-component) we consider
for simplicity two sorts of collisionless particles: cold relic
component with density $\tilde{\delta}_r$ and zero pressure which maintain
present critical density of the Universe, and light relativistic particles (ν- component) described by the Boltzman-Vlasov
equation. This model is fully specified by a single parameter

$$y = \mathcal{E}_\nu / (\mathcal{E}_\nu + \mathcal{E}_\gamma) = const \in (0, 1), \quad (1)$$

a ratio of the total number of ν – particles to the number of
all relativistic particles. At the very early times ($t \to 0$)
when relativistic components predominate cosmological perturbations are given by two gauge invariant functions of space coordinates:

$$ds^2 = dt^2 - (aR)^2 (1 + q(x)) \, d\vec{x}^2 \tag{2}$$

$$\delta_r \equiv \delta\rho_r / \rho_r = \delta(x), \quad u_r^i = (1,0,0,0)$$

where $q(\vec{x})$ determines adiabatic perturbations and $\delta(\vec{x})$ iso-thermal (isocurvature) perturbations [*], ρ_r and u_r^i is the r-component mass density and 4-velocity, R is a normalization constant, and "a" is a scale factor.

Let functions be small and Gaussian distributed

$$<q(\vec{x}) \, q(\vec{x}')> = (1/8\pi^3) \int_{-\infty}^{\infty} d^3\vec{k} \, f(kr) \, q_k^2, \quad <q(\vec{x}) \, \delta(\vec{x}')> = 0$$

$$<\delta(\vec{x}) \, \delta(\vec{x}')> = (1/8\pi^3) \int_{-\infty}^{\infty} d^3\vec{k} \, f(kr) \, \delta_k^2 \tag{3}$$

where q_k, δ_k are spectral functions, $<...>$ is the average over the field state, $f(kr) = \sin kr / kr$, $\vec{k}$ a wave vector, $k = |\vec{k}|$, $r = |\vec{x} - \vec{x}'|$. At the late times r-particles dominate the expansion and there density contrast grows as

$$<\delta_r(t,\vec{x}) \delta_r(t, \vec{x}')> = (10^{-2} \tau^4 / 32\pi^3) \int_{-\infty}^{\infty} d^3\vec{k} \, k^4 \, f(kr) \, (q_k^2 \, C_q^2(k) +$$

$$+ \, \delta_k^2 \, C_\delta^2(k)) \tag{4}$$

where $\tau \equiv R^{-1} \int dt/a \sim t^{1/3}$ is the auxiliary time, $C_q(k)$ and $C_\delta(k)$ are the transfer functions of the adiabatic and isocurvature perturbations, respectively. The astrophysical meanings of the

---

[*] Baryonic perturbations are considered separately in the next. Section, Eq. (2) takes into account only "growing" modes of perturbations, since only they are responsible for galaxy formation

variables in eqs. (3,4) are:

$$k = 360 \ h^{-2}(1-\nu)^{-\frac{1}{2}}/\lambda \quad , \quad kr = 2\pi \ 1/\lambda \ ,$$

$$\tau^2 = 10^4 \ h^2(1-\nu)(1+z)^{-1}$$

(5)

where $\lambda = 2\pi \ a_0 R/k$ and $1 = a_0 Rr$ are the present $(t = t_0)$

perturbation wave length and the distance between two spatial points (galaxies) in Mpc, z is a redshift, h is the Hubble constant in the units 100 km/s/Mpc and the background temperature $T_\gamma = 2.7$ K.

Figures demonstrate the transfer functions $C_{q,\delta}$ (k) for the two limit values of parameter $\nu = 0,1$ [1,2]. (For convenience, the calculations are briefly sketched in the Appendix). The rest of the transfer functions with the intermediate values of $\nu \in$ (0,1) are between the limit curves. Each of the functions monotonically dies with k increasing and has the following asymptotics:

$k \ll 10$:
$$C_q(k) = 1 + O(k^2 \ln k)$$
$$C_\delta(k) = 2/3 + O(k^2 \ln k) \quad ,$$

(6a)

$k \gg 100$:
$$C_q(k) = 360\alpha k^{-2} \ln (k/k_0)$$
$$C_\delta(k) = 120 \ k^{-2}$$

(6b)

So, in the short wavelength asymptotic the adiabatic transfer function depends on the ν - parameter and the isocurvature transfer function does not depend on ν at all. Coefficients $\alpha = \alpha(\nu)$ and $k_0 = k_0(\nu)$ vary slowly with ν taking the limit values $\alpha = 1$, $k_0 = 32$ for $\nu = 0$ [3] and $\alpha = 4/$

$k_0 = 55$ for $\nu = 1$ [4] . For the very small scales
($k \gg 500$) the difference between $C_q(k)$ for $\nu = 0$ and $\nu =$
reaches ~ 30 %, but physically interesting are the larger scales
($k < 500$, $\lambda > 1$ Mpc).

We have found a puzzling property of the transfer functions
[1,2,5] : $C_{q,\delta}(k)$ are practically independent of the ν -
parameter in the region of $k \in (0,500)$ for the adiabatic perturbations and in the whole region of $k \in (0, \infty)$ for the isocurvature perturbations. As it is seen from the Figures, the maximum
discrepancy between the limit curves is obtained at $k \sim 10$ and
is about 8% for $C_q(k \sim 10)$ and ~ 15 % for $C_\delta (k \sim 10)$. The
smaller deviation for the adiabatic curves C_q is explained by
the following fact: the adiabatic functions overlap at $k \sim 100$
(and then start to diverge again reaching discrepancy $\sim 8\%$ at
$k \sim 500$ and further on ~ 30 % for $k \rightarrow \infty$), while the isocurvature functions overlap only at infinity $k \longrightarrow \infty$.

Such a feeble dependence of $C_{q,\delta}(k)$ on ν in no way means
that the light particles do not affect the spectrum of the cosmological perturbations. This "scaling" effect (independence of
the spectrum shape from ν) was revealed due to a lucky choice
of the k - parameter (see eq. (5)). In the physical space all
the structural scales strongly increase with ν increasing, i.e., whil
practically invariant in the shape, the transfer function as a
function of the physical scale Mpc^{-1} shrinks or stretches uniformly when the relativistic particles increase or decrease in
number , respectively.

The transfer function scaling allows for several effective
observational predictions about large structures in the Universe

to be done without any special calculations. A comparison with the observational data serves as a new cosmological test of the total number of species of the light weakly interacting particles in the Universe (including gravitons, hypothetic "...inos" and other exotic cosmions). By light particles we mean any particles which were relativistic near the equality epoch $t_{eq} \simeq 10^{10} h^{-4}$ $(1-\nu)^{-3/2}$ s , i.e. with the restmass $m \ll 10$ eV .

First test comes from the large scale angular anisotropy of the microwave background. General dependence $\triangle T/T \sim (1-\nu)^{-1}$ and a quadrupole amplitude $\triangle T/T$ $(1 = 2) \simeq 4 \cdot 10^{-6}$ [6-8] for the standard cold dark matter model allow for the upper limit of the ν - parameter [1] :

$$\nu < 0.8 \qquad\qquad (7)$$

Here we used the RELIC data [9] developed for the standard flat spectrum of cosmological perturbations $\triangle T/T$ $(1=2) <$ $1.6 \cdot 10^{-5}$. The number C.8 in eq. (7) is only twice as great as that for the standard model with the three sorts of massless neutrinos ($\nu = 0.4$). Remind, that similar inequalities (7) come from the primordial nucleosynthesis [10, 11], z - and w-bosons' decay [12] , supernova 1987A [13] etc. However, unlike the other tests, our cosmological test of light particles gives the information not only about particles which were in the thermal equilibrium at high temperatures, but it also counts graviton and other non-equilibrium particles whose revealing by any other means seems to be improbable now.

Another test deals with a characteristic scale of large structures in the Universe with light particles (the spectrumcutof scale of the multipole $\triangle T/T(1)$ - harmonics, supercluster, cosmic void scales etc.). It approximately corresponds to the scale of the transfer functions $C_{q,\delta}(k)$ when their shape changes, $k \sim k_{eq} = 10$ (see Figures), where $C_{q,\delta}(k_{eq}) \simeq 0.5$, $\lambda_{eq} \simeq 36h^{-2}(1-\nu)^{-\frac{1}{2}}$ Mpc. This scale of the large wavelength spectrum cutoff grows substantially when relativistic particles increase in number ($\lambda_{eq} \longrightarrow \infty$ with $\nu \longrightarrow 1$). Note in conclusion, that a relation of the supercluster scale to λ_{eq} depends on the primordial perturbation spectrum: they are closely related in the models like the standard neutrino model or unstable dark matter models where superclusters condense primarily (before smaller objects, galaxies and groups). In the cold dark matter models, super clusters originate in result of the hierarchial merging of less massive clusters, and the determination of the final scales needs additional analysis

II. How many baryons are in cosmic voids and islands?

Here we consider a possibility for large perturbations in the non-dominating dark matter components.

Amongst different schemes for biased galaxy formation proposed recently there are few that speculate on the idea of large baryonic perturbations: from hypothesis about the deficit of baryons in large cosmic voids embracing scales 30-200 Mpc (cosmic bubbles) [14], and up to the assumption than all the visible matter in the Universe was born in a gigantic barionic island extending to the redshifts $z \sim 4$ and barions vanish beyond the cosmic islands [15,16]. These and similar suggestions originate from a more general assertion that nondominating mediums do not manifest themselves dynamically now. It means that even if they were highly perturbed now (but the dominating dark matter remains homogeneous) they would not contribute to the gravitational potential, i.e. to the "growing" adiabatic mode of perturbations.

However, this idea misses one point: since these large perturbations were created very early (e.g., baryons were produced at GUT's times, $t \sim 10^{-35}$ s), they did inflience the expansion dynamics during the equality epoch and as a result, they induced the "growing" adiabatic mode with a high amplitude which, in turn, caused later large background $\triangle T/T$ - fluctuations due to the Sachs-Wolfe effect and as a consequence, - contradictions with the observations [*] .

Below, we briefly outline the proof [1] .

Let us consider an early multicomponent Universe consisting of the relativistic components (which include γ- and ν -particles, see Sect. 1) and of the non-relativistic "dust" r-particles. The latter include "cold" dark matter (CDM) and "baryons" (b) whose portion is fixed by the parameter

$$\Omega_b \equiv \Omega_b(\vec{x}) = \rho_b/\rho_r \in (0,1) \tag{8}$$

For simplicity, we neglect the baryonic pressure and therefore baryons and cold particles move together[**]. Thus, Ω_b is the integral of motion, i.e. it depends only on space coordinates in the synchronous comoving frame.

---

[*] In principle, a special geometry (e.g., high degre spherical symmetry of the cosmic void or island, a certain position of the observer e.t.c.) could conceal these temperature fluctuations from the observer, but here we consider arbitrary density configuration (for discussion see ref.[16]).

[**] This simplification allows one to generalize the problem: instead of "baryons" one may consider any heavy particles, e.g. the CDM-particles which inverse the problem. (Note that Ω_b takes any values from zero to unity).

Now, we consider early evolution when the cosmological horizon was much less than a characteristic scale of variation of the Ω_b function:

$$t \ll 1 \cong |\Omega_b / \nabla_i \Omega_b| \qquad (9)$$

Let all the components be initially at rest $(t \to 0)$. Then under condition (9) the metric is locally isotropic

$$ds^2 = dt^2 - (aR)^2(dx^2 + dy^2 + dz^2) \qquad (10)$$

where $a = a(t, \vec{x})$, $R = \text{const} \sim t_{eq}$ [*) ; and the only non-trivial equations are

$$\mathcal{E} = 3C_1^2 / R^2 a^4, \quad \rho_r = 12C_1 C / R^2 a^3$$
$$\dot{a}^2 = C_1^2 + 4aC_1 C , \quad (\cdot) = aR\partial/\partial t \qquad (11)$$

where $C_{(1)} \cong C_{(1)}(\vec{x})$ - space functions fixed by initial conditions. The solution for the scale factor is the following (cf. eq.(A.1))

$$a = C_1 \mathcal{T}(1 + C\mathcal{T})$$
$$t = R\int a \, d\mathcal{T} = \tfrac{1}{2} RC_1 \mathcal{T}^2(1 + \tfrac{2}{3} C\mathcal{T}) \qquad (12)$$

Let now set the baryon excess to be produced extremely perturbed initially while the CDM and the dominating relativistic component were spatially homogeneous (for physical mechanisms se ref.[17]). This means the following choice of $C_{(1)}$-functions (see (8)):

$$C_1 = 1, \quad C = 1/(1 - \Omega_b) \qquad (13)$$

---

*) a-function and R-parameter are gauge - invariant since the reference system (10) is fixed unambiguously by the comdition $u_r^i = (1,0,0,0)$.

So that the equality time

$$a_{eq} = 0.25(1 - \Omega_b), \quad \tau_{eq} = 0.2(1 - \Omega_b) \tag{14}$$

For further estimates we shall substitute $\delta\Omega_b \simeq \Omega_b \approx 0.1$ for the baryon contrast on scales ~ 1.

At $\tau \ll \tau_{eq}$ when relativistic particles predominate, we have

$$a = (2t/R)^{\frac{1}{2}}, \quad \varepsilon = 3/4 \, t^2,$$

$$\rho_r = \frac{6}{1 - \Omega_b} \, (2R)^{-\frac{1}{2}} \, t^{-3/2} \ll \varepsilon \tag{15}$$

in accordance with the initial conditions. At $\tau \gg \tau_{eq}$ eqs. (11,12) yield:

$$a = (1 - \Omega_b)^{-1/3}(3t/R)^{2/3}, \quad \rho_r = 4/3t^2$$

$$\rho_b = 4\Omega_b /3t^2, \quad \varepsilon \ll \rho_r \tag{16}$$

which, in fact, is a sum of two perturbation modes. The upper line of eqs. (16) presents the first expansion term over the parameter $(t/1)^2 \ll 1$ of the "growing" adiabatic mode (= the quasi-isotropic solution [18]). The second line in (16) describes the isocurvature perturbations in the baryons and in the non-dominating relativistic component.

Eqs. (15,16) display an important conclusion: gravitational field perturbations (= scale factor perturbations here), being initially vanished, arise at the non-relativistic matter dominating era:

$$\delta a/a \simeq \frac{1}{3} \delta\Omega_b \simeq 0.03.$$

These large perturbations of metric bring about cosmic microwave background fluctuations $\Delta T/T \sim 10^{-3} - 10^{-2}$ at angular scales $\sim 10°-90°$ which depend on the value of the variation scale of baryon density contrast 1 .

Conclusion

1. We proposed a new test of the total number of weakly interacting particles with small restmasses (m≪10 eV) in the Universe.
2. We showed the impossibility of baryon deficit in cosmic voids.

Appendix

In the Early Universe interacting particles may be described in terms of a thermal bath (ideal fluid) with the equation of state $p = \mathcal{E}/3$. Thus we deal with the 3-component model and the background quantities can be chosen as ($c = \hbar = 8\pi G = 1$):

$$\mathcal{E}_{\gamma} = 3\frac{1-\nu}{R^2 a^4} \quad , \quad \mathcal{E}_{\nu} = 3\nu/R^2 a^4 \quad , \quad \rho_r = 12/R^2 a^3,$$

(A.1)

$$a = \tau(1 + \tau), \quad t = \tfrac{1}{2}R\tau^2(1 + \tfrac{2}{3}\tau)$$

In our normalization $a_{eq} = 1/4$ at the equality time ($\mathcal{E}_{\gamma} + \mathcal{E}_{\nu} = \rho_r$ at $\tau = \tau_{eq} = (\sqrt{2} - 1)/2 = 0.2$), and $R = 45t_{eq} = 5.4(1 - \nu)^{-3/2} h^{-4}$ Kpc. The up-to-date scale factor and the auxiliary time are given as

$$a_o = 10^4(1 - \nu)h^2, \quad \tau_o = 10^2 h (1 - \nu)^{1/2} \gg 1 \qquad (A.2)$$

For instance, if ν-component includes only three sorts of massless neutrinos then $\nu = 0.4$ and $\tau_o = 79 h$.

Scalar perturbations of the γ- and ν-components are determined by the velocity potential $u^i_{\gamma} = (1; - \mathcal{V}_{,\alpha}/2a)$ where $\mathcal{V} = \mathcal{V}(\tau, \vec{x})$, and by the function $F = F(\tau, \vec{x}, \vec{e})$ related to the integral [19] of the distribution function $f = f(\tau, \vec{x}, \vec{e}, q)$ of ν-particles over the particle momentum modulus $q = |\vec{q}|$ where $\vec{e} = \vec{q}/q$. This transformation from f to the function F is reasonable only for the relativistic ν-particles, otherwise there appear effects dependent on the distribution of the particles over their momenta q. Note, that functions $\mathcal{V}$ and F as well as $A = A(\tau, \vec{x})$ and $B = B(\tau, \vec{x})$ (see eq.(A.3)) are gauge-invariant since we specify them in the

co-moving with the r-component synchronous reference system which
is unique (cf. eq. (2)):

$$ds^2 = dt^2 - (aR)^2((1+A)\widetilde{\delta}_{\alpha\beta} + B,_{\alpha\beta})dx^{\alpha}dx^{\beta} \tag{A.3}$$

When $\tau \longrightarrow 0$, $A \longrightarrow q(\vec{x})$ and $B,_{\alpha\beta}/A \longrightarrow 0$. For the Fourier
harmonics $(\sim e^{i\vec{k}\vec{x}})$ the equation system is as follows (index
$\vec{k}$ for the Fourier components is omitted):

$$\dot{F} - ik_{\mu}F + \dot{A} - (k_{\mu})^2\dot{B} = 0 \tag{A.4a}$$

$$\ddot{\upsilon} + \frac{1}{3}k^2(\upsilon - \dot{B}) + \dot{A} = 0 \tag{A.4b}$$

$$\dot{A} + a^{-2}(2(1-\gamma)\upsilon - 3i\gamma k^{-1}\int_{-1}^{+1}d\mu\,\mu\,F) = 0 \tag{A.4c}$$

$$\ddot{B} + 2\frac{\dot{a}}{a}\dot{B} - A + 3\gamma(ka)^{-2}\int_{-1}^{+1}d\mu\,(3\mu^2 - 1)F = 0 \tag{A.4d}$$

where dot demotes derivative over τ and $\mu = \vec{k}\,\vec{e}/k$. The density contrast of r-particles can be easily derived if eqs. (A.4)
are solved:

$$\widetilde{\delta}_r \equiv \widetilde{\delta}_{\rho_r}/\rho_r = \tfrac{1}{2}(k^2B - 3A) + \widetilde{\delta}_{\vec{k}} \tag{A.4e}$$

where $\widetilde{\delta}_{\vec{k}} = $ const is the Fourier-image of function $\widetilde{\delta}(\vec{x})$ (see
eqs. (2,3), $<\widetilde{\delta}_{\vec{k}}\;\widetilde{\delta}_{\vec{k}'}> = \widetilde{\delta}^2\,\delta(\vec{k} - \vec{k}')$). Eqs. (A.4) have a
first integral for the total perturbation energy:

$$4\widetilde{\delta}_r + a^{-1}((1-\gamma)\widetilde{\delta}_{\mu} + \gamma\widetilde{\delta}_{\gamma}) = \dot{a}(\dot{A} - \frac{1}{3}k^2\dot{B}) + \frac{1}{3}k^2aA \tag{A.4f}$$

where

$$\widetilde{\delta}_{\mu} \equiv \delta\varepsilon_{\mu}/\varepsilon_{\mu} = 2\dot{\upsilon}\;, \quad \widetilde{\delta}_{\gamma} \equiv \delta\varepsilon_{\gamma}/\varepsilon_{\gamma} = \int_{-1}^{+1}d\mu\,F$$

Transfer functions (4) are found by the direct integration of
eqs. (A.4) for $\tau \geqslant 0$ with the initial conditions (2):

$$A = 1 , \quad B = \dot{B} = F = \mathcal{V} = \dot{\mathcal{V}} = \delta_{\underset{k}{\to}} = 0 \quad (\dot{A} = 0) \qquad (A.5a)$$

for the adiabatic mode (C_q) ; and

$$\delta_{\underset{k}{\sim}} = 1, \quad A = B = \dot{B} = F = \mathcal{V} = \dot{\mathcal{V}} = 0 \quad (\dot{A} = 4/3) \qquad (A.5b)$$

for the isocurvature mode (C_δ). (Conditions in parentheses are necessary for the formal definition of the (A.4) terms at $\mathcal{C} = 0$) Then, at large $\mathcal{C} \gg 1$ (just before the beginning of the nonlinear evolution of perturbations):

$$A = C\left(1 + \frac{1}{5a}\right) , \quad B = \frac{a}{10} C , \quad \mathcal{V} = \frac{\dot{a}}{10} C,$$

$$\delta_r = \frac{ak^2}{20} C , \quad \delta_\mu = \delta_y = \frac{2}{5} C$$

$$F = ik_\mu \mathcal{V} + \frac{C}{5} + F_k \exp(ik_\mu \mathcal{C}) \qquad (A.6)$$

where $C = C_q(k)$ or $C = C_\delta(k)$ depending on the initial conditions (A.5). Note, that the time independent functions $C = C(k)$ and $F_k = F(k,\mu$ are the functionals of the (A.4) solutions at $\mathcal{C} \geqslant 0$. F_k determines the decaying part of the density contrast in y -component $(\delta_y = \frac{2}{5} C + 0(\mathcal{C}^{-1}))$.

The results of the analytical and numerical analysis can be approximated by the following simple formulae [2].

For the adiabatic perturbations

$$C_q(k) \approx \left(1 + \frac{k^2}{120(\mathcal{V} + 3)\ln(k/k_0 + 1.08)}\right)^{-1} , \quad k_0 = 32 + 23\mathcal{V}. \qquad (A.7a$$

For the isocurvature perturbations

$$C_\delta(k) = \frac{120}{180 + (27 + 10\mathcal{V})k + k^2} \qquad (A.7b$$

Eqs. (7) describe real asymptotics for $k \ll 1$ and $k > 100$, and give good approximations of the true functions up to an accuracy $\lesssim 3\%$ for $0.1 < k < 20$, and less than 1% for $k > 20$.

References

1. V.N.Lukash. Proc. IAU Symp. 130 "Evolution of Large Scale Structures in the Universe", eds. J.Audouze & A.Szalay, Hungary, 1987

2. T.A.Kahniashvili, V.N.Lukash, K.A.Manukyan, G.V.Meladze, Preprint Space Res. Inst. Пp-1297, 1987

3. A.A.Starobinski, V.Sahni, in: Abstracts of contr. papers Soviet GRG Conf. 6, Moscow,79, 1984

4. T.A.Kahniashvili, V.N.Lukash, Proc. Plasma Astrophysics Workshop, ed. T.D.Guyenne (ESTEC), 41, 1986.

5. T.A.Kahniashvili, K.A.Manukyan, Preprint Space Res. Inst. Пp-1340, 1987

6. P.J.E.Peebles, Ap. J. Lett. $\underline{263}$, L1, 1982

7. J.R.Bond, G.Efstathiou, Ap. J. $\underline{285}$, L45, 1984

8. V.N.Lukash, P.D.Naselskij, I.D.Novikov, Proc. Quantum Gravity -3, eds. M.A.Markov et al. (World Scientific), 675, 1984

9. I.A.Strukov, D.P.Skulachev, A.A.Klypin, Proc. IAU Symp. 130 "Evolution of Large Scale Structures in the Universe", eds. J.Audouze & A.S.Szalay, Hungary, 1987.

10. V.F.Shvartzman, JETP Lett. $\underline{9}$, 315, 1969

11. D.N.Schramm, R.V.Wagoner, Physics Today $\underline{27}$, 41, 1974

12. D.B.Cline, D.N.Schramm, G.Steigman, FERMILAB-Pub-87/28-A, January, 1987

13. D.N.Schramm, FERMILAB-Conf. 87/166-A, October, 1987

14. L.A.Kofman, E.Saar,J.Einasto, Nature, 1987

15. N.S.Kardashev, H.J.Blome, W.Priester, Preprint Space Res. Inst Пp-1259, Moscow 1987

16. A.D.Dolgov, A.F.Illarionov, N.S.Kardashev, I.D.Novikov, Preprint Space Res. Inst. Пp-1291, 1987

17. A.Dolgov, Proc. Quantum Gravity-4, eds. M.A.Markov et al. (World Scientidic) 1987

18. E.M.Lifshitz, I.M.Khalatnikov, Usp. Fiz. Nauk $\underline{80}$, 391, 1963

19. A.V.Zakharov, JETP $\underline{77}$, 435, 1979

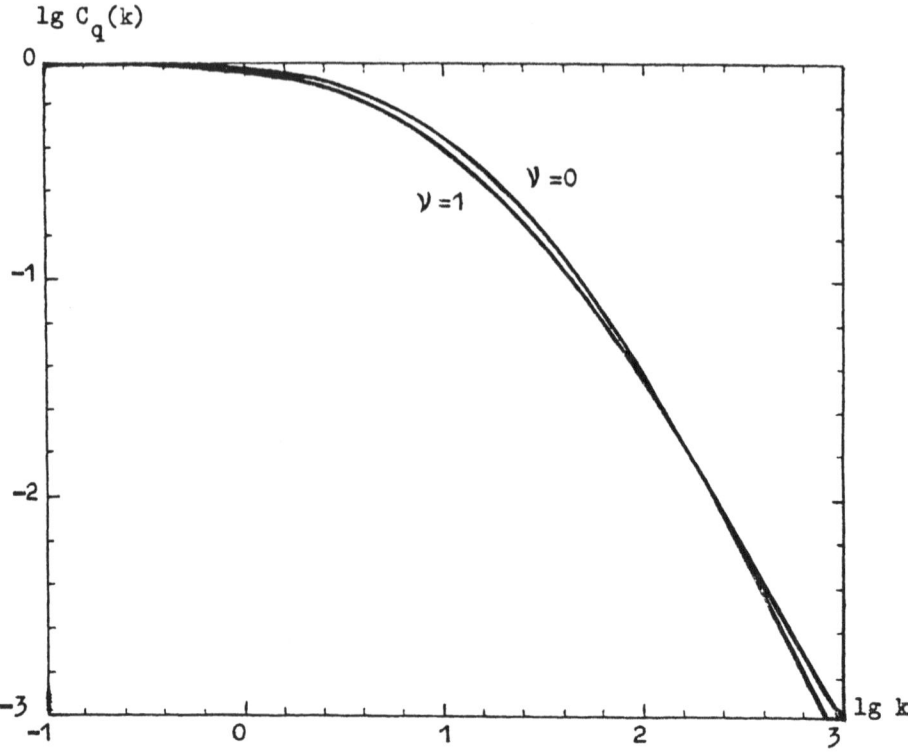

Fig. 1. Transfer function $C_q(k)$ of the adiabatic cosmological
perturbations in the cold dark matter Universe for $y = 0$
and $y = 1$

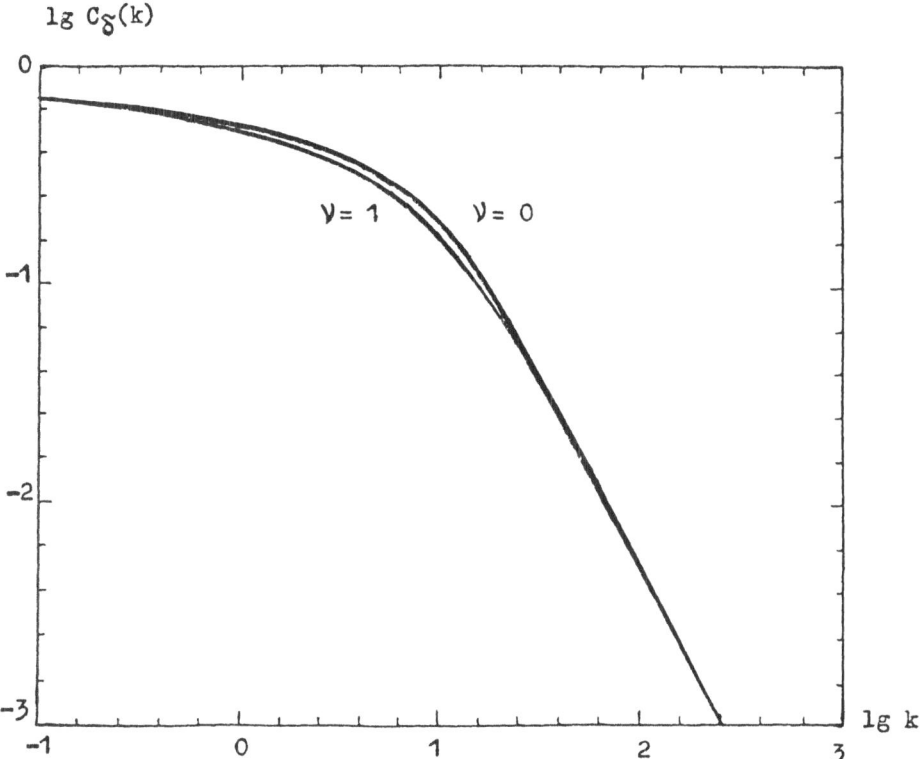

Fig. 2. Transfer function $C_\delta(k)$ of the isocurvature cosmological perturbations in the cold dark matter Universe for
$\nu = 0$ and $\nu = 1$

DISCUSSION

SCIAMA: Does your result on the absence of large-scale Baryon fluctuations contradict Tully proposal?

LUKASH: It depends on numbers. Tully's observations deal with the luminous matter (Abell clusters). The question of contradictions with $\Delta T/T$ is the question about the real total mass density contrast on such large scales. I think it cannot be large, otherwise we would observe high peculiar velocities and high $\Delta T/T$ on scales $\theta \gtrsim 1°$.

PEEBLES: I think your demonstration of the impossibility of large primordial baryon inhomogeneity depends on the assumption that the primordial radiation distribution is homogeneus. In the isocurvature picture, where fluctuations in the mass densities of baryons and radiation cancel, $\delta\rho_r(\bar{x}) = -\delta\rho_b(\bar{x})$ large primordial inhomogeneties in the baryon distribution can exist without violating any observations I am familiar with.

LUKASH: I think our definitions coincide: isocurvature initial conditions mean the absence of adiabatic "growing" mode, in the beginning, i.e. the total density is imperturbed when speaking about homogeneus distribution of radiation. I just had in mind that eq. $\delta\rho_r(\bar{x}) = -\delta\rho_b(\bar{x})$ means $(|\delta\rho_r/\rho_r|) << (|\delta\rho_b/\rho_b|)$ since $\rho_r >> \rho_b$ for t->0.

THE MODEL OF BARYONIC ISLANDS OF LARGE STRUCTURES

I.D. Novikov
Space Research Institute
Academy of Sciences of the USSR

ABSTRACT

This paper discusses the astronomical consequences of the hypothesis of the baryonic islands in the universe [1].

1. INTRODUCTION

The purpose of this paper is to discuss the possibility of the existence of the baryonic islands in the Universe [1]. A red-shift dependence of any large scale property of the universe is usually ascribed to a selectional effect of the observations or to the evolutionary effect. The well known example of this kind is the absence of quasars at large Z [2]. The latest astronomical data however give evidence in favour of real absence of a visible matter for spatial distances exceeding $Z \cong 5$. Beside no angular small-scale fluctuations in the temperature of microwave background were found despite the extensive searches [3,4]. It is difficult to explain this circumstance in the scope of the standard cosmological model [5], where anisotropy $\Delta T/T$ is determined by the matter density fluctuations at large Z. The following consideration shows that the absence of $\Delta T/T$ can be explained in the scope of a model, where the usual matter at large distances is absent. Of course, the interpretation of the observational data is by no means unambiguous and more accurate observations are needed. Nevertheless, if this trend does not disappear and the accuracy of the observations is improved, one will be pretty sure that the

M. Mezzetti et al. (eds.), Large Scale Structure and Motions in the Universe, 159–168.

density of hydrogen and heavier nuclei at large distances is considerably lower than that in the nearer part of the universe. In other words it is suggestive that we live in a baryonic island [6] submerged into an invisible matter.

Such baryonic (and antibaryonic) islands can be formed in a theory with spontaneous C (and CP) violation [1,7]. The phase transition to the CP-odd phase should take place during the inflationary epoch not long before it ends. This provides sufficiently large island size and possibly with the border inside the present-day horizon.

Because of the physical processes, which are considered in detail in [1], in the course of this phase transition there were formed bubbles with a small excess of baryons over antibaryons or vice versa, outside the bubbles the universe being absolutely charge symmetric. When the universe expanded and cooled down baryons and antibaryons in the charge symmetric region almost completely annihilated. The amount of baryons and antibaryons which survived the annihilation is very small, $N_B = N_{\bar{B}} \sim 3.10^{-19}.N_\gamma$ [8] where N_B, $N_{\bar{B}}$ and N_γ are the number densities of baryons, antibaryons and photons respectively. On the other hand the baryonic (and corresponding electronic) excess inside the bubbles does not dissapear and, as observations show, is equal to $N_B = 3.10^{-10}.N_\gamma$. Thus separate baryonic (and antibaryonic) islands are formed in the background of a charge symmetric matter. The island sizes and the distance between them are very much model dependent. In a favorable case the bank of our island can be inside the present-day horizon.

If the proposed model is indeed true then the relic background radiation observed today comes from the regions of the universe where there are no baryons, and so the radiation temperature fluctuations is not connected with the initial baryonic density perturbations in contrast to the standard model of the large scale structure formation in the Universe. Due to lack of baryons in the interisland space the radiation decoupled from the matter not at $Z \simeq 10^3$ as in the standard model but at $Z \simeq 10^7$. To this moment the frozen electron and positron number density in the interisland space being equal to $Ne+=Ne- \sim 10^{-16}.N_\gamma$, the optic depth of e+, e- for the Thompson scattering gradually (at $Z<10^7$) became small($\tau_T < 1$). See [9] for detailed discussion of these processes.

Noncentral position of an observer in our island would result in a large scale anisotropy of the microwave background. We will show however that this anisotropy is small for a considerable range of values of the model parameters.

The best way to test the model is to investigate the space distribution of astronomical objects near the island border and possibly to find other island if we were lucky enough. The latter is very much improbable however.

2. THE EVOLUTION OF THE BARYONIC ISLAND

Let us consider the dynamical evolution of matter inside and outside the baryonic island. The value $\Omega = \rho/\rho \text{crit}$ is assumed to be unity. For symplicity let us also assume that the island is of globular form and that the baryonic density is uniform inside the island. The island size today is comparable to the distance to the horizon. In the past the island was much larger than the horizon. At the beginning in the first approximation we neglect all the boundary effects we shall see afterwards.

The matter expansion inside and outside the island became different after baryon-antibaryon annihilation when extra baryons inside the island, which survived the annihilation, became nonrelativistic.

Let us denote this moment t_o. By the order of magnitude $t \cong 10^{-6}$ sec and the corresponding temperature is about baryon mass m_B. Total energy density of all forms of matter inside as well as outside the island at this moment is [8]:

$$(2.10) \qquad \rho_o = 3/32 \pi G t_o^2$$

After this moment there are three different forms of matter in the universe: baryons (with about the same number of electrons), invisible matter of light (but massive) weakly interacting particles, conventionally called neutrinos, and photons. For numerical estimates we will use $m_\nu = 10$ eV. The energy distribution between these species is the following:

$$(2.2) \qquad \rho_b = \alpha \rho_o, \quad \rho_\nu = A \rho_o, \quad \rho_\gamma = B \rho_o$$

$$A + B + \alpha = 1$$

Inside the island $\alpha \cong N_B/2N_\gamma \cong 2.10^{-10}$ and is negligible ($\alpha = 0$) outside; $A \cong B \cong 0.5$ everywhere. The neutrino component is relativistic till moment t, when the temperature is about

m_ν, t_1 is approximately equal to $t_1 \cong (m_b/m_\nu)^2 t_0 \cong 10^{10}$ sec (eq. (2.4) for a more accurate value). The solution of the Einstein equations for the mixture of pressurless baryons ($p_b = 0$) and the relativistic component with $p = \rho/3$ gives

$$(2.3) \qquad a = (t/t_0)^{1/2}[1 + \alpha/6 \cdot (t/t_0)^{1/2}]$$

where t is the proper time of the comoving observer and a is the scale factor. It is assumed that $(\alpha/6)(t/t_0)^{1/2} \ll 1$. Expression (2.3) is valid in the time interval $t_0 < t < t_1$ both inside the island where $\alpha = 2.10^{-10}$ and outside it where $\alpha = 0$. The rate of variaton of the scale factor is somewhat different inside and outside the island so temperature $T_1 \sim m_\nu$ is reached at different moments

$$(2.4) \qquad t_1 = t_0 (m_b/m_\nu)^2 [1 - (\alpha m_b/3 m_\nu)]$$

After this time the neutrino component became nonrelativistic ($p_\nu = 0$) and the gravitational influence of the photons on the further evolution can be neglected. The scale factor for $t > t_1$ can be written as follows

$$(2.5) \qquad a = [6\tilde{\Pi}GA\rho_0(m_\nu/m_b)]^{1/3}(1 + \alpha m_b/3Am_\nu) \cdot$$
$$[t - t_0(m_b/m_\nu)^2(1 - 4/3\sqrt{A} - (\alpha m_b/3m_\nu)(1 - 2A^{-3/2}))]^{2/3}$$

As above $\alpha \cong 2.10^{-10}$ inside the island and $\alpha = 0$ outside. Nonrelativistic matter with energy density ρ_* inside the island consists of baryons and neutrinos and outside the island it consists only of neutrinos. For $t \gg t_1$ difference of energy densities and the corresponding difference of the Hubble parameter inside and outside the island are respectively

$$(2.6) \qquad \Delta\rho_*/\langle\rho_*\rangle \cong 2/3(m_b/m_\nu)^3\alpha(t_0/t)(2A^{-3/2} - 1)$$

$$(2.7) \qquad \Delta H/\langle H\rangle = (\Delta\rho_*/\langle\rho_*\rangle)/2.$$

As above the boundary zone (where the evolution differs from (2.6), (2.7)) is assumed to be small. The difference of the background radiation energy densities is

$$(2.8) \qquad \Delta\rho_\gamma/\langle\rho_\gamma\rangle \cong -4\alpha m_b/3Am_\nu \cong -0.04.$$

This expression is valid when $t \gg t_1$ and when the island size is still much larger than the horizon.

Let us make some remarks about the boundary effects. During the period $t_0 < t < t_1$ the island border expands faster than the surrounding matter (see eq. (2.3)). The relative velocity of the border and of the surrounding matter is of the order of that of light. This results in an excess of matter on the boundary. Neutrinos can freely stream through the matter but baryons and photons are subjected to the hydrodynamical forces. The excess of mass on the island border is developed mainly at $t \simeq t_1$ when the equation of state changes. It can be demonstrated that excess of the mass $\Delta M/M$ is of the order ρ_b/ρ_ν, if we do not take into account nonlinear processes. These processes could diminish $\Delta M/M$. Thus the estimate:

(2.9) $\Delta M/M \simeq 10^2$

should be correct in order of magnitude.

In the narrow spherical shell the density excess is large, $\Delta \rho/\rho \simeq 1$ and the matter behaviour needs a special investigation. It's clear that the conditions for the large scale structure formation in the shell differ from those outside. Note that the picture we just described relies on the assumption of the sharp discontinuity in the baryon density at the island boundary, for a flat distribution of baryons the behavior would be different.

If initially at $t=t_0$ the matter density everywhere is equal to the critical one then the excess of mass inside the baryonic island will turn the expansion into contraction.

3.   OBSERVATIONAL TESTS

Possible tests include an observation of the island border, determination of our position relative to its center and attempts to discover other island.

We start from the consideration of the different effects which give rise to the angular anisotropy of the relic radiation. Let us assume that the observer is not situated at the island centre and the distance to the farthest side of the island corresponds to the redshift Z_* whereas the distance to the nearest side corresponds to Z_{**}. Let us start with the physical effects which are not connected with the matter excess on the boundary. The expansion rates inside and outside of the island are slightly different (see preceeding section). Thus when the light ray coming to the observer from the horizon in the direction to the farthest side of the island already reached the island the ray coming from the opposite direction was yet outside the island. So their reddening would be different. This would lead to a large scale anisotropy in the temperature of the relic background. It follows from eq. (2.5) that the temperature difference in these directions is

$$(3.1) \qquad \Delta T/T = -(2t_0\alpha/9t_{Now})(m_b/m_\nu)^3(2A^{-3/2}-1).$$

$$[(Z_*+1)^{3/2} -(Z_{**}+1)^{3/2}],$$

where t_{Now} is the present day moment. This expression is valid for $t_* \gg t_1 \cong 10^{10}$ s (i.e. $Z_* \ll 10^5$).

If just for the sake of the estimate we put $Z_* \cong 8$ and $Z_{**} \cong 3$ then $\Delta T/T \cong -3.10^{-8}$. The universe expansion together with the matter excess on the boundary leads to a much larger anysotropy $\Delta T/T$. If Z_* and Z_{**} are not much larger then unity and the Newton approximation is valid we obtain the following estimate:

$$(3.2) \qquad \Delta T/T = (1/2)(\Delta M/M)(Z_*-Z_{**})(2-1/\sqrt{Z_*+1}-1/\sqrt{Z_{**}+1})$$

$$1/200(Z_*-Z_{**})$$

It is noteworthy that the dominant spherical harmonic in $\Delta T/T$ is the dipole. Hence from the observed anisotropy it follows that $(Z_*-Z_{**})<0.3$. Note that the results presented are sensitive to the degree of the island nonsphericity and to the distribution of baryons inside the island.

Let us also estimate the anisotropy of the relic radiation due to scattering on the intergalactic plasma for a noncentral position of the observer. The optic depth corresponding to the scattering till the island border situated at the redshift Z is (see e.g. [8]):

$$(3.3) \qquad \tau = 0.03(H_0/75 \text{ km/s} \cdot \text{Mpc})\Omega_{pl}[(1+Z)^{3/2}-1],$$

where H_o is the Hubble constant and Ω_{pl} is the ratio of plasma density relative to the critical density; $\tau \ll 1$. Due scattering on the plasma, mainly near the island border, where the density at the moment of scattering was higher, the intensity of the radiation directly coming to the observer from outside the island is $(1-\tau)$ times weaker. On the other hand the part τ of the rays going primarily at the island outskirts in other directions changes it due to the scattering and comes to the observer. About one half of these rays comes from internal parts of the island which have a smaller temperature. Consequently the radiation temperature measured by the observer is approximately:

$$(3.4) \qquad T - (1-\tau)T_1 + \tau T_2/2 + \tau T_1/2 =$$

$$- (1-\tau/2)T_1 + T_2/2 \;,$$

where T_1 and T_2 are the radiation temperature outside and inside the island respectively. Hence the anisotropy born by the scattering is

$$(3.5) \qquad \Delta T_{scat}/T \cong (\tau_* - \tau_{**})/2 \; (\Delta T/T)_{max}.$$

Here τ_* and τ_{**} are the optic depths in the directions on the farthest and the nearest sides of the island respectively (It is assumed of course that $\tau < 1$); $(\Delta T/T)_{max} - 1/4(\Delta\rho_\gamma/\langle\rho_\gamma\rangle) \cong 10^{-2}$. Substituting expression (3.3) into (3.5) we get:

$$(3.6) \qquad \Delta T_{scat}/T \cong 10^{-4}\Omega_{pl}[(Z_*+1)^{3/2} - (Z_{**}+1)^{3/2}]$$

(for $H_o = 75$ km/sec.Mpc). So in the direction on the farthest side of the island a wide spot with a smaller temperature should be observed. The angular size of the spot is of the order of a steradian and the spherical harmonics (dipole, quadrupole, etc) should be correlated in the direction and have a comparable temperature. Their absence at the level $\Delta T/T \leq 10^{-4}$ leads (for $Z_* \gg Z_{**}$) to the constraint on the plasma density in the island:

$$(3.7) \qquad \Omega_{pl} < Z_*^{-3/2} \cong 1/20(Z_*/8)^{-3/2}.$$

Now let us turn to the small scale anisotropy of the background radiation which is connected with the density perturbation $\Delta\rho/\rho$. In our model, as already mentioned, there are at least two possibilities for the large scale.

The first one is completely analogous to the standard model. In this case there existed adiabatic density perturbations $\delta\rho/\rho$ at early stages everywhere inside and outside the island. The perturbations outside the island result in the temperature fluctuations, $\Delta T/T$, due to the Sachs-Wolfe effect [10]. Silk [11] and Sunyaev-Zeldovich [12] effects are negligible for every observable angular scale because the universe is transparent up to $Z\cong10^7$ instead of $Z\cong10^3$ as is in the standard model. The increase of the transparency period leads to the logarithmic rise of $\Delta T/T$ for small angular scale (of the order of tens of minutes) because of the Sachs-Wolfe effect, but due to the absence of any other sources of perturbation (which give $\Delta T/T$ of the comparable magnitude in the standard model) the total anisotropy is about the same as in the standard case.

Another possibility is that the fluctuations $\delta\rho/\rho$ outside the island are arbitrary small and the large scale structure inside the island is formed due to space variations of the baryonic charge density. In this case the small scale anisotropy of $\Delta T/T$ may be very small because the radiation comes from outside of the island where the fluctuations $\delta\rho/\rho$ are practically absent. The analysis [1.6] demonstrates that there is not any considerable distorsion of the Planck equilibrium spectrum of the relic radiation in the island model due to processes of electron-positron annihilation of NN pairs annihilation in the charge symmetric universe outside the island.

Probably evolutionary processes inside the homogeneous island do not permit to find anything about its size or about our position relative to the centre because an observer gets all the information simultaneously from different directions in the spherically symmetric part of the universe if only the sources of the radiation are not observed up to the boundary of the island. So the most interesting is a direct search of objects with the largest redshifts in different derections around the sky. The limiting values of Z would give the island size and the distribution of the farthest sources on the sky would permit to find the position of the observer relative to the centre. The most promising object are quasars, young galaxies, clusters of galaxies and clouds of interstellar gas.

One may hope that the search for quasars with the largest redshifts based on CCD detector in the near infrared region can answer the questions we pose here. If the island boundary indeed exists, a sharp cut off in the distribution of the cosmic bodies at large Z is expected. It is important for such searches to be made for several (4-6) areas on the sky with large angle separation between them.

The presented results show that the island universe model does not contradict the existing observational data with some restrictions on our position inside the island being imposed.

REFERENCES

1. Dolgov A.D., Illarionov A.F., Kardashev N.S., Space Research Institute preprint, IIp-1291, (1987).
2. Sortell W.H., Monthly Notices Roy. Astr. Soc. 213, 389 (1985); Crampton D., Cowley A.P., Hartwick F.D.A. Astrophys. J. 314, 129 (1987).
3. Korolkov D.V., Parijskii Yu.N., Communications of Special Astrophysical Observatory, 41,42,43 (1985).
4. Uson J.M., Wilkinson D.T., Astrophys. J. Lett. 277, L1,(1984).
5. Zabotin N.A., Naselskij P.D., Astr. Zh. 59, 447 (1982); 60, 467 (1983).
6. Kardashev N., Blome H., Priester W. (in press).
7. Dolgov A.D., Kardashev N.S., Space Research Institute preprint, IIp-1190, (1986); Nature (to be published).
8. Zeldovich Ya. B., Novikov I.D., Relativistic Astrophysics, v. 2, The Structure and Evolution of the Universe. Chicago: University of Chicago Press (1983).
9. Illarionov A.F., Space Research Institute preprint, IIp-127, (1987).
10. Sachs R.K., Wolfe A.M. Astrophys. J. 147, 73 (1967).
11. Silk J. Astrophys. J. 151, 459 (1968).
12. Zeldovich Ya. B., Sunyaev R.A., Astrophys. Space Sci., 6, 358 (1970).

DISCUSSION

PEEBLES: I think your baryon island model is already strongly constrained by the precise isotropy at the X-ray background.

NOVIKOV: In the simplest model the visible matter distribution is more or less uniform inside the island, there is not any radial gradient, the boundary of the island is sharp enough. Thus any observations of the sources or the background radiation from the matter inside the island (when we can not see the boundary) probably do not permit to find anything about the island's size ore about our position relative to the centre. Of course, the most complex model of the island with the radial gradient is possible.

THEORETICAL FRAMEWORK ON THE FORMATION OF THE LARGE SCALE STRUCTURE IN THE UNIVERSE

Satoru Ikeuchi
Tokyo Astronomical Observatory
University of Tokyo
Mitaka, Tokyo 181
Japan

ABSTRACT: It seems that the galaxy formation theory is now confusing. In order to resolve this situation the theoretical framework on the formation of large-scale structure in the Universe is reviewed, especially paying attention to the first-ranked observational facts. Then, the conditions which the theory must satisfy with minimum parameters are summarized in relation to additional observational facts. Finally, a possible framework of the galaxy formation theory which looks well consistent with the large-scale structure is presented.

1. A SHORT REVIEW OF HISTORY

In order to make clear of the present situation we should know the progress of research in the past and learn how the problem has been studied deeply. Then, I would like to begin my lecture from a short review of history of the galaxy formation theory (Peebles 1980).

The history of these four decades in the study of galaxy formation theory may be divided into, at least, three epochs. They are :

(I) Dawn epoch (1944 ∼ 1965)

In this epoch, the gravitational instability in the expanding universe was studied based upon the general relativity (Lifshitz 1946) and later the Newtonian mechanics (e.g. Bonnor 1957). This picture is kept unchanged till the present epoch, even though the predominant matter for the gravity changes. The galaxy formation theory began to be attacked as science, although the controversy to the steady universe was not so fruitful.

(II) Happy Days (1965 ∼ 1980)

The discovery of cosmic microwave background radiation (CBR) gave rise to the victory of big bang universe, and the fundamental framework of galaxy formation theory were presented by Peebles (1967), Zeldovich (1970) and their colleagues. Although it was rather controversial that both the up-down (or fragmentation) theory and the bottom-up (or clustering) theory were equally insisted, a new field of cosmology was certainly developed and many important studies were presented.

169

M. Mezzetti et al. (eds.), Large Scale Structure and Motions in the Universe, 169–178.
© *1989 by Kluwer Academic Publishers.*

(III) Confusing Days (1980 ∼ present ∼ 199 ?)

Many observational facts have been accumulated and correspondingly many (ad-hoc) theories are also proposed. It seems very fruitful, but I am afraid that the galaxy formation theory downs to a parameter physics. Then, we must take pause and look around the present situation.

Here, I try to look for the *Required Minimum Conditions* for the theory and to suggest a possible way to find a consistent scenario. The first step is to discriminate the observational facts as the first-ranked and additional ones. Then, the fundamental conditions for the theory is summarized. Finally, I present a probable scenario on the evolution of the universe by allocating physical roles to structures of each scale.

2. THE FIRST-RANKED OBSERVATIONAL FACTS

"The First-Ranked" means that the theory must automatically satisfy or naturally explain those observational facts. This requirement is most strict. Here, I indicate the following four as the first-ranked ones.

(I) Stringent upper limit to the anisotropy of CBR

The observational upper limits to the anisotropy of CBR have been accumulated in the angles of 30" to 90° and all results claim $\Delta T/T <$ several $\times 10^{-5}$ except for the result by Davies *et al.* (1987). From these facts, the amplitude of temperature fluctuations at the recombination epoch z_d is severely constrained.

Schematically, the temperature fluctuation at z_d is illustrated in Figure 1 (Kodama 1988). The solid lines and dashed lines, respectively, indicate the dark matter (DM) dominated universe and the baryon dominated universe. Since this spectrum is strongly coupled with the spectrum of density fluctuations as in Figure 2 (Kodama 1988). The stringent upper limit to $\Delta T/T$ also constrains $\Delta \rho/\rho$.

(II) Missing Mass Problem

The M/L of galaxies indicates the density parameter of luminous mass to be $\Omega_l(0) \sim 0.01$, while the virial theorem of galaxies and clusters of galaxies leads to $\Omega_v(0) \sim 0.2$. Then, the dynamical mass must be more abundant than the luminous mass by a factor of 10. On the other hand, the nucleosynthesis at the early universe gives rise to the upper limit of baryon density as $\Omega_B(0) < 0.1$. This means that non-baryonic matter exceeds baryonic matter. Recent discoveries of dipole distribution of IRAS galaxies and number count of distant galaxies claim $\Omega(0) = 0.8 \sim 1.0$. We must introduce some non-baryonic matter of $\Omega_{NB}(0) = 0.1$ to 0.9, depending upon the cosmological model.

(III) Fractal nature of structures

In the scale ranging from 0.7 Mpc to 70 Mpc the two-point correlation functions of galaxies and clusters of galaxies indicate the same power law form with the index of -1.8, but there is a big difference in amplitudes. It is highly probable that the difference in amplitudes may be artifact due to the difference of surveyed regions between galaxies and rich-clusters of galaxies (Ruffini *et al.* 1987). If so, the structures of this range have a fractal nature characterized by the scale-invariant similarity and the non-integer dimension. This gives an important implication on the origin of structures.

(IV) Bubbly universe

The characteristic form of structures is an assembly of bubbles, although others claim it is a sponge or a Swisscheese. The important point is the presence of voids surrounded by superclusters of galaxies, which are connected long. Such a geometry of structures must be naturally reproduced.

In the usual gravitational instability picture, all these first-ranked observational facts are essentially reduced to the spectrum of density fluctuations at the decoupling epoch. The spectrum is expressed by two parameters, amplitude A and power index n as $\Delta = \delta\rho/\rho = AL^n$ where L is the wavelength of a perturbation. The observational facts (I) and (II) necessitate the introduction of weakly interacting dark matter, and (III) and (IV) seem to require (an)other parameters. The most economical model will be that the assumed dark matter automatically makes the structures characterized by (III) and (IV).

3. PROPOSED MODELS

3.1 HDM scenario

In order to resolve the difficulty in the adiabatic perturbation model of baryons the neutrino dominated universe was proposed (Doroshkevich 1981, Sato 1981), in which the perturbations of neutrinos can grow from the earlier epoch without perturbing the CBR. If the neutrino mass is as light as several tens eV neutrinos decoupled with photons at the relativistic stage. In this case we call hot dark matter (HDM).

In this HDM scenario, a characteristic scale length is naturally introduced due to the free streaming nature of neutrinos. A simple estimate of this length gives $l \sim$ (random velocity of neutrinos) $\times$ (Hubble time) $\sim 19(20\text{eV}/m_\nu)(1+z)^{-1}$ Mpc, which nearly corresponds to the scale of superclusters of galaxies. Below this length the perturbations are erased. Then, the HDM scenario is one of the top-down theories.

This nature causes a difficulty in the distribution of galaxies. Almost all galaxies are so concentrated into superclusters that the two-point correlation function becomes too steep and too big in comparison with observations (III). In order to overcome this difficulty some unknown processes to disperse galaxies into voids must be considered. No one knows whether such an appropriate *Anti-Biasing* mechanism works or not.

3.2 CDM scenario

As an opposite model the cold dark matter (CDM) dominated universe is examined in accordance with the scale-invariant perturbations. The CDM decoupled with photons in non-relativistic stage because its mass is heavier than 1 GeV like gravitinos, photinos and so on, or its temperature is exactly zero like axions. In this case, no special length scale can be introduced, and the fractal nature of CDM distribution is naturally expected. Since a standard spectrum is a little bit higher in smaller scale perturbations, the less massive objects are formed earlier. This means that the CDM scenario predicts the bottom-up formation of structures.

In this CDM scenario, the CDM distribution becomes too smooth and the bubbly universe (IV) does not arise. The formation of bubbles in the distribution of baryons (galaxies) necessitates a new parameter which describes the efficiency of galaxy formation within smoothed potentials. The biasing galaxy formation model is representative. For example, the galaxy formation proceeds efficiently at a deeper potential depth than the threshold one. (Davis *et al.* 1985). In this case, the spherical nature of void structures can not be expected, but the CDM plus biasing are very popular at present.

3.3 Coexistence of HDM and CDM

One way to consider the biasing effect on galaxy formation is to suppose the coexistence of HDM and CDM (Umemura and Ikeuchi 1985, Ikeuchi, Norman and Yin 1988). The HDM produces a large scale bubbly universe while the CDM maintains the fractal nature. The interacting regions are dwarf galaxies, where the HDM and CDM contribute equally.

The important point of this model is that we can exactly calculate the time behavior of perturbations without other parameters because we know the dynamics of both components. In the biasing model we must introduce some unknown biasing parameters which control the efficiency. Moreover, the two-component dark matter can be naturally expected from one of models of particles physics, in which the Pecci-Quinn symmetry and resonant neutrino conversion are assumed. One fatal difficulty is that it is very difficult to check this model by numerical simulations.

3.4 Cosmic strings and others

Under the CDM dominated universe other exotic models are also proposed like cosmic strings, decaying particles and so on. The common nature is that they trigger the formation of perturbations of galactic scales and then they decay into gravitational waves, relativistic particles and low-frequency electromagnetic waves. Therefore, the distribution of galaxies depends upon the physical nature of triggering matter, on which we do not know the details. The cosmic string model is very popular and has ample convenient natures, but it looks a little artifact and no sound physical foundations are confirmed.

Other sequence of exotic models are non-gravitational theories in which the objects are not formed from primordial perturbations of matter or strings. A naive model is the explosion amplification hypothesis (Ostriker and Cowie 1981, Ikeuchi 1981) and its modified version, the superconducting cosmic strings hypothesis (Ostriker, Thompson and Witten 1986). These models suppose that some pre-galactic objects trigger the formation of galaxies due to gigantic energy injection similar to the cloud formation at the shock front of supernova remnants. The bubbly universe is naturally expected because the energy injection makes a spherical void and the surrounding dense shell is transformed into superclusters of galaxies. The fatal problem is that the inverse Compton scattering of CBR by high energy electrons or the photon production by pulsar mechanism may give rise to the anisotropy of background radiation.

Other non-gravitational theories are non-zero cosmological constant model and non-Newtonian gravitation model. These non-standard models have been proposed repeatedly when the standard model seemed unsuccessful. The past experience has taught us that non-standard or unconventional model stimulated us to find hidden parameters or new fields within the standard model. It is valuable to reexamine the standard gravitational theory which may implicitly conceive new developments. We should not take non-standard models as an easygoing way. Otherwise it becomes only a game.

4. ADDITIONAL OBSERVATIONAL FACTS

Here, I would indicate several additional observational facts to which the theory must have implications, because they are to be strongly correlated with the galaxy formation.

(I) Flat rotation curve

The isothermal distribution of stars in elliptical galaxies and the flat rotation curves in spiral galaxies denote that the density distribution in galaxies is $\rho(r) \propto r^{-2}$ irrespective of their types.

This fact claims either of following possibilities : the stellar core is formed in the first, and then the mass accretion to the core occurs (Gunn 1982), or the spectrum of density fluctuations has a positive power with respect to the mass, i.e., $\alpha = 0 \sim 1/3$ for $\Delta = AM^\alpha$ (Quinn et al. 1986). Anyway, an important implication to the origin of galaxies is indicated. The CDM scenario with cosmic strings claims the former possibility, and the two-component DM scenario does the latter one.

(II) Lyman α forest

The most important characteristics of the Lyman α forest is that it does not exactly show the tendency of spatial correlation. In other words, we can not recognize any voids of trees in the forest in contrast to the galaxy distribution.

This fact insists that the origin of the Lyman α forest is completely different from that of galaxies. The scenario of galaxy formation must also have an episode on the Lyman α forest. Representative models are the minihalos model (Ikeuchi 1986, Rees 1986) based upon the CDM scenario and the intergalactic clouds model (Sargent et al. 1980, Ostriker and Ikeuchi 1983) in relation to the quasar activity. I believe the Lyman α forest may bring important informations at $z = 2 - 4$ of the universe, if more data with higher resolution are accumulated.

(III) Hot intergalactic medium

No sign of the Lyman α trough at the shorter wavelength side of the Lyman α emission line of QSO at $z = 3.78$ (Murdoch et al. 1986) indicates that the intergalactic medium (IGM) has been already ionized at that epoch (Gunn-Peterson test). Accordingly, the temperature of IGM exceeds 10^4 K at $z = 3.8$. Moreover, if the Lyman α forest is originated in pressure-confined intergalactic clouds, the temperature of IGM must exceed 10^5 K, which is obtained from the detectability condition of clouds (Ikeuchi and Ostriker 1986).

It is generally accepted that the IGM can not be wholly ionized till $z \sim 3.5$ if the UV sources are QSOs. Other UV sources are required even if the IGM temperature is $\sim 10^4$ K at $z = 3.8$. If only the UV heating is considered it is impossible to raise the temperature higher than 3×10^4 K. Then, some collisional heating like shock phenomena must be included if it exceeds 10^5 K. Such a violent event in the intergalactic space can not be expected except for the explosion amplification hypothesis.

(IV) Deviation from the Planck spectrum of CBR

Recently, the Nagoya-Berkely group (Matsumoto et al. 1988) found the deviation from the Planck spectrum of CBR at the Wien part. According to their result, the degree of deviation is $\Delta T/T \sim 0.2$ at $\lambda = 0.5 \sim 1$ mm, and it corresponds to about 10 percent of energy fluctuation of CBR. This amount is tremendously large if we compare this with the nuclear energy of baryons.

Two possibilities on the origin of this deviation are also conceivable. One is the inverse Compton scattering of photons at the Rayleigh-Jeans part of CBR. In this case, high energy electrons with the temperature higher than 10^8 K must be injected at $z \sim 30$ in order not to produce detectable anisotropy of CBR (Yoshioka and Ikeuchi 1987). For satisfying this condition the converted mass to energy should be $\Delta \rho_B \sim \rho_{crit}/1000$, which is comparable to the helium abundance. This is implausibly high. The explosion amplification hypothesis predicts such an event although the injection energy is smaller than the above.

The other is the dust emission from population III objects at $z > 10$ in which the dust temperature is as high as $3(1 + z)$ K. (Hayakawa et al. 1987) The necessitated energy is smaller than the inverse Compton model because the conversion efficiency of the injected energy to the infra-red region is very high. The existence of hypothetical population III objects is plausible in the CDM scenario.

(V) Dwarf galaxies

In comparison with normal galaxies the dwarf galaxies with the mass $10^9 \sim 10^{10} M_\odot$ show a distinctly different structure from elliptical galaxies and globular clusters (Kormendy 1986). They have similar core radii to ellipticals but the surface brightness is smaller by a factor of 100. Although their morphological features are formed during their evolution by, for example, mass loss and merging, it seems that the mass sequence such as globular clusters ($10^6 M_\odot$), Lyman α clouds ($10^7 M_\odot$), dwarfs ($10^9 M_\odot$) and normal galaxies ($10^{11} M_\odot$) is not an accidental but logical consequence. The contrast between the mass sequence and structural difference is a challenging problem. The existence of many kinds of subgalactic objects may be preferable to the CDM scenario.

(VI) Bulk flow to the CBR

Many evidence for the presence of a bulk flow of galaxies within 60 Mpc with respect to CBR is now accumulating, and the great attractor hypothesis is now popularly discussed (Dressler et al. 1987). The mass of great attractor, if it exists, must be greater than $10^{16} M_\odot$. Such a massive inhomogeneity will be very rare even in the HDM scenario, and will declare the death of the CDM scenario, if finally confirmed.

(VII) X-ray background radiation

At present we have no definite idea on the origin of cosmic X-ray background radiation (CXBR). It may be explained by evolving AGNs and QSOs if they evolve very rapidly, but there remain residuals at $E > 100$ keV (Morisawa and Takahara 1988). Then, we must examine more detailed evolution of massive pancakes, young galaxies and QSOs how and when they contribute to CXBR.

5. ALLOCATION OF PHYSICAL ROLES

Here, I try to divide structures from subgalactic objects to superclusters into three categories and allocate an important matter or physical process to each category. This will be the most economical arrangement of physical roles to complex structures.

5.1 CDM for mass sequence

The fundamental characteristics of structures from population III objects to superclusters of galaxies is the mass sequence as population III objects, globular clusters, minihalos or intergalactic clouds, dwarfs, normal galaxies, clusters of galaxies and superclusters. From less massive to massive objects the spatial correlation increases profoundly and the scale-free nature appears at the massive side. Such a sequence should be attributed to the simple spectrum of primordial density perturbations. This is easily guaranteed by the scale-free perturbation of CDM. However, this massive sequence had been modified at the largest and intermediate scales by other factors.

5.2 Modification at the largest scale

The bubbly structure greater than 50 Mpc and the great attractor indicate that at the massive end another action modifies a simple structure. This may be supposed to be the HDM or some biasing mechanism. The bubbly nature claims that it may be gravitational and the HDM is most promising. The introduction of collisionless particles with the mass less than 10 eV is the easiest way. This also brings interesting results at the mass scale of dwarfs $\sim 10^{10} M_\odot$ because of the interaction with the CDM.

5.3 Substructures at smaller scales

The shocks from pre-existing objects, which are expected in the CDM scenario, might modify the smaller scales by producing new objects at the shock fronts. This action also affects much the thermal history of the universe. At present, we do not know what mass scales are most effective to this action.

Summing up the above, I propose to try to make all structures by means of the above three fundamentals, CDM, HDM and shocks. The important point is that we know the essential physics of these three. Arranging the mass of collisionless particles and the spectrum of their density perturbations, we can investigate without ambiguity what structures will appear in the universe. Such a rigorous investigation is now necessitated, not introducing appropriate parameters.

I would like to thank Professors C. A. Norman and J. P. Ostriker, Drs. M. Umemura and S. Yoshioka for enjoyable collaborated works. I also thank H. Suzuki for her careful arrangement for my research.

176

References

Bonnor, W. B. 1957, *M. N. R. A. S.* **117**, 111.
Davies, R. D. *et al.* 1987, *Nature* **326**, 462.
Davis, M. *et al.* 1985, *Ap. J.* **292**, 371.
Doroshkevich, A. G. 1981, in *10-th Texas Symposium on Relativistic Astrophysics*, ed. by R. Ramaty and F. C. Jones, New York Academy of Sciences, p. 32.
Dressler, A. *et al.* 1987, *Ap. J. Lett.* **313**, L37.
Gunn, J. E. 1982 in *Astrophysical Cosmology* ed. by H. Bruck *et al.*, Pontificiae Academiae (Vatican).
Hayakawa, S. *et al.* 1987, *Pub. Astron. Soc. Japan* **39**, 941.
Ikeuchi, S. 1981, *Pub. Astron. Soc. Japan* **33**, 211.
Ikeuchi, S. 1986, *Astrophys. Sp. Sci.* **118**, 509.
Ikeuchi, S. and Ostriker, J. P. 1986, *Ap. J.* **301**, 523.
Ikeuchi, S., Norman, C. A. and Yin, Z. 1988, *Ap. J.* **324**, 35.
Kodama, H. 1988, *Journal of Physical Society of Japan (in Japanese)*, **43**, 201.
Kormendy, J. 1986, *Ap. J.* **295**, 73.
Lifshitz, E. M. 1946, *J. Phys.* **10**, 116.
Matsumoto *et al.* 1988, *Ap. J.* **329**, in press.
Morisawa, K. and Takahara, F. 1988, in *Proceedings of The Yamada Conference XX on Big Bang, Active Galactic Nuclei and Supernovae*, ed. by S. Hayakawa and K. Sato, Universal Academy Press, in press.
Murdoch, H. S. *et al.* 1986, *Ap. J.* **309**, 19.
Ostriker, J. P. and Cowie, L. L. 1981, *Ap. J. Lett.* **243**, L127.
Ostriker, J. P. and Ikeuchi, S. 1983, *Ap. J. Lett.* **268**, L63.
Ostriker, J. P. Thompson, C. and Witten, E. 1986, *Phys. Lett.* **B 180**, 231.
Peebles, P. J. E. 1967, *Ap. J.* **147**, 859.
Peebles, P. J. E. 1980, *The Large-Scale Structure of the Universe*, Princeton Univ. Press.
Quinn, P. J. *et al.* 1986, *Nature* **326**, 329.
Rees, M. J. 1986, *M. N. R. A. S.* **218**, 25p.
Ruffini, R. *et al.* 1987, *Ap. J.*, in press.
Sargent, W. L. W. *et al.* 1980, *Ap. J. Suppl.* **42**, 41.
Sato, H. 1981 in *10-th Texas Symposium on Relativistic Astrophysics*, ed. by R. Ramaty and F. C. Jones, New York Academy of Sciences, p. 39.
Umemura, M. and Ikeuchi, S. 1985, *Ap. J.* **299**, 583.
Yoshioka, S. and Ikeuchi, S. *Ap. J. Lett.* **323**, L7.
Zeldovich, Ya B. 1970, *Astron. Astrophys.* **5**, 84.

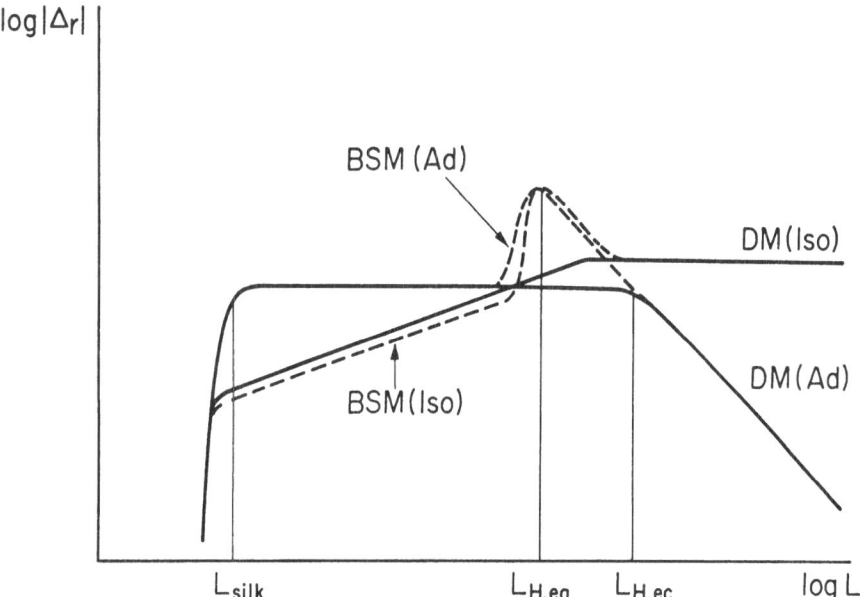

Figure 1 :   Expected spectrum of temperature fluctuations, $\nabla_r = \delta T/T$, at the recombination epoch with respect to the size L. Solid and dashed lines indicate, respectively, the dark matter dominated and baryon dominated universe (Kodama 1988).

178

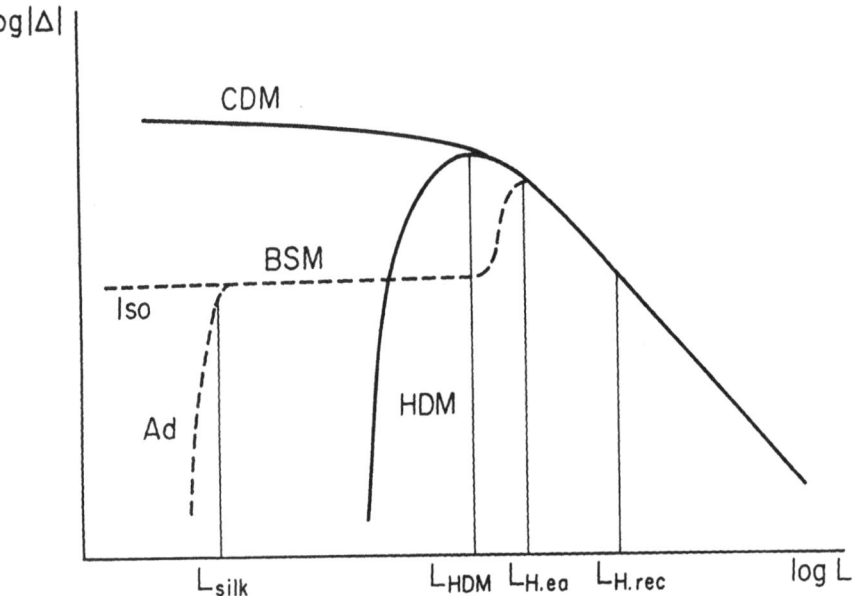

Figure 2 :   Corresponding spectrum of density fluctuation, $\nabla = \delta\rho/\rho$
(Kodama 1988).

LARGE-SCALE MOTIONS IN THE NEARBY UNIVERSE

David Burstein, Dept. of Physics, Arizona State University
Roger L. Davies, NOAO, Kitt Peak National Observatory
Alan Dressler, Mt. Wilson and Las Campanas Observatories
S.M. Faber, Lick Observatory, U.C. Santa Cruz
Donald Lynden-Bell, Institute of Astronomy, Cambridge, England
Roberto Terlevich, Royal Greenwich Observatory
Gary Wegner, Dept. of Physics and Astronomy, Dartmouth College

ABSTRACT. The velocity field of galaxies relative to the cosmic microwave background has been investigated to a distance of 3000 km/s, using
two samples of spiral galaxies together with elliptical galaxies. The
motions of these galaxies agree well with an optimized velocity field
model that includes: a) A spherically-symmetric 'Great Attractor' whose
velocity field dominates this region; b) a Virgocentric flow, which
dominates only within 700 km/s of the center of Virgo; and c) a Local
Anomaly motion which all galaxies within a distance of 700 km/s of the
Local Group share. Systematic properties of the velocity field not fit
by this model are also discussed.

1. INTRODUCTION

Our group originally investigated non-uniformities in the local Hubble
flow using the motions of elliptical galaxies (Dressler et al. 1987;
Lynden-Bell et al. 1988, hereafter LFBDDTW). Recently we have expanded
that investigation to include the motions of nearby spiral galaxies as
determined from the studies of Aaronson et al. (1982a) and Bothun et al.
(1984; collectively referred to as Aaronson et al.) and de Vaucouleurs
and Peters (1984). A detailed description of this analysis is given in
Faber and Burstein (1988; hereafter FB) and will not be repeated here.
Rather, in the space available, aspects of that analysis are given that
illustrate both the successes and the problems that we have had with our
current model of the nearby velocity field.

2. ACCURACY OF DATA

The measurement of peculiar motions requires adoption of a standard
velocity reference system (here taken to be that defined by the Cosmic
Microwave Background [CMB]), and an independent measure of relative
distance. Not all distance predictions are of the same accuracy, even

179

M. Mezzetti et al. (eds.), Large Scale Structure and Motions in the Universe, 179–196.
© 1989 by Kluwer Academic Publishers.

within a given data sample. In particular, the Aaronson et al. data can be objectively divided into two categories: galaxies with diameters, blue magnitudes and axial ratios given the Second Reference Catalog of Bright Galaxies (de Vaucouleurs, de Vaucouleurs and Corwin 1976), and galaxies without one or more of these parameters. Galaxies that fall into the first category can be shown to have signifcantly smaller observational errors in the near infrared Tully-Fisher (IR-TF) relationship: From an analysis of the velocity field, FB estimate the observational error in predicted distance for the Aaronson et al. 'good' sample corresponds to a magnitude error of 0.37 mag. The Aaronson et al. 'fair' sample has an adopted average observational error of 0.50 mag. Pierce and Tully (1988), among others, have confirmed that errors in diameters and axial ratios are a significant source of observational error in predicted distances in the IR-TF relationship.

The observational error for a predicted distance to an individual elliptical galaxy is equivalent to a magnitude error of 0.45 mag (LFBDDTW), but many galaxies are in groups and clusters that, as a unit, have smaller observational errors. Only those galaxies in de Vaucouleurs and Peters (1984) that have blue magnitude Tully-Fisher distances are used in the following analysis, with an adopted error of 0.56 mag. As this last sample has the largest observational error, if a galaxy is observed by both Aaronson et al. and de Vaucouleurs and Peters, the data from the former are used.

All distances in this analysis have been corrected for Malmquist bias, along the lines discussed in LFBDDTW. The problematic issues of Malmquist bias correction that arise from a clumpy distribution of galaxies are minimized when dealing with accurate data for nearby galaxies. As such, the analysis in FB, and the analysis here, uses the best data to establish trends, and the poorer data as a consistency check.

3. THE G.A.+Vinf+L VELOCITY FIELD MODEL

The velocity field model that we have investigated is a revised version of the model presented in LFBDDTW, with an additional component. The three components of the current model are:

3.1 The Great Attractor (G.A.)

The peculiar velocities of galaxies in each data sample require the presence of a dominant, coherent differential velocity field across our volume of space. We partake of this flow, and observe galaxies in one direction moving away from our location and those in the opposite direction moving towards our location. Galaxies in directions perpendicular to the flow direction show a compressional motion of sizeable amplitude towards the flow direction. Both kinds of motions (flow and compression) combine to yield the large quadrupole terms in the solutions of Lilje, Yahil and Jones (1986) and LFBDDTW.

A gravitational mass has been postulated as the source of this velocity field, the center of which has been termed the 'Great Attractor.' The adopted G.A. velocity field model is similar mathematically to a standard, spherically-symmetric Virgocentric flow

model (e.g. Aaronson et al. 1982). The center of the G.A. (according to
FB) is found to be at a distance of 4200+350 km/s towards a direction
(in Galactic coordinates) of l=309°, b=18° (+ 10° in both coordinates).
The flow velocity of the Local Group $_{-1.7}$ due to the G.A. is found to be
535+60 km/s, velocity decreases as $r^{-1.7}$ from the center of the G.A.,
and the G.A. is given a 'core' radius of 1350 km/s. Compared to the
previous model of LFBDDTW, the radial profile of the new model is
substantially altered, but the location of the G.A. and amplitude of the
flow at the Local Group are essentially unchanged.

3.2 Virgocentric Infall (Vinf)

Previous estimates of the amplitude and extent of Virgocentric infall
have been heavily influenced by the more dominant G.A. motion, as well
as by mis-estimates of the errors of observation (see FB for details).
With the same data as used by Aaronson et al. (1982), our current best
estimate of the size of Virgocentric infall at the position of the Local
Group is between 85-133 km/s (cf. Peebles 1988). Moreover, our graphs
show that the Virgo velocity field is apparent only within a distance of
700 km/s of the cluster center.

3.3 The Local Anomaly (L)

We (FB) and others (e.g. Peebles 1988) have shown that there is a
coherent velocity flow associated with the Local Group that extends to a
region of about 700 km/s in radius around it. If we remove the
components of the Local Group's CMB motion that are due to the G.A. and
Virgo infall, the remaining vector has an amplitude of 360 km/s towards
l=199°, b=0°. Thus, this 'Local Anomaly' is 2/3 the amplitude of the
G.A.-induced flow, and three times the amplitude of Virgo infall at the
position of the Local Group. Lynden-Bell and Lahav (1988) and Lahav
et al. (1988) point out that the direction of (199,0) is nearly opposite
in the sky to the 'Local Void' identified by Tully and Fisher (1987).
Lower mass density in this void would result in an increased motion in
the opposite direction, which leads these authors to propose that the
existence of this void is the cause of the Local Anomaly motion.

    It is shown below (and detailed in FB) that the Local Anomaly
appears to be typical of motions relative to the G.A. in other regions
of space. Our velocity field model includes a 'switch' that assigns
this region a bulk-velocity correction to make its motion blend smoothly
with its surroundings.

4. 'PICTURES' OF THE VELOCITY FIELD

4.1 Picturing the Flow within the Supergalactic Plane

Fig. 1 presents the peculiar motions of galaxies located within +22.5°
of the Supergalactic plane, in a format similar to that used by LFBDDTW.
The peculiar motions of galaxies from all three data sets are included
As shown in FB, the predicted peculiar motions of galaxies from all

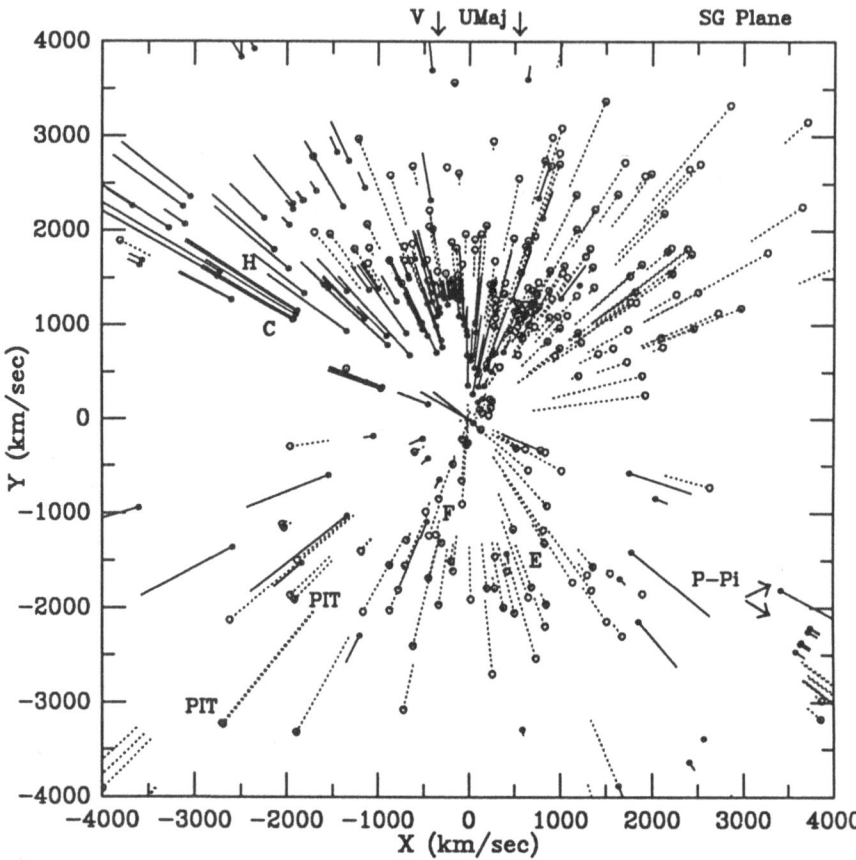

Fig. 1. A velocity field map like those presented in LFBDDTW. The peculiar motions for all galaxies within 22.5° of the Supergalactic plane are plotted (i.e., the plane perpendicular to the direction l=227°, b=47°). Peculiar velocities are measured with respect to the cosmic microwave background frame. Data from all three samples are used: Elliptical galaxies, Aaronson et al. IR–TF spirals, de Vaucouleurs–Peters blue–TF spirals. The general positions of the major regions in this volume of space are noted: V = Virgo cluster; UMaj = Ursa Major cluster; P–Pi = Perseus–Pisces clusters; F = Fornax cluster; E = Eridanus cluster; PIT = Pavo–Indus–Telescopium region; C = Centaurus cluster and H = Hydra cluster.

three data sets are in good agreement in all directions. The locations of major clusters in this volume of space are noted. This diagram is useful in showing the two major features of the velocity field: a) G.A.-induced flow. Galaxies in the direction of the G.A. are moving away from our location, galaxies opposite to the direction of the G.A. are moving towards our location. Note that galaxies as far as 1500 km/s further away from the G.A. are moving towards the G.A. b) G.A.-induced tidal compression. Galaxies located at large angles relative to the G.A. flow, on either side of the flow, are moving towards the flow direction. As shown in LFBDDTW (Fig. 6), this compression is consistent with a tidal field generated by the G.A.. The compresion gives rise to the strong quadrupole terms in the velocity field which, in turn, imply the presence of a central, gravitational source of this motion (cf. Lilje, Yahil and Jones 1986).

4.2 'X-Slices': Picturing the Tidal Field of the G.A.

The distribution of galaxies relative to the Local Group in the volume of space out to 3000 km/s permits one to picture the velocity field in two dimensions using relatively simple geometric 'cuts'. In particular, as shown in Fig. 1, the four major concentrations of mass closest to the Local Group are situated in each of the four quadrants defined by Supergalactic X,Y coordinates: the Ursa Major cluster is at positive X, positive Y; the Virgo cluster is at negative X, positive Y; the Eridanus cluster is positive X, negative Y; and the Fornax cluster is at negative X, negative Y. As shown in FB, 'slices' of this volume along lines of constant X or Y yield useful views of the velocity field.

One 'slice' that is particularly useful in picturing the tidal field of the G.A. uses values of X between 0 and 1500 km/s, parallel to the Y axis and perpendicular to the Supergalactic plane. This slice includes the Ursa Major cluster at positive values of Y, and the Eridanus cluster at negative Y values. Fig. 2 (upper panel) plots the observed peculiar velocities of Aaronson et al. 'good' galaxies in this 'X-slice' versus the Y component of their predicted distances. Only the Aaronson et al 'good' data are used in this diagram so as to minimize the effects of observational error.

This 'X-slice' is particularly illustrative since it includes galaxies located at large angles relative to the direction of the G.A. flow. It is the net negative peculiar motion of galaxies in this 'slice' that is primarily repsonsible for the quadrupole term in the analyses of Lilje, Yahil and Jones, LFBDDTW and FB. An amplitude of ~400 km/s for both sides of this compression, relative to the flow direction, can be estimated from the upper panel in Fig. 2. The relative independence of this amplitude with distance from the Local Group places a signficant constraint on possible sources of systematic errors in these peculiar velocities, as described in Section 6.

Fig. 2 (lower panel), in which the peculiar motions of these galaxies relative to the G.A.+Vinf+L model are plotted, illustrates how well the compressional motion is removed by the G.A. tidal field. Although the Ursa Major cluster itself has a small residual net motion

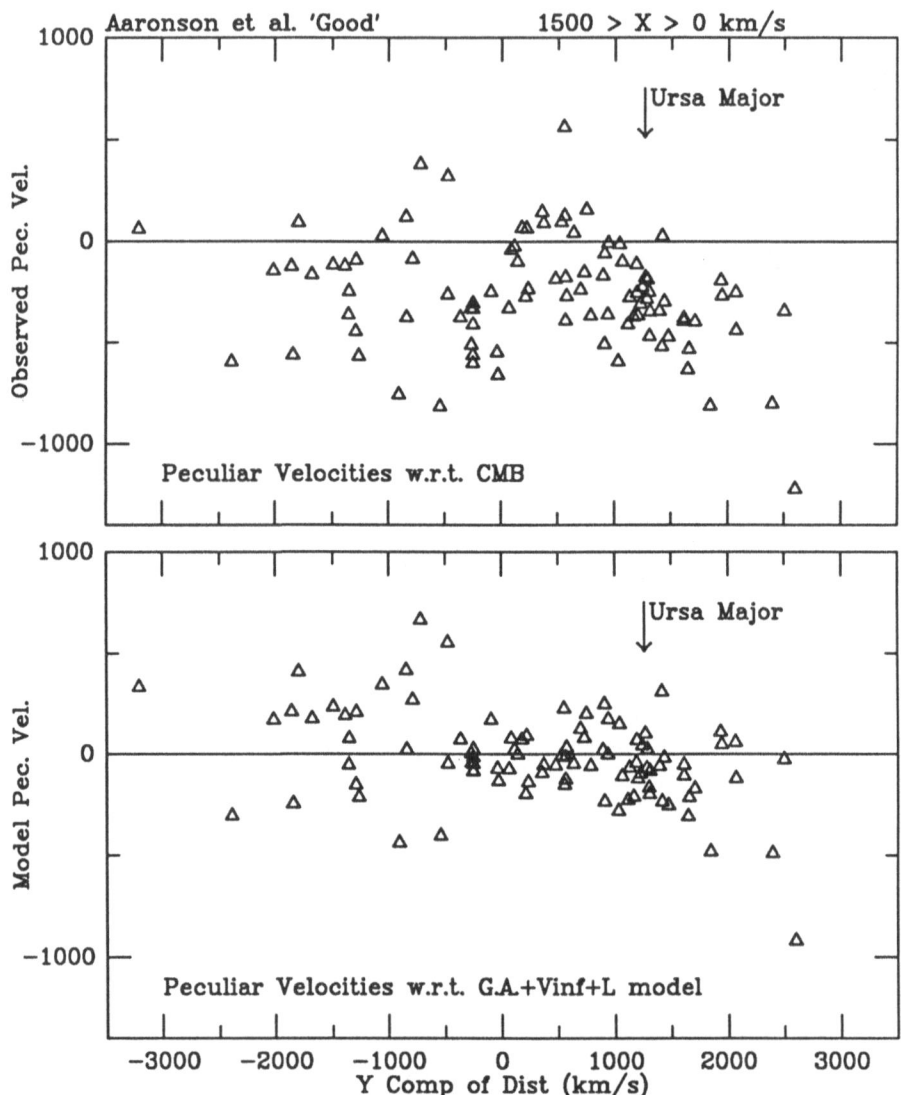

Fig. 2. Peculiar velocities within a 'slice' of the local volume.
Upper panel: The observed peculiar velocities of Aaronson et al. 'good'
galaxies in a region bounded by 1500<X<0 km/s, plotted versus Y
distance for predicted distances less than 4600 km/s. This slice
includes the Ursa Major cluster, whose Y distance is indicated. Lower
panel: The peculiar velocities of these galaxies relative to the
G.A.+Vinf+L model, plotted versus Y distance. Note that the strongly
negative observed peculiar velocities of galaxies in this slice (which
is nearly perpendicular to the direction of the G.A.) are adequately
modeled by the tidal field generated by the G.A.

(see Fig. 3b below), there is little remaining net compressional motion relative to the model.

4.3 Picturing the Characteristic Motions of 'Regions' of Space

Tully and Fisher (1987) have subdivided the volume of space within 3000 km/s into separate clusters and 'clouds', following the lead of de Vaucouleurs (1975). Using their maps as a guide, we have divided much of the sky into 'regions' according to limits in either Galactic coordinates or Supergalactic coordinates, and have examined the peculiar motions of galaxies within these regions as a function of distance (FB).

We found that most galaxies tend to lie in one of these clusters or 'clouds' that have a typical size scale of ~1500 km/s. Galaxies within a given region tend to share a common motion, relative to the G.A. flow, of one of three types: 'quiet', 'cluster-dominated' or 'noisy'. Examination of motions relative to the G.A. flow is important, as the G.A. gravitational field so dominates this volume of space that its effects must be removed before other, smaller-scale motions can be easily seen.

4.3.1 'Quiet' motions. Four regions of space — the Local Anomaly, the Leo Cloud, N1023-Cetus and the near region of Camelopardolis — appear to have 'bulk' velocity motions that have a low internal velocity dispersion (less than 100 km/s). The motions of galaxies in two of these regions (Leo and N1023-Cetus) are illustrated in Fig. 3(a). Data from all three data samples are used in Fig. 3, with the Aaronson et al. data being divided into 'good' and 'fair' sub-samples. (Distance errors in this kind of diagram lead to correlated errors, as noted by the line drawn in each figure. Moreover, scatter increases with distance for a constant observational error.)

The observed peculiar motions of the Leo Cloud with respect to the CMB are particularly dramatic (Fig. 3(a), upper panel, left-hand (LH) side): These galaxies exhibit a range of over 1000 km/s in peculiar motion that is strongly correlated with distance. As shown in the right-hand (RH) panel, almost all of this motion can be attributed to the G.A.-induced flow. Net motions relative to the model, of one-dimensional amplitude 100-200 km/s, are evident for the two segments of the Leo Cloud.

In contrast, the peculiar motions of the N1023-Cetus region are more uniform as seen from our vantage point (Fig. 3(a), lower panel, LH side): Galaxies in this region share a net -500 km/s motion relative to the CMB, which is nearly all removed by the model (RH side).

4.3.2 'Cluster-Dominated' Regions. Seven of the regions investigated by FB are dominated by centralized clusters: Virgo, Ursa Major, N5846, N1549, Crater, Eridanus and Fornax. The motions of galaxies relative to two of these clusters are illustrated in Fig. 3(b). Essentially all of the galaxies in the direction of N5846 appear to be located near the cluster (upper panel, LH side). The N5846 cluster itself has little observed peculiar motion relative to the CMB, but may have a small net motion relative to the model (RH side). The evidence for infall of galaxies into the cluster center is marginal with the present data.

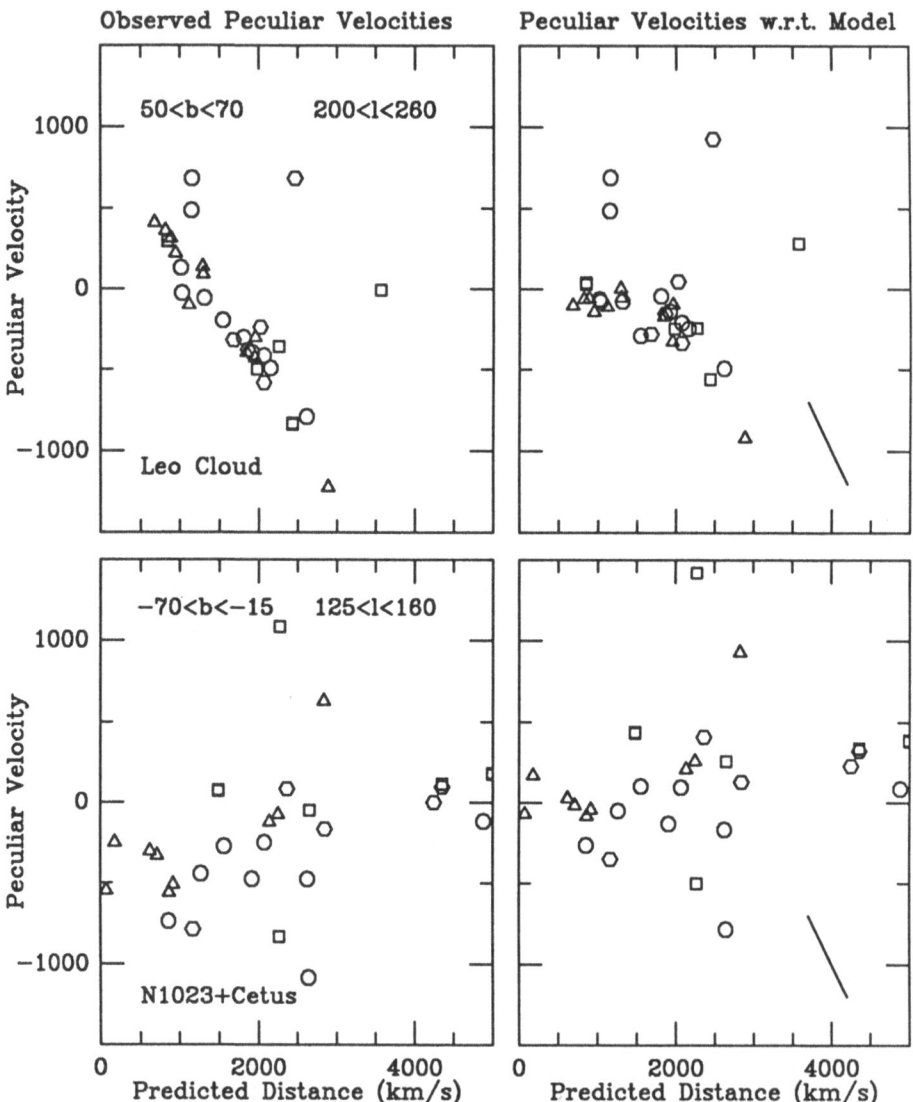

Fig. 3. Examples of peculiar velocity vs. predicted distance for galaxies in different kinds of 'regions' in the sky. Data from all three samples are used: Squares = Ellipticals; Triangles = Aaronson 'good'; Hexagons = Aaronson 'fair'; Circles = de Vaucouleurs–Peters. Left–hand panels plot observed peculiar velocities relative to the CMB; right hand panels plot peculiar velocities relative to the G.A.+Vinf+L model. The line drawn represents a 'distance–error' slope of ±250 km/s.

(a) Two examples of 'quiet' regions. Upper panels: Leo Cloud, defined here as within 50°<b<70°, 200°<l<260°. Lower panels: N1023–Cetus region, defined here as within −70°<b<−15°, 125°<l<160°.

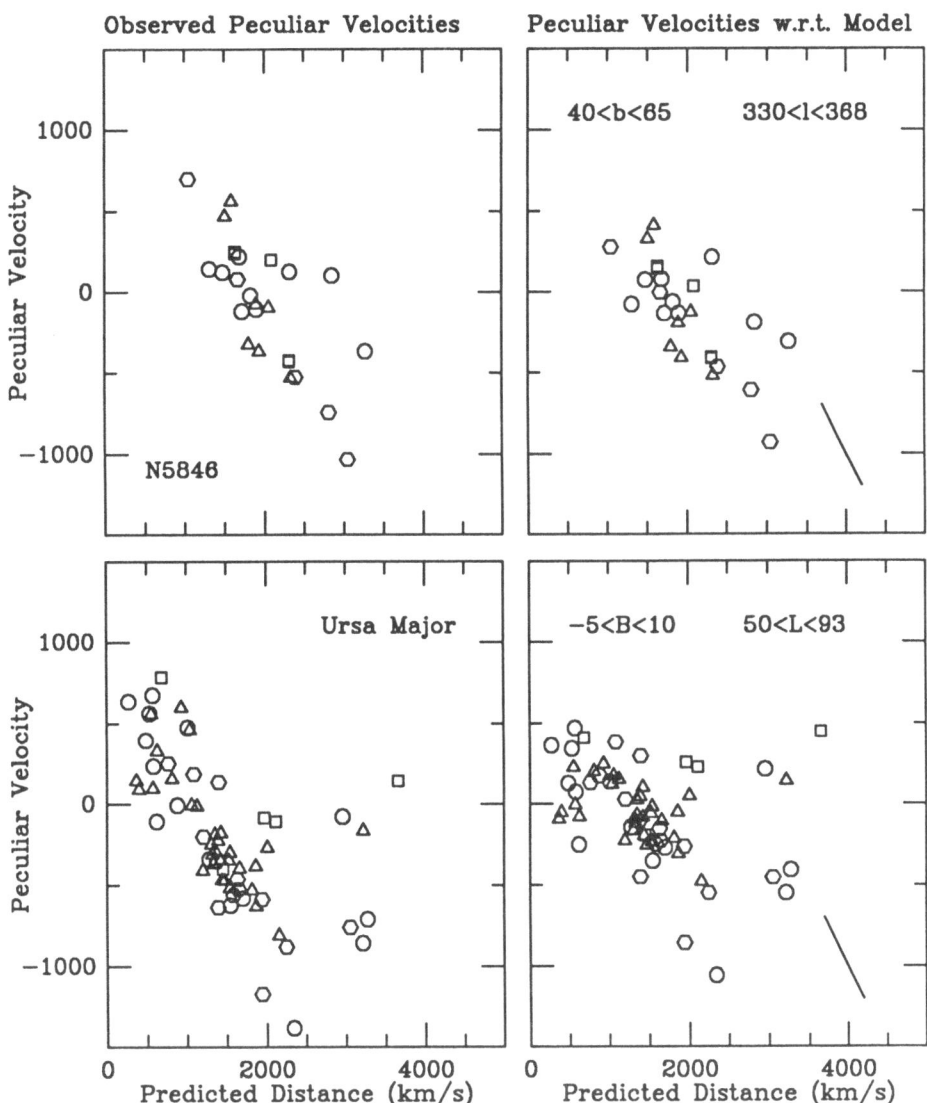

Fig. 3(b). Two examples of 'cluster'–dominated regions.
Upper panels: The N5846 region, defined here as within 40°<b<65°,
330°<l<368°. Lower panels: The Ursa Major region, defined here as
within supergalactic coordinates −5°<B<10°, 50°<L<93°. The
peculiar velocities of galaxies relative to the model, on the near
side of the Ursa Major cluster, show evidence of infall relative
to the cluster center. Too few data are available to reliably
detect infall on the far side of the cluster.

188

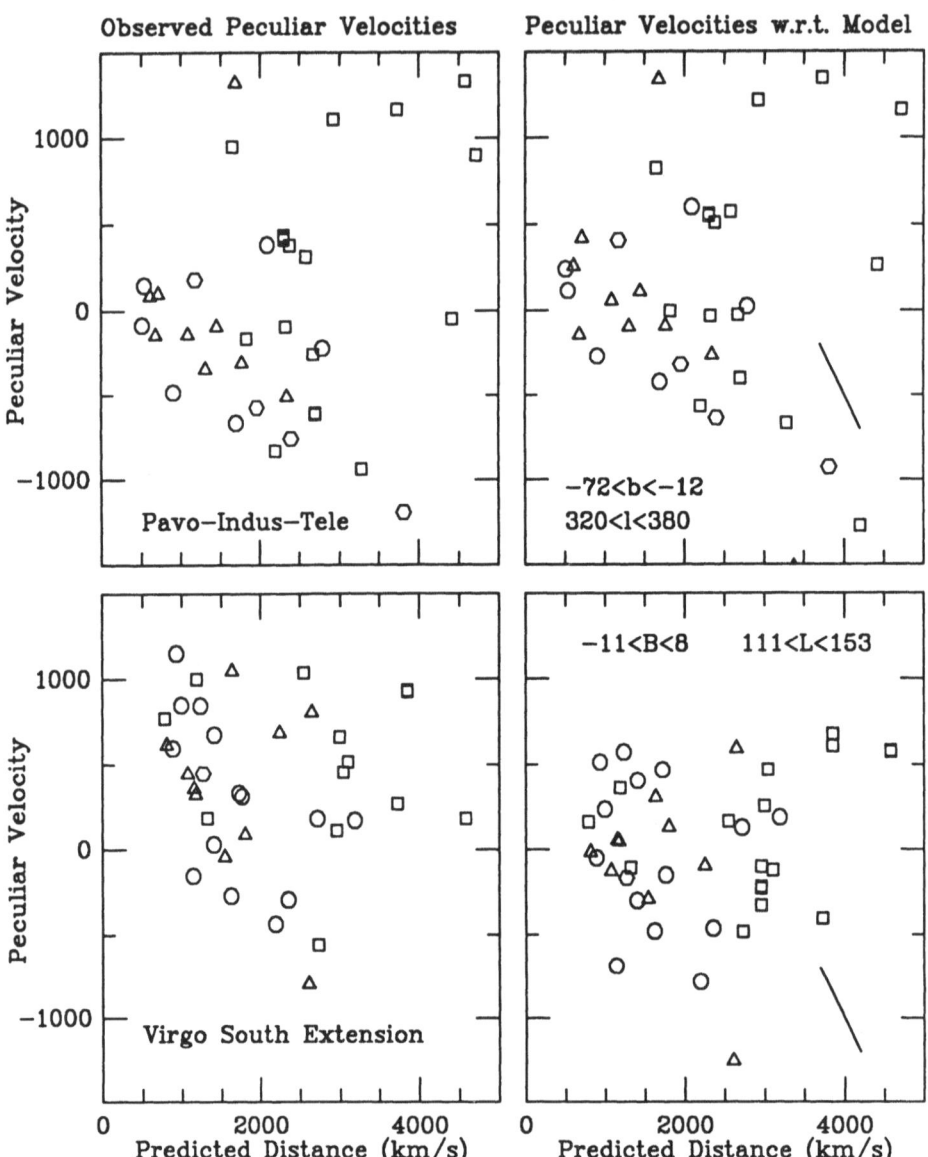

Fig. 3(c). Two examples of 'noisy' regions. Upper panels: Pavo–Indus–Telescopium region, defined here as within $-72°<b<-12°$, $320°<l<380°$. Lower panels: The Virgo South Extension, defined as within supergalactic coordinates $-11°<B<8°$, $111°<L<153°$.

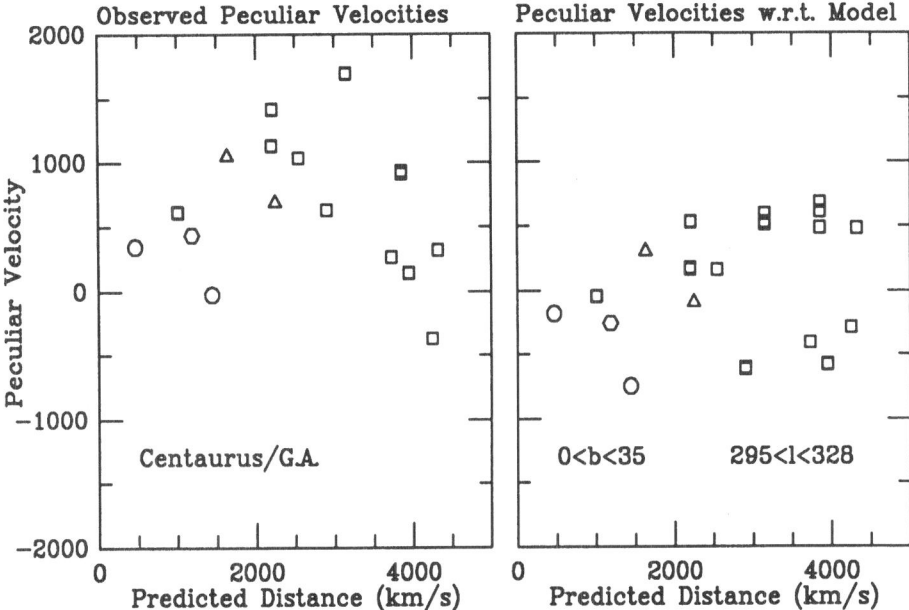

Fig. 3(d). The Great Attractor region north of the galactic plane, centered on the Centaurus cluster and defined here as within 0°<b<35°, 295°<l<328°. Note the change of vertical scale with respect to the previous figures. The decline of peculiar velocity from a distance of ~3000 km/s to 4000 km/s suggests that the G.A. may have a core of about 1000 km/s in radius. If roughly spherical in shape, this projects to an apparent size of ~1 steradian at a distance of 4000 km/s; comparable in size to the region observed by Dressler to be dominated by galaxies with radial velocites of ~4000 km/s.

In contrast, the observed motions of galaxies relative to the center of the Ursa Major cluster are dominated by the G.A. flow (Fig. 3b, lower panel, LH side), much for the same reasons as in the case of the Leo Clouds. Residual infall motion towards this cluster is evident on the near side of the cluster (lower panel, RH side), with an amplitude of ~300 km/s. These data require further detailed analysis to determine the extent of infall from the back side. The cluster itself, however, as a net motion of ~200 km/s relative to the model.

4.3.3 'Noisy' Regions. In three of the regions investigated, galaxies appear to have a signficantly larger velocity dispersion than within the 'quiet' regions. Fig. 3(c) and 3(d) present the data for these three regions: Pavo-Indus-Telescopium (PIT), the Virgo South Extension, and the direction towards the Centaurus cluster and the Great Attractor.

The galaxies in PIT have a large negative peculiar motion relative to the CMB, which is largely removed by the model (Fig. 3(c), upper panels). Galaxies in the Virgo South Extension have a large positive

peculiar motion relative to the CMB that is also largely removed by the
model (Fig. 3(c), lower panels). In the case of PIT, Lynden-Bell and
Lahav (1988) note that this might be a nearby supercluster that is only
sparsely-sampled with the present data, and is seen along the
line-of-sight. The Virgo South Extension samples galaxies that cut
across the line-of-sight towards the G.A. region. A more detailed
investigation should be made of the motions in this region.

The motions of galaxies towards the G.A. itself are shown in
Fig. 3(d). Motions of galaxies relative to the CMB in this region show
a rapid rise in peculiar velocity, to a value of ~1000 km/s at a
distance of ~3000 km/s, and then a decrease to near zero at a distance
of ~4500 km/s. It is this latter fact that led FB to include a 'core'
radius in the mathematical model for the G.A. velocity field. From the
remaining large peculiar velocities of galaxies relative to the model,
the current spherically-symmetric G.A. model may not fit well the
motions within 1500 km/s of the center of the G.A..

5. KNOWN PROBLEMS WITH THE G.A.+Vinf+L VELOCITY FIELD MODEL

5.1 A 'Clumpy' Core

As illustrated above, the current model for the G.A. is too symmetric to
fit the motions of galaxies within the core of the G.A. This fact is
also clear from the observations of Dressler (1988), who shows that
galaxies covering a solid angle of nearly a steradian around $l=310^{o}$,
$b=10^{o}$ have radial velocities between 4000-6000 km/s. The observed
luminous mass distribution within this region is lumpy, and the
indication is that the Centaurus region itself is one of the major
sub-condensations whose gravitational influence on the Local Group is
considerable. An obvious modification to the present model is to permit
the core region to be asymmetric. This is a move, however, that will
add more free parameters to a model that already has 6 parameters to
describe motions due to the G.A..

5.2 'Regional' Motions

The analysis in FB shows that the motion of the Local Anomaly is fairly
typical of the motions of other regions of space: Galaxies within a
region of ~1500 km/s in size appear to share a common motion, be it
'quiet' bulk, 'cluster-dominated' or 'noisy.' In addition, many regions
have a net motion relative to the larger-scale G.A. flow. Indeed, two
such 'regional' motions, namely Virgo infall and the Local Anomaly, are
explicitly incorporated into our general velocity field model, since
they directly affect the motion of the Local Group.

The Local Anomaly is the only region for which we can know the
full three-dimensional motion. Nonetheless, the 3-D motion of 360 km/s
(= 210 km/s in one-dimension) for the Local Anomaly relative to the G.A.
flow, is consistent with the 1-D velocities of 100-250 km/s of other
regions (e.g., Leo Clouds, Ursa Major). In order to fully model all of
the motions in this volume of space, on all size-scales, one will

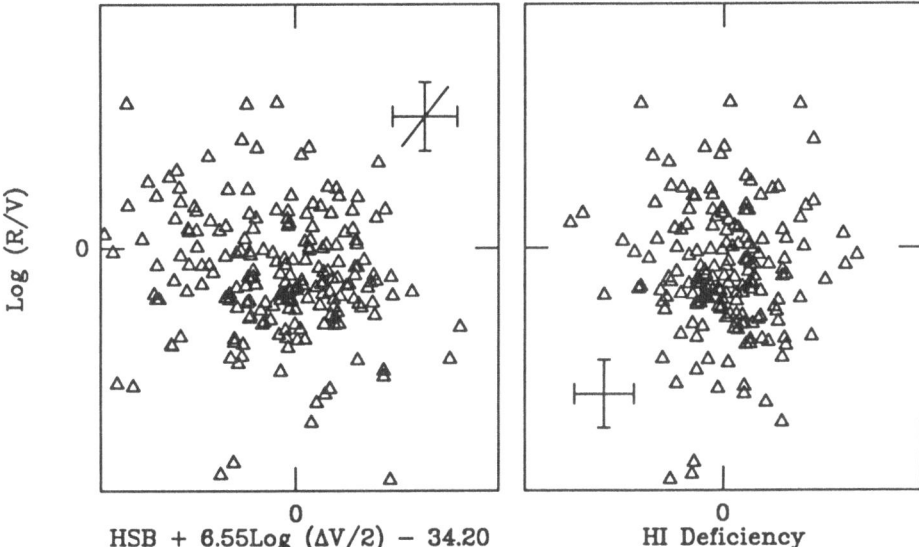

Fig. 4. Two tests for systematic effects in predicted distances of spiral galaxies, using the IR–TF distances for the Aaronson 'good' sample. Left–hand panel: Do differences in stellar population affect predicted peculiar velocities? The log of the ratio of predicted distance (R) to CMB radial velocity (V), plotted versus the residual of H mag surface brightness (HSB) w.r.t. rotation velocity for 204 galaxies. Random errors in rotation velocity lead to correlated errors in ΔHSB, as shown. Right–hand panel: Does environmental density affect predicted peculiar velocities? Log (R/V) plotted versus 21 cm HI deficiency for 179 galaxies. No significant correlation is seen in either test.

need a much more detailed map of the distribution of matter than can be obtained by the methods presented here. In this regard, combining the observed peculiar velocity data with the mass density predictions of the IRAS surveys (Strauss and Davis 1988; Yahil 1988) should yield very interesting results.

6. TEST OF SYSTEMATIC EFFECTS

6.1 Using the Intrinsic Properties of Spiral Galaxies

Fig. 4 presents two tests of systematic errors that might affect predicted distances, using the Aaronson et al. 'good' sample of spiral galaxies. Analogous tests can be made of elliptical galaxies, which will be presented elsewhere (Burstein 1988). In both tests systematic errors would lead to an error in predicted distance that would scale with distance. As such, the quantity to be compared is taken to be the logarithm of the ratio of predicted distance (R) over observed CMB

radial velocity (V). The LH side of Fig. 4 compares log (R/V) against a
measure of the dispersion of stellar population in these spiral
galaxies, the residual of H mag surface brightness (HSB) as a function
of rotation velocity (V/2; Burstein 1982; Burstein, Condon and Yin
1987). Although an intrinsic range in stellar population is known to
exist for this sample (Burstein, Condon and Yin), there is no indication
that this range in stellar population affects the predicted distances in
any systematic manner.

The deficiency of 21 cm HI emission from spiral galaxies is
correlated with distance from the center of a cluster (e.g. Giovanelli
and Haynes 1983). Fig. 4 (RH side) plots log (R/V) versus HI deficiency
for the subset of these same galaxies with reliable HI fluxes. Again,
no significant correlation is seen, implying that the predicted distances for these galaxies are not a systematic function of environment.

6.2 Using the Tidal Velocity Field

The peculiar motions of galaxies in this volume that give rise to the
quadrupole term in the velocity field are observed at distances ranging
from 500 km/s to over 2500 km/s from the Local Group. Galaxies over
these distances show a negligible systematic trend of peculiar velocity
with distance, with an average value of about -400 km/s. Any systematic
effect that could produce this constancy of motion would have to be both
large and concentric with the Local Group: A peculiar motion of 400
km/s at a distance of 500 km/s is a very large perturbation, much larger
than that produced by any error in Malmquist bias estimate. A motion of
400 km/s at a distance of 2500 km/s is a much smaller perturbation.
Both the required amplitude and Galactocentric behavior of such a
systematic effect make it extremely improbable.

In sum, we take these comparisons to indicate that systematic
errors in the present Aaronson et al. 'good' distances (Malmquist bias
errors excluded) affect the derived peculiar velocities at a level that
is currently undetectable.

7. A 'STRAIGHTFORWARD' VIEW AND A 'MINIMALIST' VIEW
   OF THE VELOCITY FIELD

The volume of space dominated by the Great Attractor can be described in
a number of ways, the most informative of which (from an observer's
point of view) can be described as 'straightforward' and 'minimalist'
(FB).

In the 'straightforward' view, one simply asks over what distances
are the peculiar motions of galaxies dominated by a motion towards the
Great Attractor. The current observations indicate that the G.A.
dominates the motions of galaxies near its center, at a distance of
4200 km/s, and galaxies that are at distances of 1500 km/s in the
opposite direction from the G.A., or a total volume of about 6000 km/s
in length. The volume pictured in this way could conceivably increase
by a few thousand km/s if it is found that galaxies on the other side of
the G.A. from us are also falling into the core.

In the 'minimalist' view, one asks what is the actual volume of space covered by the observations, in which the motions of galaxies are coherently dominated by the G.A. flow. As discussed in FB, the quadrupole term in the motions of the local galaxies suggests that a coherent flow exists in a cylinder-shaped region of dimension 4500 km/s long and 3500 km/s wide. The long axis of this cylinder is oriented towards the G.A., with one end 1500 km/s in distance on the other side of the Local Group, and one end 3000 km/s in distance towards the G.A.. This latter distance does not include the 1500 km/s radius of the core of the G.A., in which the motions appear to be more complicated.

Finally, it should be noted that, at the time of this writing, Dressler and Faber have made further observations of ellipticals in the direction of the Great Attractor which, they expect, will sample the velocity field on the back side of the G.A. It is hoped that, by the time the proceedings of this conference are published, these new data will have been analyzed.

8. ACKNOWLEDGEMENTS

We wish to thank again the many institutions and observatories who have made this research possible, and who are too numerous to name here individually. We also wish to thank Ofer Lahav for his continuing efforts on our behalf.

8. REFERENCES

Aaronson, M., Huchra, J., Mould, J., Schechter, P. and Tully, R.B. 1982, Ap. J. 258, 64.
Bothun, G.D., Aaronson, M., Schommer, B., Huchra, J. and Mould, J. 1984, Ap. J. 278, 475.
Burstein, D. 1988, in Le Monde des Galaxies, a Symposium honoring G. and A. de Vaucouleurs, to be published.
Burstein, D. 1982, Ap. J. 253, 539.
Burstein, D., Condon, J.J. and Yin, Q.F. 1987, Ap. J. Lett. 315, L99.
de Vaucouleurs, G.A. 1975, in Stars and Stellar Systems, Vol. 9, Galaxies and the Universe, ed. A. Sandage, M. Sandage, J. Kristian, Chap. 14. (Univ. of Chicago Press).
de Vaucouleurs, G.A. and Peters, W.L. 1984, Ap. J. 287, 1.
de Vaucouleurs, G.A., de Vaucouleurs, A. and Corwin, H.G. 1976, Second Reference Catalog of Bright Galaxies (Univ. of Texas: Austin).
Dressler, A. et al. 1987b, Ap. J. Lett. 313, L37.
Dressler, A. 1988, Ap. J. in press.
Faber, S.M. and Burstein, D. 1988, in Large-Scale Motions in the Universe, proceedings of Vatican Study Week #27, ed. G. Coyne and V.C. Rubin (to be published by Princeton University Press). (FB)
Giovanelli, R. and Haynes, M.P. 1983, Astron. J. 88, 881.
Lahav, O., Rowan-Robinson, M. and Lynden-Bell, D. 1988, M.N.R.A.S., in press.
Lilje, P.B., Yahil, A. and Jones, B.J.T. 1986, Ap. J. 307, 91.

8. REFERENCES (continued)

Lynden-Bell, D. and Lahav, O. 1988 in <u>Large-Scale Motions in the Universe</u>, proceedings of Vatican Study Week #27, ed. G. Coyne and <u>V.C. Rubin</u> (to be published by Princeton University Press).

Lynden-Bell, D. et al. 1988, Ap. J. 326, 19. (LFBDDTW).

Peebles, P.J.E. 1988, preprint.

Pierce, M. and Tully, R.B. 1988, preprint.

Strauss, M.A. and Davis, M. 1988, in <u>Large-Scale Motions in the Universe</u>, proceedings of Vatican Study Week #27, ed. G. Coyne and <u>V.C. Rubin</u>. (to be published by Princeton University Press).

Tully, R.B. and Fisher, J.R. 1987, <u>Nearby Galaxies Atlas</u>, (Cambridge Univ. Press: New York).

Yahil, A. 1988, in <u>Large-Scale Motions in the Universe</u>, proceedings of Vatican Study Week #27, ed. G. Coyne and V.C. Rubin. (to be published by Princeton University Press).

DISCUSSION

TULLY: You recall from our discussions at the Vatican Study
Week that I have been talking about a 'local velocity
anomaly', too. You propose that our Coma-Sculptor Cloud has
a bulk translation motion while I suggest that within the
same cloud there is a retardation of the Hubble expansion,
probably to be explained simply by the concentration of mass
within the cloud. In retrospect, it seems to me to be quite
possible that both these circumstances could be correct.

BURSTEIN: Yes, of course I remember, but I did not find a
direct reference to quote. I'm sure you also remember our
discussions on your interpretation of "retardation" motions
within the local anomaly, and how your interpretation is
directly tied to your handling of Malmquist bias.   Although
it is certainly possible that both circumstances could be
correct, until we can agree on how to correctly handle
Malmquist bias, I must still view your interpretation of
"retardation" with some scepticism.

ROWAN-ROBINSON: My question is where is the great
attractor?   In the distribution of IRAS sources I showed
yesterday, I see the Centaurus Cluster at (l, b) $\sim$ (313,34)
with $\bar{v} \sim 3000$. I see a prominent cluster at (l, b)$\sim$
(304,20) also with $\bar{v} \sim 3000$. And I see the Pavo-Indus
complex at (l, b)$\sim$(336,-27) with $\bar{v} \sim 4500$. The only place
for the "Great Actractor" appear to be behind the Galactic
plane. (Incidentally, rich clusters like Virgo and Coma are
also clearly see, though of course not with the same
contrast as elliptical-rich surveys).

BURSTEIN: As we discuss in the paper, it is entirely
possible that as much as half of the mass of the G.A. is
hidden behind the galactic plane. Regarding your
interpretation of your version of the IRAS map: your map
has very little contrast in it, much less than in maps made
from the UGC and the ESO catalogues. I think there are
plausibly 3 reasons for this:
a) You use a flux limit of 0.6 Jy, thus looking at more
distant galaxies. This will tend to 'wash-out' nearby
structures.
b) IRAS galaxies tend to be late-type galaxies, which tend
to cluster less in optical catalogues.
c) The luminosity function of IRAS galaxies is intrinsically
wide, again washing out structures in a map such as yours.
Have you tried to make IRAS maps cuts at different flux
levels?

LILJE: a) What is the number of free parameters in your
model? b) Do you have a quantitative measure of the

significance of the reduction in peculiar motions relative to the model?

BURSTEIN: a) 13. b) Using only the Aaronson et al. 'good' data, we find that the likelihood increase +88.5 for the G.A+VINF+L model.

GUDEHUS: In your solution number 1 you have the Hubble constant as one of the free parameters. This would mean that your peculiar velocities are a function of the Hubble parameter. Isn't the free parameter a combination of the Hubble and an absolute magnitude?
The average direction of the peculiar velocity vectors derived from the m* method is consistent with your direction of the Great Attractor.

BURSTEIN: The Hubble constant is not a "free" parameter, per se, in that it's value does not matter in determining peculiar velocities. Expressing all distances in units of km/s for galaxies is analogous to express distances of solar system objects in units of A.U. We just have to agree on what an "A.U." is (in this case, the distance to the Coma cluster).

LOCAL LARGE SCALE STRUCTURE vs COLD DARK MATTER

Nick Kaiser
Canadian Institute for Theoretical Astrophysics
University of Toronto
60 St. George St.
Toronto, Ontario M5S 1A1
Canada

ABSTRACT The cold-dark-matter model makes very specific predictions for large scale structure. Here I compare these predictions with some observations of large scale structure in the nearby universe. I review a number of tests described in more detail elsewhere (Kaiser and Lahav, 1988a,b). These comprise a comparison of the cold-dark-matter predictions with angular dipoles of IRAS galaxies and bulk flow solutions; an improved determination of the bias parameter from a variant of Gott's method; and a comparison of both angular dipoles and acceleration vector derived from the IRAS 2Jy redshift survey with the density field predicted from modelling of the peculiar velocity field. A new method for estimating the peculiar velocity autocorrelation function is described and applied to a selection of available catalogues. I conclude with a discussion of the connection between biased theories for galaxy formation and the distance estimates used to obtain peculiar velocities.

1 THEORETICAL MODEL

The theoretical model we shall assume throughout is the 'standard' cold-dark-matter model, hereafter CDM, with Gaussian adiabatic initial fluctuations with the Harrison-Zel'dovich spectrum. We assume that $\Omega = 1$, $h = 0.5$, and work entirely in linear theory. The density field can be written as $\rho(\mathbf{r}) = \overline{\rho}(1 + \Delta(\mathbf{r}))$, where $\Delta(\mathbf{r})$ is a random Gaussian scalar field: $\Delta(\mathbf{r}) = \sum \Delta(\mathbf{k}) \exp(i\mathbf{k} \cdot \mathbf{r})$, where the Fourier components have random phases. At late times, when the universe has become matter dominated, the power spectrum $P(k) \equiv \langle \Delta(\mathbf{k})\Delta^*(\mathbf{k}) \rangle$ has the form given by Bond and Efstathiou (1984). The peculiar velocity field $\mathbf{v}(\mathbf{r})$ is a random Gaussian vector field, and is linked to $\Delta(\mathbf{r})$ by the equation of continuity: $\mathbf{v}(\mathbf{k}) = iH\mathbf{k}/k^2 \, \Delta(\mathbf{k})$. The treatment of these continuous fields is essentially exact, subject only to the restriction of linear theory. What is less certain is how one should model the spatial distribution of galaxies, which we shall need in order to predict the angular dipoles for instance. We treat the galaxies as a random point process (i.e. we describe the galaxies simply by occupation numbers $n(L_i, \mathbf{r}_i)$ for cells in $L, \mathbf{r}$ space). These galaxies sample a universal luminosity function $\Phi(L)$ and a *biased* density field $\propto (1 + b\Delta(\mathbf{r}))$, so the mean occupation number for the cell with labels i, j is $\langle n(\mathbf{r}_i, L_j) \rangle \propto (1 + b\Delta(\mathbf{r}_i)) \cdot \Phi(L_j)$. Last, but not least, the observer is assumed to be a randomly chosen point in space.

M. Mezzetti et al. (eds.), Large Scale Structure and Motions in the Universe, 197–213.

In addition to these assumptions, two normalisation parameters are needed to fully specify the theoretical model. The first of these we denote by σ_ρ, which is the rms mass fluctuation in a sphere radius $a = 8$ Mpc/h, and which can be exressed as an integral over the power spectrum. The second parameter b is the ratio of the fluctuations in galaxy counts to mass fluctuations. Considerations of galaxy clustering and cluster dynamics suggest $\sigma_\rho \simeq 0.7$, and $b\sigma_\rho \simeq 1.0$, though these determinations are quite uncertain. This normalised model has great predictive power. We will use observations of large scale structure in our vicinity to independently constrain the normalisation parameters and thus test the theory.

2 ANGLE-FLUX DIPOLES AND BULK FLOWS

The statistics we shall consider here are dipole moments on the sky of either the number of galaxies

$$\mathbf{D} = \frac{3 \sum w(S_q) \, \hat{\mathbf{r}}_q}{\int dS \, N(S) \, w(S)}$$

with some flux-weighting scheme $w(S)$, or of line of sight peculiar velocities u

$$\mathbf{U} \equiv \frac{3 \sum w(r_q) \, \hat{\mathbf{r}}_q \, u_q}{\int d^3r \, n(r) \, w(r)},$$

again with some weighting scheme; now a function of distance.

One can write such a dipole as a linear convolution of $\Delta(\mathbf{r})$ plus a 'noise' due to discreteness of galaxies or distance errors in the case of the bulk-flow dipoles $\mathbf{U}$. If we imagine observers throughout the universe all calculating these statistics, then the results form a random Gaussian vector field, which is a filtered version of the peculiar velocity field $\mathbf{v}(\mathbf{r})$. The main problem here is calculating the appropriate filter characteristics. For example, for the angular dipole $\mathbf{D}$ we have

$$\mathbf{D}(\mathbf{r}_0) = 3b \int d^3r \, W(r) \Delta(\mathbf{r}_0 + \mathbf{r})\hat{\mathbf{r}} + \text{noise}$$

where

$$W(r) \propto \int dL \, w(L/4\pi r^2) \, \Phi(L).$$

so the total variance is

$$\langle \mathbf{D} \cdot \mathbf{D} \rangle = \frac{b^2}{(2\pi)^3} \int d^3k \, P(k) \, \dot{W}(k) \, W^*(k)$$
$$+ \frac{9}{N_g} \frac{\int dS \, S^{-5/2} \int dS \, w^2(S) \, S^{-5/2}}{\left(\int dS \, w(S) \, S^{-5/2} \right)^2}$$

where the filter function (in k-space) is $W^*(k) \equiv -12\pi i \int dr \, r^2 \, W(r) \, j_1(kr)$, and the second term is the 'shot noise' due to the discreteness of galaxies.

Modelling of a bulk flow solution $\mathbf{U}$ is formally very similar.

$$\mathbf{U}(\mathbf{r}_0) = 3 \int d^3r \, W(r) \, (\hat{\mathbf{r}}.\mathbf{v}(\mathbf{r}_0 + \mathbf{r})) \, \hat{\mathbf{r}} + \text{noise}$$

where

$$W(r) = \frac{n(r)w(r)}{\int d^3r\, n(r)\, w(r)}.$$

with its pair in Fourier space

$$W(k) = \frac{12\pi}{k} \int dr\, r^2\, W(r) \left(\frac{1}{3}j_0(kr) - \frac{2}{3}j_2(kr)\right).$$

We have assumed here that the survey region is spherically symmetric. For the more general analysis for surveys with patchy sky coverage see Kaiser (1988a).

Generalising, one can form the probability distribution for a combination of N vectors $\mathbf{V}_i$ of this form. The procedure is to first calculate the window functions $W_i(k)$; then form the covariance matrix elements $C_{ij} \sim \int d^3k\, P(k)W_i(k)W_j(k) +$ noise, and the multivariate pdf is then

$$P(V_{i\alpha})d^{3N}D = \frac{d^{3N}D}{(2\pi)^{3N/2}|C|^{3/2}} \exp -\frac{1}{2}\sum\sum C_{ij}^{-1}\sum_\alpha V_{i\alpha}V_{j\alpha}$$

We can use this either to test our normalised CDM model via the χ^2 statistic which appears in the exponent, and which has the usual distribution, or to calculate likelihood $L = P(\text{data}|\text{CDM})$ as a function of normalisation parameters.

2.1 CDM vs IRAS Dipoles

Kaiser and Lahav (1988a,b) have extracted 5 dipoles from the Meurs and Harmon (1987) IRAS catalogue (5 equal log intervals of flux from 0.7 to 10 Jy).

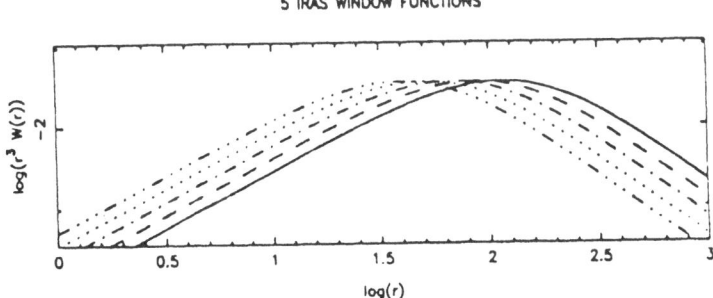

Figure 1: Window functions for 5 IRAS flux slices

S_{min}	S_{max}	D_x	D_y	D_z	D_{obs}	D_{pred}	D_{th}	D_{sn}
0.7	1.2	-0.008	-0.131	0.049	**0.140**	**0.110**	0.103	0.038
1.2	2.0	-0.003	-0.103	0.089	**0.136**	**0.154**	0.143	0.058
2.0	3.5	0.053	-0.179	0.147	**0.238**	**0.212**	0.195	0.083
3.5	5.9	-0.102	-0.182	0.081	**0.224**	**0.295**	0.265	0.129
5.9	10.0	-0.177	-0.090	0.168	**0.260**	**0.401**	0.353	0.190

Table 1: Individual dipoles vs theoretical predictions for the rms.

Applying the χ^2 test with our preferred normalisation gave 15.0 with 15 degrees of freedom. This was clearly a bit of a fluke to get a value so close to the expectation value. Likelihood analysis gave a maximum likelihood estimate $b\sigma_\rho = 1.1\pm^{0.8}_{0.4}$. We conclude that there is no conflict between CDM theory (with our preferred normalisation at least) and the IRAS dipoles.

2.2 CDM vs BULK FLOWS

In a very similar analysis, Kaiser and Lahav have used 3 bulk flow solutions comprising the Local Group motion; a 'near field' solution for elliptical galaxies (Lynden-Bell et al., 1987) and the AHMST (1982) spirals in the distance range 0-3200 km/s; and a 'far field' solution for the elliptical galaxies in the distance range 3200-8000 km/s. These data can be used to constrain σ_ρ alone.

sample	U_x	U_y	U_z	U_{obs}	U_{pred}	U_{th}	U_{sn}
1. Local Group	-20	-532	277	**600**	**990**	990	—
2. E's 0-3200	362	-486	112	**616**	**440**	431	90
3. E's 3200-8000	194	-132	-3	**235**	**376**	268	251
4. S's 0-3000	61	-349	-13	**355**	**628**	627	31
5. E's + S's 0-3200	139	-398	15	**420**	**591**	590	29

Table 2: Individual bulk flows vs predicted rms values

The χ^2 test gave 8.7 with 9 degrees of freedom (another fluke!), and likelihood analysis gave $\sigma_\rho = 0.6\pm^{0.2}_{0.1}$. Again, we conclude there is no conflict between theory and these bulk flow solutions.

Various previous discussions (e.g. Vittorio et al., 1986; Bond, 1987) of the implications of the same data have reached quite different conclusions. For the main part, the same normalisation and choice of Hubble constant have been used. The discrepancy can be traced to i) assuming a window function which is much broader (in real space) than that which emerges from the calculations shown here, and ii) a failure to include the noise in the observations. I would stress that there is no ambiguity in the choice of window, but, for what its worth, the large windows which have been assumed in other studies do provide a reasonable approximation to the window calculated for the deep field elliptical sample. Since for that sample no motion is detected above the noise level (see table 2 above), this is clearly not in conflict with CDM, which actually predicts an rms value somewhat larger than that observed.

A useful graphical indication of the effective depth of the elliptical sample is given by plots of the number and weight distribution. The weight assigned to a galaxy in ML bulk flow solutions varies as the inverse square of the distance, reflecting the constant *fractional* error in distance estimates.

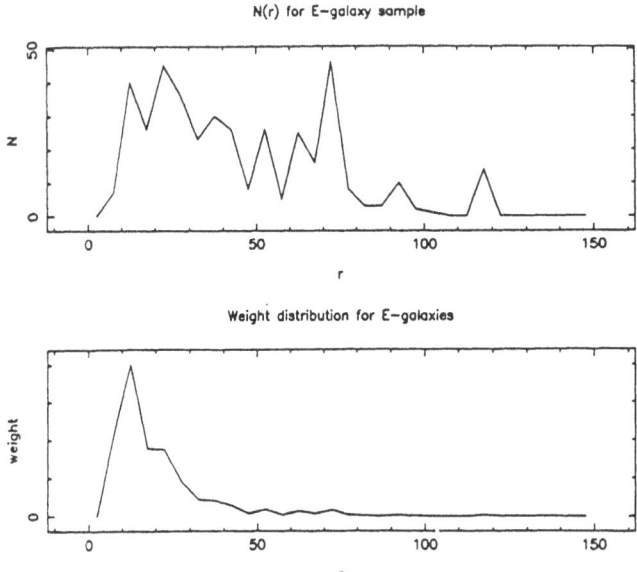

Figure 2: Number and weight distribution for elliptical galaxy sample

It should be apparent that the effective depth of the survey is much smaller than, say, a top hat of radius 50 Mpc/h; in which case the weight distribution would peak at the outer radius. Since the predictions depend quite sensitively on the assumed window function it is imperative that this be calculated from the geometry of the survey and the weighting scheme used by the observer.

2.3 Maximum Likelihood Determination of the Bias Parameter

The angular dipoles and bulk flows can be used together to determine the bias parameter. In this application the assumed power spectrum takes a subservient role. The results are not particularly sensitive to the precise form adopted, but some assumption about the depth distribution of the fluctuations is necessary to give a realistic estimate of the uncertainty in the method.

Gott's original idea was to use flux weighted dipole $\mathbf{D}_S$ as a measure of $(b$

times) the net acceleration acting on the galaxy due to all galaxies outside the Local Group, and to compare this with local group motion $\mathbf{U}_{LG}$ and thus determine the bias parameter b. (Actually it was assumed that light-traces-mass, and that the constant of proportionality was not b but $\Omega^{-0.6}$. Here we are assuming $\Omega = 1$, and allowing a linear bias, but the calculation is still the same.) A ML estimate of b can be obtained from the likelihood function $L(b) = P(\mathbf{U}_{LG}, \mathbf{D}_S | b....)$.

An useful way to quantify the uncertainty in the method is to use the conditional probability $P(\mathbf{U}_{LG} | \mathbf{D}_S, ...)$ which is Gaussian with mean

$$\overline{\mathbf{U}}_{LG} = \frac{\langle \mathbf{U}_{LG}.\mathbf{D}_S \rangle}{\langle \mathbf{D}_S^2 \rangle} \mathbf{D}_S,$$

and with fractional uncertainty E_U where

$$E_U^2 \equiv \frac{\langle (\mathbf{U}_{LG} - \overline{\mathbf{U}}_{LG})^2 \rangle}{\langle \mathbf{U}_{LG}^2 \rangle} = \left[1 - \frac{\langle \mathbf{U}_{LG} \cdot \mathbf{D}_S \rangle^2}{\langle \mathbf{U}_{LG}^2 \rangle \langle \mathbf{D}_S^2 \rangle} \right].$$

and each $\langle ... \rangle \sim \int d^3k \, P(k) \, W^*W$ with the relevant window functions inserted and noise terms added where appropriate.

For the traditional $\mathbf{U}_{LG}, \mathbf{D}_S$ combination. the uncertainty is quite large: $E_U \simeq$ 35%. The uncertainty derives in part from the graininess of the nearby galaxies, and in part because of the flux limits which mean that the ratio of the dipole to the acceleration becomes wavelength dependent for a realistic survey. Kaiser and Lahav (1988a,b) have constructed and applied an improved test which uses 5 flux dipoles and 3 bulk flows via likelihood analysis. The basic idea is that just as $\mathbf{U}_{LG}$ correlates well with an idealised $\mathbf{D}_S$, other $\mathbf{U}$ vectors which measure the motion of larger regions correlate well with $\mathbf{D}$ vectors which give lower weight to bright nearby galaxies, and which therefore have much reduced statistical uncertainty. From the likelihood function

$$L(b) = P(\mathbf{D}_1, ..., \mathbf{D}_5, \mathbf{U}_1, ..., \mathbf{U}_3 | b....)$$

we obtained $b = 1.6^{+0.3}_{-0.2}$, quite similar to our preferred value, and with much smaller uncertainty than previous studies. This corresponds to an 'Ω' for IRAS galaxies of about 0.45.

2.4 IRAS Dipoles vs 'Great Attractor' Model

One can also compare IRAS dipoles to predictions for models comprising one or more spherically symmetric $\Delta \propto r^{-2}$ perturbations. For the 5 dipoles described above, the best fitting single attractor is centred at 40 Mpc/h, though the distance to the centre is poorly determined. Acceptable χ^2 values were obtained for either single or multiple attractor models derived from velocity studies.

We have found that CDM is compatible with IRAS dipoles; CDM is compatible with bulk flows; and the IRAS dipoles are compatible with those predicted in the 'Great Attractor' model. It is interesting therefore to ask; Is CDM compatible with the 'Great Attractor'? This is a tricky question and must be phrased very carefully. If we take 'Great Attractor' to denote a perfectly spherical monolithic structure, then the probability of getting something 'just like that' in CDM is zero. However, the data never said the world was monolithic; rather this symmetry was imposed in the form of the model fitted. Moreover, it should be clear from the very local distribution of weight in the elliptical galaxy sample that this model fitting rests essentially on a fit to the local tidal distortion. One can loosely interpret the results of this model fitting as indicating that a substantial part of our acceleration (about 50% in the single attractor model) derives from distances greater than 40 Mpc/h. Now in CDM there are certainly substantial long wavelength perturbations - the rms acceleration from $r > 40$ Mpc/h amounts to $300\sigma_\rho$ km/s, or about 200 km/s with our preferred normalisation. This should be compared with $\simeq 300$ km/s in the 'Great Attractor' model, so looked at this way, the discrepancy does not seem to be unacceptable. It may be that this comparison has thrown out some important ingredient which may be crucial for rejecting the CDM model. I shall return to consider the peculiar velocities in §4 where I develop a statistic to measure the autocorrelation function of the peculiar velocity field.

3 GRAVITY VECTOR FROM THE IRAS RESHIFT SURVEY.

Strauss and Davis (1987) have extracted from the 2 Jy IRAS redshift survey the statistic

$$\mathbf{v}_s = \frac{H_0}{4\pi n_1} \sum_i \frac{1}{\phi(s_i)} \frac{\hat{\mathbf{s}}_i}{s_i^2}.$$

Like the angular dipole, this quantity also measures acceleration if redshift is equated with distance. They found the acceleration converged by $r \simeq 40$ Mpc/h with little pull from larger r. This seems to conflict with the 'Great Attractor' view of the world. This apparent discrepancy is resolved once allowance is made for peculiar velocities. With our model,

$$\mathbf{v}_s = \frac{1}{4\pi} \int d^3 s \, \frac{\hat{\mathbf{s}}}{s^2} \Delta_s(\mathbf{s}) + \text{noise}$$

but with

$$\Delta_s = b\Delta(\mathbf{r}) - \left(2 + \frac{d\ln\phi}{d\ln r}\right) \frac{\hat{\mathbf{r}}.(\mathbf{v}(\mathbf{r}) - \mathbf{v}(0))}{r} - \hat{\mathbf{r}} \cdot \frac{d\mathbf{v}(\mathbf{r})}{dr}$$

The (unwanted) terms involing peculiar velocities arise from the distortion of the clustering pattern (Kaiser, 1987) The details of what effect these corrections have in the present instance are given elsewhere (Kaiser and Lahav 1988a). The main conclusions are: i) If there was really no acceleration from beyond some radius then our motion would *induce* an apparent acceleration from that region

$$\mathbf{v}_s = \frac{1}{3}\mathbf{v}(0) \int \frac{dr}{r} \left(2 + \frac{d\ln\phi}{d\ln r}\right)$$

so from the distance range 40-80 Mpc/h one would have expected to see an apparent repulsion of amplitude 150 km/s. In so far as this is *not* seen, we can discount the possibility that the true acceleration converges within 40 Mpc/h. ii) the $\mathbf{v}_s$ predicted for a single 'Great Attractor' model gives an overall apparent acceleration of about the right magnitude, but like the data, shows a small apparent acceleration from large r. One concludes that with peculiar velocity distortion taken into account, $\mathbf{v}_s$ also indicates significant long range power, and appears to be consistent at least with the 'Great Attractor' picture.

4 VELOCITY AUTOCORRELATION FUNCTION

I have shown that the bulk flow solutions, when realistically modelled, are compatible with CDM, but they are also compatible with models with more power at large scales. It may be that by using this highly reduced representation of the data we have discarded some important information which may rule out CDM as a viable model. One potentially fruitful approach which suggests itself is to use the velocity autocorrelation function $\xi_{vv}(r) \equiv \langle \mathbf{v}(\mathbf{r}_0 + \mathbf{r}) \cdot \mathbf{v}(\mathbf{r}_0) \rangle$. In a previous study (Kaiser 1988a) I applied likelihood analysis, using a family of theoretical models characterised by a coherence length and a total variance $\xi_{vv}(0)$, and confronted these with the elliptical galaxy data. This approach has the benefit that the data are used in an essentially unreduced form. Unfortunately, while the total variance seemed to be well constrained, the data appeared to be quite ambivalent as to the coherence length, particularly if some tolerance is allowed in the real uncertainty. Here I shall develop a method to obtain a more direct estimate of ξ_{vv}. If we had access to the 3-dimensional peculiar velocity $\mathbf{v}$ then it would be straightforward to construct an estimator: simply sum the dot product $\mathbf{v} \cdot \mathbf{v}$ for all pairs of galaxies with a given separation and divide by the number of pairs. (There is a slight technical problem here in that one would have to use either a noisy distance estimate or a recession velocity to determine the separation. Neither of these is ideal, but the latter choice should give reasonable results, at least for large separations.) Unfortunately, $\mathbf{v}$ is not observable, only the line of sight component $u \equiv \hat{\mathbf{r}} \cdot \mathbf{v}$. From this one can construct a correlation function $\phi(r, \theta_1, \theta_2) \equiv \langle u_1 u_2 \rangle$, but this is a function of the three variables r, θ_1, θ_2 which define the geometry of a pair of galaxies relative to the observer.

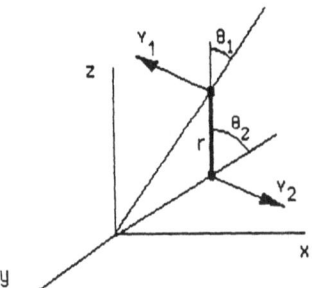

Figure 3: Geometry of a pair of galaxies

If we introduce a little theoretical prejudice at this point and assume that the velocity field is curl free, as predicted in the gravitational instability picture, and is statistically isotropic (i.e. we assume that the power spectrum is a function only of the modulus of the wave vector), then we find (Kaiser 1988a)

$$\langle u_1 u_2 \rangle = f(\theta_1, \theta_2)\xi_{vv}(r) + g(\theta_1, \theta_2)\zeta_{vv}(r)$$

where

$$f = \cos \theta_1 \cos \theta_2; \quad g = 2 \cos \theta_1 \cos \theta_2 - \sin \theta_1 \sin \theta_2,$$

$$\xi_{vv}(r) = \int d^3k \, P_{\mathsf{v}}(k) \frac{\sin kr}{kr}$$

and the function ζ_{vv} is another integral over the power spectrum:

$$\zeta_{vv}(r) = \int d^3k \, P_{\mathsf{v}}(k) \left(\frac{\cos kr}{(kr)^2} - \frac{\sin kr}{(kr)^3} \right).$$

The separation of the angular and radial dependence dependence of $\langle u_1 u_2 \rangle$ admits a simple way to estimate $\xi_{vv}(r)$: First compute a raw estimate of the 3-dimensional autocorrelation function

$$\hat{\phi}(r, \theta_1, \theta_2) \equiv \sum_{\text{pairs}} u_1 u_2 / \sum_{\text{pairs}} 1$$

for bins in r, θ_1, θ_2 space, and then fit this to the theoretically predicted form $\phi = f\xi_{vv} + g\zeta_{vv}$. If the coverage of the θ_1, θ_2 plane were even then one could simply do this by Fourier analysis. In fact the coverage is highly non-uniform so it is preferable to fit by minimising a 'chi-squared' statistic

$$\chi^2 \equiv \sum_{\theta_1 \theta_2} w(r, \theta_1, \theta_2) \left[\hat{\phi}(r, \theta_1, \theta_2) - f\hat{\xi}_{vv}(r) - g\hat{\zeta}_{vv}(r) \right]^2$$

resulting in a final estimator

$$\hat{\xi}_{vv}(r) = \frac{\sum wg^2 \sum wf\hat{\phi} - \sum wfg \sum wg\hat{\phi}}{\sum wf^2 \sum wg^2 - (\sum wfg)^2}$$

where the sums are over the cells at fixed r.

The choice of weighting function is non-trivial. One can construct a weighting scheme which minimises the 'formal' errors (i.e. those arising due to the errors in the distance estimates alone). However, this is rather pathological as it gives nearly all the weight to very nearby galaxies, and thus results in a realistic error which is enormous as one is then sampling only a tiny, and probably unrepresentative region. I have chosen to weight pairs equally. No doubt one can improve on this, but it seems a useful starting point.

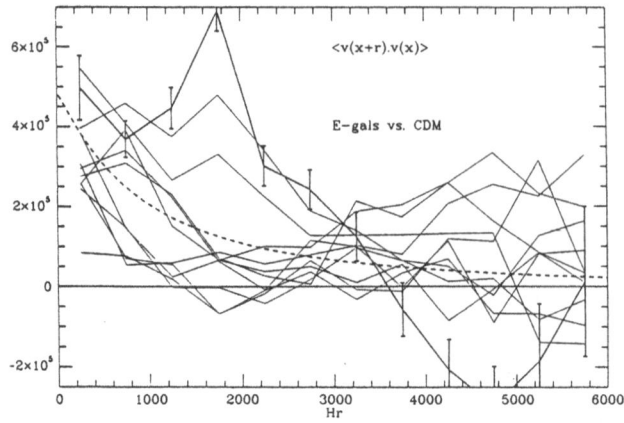

Figure 4: Velocity autocorrelation function estimated from the 'old' elliptical galaxy sample (heavy line). Formal error bars shown. The dashed line is the ensemble average for CDM with our preferred normalisation. The thin lines show an ensemble of CDM realisations with mock galaxies with the same spatial distribution as the real elliptical galaxies.

The result of applying this analysis to the sample of elliptical galaxies given to me some time ago by Lynden-Bell is shown in figure 4. It became immediately apparent, by dividing the survey into subsamples and analysing these separately, that the survey is too small to be a 'fair sample'. Nonetheless, it is still interesting to compare with the theoretical prediction. The coherence length for this sample is considerably greater than the ensemble value for CDM. I made a set of 10 realisations of CDM using linear theory and a spatial distribution of 'galaxies' just like those observed and analysed them in an identical fashion. The results are shown as the thin solid lines in figure 4. One of these had a coherence length as large as the observed sample, so this feature of the data is not so exceptional, but it is notable that none of the realisations had as large fluctuations at large separations as the actual estimate.

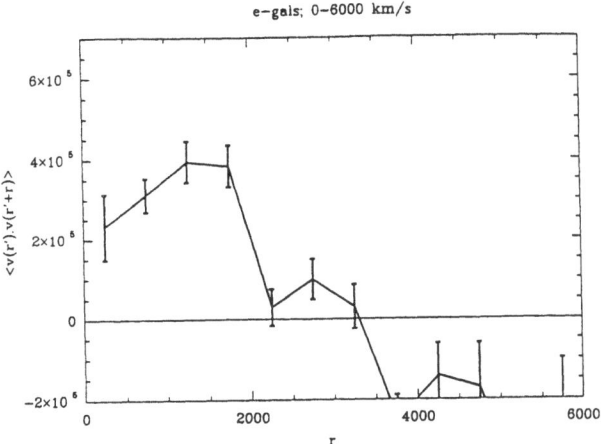

Figure 5: Velocity autocorrelation function for the new elliptical galaxy sample.

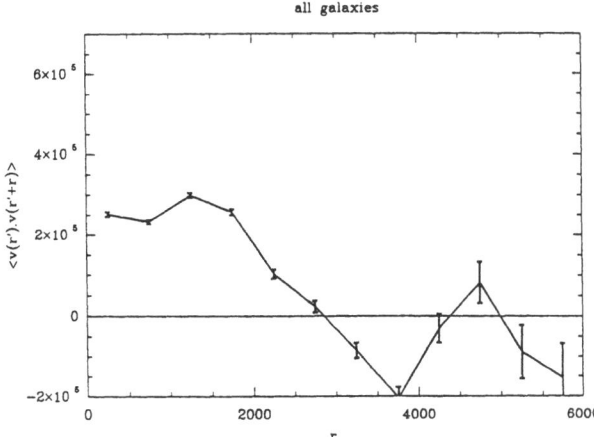

Figure 6: Velocity autocorrelation function for combined elliptical and spiral sample.

Estimates of ξ_{vv} for the revised elliptical sample and the combined elliptical-spiral compilation (Burstein, these proceedings) are shown in figures 5 and 6. The amplitude is a little lower than for the old elliptical galaxy sample, but the general features persist.

There are several major problems in interpreting these estimates that are perhaps worth mentioning. In addition to the apparent lack of representativeness of the samples available, the procedure for making the artificial catalogues is less than perfect in many ways. Numerical simulations would be a great help here to quantify the effect of non-linearity and to generate a more realistic set of 'galaxies', and this work is currently under way. One important difference between this statistic and the bulk flows is that the bulk flow solutions are insensitive to Malmquist bias to zeroth order in the perturbation amplitude; in a uniform universe an uncorrected (or incorrectly corrected) Malmquist bias does not couple to a dipole moment. For the ξ_{vv} estimator this is unfortunately not the case, and I have found that if I split the sample into shells in radius the estimates become very large. Perhaps this is reflecting a Malmquist effect. Another potential source of bias arises because the mean Hubble parameter will typically have been determined from the same data set. This is analogous to the 'integral constraint' problem familiar in conventional spatial correlation studies, and this may result in spurious apparent long-range correlations.

While there are many potential problems that need to be explored here, I think it is exciting that this statistic is showing the first quantitative (rather than anecdotal) evidence that all might not be well with CDM and large scale structure in the local universe. One very nice feature which emerges is that the 'formal' error bars are very small. If it became feasible to extend the samples to greater depth with similar precision in the distance estimators, then, even though the formal errors actually *increase* with more data, they do so only as the square root of the depth for magnitude limited samples, so one could hope to obtain a reasonably representative sample before the formal errors overwhelm the signal.

5 DISTANCE ESTIMATORS AND GALAXY FORMATION

So far, I have interpreted the estimated peculiar velocities literally and asked whether these are compatible with predictions of CDM at large scales. There is another connection with the theory at much smaller scales, since a viable theory for the formation of structure should also be able to account for the internal properties of galaxies which are used to estimate distances. There are various ways one can exploit this interconnection: From the theoretical side, one might ask whether, in a particular theory we should expect to find good distance indicators; what are the likely limitations on these, and how far can we reasonably take the distance estimates at face value? From the observational side we can ask; what does the existence of good distance indicators tell us about the process of galaxy formation?

I shall now argue that, at least in the context of theories for galaxy formation like CDM, in which the initial fluctuations have a Gaussian distribution, one would *not* have expected *a priori* to have good Tully-Fisher or Faber-Jackson distance estimators; rather one would most naturally expect much noisier estimators with systematic shifts of the offset giving rise to spurious peculiar velocities. The existence of what appear to be at least reasonably good distance indicators seems to indicate that galaxy formation was biased in a rather special way, and the challenge to galaxy formation theory is to give an astrophysical explanation of how this came about.

I shall work with the fairly conventional, though of course questionable, assumptions that the non-linear dark matter concentrations are the potential sites of galaxy formation, and that the final stellar velocity dispersion or rotation velocity reflects the depth of the confining potential well. In this context, a 'theory for galaxy formation', is just a rule for assigning luminosities to these sites. A simple rule, and arguably the most natural one, is that the luminosity is just proportional to the mass; this gives rise to a distribution of stellar luminosity which traces the mass distribution on large scales. According to the conventional wisdom, the observed correlations between luminosity and internal velocity are then simply attributed to the spectral index of the CDM spectrum, which is close to $n = -2$ on galactic scales (e.g. Blumenthal *et al.*, 1984). However, in theories like CDM, it turns out that mass and velocity are rather weakly correlated. An estimate (Kaiser and Lahav, 1988a) for the intrinsic scatter about this mean line due to the Gaussian distribution of the initial fluctuations alone gives about 2 magnitudes uncertainty, much larger than that observed. Faber (1982) argued that the ν-value for a perturbation (i.e. the amplitude relative to the rms for that mass scale) determines the morphology through the correlation between ν and shape of the primordial perturbation. This would act to decrease the scatter about the luminosity-velocity relation for any particular morphological type, though it now seems somewhat doubtful whether the ν-shape correlation is strong enough to resolve this problem.

Another very interesting feature of this 'light-traces-mass' model emerges when we consider the modulation of the galactic-scale haloes by density fluctuations on much larger mass scales. The effect of a positive long-wave perturbation on an individual proto-galactic perturbation is to enhance the redshift of collapse, and therefore the internal velocity. By assumption though, the luminosity is unchanged, so we would expect to find the galaxies in dense regions to have a $L - v$ relation which is offset to higher v (or lower L at a given v), and we would overestimate the distance to galaxies in dense regions and *vice versa*. If the mean $L - v$ relation has slope 4, then the fractional systematic error in the distance is the ratio of the long wave perturbation amplitude Δ_B to that of the galactic scale perturbation, or

$$\Delta r = \frac{\Delta_B}{1.68(1 + z_f)} r,$$

if we assume that the halo containing the galaxy collapses at redshift z_f. On the other hand, the real peculiar velocity due to a spherical perturbation of radius R and amplitude Δ_B is $v_{\rm pec} = (1/3)\Delta_B H R$, so the spurious 'peculiar velocity' $H\Delta r$ would overwhelm the true peculiar velocities for perturbations at distances $r > 1.68(1 + z_f)R/3$. In the CDM picture, galaxies form at quite recent epochs $(1 + z_f) \simeq 3$, (more precisely, the comoving number density of halos with rotation velocities like galaxies peaks around that epoch (Cole and Kaiser, in preparation)) so one would expect the spurious velocity to dominate, except for the nearest perturbations.

It is easy to construct a theory for galaxy formation in which all galaxies lie along a universal $L - v$ relation with no intrinsic scatter, and in which inferred peculiar velocities are unbiased; simply postulate that the efficiency of conversion of baryons to luminous stars varies as the 1/3 power of the mass and with collapse time as $(1 + z_f)^2$. With this prescription all 'galaxies' have $L = v^4$ exactly, and it is easy to see that this implies a positive bias in the net luminosity density; in a region subject to an positive long wave perturbation, the protogalaxies collapse earlier, and therefore the net L/M will be enhanced (Kaiser 1988b)

$$\left(\frac{L}{M}\right)' = \left(1 + \frac{\Delta_B}{1.68(1 + z_f)}\right)^{\gamma/2} \left(\frac{L}{M}\right),$$

where γ is the slope of the $L - v$ relation. This bias is seen in numerical simulations where 'galaxies' above a luminosity limit are identified with the dark matter halos with rotation velocities above a corresponding threshold (White *et al.*, 1987; see also Carlberg and Couchman, 1988). This has been dubbed 'natural' bias, though how the non-trivial dependence of star formation efficiency on the properties of the parent haloes that is implicitly assumed here actually comes about is still an unanswered question.

It is possible to construct models with more or less bias than this $L \equiv v^4$ model, by varying the assumed dependence of L/M on collapse time, and the mass dependence of L/M can always be adjusted to give an $L - v$ relation with the right overall slope (Kaiser, 1988b). We have already seen one example of this in the light-traces-mass model. Also of interest are models which predict a stronger 'super-natural' bias; particularly in the context of elliptical galaxies, which seem to be very strongly biased. Just as the 'sub-naturally' biased light-traces-mass model predicts systematic variations in the offset of the $L - v$ relation driven by long wavelength modes, spurious peculiar velocities will also appear in a 'super-naturally' biased model, though in this case the prediction is that galaxies with a given internal velocity dispersion will be *brighter* in dense regions and *vice versa*.

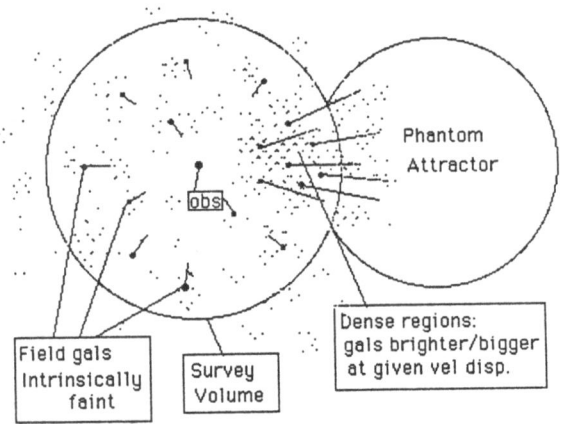

Figure 7: Spurious peculiar velocities in a strongly biased universe

This kind of biasing, if relevant to real galaxies, would have very important consequences for the interpretation of 'peculiar velocities'. Consider the situation sketched in figure 7, in which the observer's peculiar velocity is driven predominantly by the overdensity lying to the right, but which lies inside the region for which the observer has good distance estimates. In the super-naturally biased model the galaxies in this attractor would be anomalously bright (or have larger D_n) at a fixed velocity dispersion. Applying an assumed universal $L - \sigma$ relation one would infer that this attractor has a peculiar velocity in the same direction as the observer, and one might well be tempted to infer the existence of a 'phantom' great attractor at much greater depth. This simple picture has some similarity to our own situation if we identify the real attractor with the Centaurus region, and it is interesting that this type of galaxy formation theory predicts that in a situation like this, one would tend to overestimate the coherence length of the peculiar velocity field.

We are clearly a long way from a convincing theory for galaxy formation that can predict with any certainty the internal properties of galaxies. In the models discussed here, it is assumed that the stellar velocity dispersion is tied to the halo potential, but that the dependence of star formation efficiency on halo properties is left at the disposal of the theorist. I have argued that the simplest light-traces-mass model has difficulty accounting for the small scatter seen in the distance estimates used in practice, and that these observations seem to be pointing towards a positive bias. One biased model which is automatically compatible with the small observed scatter is the 'naturally' biased model in which all galaxies are placed on a universal $L - v$ relation by fiat. While empirically motivated, this identification of L and v^4 disguises a highly non-trivial dependence of star forming efficiency on collapse epoch. It is possible that this is the correct theory (at least for some class of galaxy); in which case astrophysics has contrived to produce distance indicators which are unbiased, but it seems equally plausible that there are some biases in the offset of the $L - v$ relation driven by long-wavelength modes, in which case the inferred 'peculiar velocities' will in general be a combination of true velocities and systematic offsets. The most convincing way to test between these hypotheses is to have much deeper samples in which any systematic offsets will dominate over true peculiar velocities. Unfortunately, such surveys will require prodigious amounts of telescope time and dedication, but the end results would have profound implications for our understanding of the formation of both large-scale and galactic-scale structure.

6 CONCLUSIONS

The CDM theory (or any other model invoking Gaussian fluctuations) permits detailed modelling of linear observations such as the dipole moments of galaxy counts; bulk flow solutions, and estimates of the apparent acceleration vector in redshift space. With our preferred normalisation, CDM predicts IRAS dipoles and bulk flows of about the right amplitude. Applying an improved 'Gott' test gives 'Ω' $\simeq 0.45$ for IRAS galaxies. The 'gravity' vector from the IRAS 2Jy redshift survey also indicates structure on quite large scales, and, while each line of evidence is individually quite questionable, these various statistics all seem to be consistent with about the amount of large-scale power implied by the 'Great Attractor' picture.

Velocity autocorrelation analysis indicates a more coherent velocity field than the theory predicts. The sample is probably unrepresentative, and there may be other problems, but this provides the strongest quantitative evidence for more power on large scales than is supplied by CDM.

I have argued that the very tight correlations between internal properties required for good distance estimates would not naturally arise in Gaussian models, and that the light-traces-mass model in particular seems unacceptable. The small observed scatter and lack of very large spurious peculiar velocities seems to indicate that the galaxies are biased - at least if internal stellar velocities are tied to the depth of the potential of the parent dark halo. Only in one specific biased model would galaxies have a truly universal luminosity velocity dispersion relation, and it seems not implausible that what we are seeing in the derived peculiar velocity field is actually a combination of true velocities and systematic offsets driven by large scale density perturbations.

REFERENCES

Kaiser N., and Lahav, O., 1988a, in "Large Scale Motions in the Universe", Proc. of the Vatican Study Week, 1987.

Kaiser, N., and Lahav, O., 1988b, submitted.

Bond, J.R. and Efstathiou, G., 1984. Mon. Not. R. astr. Soc., **285**, L45.

Meurs, E.J.A. and Harmon, R.T., 1987. preprint.

Lynden-Bell, D., Faber, S.M., Burstein, D., Davies, R.L., Dressler, A., Terlevich, R.J. & Wegner, G., 1987. Astrophys. J., in press.

Aaronson, M., Huchra, J., Mould, J., Schechter, P.L. & Tully, R.B., 1982. Astrophys. J., **258**, 64. (AHMST)

Vittorio, N., Juszkiewicz, R. & Davis, M. 1986. Nature, **323**, 132.

Bond, J.R., 1987. in "Nearly Normal galaxies", the 8th Santa-Cruz summer workshop, ed. S.M. Faber, New York: Springer Verlag.

Strauss, M.A. & Davis, M., 1987. in "The Large Scale Structures of the Universe", IAU Symp No. 130, Hungary. eds. J. Audouze & A. Szalay, Dordrecht: Reidel.

Kaiser, N., 1987. Mon. Not. R. astr. Soc.**227**, 1.

Kaiser, N., 1988a. Mon. Not. R. astr. Soc.**231**, 149.

Blumenthal, G.R., Faber, S.M., Primack, J.R. and Rees, M.J., 1984. Nature, **311**, 517.

Faber, S.M., 1982. In "Astrophysical Cosmology" proceedings of the Vatican Study Week.

White, S.D.M., Frenk, C.S., Davis, M. and Efstathiou, G., 1987. Nature, **330**, 451.

Carlberg, R. and Couchman, H., 1988. Preprint.

Kaiser, N., 1988b. in "The Large Scale Structures of the Universe", IAU Symp No. 130, Hungary. eds. J. Audouze & A. Szalay, Dordrecht: Reidel.

DISCUSSION

PEEBLES: Your move to bias parameter close to unity, $b \cong 1.5$; is, I think, physically sensible but may cause problems with the dynamical estimates of Ω. I wonder if you would state what you consider to be values for our virgocentric flow speed, $v_{\surd}$, and the galaxy density contrast in the local supercluster within our position, δ, consistent with the observations and your preferred value of b.

KAISER: When I looked into Virgo infall, I found that with an infall of 350 km/s, the apparent density contrast seen in the RSA catalogue would be compatible with a unbiased distribution once allowance is made for peculiar velocity distortion. If the infall is around 100 km/s as Burstein had argued then a considerable bias is required. The b value given here is for IRAS galaxies. When Lahav and I analyzed the optical dipole we found a larger bias $b \cong 2.0$. A very low infall ≤ 100 km/s' would still be a problem, but given the various uncertainties I wouldn't write off the model just yet.

FAIRALL: Is CDM in any way compatible with bubbles?

KAISER: Yes. I see no obvious discrepancy between the CDM predictions and the bubbles seen in the CfA slice. With the moderate bias indicated here, one would expect the distortion of the clustering pattern in redshift space to be quite marked, and I have argued that the crowding of galaxies behind velocity caustics might well account for the sharp features seen.

SCIAMA: Would hot dark matter survive your tests?

KAISER: Yes. The tests shown here are rather one sided. I have found CDM to be consistent with the bulk flows, but it is always easier to reconcile a high prediction (as, e.g. in the hot dark matter model) with a low observation than vice versa.

VELOCITY CORRELATIONS AND LARGE SCALE MOTIONS

ALEXANDER S. SZALAY

Dept.of Physics and Astronomy, J.Hopkins University, Baltimore, and
Dept. of Atomic Physics, Eötvös University, Budapest

ABSTRACT   For a given shell of galaxies the radial velocity as a function of direction can be expanded in terms of multipoles, the $l = 1$ term corresponding to the bulk motion, the $l = 2$ to a quadrupole anisotropy of the velocity field. The $l = 0$ monopole, even if present is absorbed into the local value of the Hubble constant. However, its absence strongly affects the velocity correlations. It is shown, that from all multipoles only the dipole carries large scale contributions, the other terms arise from local effects. We also present general properties of the velocity correlation functions that can be associated with the large scale velocity fields. The current data indicates that the correlation function is unstable against small perturbations in the radial distances. Optimal observing strategies are also discussed.

1. Introduction

There has been a rapid development in the large scale motions in the Universe. This is a particularly challenging area, since it provides a way to measure the perturbations in the gravitational potential directly, without having to worry about biasing, or how well light traces the mass distribution. If the presence of large scale velocities is confirmed beyond a doubt, we have an ultimate normalization of the primordial fluctuations. The velocity correlation functions carry a lot of information on the shape of the fluctuation spectrum.

Peculiar velocities are hard to determine, for a review see Burstein, this volume. One needs secondary distance indicators, which enable us to get the distances. This is the weakest point, the errors (in spite of the great effort) are still quite large, about 23 %. Substracting the Hubble flow from the redshifts will give us the peculiar velocity. However, we do not know the precise value of the Hubble constant either. This is determined from the sample, and the way this fit is done will affect all peculiar velocities. For the elliptical galaxy sample all distances are actually distance moduli relative to Coma. The parameter affecting the magnitude of the Hubble velocities is the peculiar velocity of Coma. Some of the consequences will be discussed below.

The theoretical significance is clear, the question is how to extract meaningful information from the data, that is also robust. There are several possible strategies for analysing the large scale velocity field. They all have advantages and disadvantages, and represent a compromise. We discuss some of the possibilities here.

M. Mezzetti et al. (eds.), Large Scale Structure and Motions in the Universe, 215–226.
© *1989 by Kluwer Academic Publishers.*

1.1 BULK VELOCITY

The simplest quantity to estimate is the bulk velocity. It is the center of mass motion of the galaxies in the survey. The first such motion was observed by Rubin *et al.* (1976), using a set of spiral galaxies. Subsequent work along these lines was made by deVaucouleurs and Peters (1984), Hart and Davies (1982) and most recently Burstein *et al.* (1986), Collins *et al.* (1986). The major advantage of the bulk flow solution besides its simplicity is its straightforward connection to theory based upon the density fluctuation spectrum (Clutton-Brock and Peebles 1981, Kaiser 1983). The disadvantage is, that it may be a too simple description of the complex velocity structure seen in the data. Also, the actual value of the bulk velocity may be contaminated by contributions from other multipoles, caused by anisotropic sampling in the galaxy distribution (Regős and Szalay 1988).

1.2 FINDING THE MASS

The other major way of analysing the velocity field is to identify the sources of gravitational attraction, the dominating parts of the mass distribution. This can be complicated for two reasons: if we start from the velocities, we only see the radial components of the peculiar velocities, so we base our estimate on incomplete information. After the masses have been obtained in this way, one can of course test statistical isotropy of the velocities as a consistency check. On the other hand, starting with the visible galaxy distribution we do not know how well light is tracing mass.

The first scheme was used by Lynden-Bell *et al.* (1988) in a model that led to the Great Attractor, the largest mass concentration around, hidden by Galactic extinction. Still, the fits to the velocity field are suggestive and recent redshift surveys seem to confirm an overdensity in that region. This method, first used for the Virgocentric infall has the advantage of reproducing the full complexity of the velocity field, but it has a lot of disadvantages as well: it requires a lot of parameters, which are probably strongly correlated with each other. Also, the theoretical significance is much harder to assess, e.g. what is the probability in the different galaxy formation scenarios of observing a Great Attractor with its density profile, and having the Virgo cluster, Hydra and Centaurus where they are? These can be partially answered via Monte-Carlo simulations (Bertschinger and Juszkiewicz 1988), but the elegant simplicity is lost.

1.3 FLUX WEIGHTED SCHEMES

Kaiser and Lahav (1988) used flux weighted dipoles to estimate the contribution of different masses to the acceleration, using the fact that both the optical flux and the gravitational force decrease as $1/r^2$, and implicitly assuming that M/L is constant. The above authors have also used the IRAS catalog, which may be a dangerous venture. The luminosity function of the IRAS galaxies is a power law on the bright end, so there are much larger fluctuations in luminosities than for the optically selected catalogues, having an exponential cutoff. For the optical sample it is fairly well established, that M/L is constant to a factor of 10, but this cannot be true for the IRAS galaxies, as one can see from the difference in the bright end luminosity functions. The infrared luminosity may be correlated with the mass, but not in a simple linear relation, so the validity of the flux limited scheme is at least highly questionable.

1.4 SYSTEMATIC EXPANSIONS

One can also use some orthogonal expansion of the velocity field. This would eliminate the arbitrariness in identifying point masses, and placing them to certain locations. Using enough functions one can represent an arbitrary level of complexity. It is easy to make the connections to theory, one can derive the appropriate window function for each of the orthogonal modes and calculate the probability distribution of the coefficients for a given fluctuation spectrum. The disadvantage is that the modes are not really orthogonal for a realistic case, they mix and correlate due to the anisotropic sampling of the galaxies. Here below we outline, how such an expansion is done, and discuss some of the results.

1.5 CORRELATION FUNCTIONS

In principle one can obtain a lot of interesting information by studying the velocity correlation function. It is directly related to the fluctuation spectrum, and from its shape and amplitude one can hopefully constrain the possible cosmological scenarios. The current observational status is much worse, the data are rather noisy, contaminated by selection effects. We discuss some of the possible clues the correaltion function can offer us at this point, using the elliptical galaxy sample of Davies *et al.* (1987).

2. Multipole Expansion

Here we cut the sample into spherical shells, and calculate the multipole coefficients of the radial velocity distribution. This has the advantage of retaining all details of the flow, and it is easy to calculate the statistical distribution of these velocity multipoles, as we will show below. Furthermore, our analysis of the required window functions shows that only the dipole will carry any significant information about the large-scale part of the fluctuation spectrum, all the other multipoles measure local scales only, but they are still related to the spectrum. Besides, the velocity multipoles will not be independent, from measuring values in a few shells one can predict values in subsequent ones; one can calculate the correlation coefficients between the multipole components within and across the shells (see also Kaiser and Lahav 1988). All such data can be used together to assess the statistical significance of our guess on the shape of the fluctuation spectrum.

The fluctuation spectrum causes a large scale coherence in the velocity field, quantified by the multipole components of the projected velocities. The use of radial projection mixes these modes corresponding to different l, and a nonuniform sampling results in nonseparable errors. With an anisotropic sample the spherical harmonics become nonorthogonal, and the usual methods of least squares or maximum likelihood yield systematically biased estimations of the bulk motion. The major effect comes from the dipole and the quadrupole anisotropy: the direction of the velocity will be correlated with the dipole moment of the galaxy distribution and statistical uncertainties align perpendicular to the quadrupole (Regős and Szalay 1988). This will cause the measured bulk velocity to be preferentially close to the galactic plane and aligned with the dipole anisotropy, very close to what the actual observations show, although there is a real density enhancement in that direction.

If we select a relatively thin shell of galaxies, then the radial velocity on the surface of the sphere is a scalar function of the angles α, δ. It can be expanded in terms of spherical harmonics :

$$V_r(\Omega_s) = \sum_{l,m} V_{lm} Y_{lm}(\Omega_s) \tag{1}$$

with $V_{lm} = (-1)^m V^*_{l,-m}$. For that particular shell one can calculate the ensemble average of the multipole moment correlation matrix

$$\langle V^*_{lm} V_{l'm'} \rangle = \delta_{ll'} \, \delta_{mm'} (H_0 a f)^2 \int d^3k \, \frac{|\delta_k|^2}{k^2} |W_l(k)|^2 \equiv \delta_{ll'} \, \delta_{mm'} u_l^2 \tag{2}$$

where $W_l(k)$ is the window function determined by the selection function $\Phi(s)$:

$$W_l(k) = \int d^3s \; \Phi(s) \; j'_l(ks) \tag{3}$$

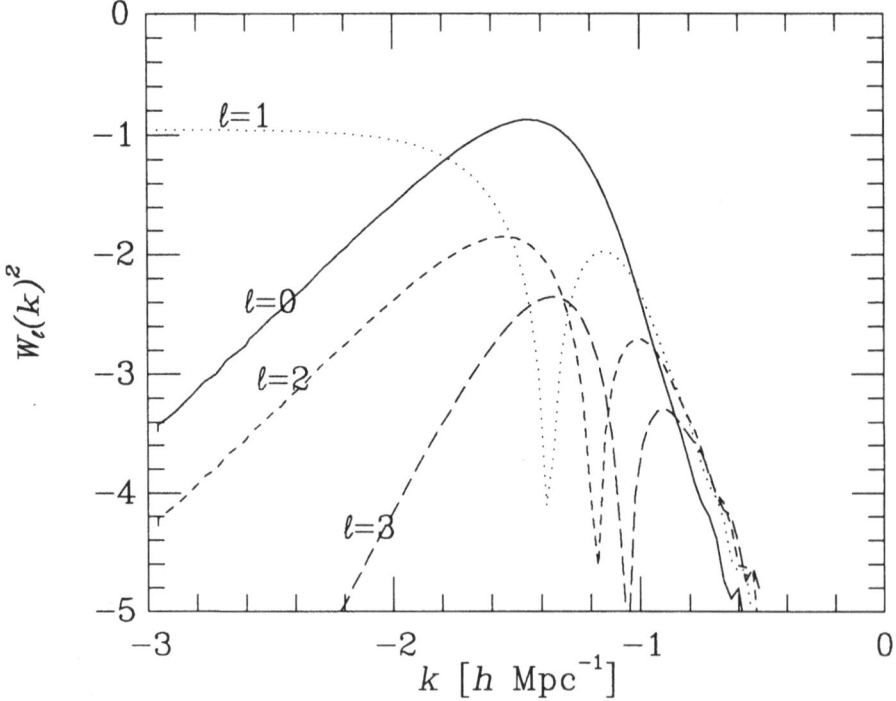

Figure 1. Window functions corresponding to various l, using a Gaussian selection with $R = 3200$ km/s, and the errors are roughly proportional to distance. The actual quantity plotted is $|W_l(k)|^2$, this appears in the integrals weighting $||\delta_k|^2|^2$. Note, that the only only window function with significant large scale weights is the $l = 1$, all the others are measuring local fluctuations.

The window functions describing the various multipole moments are integrals over $j'_l(ks)$, and therefore one can immediately see, that the only window function which does not disappear at large scales (small k) is the j'_1, corresponding to the dipole motion. All the other window functions describe local contributions to the flow, inside or at the boundary of the survey, as shown on Fig.1.

The dipole motions on the other hand carry information from scales outside the survey, so in order to determine the large-scale part of the fluctuation spectrum they seem to be the best choice, and in this respect one can probably design future deeper surveys by selecting a sample for an optimal determination of the dipole velocity. One such scheme may be to choose 6 patches on the sky each 90 degrees apart, in directions normal to the faces of a cube. The 'cube' can be oriented so that the zone of extinction is mostly avioded. By measuring an equal number of galaxies in each region one can assure, that both the dipole and the quadrupole moment of the galaxy distribution vanish. Then it becomes rather simple to obtain the dipole velocity. The above result also suggests, that the fine details of the velocity field (higher multipole moments) have relatively little influence on the determining the large scale contribution from the fluctuation spectrum.

We already discussed the $l = 1$ term corresponding to the dipole 'bulk' motion. There is a monopole $l = 0$, corresponding to an overall uniform expansion or infall, caused by the local over- or underdensity, which may be partially or fully absorbed into the local value of the Hubble constant. The presence of this term can possibly be observed by taking larger and larger shells. In the elliptical galaxy sample of Davies et $al.$ (1988) there is no monopole when one takes the data at face value. However one should be aware of the fact, that the peculiar velocities were obtained after several steps. After having determined the relative distance moduli from Coma, the actual distance of the Coma cluster will set the distance scale. One cannot assume that Coma is at rest in the comoving frame, any peculiar velocity assigned to Coma will shrink or stretch the distance scale as measured in Hubble velocities. This velocity was determined by a least squares fit, minimizing the deviations from the Hubble flow. This effectively removes the monopole velocity. The bulk velocity was determined in a next fit. If one takes anisotropies into account, the two are correlated, and the results of a simultaneous fit are different from the sequential one.

The relative magnitude of the u_l cofficients will determine the coherence, or patchiness of the large scale velocity field: a smooth velocity field has a cutoff for the higher harmonics. Here we suggest that this patchiness carries significant information about the shape of the fluctuation spectrum, independent of the normalization. On Fig.2. we plotted the coefficients u_l^2 as a function of l for the CDM spectrum, using $R = 3000$ km/s, weighted with errors, normalized to a mass-variance of 1 for a top-hat radius of 800 km/s. It is obvious, that the $l = 0$ monopole term is the largest, and the others are falling off quite rapidly. For this normalization,

$$u_0^2 = 2.147$$
$$u_1^2 = 1.032 \qquad (4)$$
$$u_2^2 = 0.188$$

in units of $(100 \text{ km/s})^2$. One should note that these are the dispersions for each respective $|V_{lm}|^2$. The rms 'monopole' velocity is about 150 km/s, about 5% distortion on the Hubble flow. The dispersion of the isotropic bulk motion is $\langle U^2 \rangle = 9u_1^2 = (310 \text{ km/s})^2$. For the quadrupole, there are 5 components of the irreducible tensor, and the dispersion of the

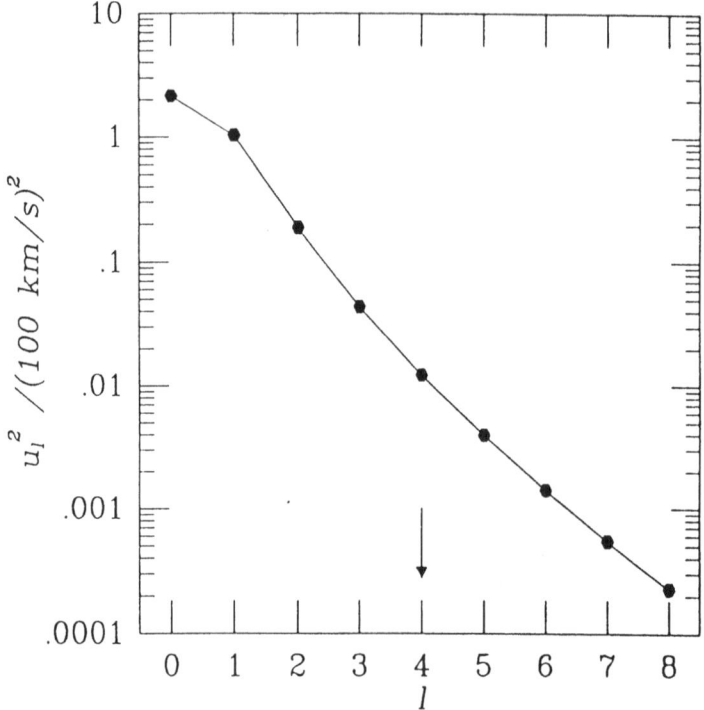

Figure 2. The dependence of the velocity multipole moments u_l^2 on the harmonic number l, using the CDM spectrum, normalized to unit top-hat mass variance at 800 km/s, no biasing. The sample is approximated by a Gaussian selection of $R = 3000$ km/s radius, and error weighting. $\sigma_f = 300$ km/s, and $\Delta = 0.23$. The arrow indicates the angular coherence scale of the CDM picture.

anisotropy in the quadrupole components can be estimated to be $(337 \text{ km/s})^2$, comparable to the bulk motion prediction. A systematic analysis of the multipole expansion will be published in detail elsewhere (Kaiser and Szalay 1988).

This may be rather close, to what the observations show. In a third, separate fit currently Lynden-Bell *et al.* (1988) found a quadrupole anisotropy of $0.1-0.2$ in the Hubble flow for the various samples (300 - 600 km/s). It is interesting to note, that this anisotropy seems to be aligned with the dipole motion. This feature seems to be present in the 'near' $0 < V < 3200$ and the 'far' $3200 < V < 8000$ subsamples as well. Besides, in these two subsamples the monopole term does not go to zero, only when we combine the two. Besides, the velocity field shows remarkable coherence, on linear scales of $500 - 2000$ km/s, which may be best characterised by using velocity correlation functions.

3. Velocity Correlation functions

We follow the notation of Monin and Yaglom (1971): velocity correlation functions will be denoted by $B(r)$. One can write the general correlation function of the isotropic random vector field $\mathbf{v}(r)$ as

$$B_{ij}(r) = \langle v_i v_j \rangle = \int d^3\mathbf{x}\, v_i(\mathbf{x}) v_j(\mathbf{x} + \mathbf{r}) \tag{5}$$

For a known power spectrum one can write this as a Fourier integral

$$B_{ij}(r) = \delta_{ij} \int d^3\mathbf{k}\, |v_\mathbf{k}^2| \frac{j_0(kr) + j_2(kr)}{3} - \frac{r_i r_j}{r^2} \int d^3\mathbf{k}\, |v_\mathbf{k}^2|\, j_2(kr) \tag{6}$$

where the expression $|v_\mathbf{k}^2| = (H_0 a f)^2 P(k) |W(k)|^2$ includes the contribution of the power spectrum $P(k) = |\delta_\mathbf{k}|^2$ to the velocities in the given survey, described by the window function $W(k)$. a is the scale factor, $H_0 = \dot{a}/a$ is the Hubble constant and $f = (\dot{D}/D)/H_0 \approx \Omega^{0.6}$ (Peebles 1980).

The two parts with the different symmetry can be decomposed into a longitudinal and a normal correlation function B_{LL} and B_{NN} :

$$B_{ij}(r) = \delta_{ij} B_{NN}(r) + \frac{r_i r_j}{r^2}[B_{LL} - B_{NN}] \tag{7}$$

The trace of the velocity correlation tensor becomes $B_{LL} + 2B_{NN}$:

$$B_{LL} + 2B_{NN} = \int d^3\mathbf{k}\, |v_\mathbf{k}^2|\, j_0(kr) \tag{8}$$

The two components are not independent, there are several contraints they have to satisfy. One can define the Fourier transform of B_{ij} :

$$F_{ij}(k) = \int d^3\mathbf{k}\, e^{-i\mathbf{k}\mathbf{r}} B_{ij}(r) = \delta_{ij} F_{NN} + \frac{k_i k_j}{k^2}[F_{LL} - F_{NN}] \tag{9}$$

This Fourier transform must be a non–negative function for any arbitrary value of $\mathbf{k}$. One can easily show, that for potential motion arising in the growing mode of self–gravitating linear perturbations $F_{NN} = 0$, whereas for rotational motion $F_{LL} = 0$. This is a rather strong constraint, as we will see. Also, if one knows B_{NN}, one can determine the fluctuation spectrum:

$$|\delta_\mathbf{k}|^2 = \frac{1}{\pi (H_0 a f)^2} \int dr\, (kr \sin kr - k^2 r^2 \cos kr) B_{NN}(r) \tag{10}$$

This inversion is unfortunately not as simple as it seems, due to the large noise in the determination of B_{NN}.

4. Using Real Data

The first complication arises from the fact, that only the radial components of the peculiar velocities are known. Let us first assume that the errors are negligible, we can measure all quantities exactly. We consider a pair of galaxies at an angular separation θ, and radial distances r_1 and r_2. The galaxies are then at a relative distance r, given by

$$r^2 = r_1^2 + r_2^2 - 2r_1 r_2 \cos \theta \tag{11}$$

From the difference of the redshift and the Hubble flow one can determine the radial peculiar velocities u_1 and u_2, which are the line-of-sight components of the velocities $\mathbf{v}_1$ and $\mathbf{v}_2$. The relative angle of the line connecting the two galaxies with the two lines of sight are γ_1 and γ_2, respectively. The two angles are not independent : $\gamma_1 = \gamma_2 + \theta$. We choose our coordinate system with the z axis connecting the galaxies. The radial velocity components can be written in terms of the longitudinal and normal ones, where v_L is along the z-axis, v_N is normal to it, but still in the plane of 012, where 0 is the observer, 1 and 2 are the two galaxies.

$$
\begin{aligned}
u_1 &= v_{L1} \cos \gamma_1 + v_{N1} \sin \gamma_1 \\
u_2 &= v_{L2} \cos \gamma_2 + v_{N2} \sin \gamma_2
\end{aligned} \tag{12}
$$

Furthermore, the various components are correlated in the following simple way:

$$
\begin{aligned}
\langle v_{L1} v_{L2} \rangle &= B_{LL} \\
\langle v_{N1} v_{N2} \rangle &= B_{NN} \\
\langle v_{L1} v_{N2} \rangle &= 0 \\
\langle v_{N1} v_{L2} \rangle &= 0
\end{aligned} \tag{13}
$$

Combining these expressions we get for the expectation value

$$B(\theta, r_1, r_2) = \langle v_{r1} v_{r2} \rangle = B_{LL} \cos \gamma_1 \cos \gamma_2 + B_{NN} \sin \gamma_1 \sin \gamma_2 \tag{14}$$

Expanding further, and using $\theta = \gamma_1 - \gamma_2$:

$$B(\theta, r_1, r_2) = B_{LL} \cos \theta + [B_{NN} - B_{LL}] \sin^2 \theta \, \frac{r_1 r_2}{r^2} \tag{15}$$

If we knew everything very accurately, it would be easy to invert this expression. However, we only know θ well, the distances are uncertain with a formal relative error of $\Delta = 0.23$.

We bin the data as a function of the radial separation r. We determine the value of the two correlation functions B_{LL} and B_{NN} in the bins by minimizing

$$\chi^2(n) = \sum_{i \neq j} \Pi(r_{ij}, n) \frac{[u_i u_j - B_{LL}(n) \cos \theta - (B_{NN} - B_{LL}) \sin^2 \theta \, r_i r_j / r^2]^2}{\Delta^2 (r_i^2 + r_j^2)} \tag{16}$$

where the sum is over all distinct pairs, and $\Pi(r, n)$ is 1 if r is in the nth bin, 0 otherwise. There is one complication: for small separations $B_{LL} = B_{NN}$, and the fitting scheme becomes slightly degenerate, but this can be easily overcome.

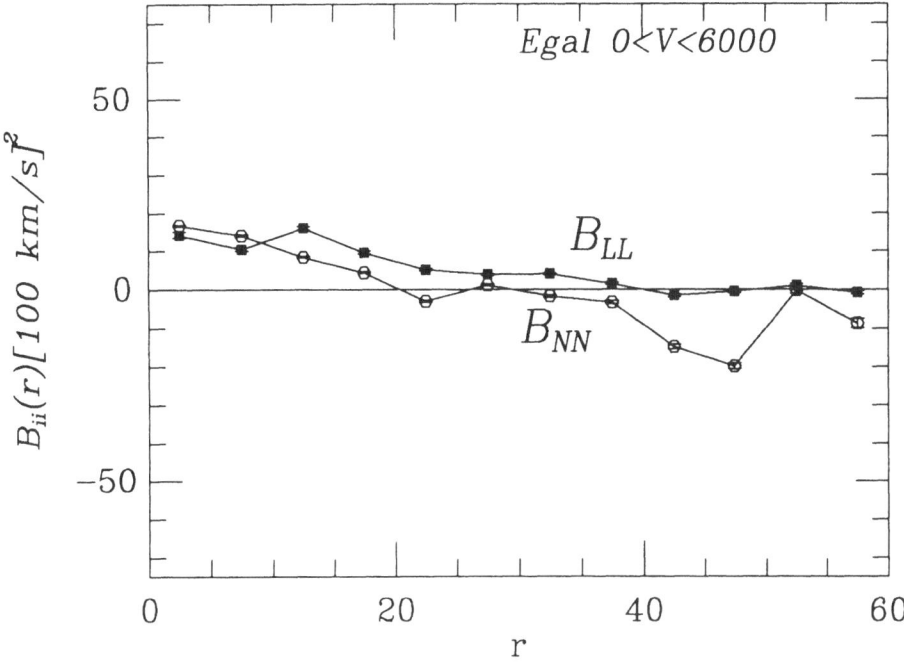

Figure 3. The longitudinal and normal velocity correlation functions B_{LL} and B_{NN} as a function of radial separation measured in h^{-1} Mpc. The data used is the elliptical galaxy sample of Burstein *et al.* (1986) within a 6000 km/s radial distance limit.

The results of the fit for the elliptical galaxy sample limited to radial distances < 6000 km/s can be seen on Fig.3. We used only the systematic error for the weighting scheme. The correlations between B_{LL} and B_{NN} are not very large (0.2-0.5). When we include a systematic field dispersion of $\sigma_f = 300$ km/s, slightly changing the weighting, the correlation function changes considerably. All this implies that the processed data show a smoother Hubble flow, than in reality!

In order to see the stability of the correlation function with respect to the radial distances, 100 Monte-Carlo catalogues were generated in a way, that each galaxy remained at the same angular position and redshift as the measured sample. Only the radial distances were drawn for each galaxy from a lognormal distribution, with the same mean as the Mahlmquist corrected distance, and the dispersion in the log distance was Δ. The result of this calculation is shown on Fig.4. The strong anticorrelation in the random catalogues is due to the emergence of the monopole term, which is not removed this time. The presence of the monopole means that the opposite edges of the survey are moving with an opposite velocity. At 40 Mpc separations this contribution is substantial, since the effective (error weighted) radius of the galaxy distribution is only about 15-20 Mpc. This result is not surprising, but it indicates the sensitivity of the correlation function to such 'details'.

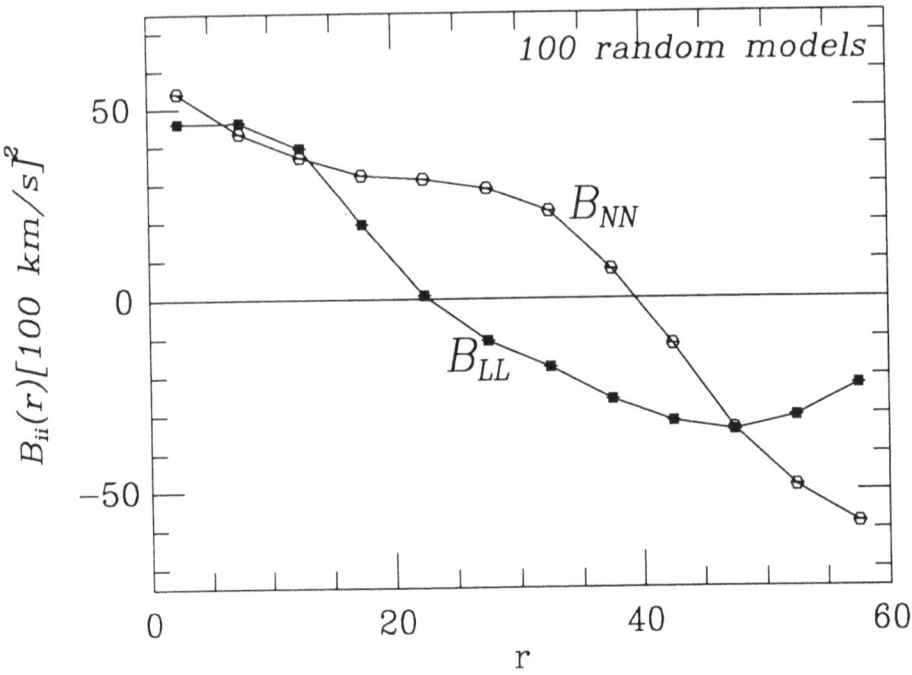

Figure 4. The effect of perturbing the radial distances for the elliptical galaxy sample. The graph shows the average over 100 random catalogues.

5. Conclusion

The radial velocity field observed over the whole sky can be also expanded in terms of spherical harmonics, and each coefficient has a different physical meaning. The monopole term describes a uniform distortion in the expansion, the dipole is the bulk motion, the quadrupole appears as an anisotropy of the Hubble flow. The shape of the correlation function carries interesting information on the shape of the power spectrum around the 10 Mpc scales. Current estimates of the velocity correlations seem to be still too unstable to be used, but a more thorough understanding of the Mahlmquist bias and the proper substraction of the Hubble flow may enable improved estimates. The analysis presented in this paper indicates that the monopole term in the data may have been too well removed, using the same data set for obtaining an effective Hubble-constant yields an integral constraint, that strongly affects the velocity correlation function.

The author would like to acknowledge useful conversations with Enikő Regős, Dave Burstein, Vera Rubin and Nick Kaiser. This work was supported by a grant of the Hungarian Academy of Sciences (OTKA) in Hungary, and by the NSF in the US.

6. References

Bertschinger,E. and Juszkiewicz,R. 1988, preprint.

Burstein,D., Davies,R.,L., Dressler,A., Faber,S.M., Lynden-Bell,D., Terlevich,R.J., and Wegner,G. 1986, Proc.*Galaxy distances and deviations from universal expansion* (NATO ASI Series) ed. B. Madore and B. Tully, Reidel p.123.

Burstein,D. and Szalay,A.S. 1988, in preparation.

Clutton-Brock,M. and Peebles, P.J.E. 1981, *Astron.J.*, **86**, 1115.

Collins,C.A., Joseph,R.D. and Roberts,N.A. 1986, *Nature*, **320**, 506.

Davies,R.L., Burstein,D., Dressler,A., Faber,S.M., Lynden-Bell,D., and Terlevich,R.J. 1987, *Ap.J.Suppl.*, **64**, 581.

de Vaucouleurs,G. and Peters,W.L. 1984, *Ap.J.*, **287**, 1.

Dressler,A., Lynden-Bell,D., Burstein,D., Davies,R.L., Faber,S.M., Terlevich,R.J., and Wegner,G. 1987, *Ap.J.*, **313**, 42.

Hart,L. and Davies,R.D. 1982, *Nature*, **297**, 191.

Kaiser,N. 1983, *Ap.J.Lett.*, **273**, L17.

Kaiser,N. 1988, *M.N.R.A.S.*, **231**, 149.

Kaiser,N. and Lahav,O. 1988,*M.N.R.A.S.*, in press.

Kaiser,N. and Szalay,A.S. 1988, in preparation.

Lynden-Bell,D., Faber,S.M., Burstein,D. Davies,R.L., Dressler,A., Terlevich,R.J. and Wegner,G. 1988, *Ap.J.* in press.

Monin,A.S. and Yaglom,A.M. 1971, *Statistical Fluid Mechanics*, The MIT Press, Cambridge.

Peebles,P.J.E. 1980; *The Large Scale Structure of the Universe*, (Princeton University Series, N.J.)

Regős,E. and Szalay,A.S. 1988, submitted to *Ap.J.*.

Rubin,V.C., Thonnard,N., Ford,W.K. Jr., and Roberts,M.S. 1976, *Astron.J.*, **81**, 687.

Rubin,V.C., Thonnard,N., Ford,W.K. Jr., and Roberts,M.S. 1976, *Astron.J.*, **81**, 719.

DISCUSSION

EVRARD: A comment rather than a question. I'm a bit pessimistic about determining the shape of P(k) from these data for the following reason. If you believe things on scales < 1000 km/s are non-linear and the data only sample reliably out to 3000 km/s, then that only leaves a factor of 3, or 1/2 decade in k to determine the shape of P(k).

SZALAY: I agree, what I had in mind is to use future data, at larger radii, with more independent volume. We should consider shells with at least 1000 km/s in thickness, so that non linear effects should average out.

RHEE: Are the peaks in the histograms coming purely from clusters or are these clusters possibly embedded in sheet-like structures?

SZALAY: As shown on some of the (XY) plots, indeed clusters or groups seem to be imbedded in larger structures.

THE IMPRINT OF LARGE SCALE STRUCTURE ON THE COSMIC BACKGROUND RADIATION

R. B. Partridge
Haverford College
Haverford, PA   19041
U.S.A.

ABSTRACT.   Searches for small scale anisotropies in the cosmic background radiation provide unique information about the large scale distribution of matter at early epochs.   This paper reviews recent observational programs to detect such anisotropies on angular scales from 12" to 8°.   That range in angle corresponds to a scale of density perturbations ranging from roughly a galactic mass to more than a supercluster mass.   In two cases, there is very tentative evidence for the existence of fluctuations in the background.

Also reviewed here are new measurements of the dipole moment of the cosmic background radiation, and upper limits on the quadrupole and other low-order moments.   These results provide constraints on the local distributon of gravitating matter, whether luminous or not.

1.   LARGE SCALE STRUCTURE AND THE COSMIC BACKGROUND RADIATION

The connection between anisotropies in the cosmic microwave background radiation (CBR) and large-scale structure in the Universe was pointed out (by Sachs and Wolfe, 1967, and Silk, 1968, for instance) within a few years of the discovery of the radiation by Penzias and Wilson (1965).   Early measurements (e.g., Partridge and Wilkinson, 1967 and Conklin and Bracewell, 1967) showed the CBR to be remarkably isotropic; more than 20 years later we still have only upper limits on anisotropies in the CBR temperature (for a recent review, see Partridge, 1988).

Nevertheless, at some level, angular variations $\Delta T/T$ must be present in the CBR, produced by one or more of the following physical processes:--

1.)   Differential gravitational redshift across a mass inhomogeneity as the Universe expands (Sachs and Wolfe, 1967).

2.)   The correlation between density and temperature expected in adiabatic density perturbations (Silk, 1968).

3.)   The motion of the observer (Peebles and Wilkinson, 1968).

4.)   Absorption and reemission by inhomogeneously distributed dust (e.g., Hogan, 1980).

227

M. Mezzetti et al. (eds.), Large Scale Structure and Motions in the Universe, 227–240.
© *1989 by Kluwer Academic Publishers.*

5.) Inverse Compton scattering of the CBR photons by hot electrons (Sunyaev and Zel'dovich, 1972; Sunyaev, 1978).

6.) Anisotropic expansion or rotation of the Universe (Hawking, 1969).

Which of these processes will dominate, and in particular what amplitude $\Delta T/T$ is expected at a given angular scale, will depend on a number of physical and cosmological parameters. That very dependence on the models is what makes the CBR so useful a test of both cosmological models and scenarios for the origin of structure in the Universe. A partial list of the parameters which determine $\Delta T/T$ includes the amplitude and mass spectrum of density perturbations on the surface of last scattering, $(\Delta\rho/\rho)_s$; the redshift z_s of the surface of last scattering; the amount and nature of "dark matter"; and the distribution of local ($z \ll 1$) matter. All of these, in turn, are related to the origin and properties of large-scale structure (LSS) in the Universe. Indeed, as I hope to show, CBR measurements provide unique information about large-scale structure, information which nicely complements what we can learn from local observations, such as the Center for Astrophysics redshift survey discussed earlier here by Geller.

2. SMALL-SCALE ANISOTROPIES

I will begin by making the assumption that the CBR photons last scattered when the primeval plasma combined at T ~ 4000 K or z_s ~ 1500. The scale and amplitude of CBR fluctuations under this assumption have been calculated by Vittorio and Silk (1984) and Bond and Efstathiou (1984) among others. Figure 1 shows an example; note the decrease in $\Delta T/T$ at $\theta \lesssim 10'$ and the (not very strong) dependence on the two cosmological parameters H_0 and Ω. In this work, Bond and Efstathiou (1984) assumed adiabatic perturbations at z_s ~ 1500 and included dark matter; the predicted level of CBR fluctuations for pure-baryon models with $\Omega = \Omega_b \approx 0.1$ is an order of magnitude <u>higher</u>. As we shall see, the currently fashionable "dark matter" models are constrained by the observations now available, and the pure-baryon models essentially ruled out (see Kaiser and Silk, 1986, and Bond and Efstathiou, 1987, for recent reviews of the theory).

What if we relax the assumption that the Universe remains transparent after the epoch when the primeval plasma combined? Then the redshift of last scattering shifts to a value much less than 1500. This scenario has been investigated by Hogan (1980, 1984) and by Ostriker and Vishniac (1986; see also Vishniac, 1987). In outline, while the primordial fluctuations produced at z ~ 1500 are fully or partially erased, new perturbations are produced. Primordial fluctuations can be smoothed away by rescattering on scales up to ~ 5° (the horizon scale). On the other hand, Ostriker and Vishniac (1986) have shown that the process of galaxy formation--independent of specific, model-dependent details--will produce fluctuations in the CBR of roughly the same amplitude (~ 10^{-5}) as those erased; in addition the angular spectrum of the fluctuations may increase with

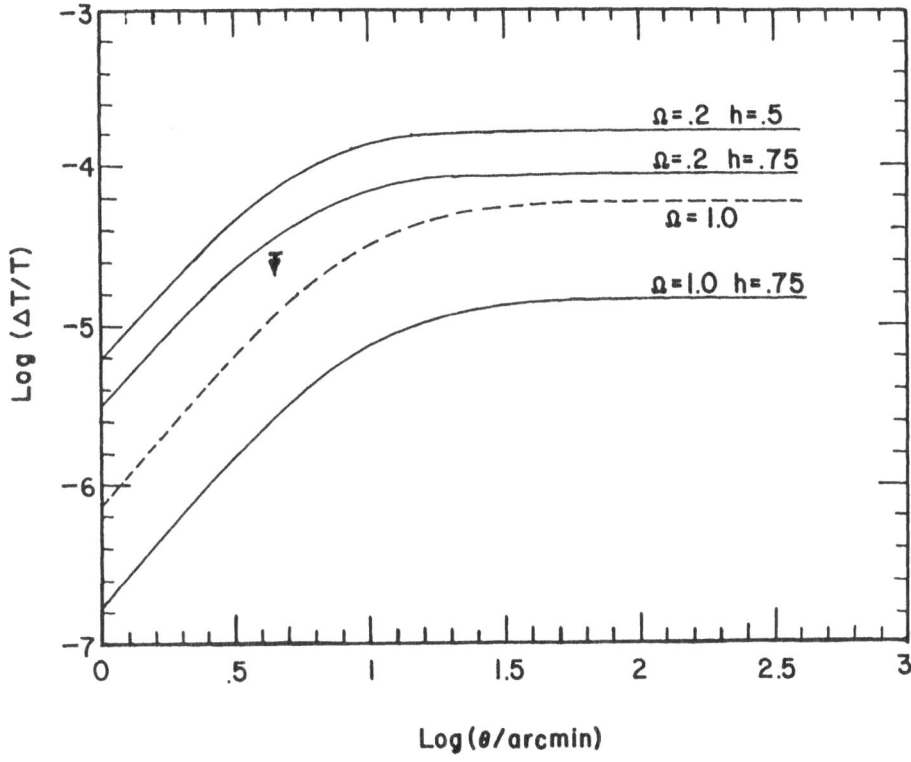

Fig. 1. Predicted values of ΔT/T for models including dark
matter, for various values of the cosmological parameters
h = H_0/100 and Ω (adapted from Bond and Efstathiou, 1984).
The single dashed line is for a hot dark matter model; the
solid lines are for cold dark matter models. The
observational upper limit is the result of Uson and
Wilkinson (1984b) discussed in the text.

decreasing θ, giving more power at lower θ than in the models shown in
figure 1. Specific cases have been studied by Vishniac (1987; see
fig. 2). To make the contrast with the models discussed above
clearer, I have dotted in the predictions of Bond and Efstathiou for a
cold dark matter model with H_0 = 75 Km/sec per Mpc and Ω_0 = 1.
Clearly, the angular spectrum of CBR anisotropies alone can tell us
much about the early evolution of structure in the Universe.

But we first need to detect such fluctuations--and I think there
is as yet no certain evidence that we have.

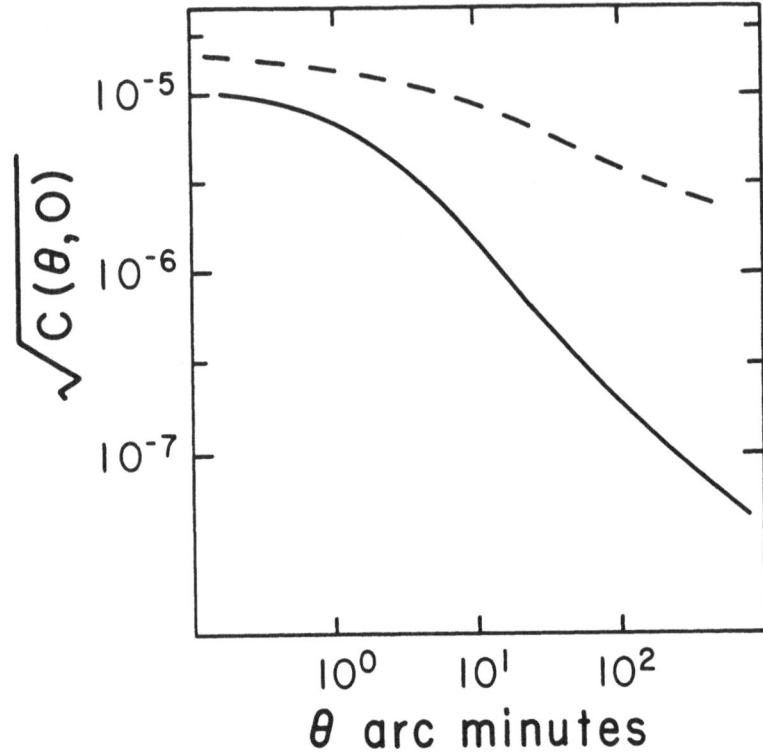

Fig. 2. Predicted values of ∆T/T for a model including
reionization at z << 1500 (adapted from Vishniac, 1987).
The square root of the angular autocorrelation coefficient
is shown. In this case, a cold dark matter model was
assumed, with h = 0.75. Dashed curve--comparable
predictions for primordial fluctuations, taken from Vittorio
and Silk (1984).

2.1. Searches on Arcminute Scales

The first of a series of recent searches for fluctuations in the CBR
on scales of arcminutes and degrees was carried out by Uson and
Wilkinson (1984a, 1984b) using the 40-meter antenna of the U. S.
National Radio Astronomy Observatory in West Virginia. The wavelength
employed was 1.5 cm, and the observers employed a triple beam pattern
on the sky to reduced atmospheric noise (see their papers and
Partridge, 1988 for a discussion). The beam switch angle was 4.5, and
it is on this approximate scale that their observations can impose the
tightest constraints on ∆T/T. Their 95% confidence upper limit is
∆T/T ≤ 2.1-2.5×10^{-5}, depending slightly on the model assumed for the
CBR fluctuations. It has been pointed out (by Boynton, private
communication, and by Kaiser and Silk, 1986, among others) that such a

low limit is unlikely given the instrumental noise present. Of course particularly low limits on ΔT/T as well as particularly high limits on ΔT/T will emerge in any collection of observational programs. But I would urge my theorist colleagues not to pin their hopes on the last decimal place of any of our reported values of ΔT/T--more on that point later. Even if the Uson and Wilkinson upper limit is only ~ 4×10^{-5}, it still is a crucial datum in cosmology.

As it happens, a new measurement may soon supplant the careful work of Uson and Wilkinson--a search for CBR fluctuations, again at λ = 1.5 cm, underway at Cal Tech (by Lawrence, Moffett and Readhead, among others). In design, it is very similar to the experiment of Uson and Wilkinson, but the beam switch angle is larger, ~ 7′. As of the time of this meeting, the results of this work were not yet available, but the probability is that the Cal Tech group can reach ~ 1.5×10^{-5} in ΔT/T.

2.2. Searches on Scales of Degrees

Next, let me turn to attempts to set limits on ΔT/T on larger angular scales. Although there has been earlier work on degree scales by Pariiskii (1977; see also Berlin et al, 1983), Fabbri et al (1980) and Mandolesi et al (1986) (see also the contribution to this volume by F. Melchiorri, reporting new and unpublished results), I will focus on the work of Davies et al (1987), not least because a fuller report of that program is given by R. A. Watson in this volume. The observations were carried out at a high altitude, dry site at a wavelength of 3 cm. Triple beam switching was achieved using a plane mirror. The published results (Davies et al, 1987) were obtained with a beam switch angle of 8°; the beams themselves had full width at half maximum of 8° also. A five-hour long strip at δ = 40° was scanned (see figure 3). Given the beam shape shown in the figure, the data are obviously oversampled; I estimate there are only six or seven independent samples in the data shown. Note also the apparent peak at RA ~ 223° ~ 15^h, discussed further in this volume by Watson.

From the data shown in figure 3, Davies and his colleagues derive a value of ΔT/T = 3.7×10^{-5} for fluctuations in the sky emission. As Watson notes here, much of the observed fluctuation arises from the peak at ~ 15^h R.A.

If these results are confirmed, they will be of extraordinary value to scientists working on the origin and evolution of LSS. Given their importance, it is not inappropriate to ask whether these results could have been influenced by systematic or instrumental problems. As the observers themselves note, patchy emission from the Galaxy might add to the observed sky fluctuation. Davies and his colleagues are checking this possibility by reobserving the same strip of the sky at a longer wavelength, where any Galactic emission is expected to be ~ 5-7 times more prominent; I understand that these additional observations do not provide an explanation for the sky fluctuations observed at 3 cm wavelength. Another potential problem is what radio astronomers call side-lobe pickup--the radiation of bright sources into subsidiary maxima of the diffraction pattern of an antenna. I

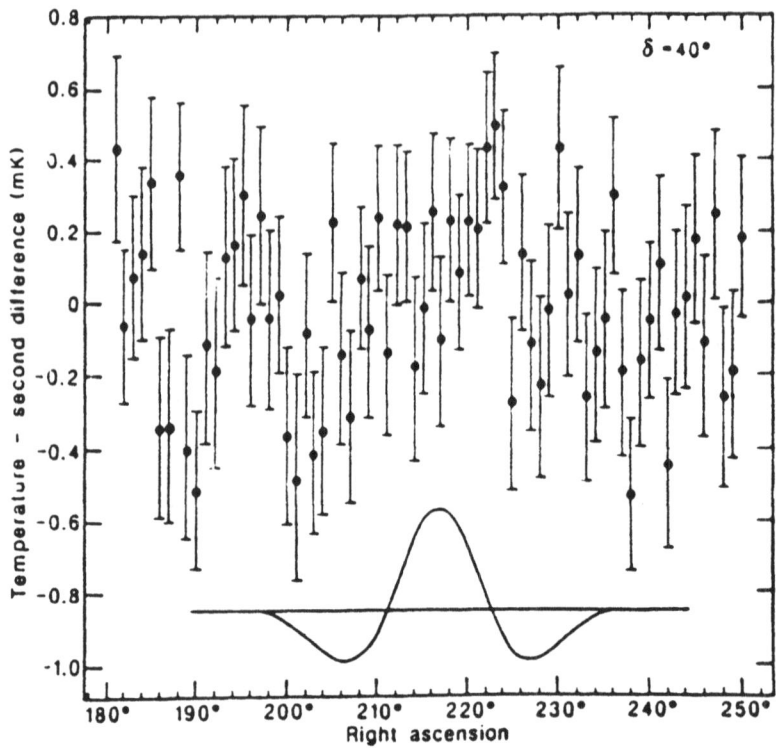

Figure 3. Scan of the sky made at $\delta = 40°$ by Davies et al (1987). The lower curve represents the beam of the instrument (triple beam switching was employed).

worry about side-lobe pickup in this experiment particularly because the plane mirror used was located so close to the mouths of their horn antennas; it was also not very large. The former raises the possibility of distortions in the beam, with an increase in side-lobe response; the latter, the possibility of beam spillover past the reflector. Could bright astronomical sources--the Moon, the Galactic plane, Cyg A, etc.--have radiated enough power into side lobes to cause the sky fluctuations Davies et al (1987) report? Finally, I note that the value $\Delta T/T = 3.7 \times 10^{-5}$ was derived by making an assumption about the angular spectrum of the temperature fluctuations: that it was Gaussian. That assumption and its consequences are discussed in detail by Vittorio, de Bernardis and others in this volume.

2.3. Searches on Subarcminute Scales

The most recently announced results concern possible CBR fluctuations on subarcminute scales. To reach scales $\theta < 1'$ at centimeter wavelengths requires the use of arrays of telescopes and aperture synthesis (see Kellermann and Verschuur, 1974), rather than conventional radio telescopes. This technique has been used by two groups (Knoke et al, 1984, and Martin and Partridge, 1988; and Fomalont et al, 1984, and Fomalont et al, 1988) at the Very Large Array (VLA), operated by the U. S. National Radio Astronomy Observatory in New Mexico. Most of the observatons were made at a wavelength of 6 cm, and the most sensitive ones span an angular scale of 12" to 60" (see Table 1).

Reference	Angular Scale	Upper Limit on $\Delta T/T \times 10^{-4}$
Knoke et al	6"	32
(1984)	12"	17
	18"	12
Martin and	18"–80"	1.7 ± 0.5
Partridge (1988)	36"–160"	1.3 ± 0.2
Fomalont	12"	8.5
et al (1988)	18"	1.2
	30"	0.8
	60"	0.6

Table 1. Upper limits on or tentative measurements of $\Delta T/T$ fluctuations on subarcminute scales. All are VLA measurements at $\lambda = 6$ cm.

A glance at the table will show that this technique has not yet caught up in sensitivity with the measurements described above using single instruments rather than arrays (see Partridge, 1988 for a discussion). It is also clear that the conclusions of the two groups are qualitatively discrepant: we appear to find evidence of fluctuations in the CBR, and Fomalont, Kellermann and their colleagues do not. Although we used rather different observing techniques and looked at different parts of the sky, the discrepancy does not lie in the observations themselves. We both see significant levels of fluctuation in the sky. Our disagreement instead hinges on the ways we treat faint radio sources in the fields we study. Can the presence of radio sources too weak to detect individually explain all of the sky variance both groups detect? Martin and I (1988) model the effect of such weak sources and find they cannot explain all the variance we see; we tentatively assign the rest to the CBR. Fomalont et al (1988) model the source contribution in a different way and find that radio sources can explain all the observed sky variance.

I discuss the details elsewhere (1989); here, let me add just two things. First, I think all these VLA results should be used with caution, since they involve modeling and subtracting a substantial contribution from weak radio sources, and that may introduce some bias in the results (see Franceschini et al, 1988, for a general discussion of radio source contributions). Second, more work is clearly needed. Each group plans to subject data from the other to its data analysis program. Further observations are planned or underway. In particular, as one way of checking on the role of weak radio sources, Craig Hogan and I (unpublished) have repeated our VLA observations at 2 cm. Since the spectrum of typical radio sources, expressed in temperature terms, is $T \propto \lambda^{2.7}$, the sources should cause much less trouble at 2 cm. Unfortunately, the receivers at 2 cm are less sensitive, so we have only a marginal result:

$$\text{at } \lambda = 2 \text{ cm and } \theta = 18", \quad \Delta T/T = 2 \pm 1 \times 10^{-4}.$$

This result, while intriguing, is not yet precise enough to decide between the interpretations of the two groups.

2.4. Small-scale Polarization of the CBR

VLA observations may also be used to set limits on the linear polarization of fluctuations on scales $\theta \lesssim 1'$. Because polarization measurements are freer of the effects of radio sources and of some instrumental effects, the limits on $\Delta T/T$ are 2-3 times more sensitive than those in Table 1 (see Partridge, Nowakowski and Martin, 1988). On the other hand, the amplitude of polarized fluctuations is also expected to be lower (Bond and Efstathiou, 1987), so these observations do not further constrain models for LSS.

2.5. What Do We Learn from the Measurements?

These results at scales from arcseconds to degrees, and other older results not mentioned here, place constraints on the inhomogeneity of matter at very early times, and thus complement studies of very local inhomogeneity such as those reported here by Margaret Geller. In addition, if the epoch of last scattering was at $z_S \sim 1500$, we can be certain that any density inhomogeneities responsible for CBR fluctuations were linear, with all the mathematical and physical simplification an absence of non-linear effects entails.

What do we learn from the observations now in hand? Models with no non-baryonic "dark matter"--pure baryon models--are essentially ruled out if the redshift of last scattering is at $z_S \sim 1500$; pure baryon models with subsequent reionization are just consistent with the observations (Vishniac, 1987). If $z_S \sim 1500$, even "dark matter" models are barely consistent with the observations unless $\Omega_0 \sim 1$--we need lots of dark matter. Another approach to reducing the predicted level of CBR fluctuations for a given model of LSS is to introduce "bias" in galaxy formation (see, for instance, Bardeen, 1986; Kaiser, 1986). To a first approximation, this lowers $\Delta T/T$ by $\sim 1/b$ where b is

the bias factor, for which currently fashionable values seem to be
~ 2. If we include bias the curves shown in fig. 1 should therefore
be lowered by ~ 2; but in the next few months we may hope to see the
experimental upper limits lowered by a like factor (by the Cal Tech
group, Readhead et al). It thus remains the case that limits on the
small scale anisotropy of the CBR are only just consistent with
current models for the origin of LSS in the Universe. To me it seems
that the dilemma presented by a very smooth early Universe and a very
structured present Universe still lies at the heart of modern
cosmology.

3. LARGE-SCALE ANISOTROPY

As it happens, the CBR also provides us with some useful data bearing
on the inhomogeneous distribution of matter in the Universe at
present. The dipole moment observed in the CBR radiation is
conventionally interpreted as a result of the motion of the observer
(Peebles and Wilkinson, 1968). The motion, in turn is normally
assumed to result from gravitational acceleration caused by some
relatively local "lump" of matter. Davis (1987) among others has
discussed the constraints the measured amplitude of the dipole
(~ 1.2×10^{-3}--see below) can place on Ω_0 and local density
inhomogeneity, and that topic appears frequently in these proceedings.
     Before proceeding to the observations, it is perhaps worth asking
how confident we are that the dipole really is gravitationally
induced. Let me offer two arguments supporting the conventional view
which so many of my colleagues implicitly assume. Both are negative
arguments, in effect shooting down other possibilities.
     First, let us consider the possibility of a dipole moment
introduced into the CBR in a purely homogeneous but anisotropic
cosmological model. Anisotropic cosmologies (see Barrow et al, 1983,
for a review) produce anisotropic expansion, which can in turn produce
a dipole moment T_1 in the measured temperature of the CBR. It is my
understanding, however, that in all such models the magnitude of the
quadrupole moment T_2 and higher order moments is larger than the
dipole. The measurements have long shown the opposite; for instance,
the Soviet results to be discussed below show $T_2/T_1 \lesssim 10^{-2}$. One can
imagine that anisotropic expansion is responsible for some small
fraction (at most a few percent) of the observed dipole, but not all
of it.
     Second, we know from studies in the IR (Meiksin and Davis, 1986;
Yahil et al, 1986) and optical (Lahav, 1987) that nearby galaxies are
distributed anisotropically. Further, the dipole moment of the
distribution of local galaxies is aligned to within several degrees of
the CBR dipole. If the CBR dipole is gravitationally induced, the
approximate alignment is no surprise. Here let us consider an
alternative possibility which Lahav and I have recently worked on
(1988): could the radio emission from the anisotropically distributed
galaxies itself contribute the observed ~ 3 mK anisotropy? The point
is that the microwave emission from galaxies has a very different

Group	Berkeley	MIT/UBC	Moscow	Princeton
Reference	Lubin et al (1983)	Halpern et al (1985)	Strukov and Skulachev (1984)	Fixsen et al (1983)
Wavelength, mm	3	1-3	8	12
Vehicle	balloon	balloon	satellite	balloon
Detector	heterodyne	bolometric	heterodyne	heterodyne
Dipole amplitude, mK	3.4 ± 0.2	3.0 ± 0.5	3.16 ± 0.12	3.1 ± 0.2
Direction of solar motion, R.A. and Dec.	$11.5^h, -6°$	(similar)	$11.3, -8°$	$11^h, -10°$
Limit on quadrupole moment, mK	0.25*	n.a.	$\lesssim 0.03$	≤ 0.19

Table 2. Results of recent measurements of the large-scale distribution of the CBR.

*Calculated from Table I of Lubin et al by the present author, by taking the quadrature sum of the coefficients Q_2, Q_3, Q_4 and Q_5; $T = 2.75$ was used for the conversion to thermodynamic temperature.

spectrum from the blackbody CBR, typically $T \propto \lambda^{2 \cdot 7}$. If the nearby galaxies were to contribute a full 3 mK at $\lambda = 3.3$ mm where the dipole measurements of Lubin et al (1983) were made, they would contribute ~ 30 K to the observed temperature at 12 cm, where Sironi (1987) has shown the background temperature is 2.5 ± 0.7. Partridge and Lahav (1988) treat this argument in somewhat more detail, and conclude that not more than 1% and probably less than 0.1% of the observed CBR dipole can be due to direct microwave emission from anisotropically distributed galaxies.*

Of course, you all knew that all along--and have been using the CBR observations to constrain models of LSS and large-scale streaming on the assumption the dipole is gravitationally induced. So let me now turn briefly to the observations summarized in Table 2.

The agreement among the various groups is impressive--we know the dipole amplitude to an accuracy of a few percent. In particular let me note that T_1 is essentially independent of wavelength over a wavelength range of ~ 20. That independence is expected if the dipole is a Doppler effect and if the CBR spectrum is blackbody--and not otherwise (Lubin et al, 1983). Hence these measurements confirm direct measurements of the spectrum of the CBR (Smoot et al, 1985; Crane et al, 1986; Johnson and Wilkinson, 1987), which show it to be blackbody over a wavelength range of ~ 100.

As is well known, when the measured dipole moment of ~ 1.2×10^{-3} in $\Delta T/T$ is converted to a velocity for the local group, we find V_{LG} ~ 600 Km/sec, an uncomfortably large speed. The dipole direction also provides a standard against which to measure other, more local, velocity fields and motions (see, for instance the paper by Burnstein here).

4. CONCLUSION

Even as upper limits, measurements of the CBR anisotropy have been of great value in constraining theories for the origin and evolution of large-scale structure of the Universe. If we have not already begun to actually detect such anisotropies (Davies et al, 1987; Martin and Partridge, 1988), we are close to being able to do so. Think for a moment about the power of actual measurements of $\Delta T/T$ on a range of angular scales.

I would like to close with a word of thanks to the organizers of this workshop, and to the city of Trieste, which I have visited many times, each time with pleasure.

---

*A related argument in the same paper shows that the maximum dipole moment in the submillimeter flux reported by Matsumoto et al (1988) is < 22% (easy to reconcile with the results reported here by Melchiorri), and the maximum quadrupole is < $\frac{1}{2}$%.

238

REFERENCES

Bardeen, J. 1986, in Inner Space/Outer Space, E. W. Kolb, M. S. Turner, D. Lindley, K. Olive, and D. Seckel, eds., University of Chicago Press.

Barrow, J. D., Juszkiewicz, R., and Sonoda, D. H. 1983, Nature, 305, 397.

Berlin, A. B., Bulaenko, E. V., Vitkovsky, V. K., Kononov, V. K., Pariiskii, Yu. N., and Petrov, Z. E. 1983, in I.A.U. Symposium 104, G. Abell and G. Chincarini, eds., Reidel Publ. Co., Dordrecht, Holland.

Bond, J. R. and Efstathiou, G. 1984, Ap. J. (Letters), 285, L45.

Bond, J. R., and Efstathiou, G. 1987, Mon. Not. Roy. Astr. Soc., 226, 655.

Conklin, E. K., and Bracewell, R. N. 1967, Nature, 216, 777.

Crane, P., Hegyi, D. J., Mandolesi, N. and Danks, A. C. 1986, Ap. J., 309, 822.

Davies, R. D., Lasenby, A. N., Watson, R. A., Daintree, E. J., Hopkins, J., Beckman, J., Sanchez-Almeida, J., and Rebolo, R., 1987, Nature, 326, 462.

Davis, M. 1987, in I.A.U. Symposium 117, Reidel Publishing Company, Dordrecht, Holland.

Fabbri, R., Guidi, I., Melchiorri, F. and Natale, V. 1980, Phys. Rev. Letters, 44, 1563.

Fixsen, D. J., Cheng, E. S., and Wilkinson, D. T. 1983, Phys. Rev. Letters, 50, 620.

Fomalont, E. B., Kellermann, K. I., and Wall, J. V. 1984, Ap. J. (Letters), 277, L23.

Fomalont, E. B., Kellermann, K. I., Anderson, M. C., Weistrop, D., Wall, J. V., Windhorst, R. A., and Kristian, J. A. 1988, submitted to Ap. J.

Franceschini, A., Toffolati, L., Danese, L., and De Zotti, G. 1988, in preparation.

Halpern, M., Weiss, R. and Benford, R. 1985, in The Cosmic Background Radiation and Fundamental Physics, F. Melchiorri, ed., Editrice Compositori, Bologna.

Hawking, S. W. 1969, Mon. Not. Roy. Astr. Soc., 142, 129.

Hogan, C. J., 1980: Mon. Not. Roy. Astr. Soc., 192, 891.

Hogan, C. J. 1984, Ap. J. (Letters), 284, L1.

Johnson, D. G., and Wilkinson, D. T. 1987, Ap. J. (Letters), 313, L1.

Kaiser, N. 1986, in Inner Space/Outer Space, E. W. Kolb, M. S. Turner, D. Lindley, K. Olive, and D. Seckel, eds., University of Chicago Press.

Kaiser, N. and Silk, J. 1986, Nature, 324, 529.

Knoke, J. E., Partridge, R. B., Ratner, M. I., and Shapiro, I. I. 1984, Ap. J., 284, 479.

Lahav, O. 1987, Mon. Not Roy. Astr. Soc., 225, 213.

Lubin, P. M., Epstein, G. L., and Smoot, G. F. 1983, Phys. Rev. Letters, 50, 616.

Mandolesi, N. et al 1986, Nature, 319, 751.

Martin, H. M. and Partridge, R. B. 1988, Ap. J., 324, 794.

Matsumoto, T., Hayakawa, S., Matsuo, H., Murakami, H., Sato, S., Lange, A. E., and Richards, P. L. 1988, Ap. J., 329, .

Meiksin, A., and Davis, M. 1986, Astron. J., 91, 191.

Ostriker, J. P. and Vishniac, E. T. 1986, Ap. J. (Letters), 306, L51.

Pariiskii, Yu. N., Petrov, Z. E., and Cherkov, L. N. 1977, Sov. Astron. Letters., 3, 263.

Partridge, R. B. 1988, Rep. Prog. Phys., 51, .

Partridge, R. B. 1989, in Proceedings of the Third ESO-CERN Symposium, "Astronomy, Cosmology and Fundamental Physics," to be published by Reidel Publ. Co., Dordrecht, Netherlands.

Partridge, R. B., and Lahav, O. 1988, submitted to Mon. Not. Roy. Astr. Soc.

Partridge, R. B., Nowakowski, J., and Martin, H. M. 1988, Nature, 331, 146.

Partridge, R. B. and Wilkinson, D. T. 1967, Phys. Rev. (Letters), 18, 557.

Peebles, P. J. E., and Wilkinson, D. T. 1968, Phys. Rev., 174, 2168.

Penzias, A. A. and Wilson, R. W. 1965, Ap. J., 142, 419.

Sachs, R. K., and Wolfe, A. M. 1967, Ap. J., 147, 73.

Silk, J., 1968, Ap. J., 151, 459.

Sironi, G., 1987, private communication.

Smoot, G. F., De Amici, G., Friedman, S., Witebsky, C., Sironi, G., Bonelli, G., Mandolesi, N., Cortiglioni, S., Morigi, G., Partridge, R. B., Danese, L., and De Zotti, G. 1985, Ap. J. (Letters), 291, L23.

Strukov, I. A. and Skulachev, D. P. 1984, Sov. Astron. Letters, 10, 1.

Sunyaev, R. A. 1978, in I.A.U. Symposium 79, The Large Scale Structure of the Universe, M. S. Longair and J. Einasto, eds., Reidel Publ. Co., Dordrecht, Netherlands.

Sunyaev, R. A. and Zel'dovich, Ya. B. 1972, Comments Astrophys. Space Sci., 4, 173.

Uson, J. M., and Wilkinson, D. T. 1984a, Ap. J., 283, 471.

Uson, J. M., and Wilkinson, D. T. 1984b, Nature, 312, 427.

Verschuur, G. L., and Kellermann, K. I. 1974, Galactic and Extragalactic Radio Astronomy, Springer-Verlag, Heidelberg.

Vishniac, E. T. 1987, Ap. J., 322, 597.

Vittorio, N., and Silk, J. 1984, Ap. J. (Letters), 285, L39.

Yahil, A., Walker, D., and Rowan-Robinson, M. 1986, Ap. J. (Letters), 301, L1.

DISCUSSION

BURSTEIN: Am I correctly interpreting your graph that plots
the   temperatures   derived   from   the   recent japanese
experiment? It appears that the only point that forces the
fitting curve to rise is that point at the shortest
wavelength.

PARTRIDGE: Certainly that last point – at   400μm is the
crucial   one.   Nevertheless,   the   longer wavelenght
measurements of Matsumoto, Richards et al.   also show an
excess – and note the small errors of the measurements. I
would agree, though, that hoping to the conclusion that the
data now favor a comptonization model is very premature.

WATSON: If there is clustering of discrete sources at high
redshift could you produce a "hot spot" at 3 cm, but not at
1 cm, so explaining the non-detection in the relic?

PARTRIDGE: Yes – one only needs a suitable spectrum to give
large fluctuations at 3 cm or a single large "blip" at 3 cm
with no visible effect on the Soviet 8 mm observations. But
one thing you can't do is to have a large blip at 3 cm, the
same blip in Melchiorri in data 0.6 mm and no trace of it in
the Soviet data.

PECULIAR VELOCITY AND GRAVITY
AS COSMOLOGICAL PROBES

NICOLA VITTORIO

Dipartimento di Fisica, Universita' dell'Aquila, Italy

ROMAN JUSZKIEWICZ

Joseph Henry Laboratories, Princeton University, U.S.A.,
and Copernicus Center, Warsaw, Poland

ABSTRACT. We summarize the theoretical implications of recent observations of peculiar velocities, represented as (1) *empirical models*, such as the bulk flow or the Great Attractor infall model and (2) velocity correlation tensor, derived directly from the *"raw data"*. Our tests have varying power to constrain theories for the growth of cosmic structure. Each of these tests has its own drawbacks and technical limitations. However, one result seems to be particularly robust: the predictions of the canonical biased cold dark matter (CDM) scenario fail all tests by a large margin. A change of the interpretation or a drastic revision of the existing fits to the observational data would be required to save this theory. The isocurvature baryon-dominated models, recently proposed by Peebles (hereafter PIB) fare much better and naturally lead to large scale flows, although some of these models may actually have too much large scale power.

We also consider the magnitude and direction of the observed dipole anisotropy of the galaxy distribution, derived from the IRAS catalogue. In contrast to the peculiar velocity data, constraints on cosmological models, based on this statistic (and the associated estimate of the local peculiar gravity) appear to be rather weak and cannot distinguish between models with a wide variation of large scale power, such as CDM and PIB. Estimates of the density parameter, Ω, based on the IRAS data may be misleading as one cannot be assured that there is indeed neglible power on scales beyond the sample depth.

1. INTRODUCTION

The theory of structure formation from scale-invariant adiabatic fluctuations in a flat Universe, dominated by cold dark matter agrees well with observations on small scales (Davis *et al.* 1985; Bardeen *et al.* 1986). However, observations of large scale streaming motions (Collins *et al.* 1986; Dressler *et al.* 1987) disagree with the CDM predictions (Vittorio *et al.* 1986; Bond 1986; Melott 1987; see also Vittorio and Silk 1985). These observations indicate that Universe on scales $\gtrsim 20$ Mpc may be more inhomogeneous than theorists ever expected. The observed rich cluster correlation function (Bahcall 1988 and references therein) and a possible detection of cosmic microwave background anisotropy at 8° (Davies *et al.* 1987) can be regarded as additional evidence of structure on very large scales (unless cluster-cluster correlations are an artifact of errors in the catalogues while the Davies *et al.* anisotropy is non-cosmological). To deal with this impasse, several new scenarios with enhanced large scale power have been constructed (Peebles 1987 and 1988; Bardeen *et al.* 1987; Vittorio *et al.* 1987; Blumenthal *et al.* 1988).

241

M. Mezzetti et al. (eds.), Large Scale Structure and Motions in the Universe, 241–258.
© *1989 by Kluwer Academic Publishers.*

In this article we focus on the problem of of "missing power" on large scales. We review the present status of two alternative families of models, using the available observational constraints on the peculiar velocity and peculiar gravity fields of galaxies with redshifts within several thousands $km s^{-1}$. These are: flat CDM models (with two different normalizations); and a set of open baryon-dominated universes with isocurvature perturbations. The latter can be regarded as a set of models with "extra power" on large scales. All calculations presented here assume gravitational instability and random Gaussian initial inhomogeneities as a source of the cosmic structure. The statistics are then fully specified by the density fluctuation power spectrum, $P(k)$, where k is the wavenumber.

2. PECULIAR VELOCITY

We consider (1) the CDM model with the transfer function of Davis et al. (1985) and $\Omega = 1$; and (2) a set of isocurvature models with transfer functions of Peebles (1987) with primordial $n = 0$ and $n = -1$ spectra ($P \propto k^n$), partial (minimum ionization fraction $x_e = 0.1$) or no ($x_e = 1$) recombination and a range of values for Ω. All spectra are derived assuming a Hubble constant $H_0 = 50 \, km s^{-1}/Mpc$. All distances are measured in $km s^{-1}$ (hence, $H_0 r$ is replaced by r in all expressions where r is the distance; wavenumbers k are expressed in $km s^{-1}$). Davis et al. advocate $\Omega = 1$ biased CDM, with rms mass fluctuation in a sphere of radius $800 \, km s^{-1}$ equalling $b^{-1} = 0.4$; this is the the "canonical CDM" – the first model we consider. We also study the case $b = 1.5$ for the CDM model. The isocurvature models are "unbiased" and their spectra are normalized as described in Peebles (1987). We will refer to these models as PIB, for "Peebles' Isocurvature Baryon-dominated models". In all cases we filter the density field with a Gaussian of radius $R_f = 400 \, km s^{-1}$, in order to smooth over small scale nonlinear structure.

2.1. Bulk motions

The statistical measure of the peculiar velocity field we discuss here, is the *bulk velocity*, or center of mass velocity of a large volume. In reality, all measurements of this kind involve averaging the velocity field over some selection function (or "effective volume"). In all scenarios considered here, bulk velocities are expected to have a Maxwellian distribution, with a dispersion

$$\sigma_v^2(R_\star) = \frac{\Omega^{1.2}}{2\pi^2} \int_0^\infty P(k) W^2(k) dk \tag{1}$$

where the $W(k)$ models the selection function. For our purposes here, we adopt $W(k) = \exp(-k^2 R_\star^2/2)$. The choice of $R_\star$, appropriate for the recent observations, is a subject of controversy. The depth of the Rubin-Ford sample of spiral galaxies (Rubin et al. 1976), re-observed by Collins et al. , is $\sim 6500 \, km s^{-1}$. The "Seven Samurai" (Dressler et al. 1987) elliptical galaxy sample extends to similar distances. For a volume-limited sample this would correspond to $R_\star \sim 4000 \, km s^{-1}$. However, Kaiser (1988) argues that the maximum likelihood analysis employed by Dressler et al. reduces the effective depth of their sample to $1500 \, km s^{-1}$. On the other hand, Górski and Hoffman (1988) suggest that Kaiser's formalism mixes *a priori* and *a posteriori* probabilities, and that when $W(k)$ is constructed in a self-consistent, *a priori* fashion, the effective depth increases to $R_\star = 2500 \, km s^{-1}$.

In any case, it seems that $1500\,\mathrm{km\,s^{-1}}$ can be regarded as a very conservative lower limit on R_* for the elliptical sample. In Table 1 we list the predicted rms bulk velocities for both $R_* = 1500\,\mathrm{km\,s^{-1}}$ and $4000\,\mathrm{km\,s^{-1}}$. These results may be compared with the bulk velocity $V = 970 \pm 300\,\mathrm{km\,s^{-1}}$ reported by Collins $et\ al.$ and $599 \pm 104\,\mathrm{km\,s^{-1}}$ found by Dressler $et\ al.$ For a randomly placed observer, there is only a 5% probability of finding a bulk flow with $V > 1.65\sigma_v$.

2.2. Are Great Attractors common?

After considering the bulk flow solution to their data, the Seven Samurai proposed what they believe is a better fit for their sample – a spherically symmetric infall toward a massive density perturbation – the "Great Attractor" (hereafter GA; Lynden-Bell $et\ al.$ 1988; Burstein 1988).

In this section we summarize the results recently obtained by Bertschinger and Juszkiewicz (1988; BJ), who have reexamined the theoretical implications of the streaming motions, using the GA infall model. The velocity field in this model is spherically symmetric about a point located $4200\,\mathrm{km\,s^{-1}}$ from the Local Group (LG). The velocity

TABLE 1: FLOW MODEL STATISTICS

| Spectrum | Bulk Velocity | | Spherical Model | | | Cylindrical Model |
	R_{15}	R_{40}	ν	N_{up}	f_6	f_6
CDM(2.5)	150	72	7.3	2.7E-6	0	0
CDM(1.5)	250	121	4.4	26	0	1
PIB(0;0.1;1)	156	110	7.6	7.0E-7	0	0
PIB(-1;0.1;1)	328	291	5.0	0.63	0	0
PIB(0;0.1;0.1)	145	93	7.5	1.2E-6	0	0
PIB(-1;0.1;0.1)	257	208	5.1	0.42	0	0
PIB(0;0.4;1)	521	348	2.3	3.2E3	74	1.0E3
PIB(-1;0.4;1)	749	559	1.8	1.7E3	827	1.1E4
PIB(0;0.4;0.1)	399	214	2.6	3.6E3	39	59
PIB(-1;0.4;0.1)	536	340	2.1	2.8E3	254	1.3E3

Table Caption: Bulk Velocity= $\sigma_v(R_*)$, in $\mathrm{km\,s^{-1}}$, calculated for $R_* = 1500\,\mathrm{km\,s^{-1}}$ (R_{15}) and $R_* = 4000\,\mathrm{km\,s^{-1}}$ (R_{40}). $N_{up} = n_{up}(c/H_0)^3$; $f_6 = f \times 10^6$. CDM(...)= CDM spectrum with parameter (b). PIB(...)= Isocurvature spectrum $(b = 1)$ with parameters $(n; \Omega; x_e)$. $\nu =$ Great Attractor density contrast in standard deviations.

244

profile is

$$u_A = -v_A(r_A/d_A)[(d_A^2 + c_A^2)/(r_A^2 + c_A^2)]^{(n_A+1)/2} , \tag{2}$$

where r_A is the distance from the flow center, d_A is the distance of the Great Attractor from the LG, $v_A = 535\,\mathrm{km\,s^{-1}}$ is the flow velocity at $r_A = d_A$, and $n_A = 1.7$. The parameter c_A is the "core radius", $c_A = 0.34d_A$. This is a version of the Lynden-Bell et al. model, updated by Faber and Burstein (1988; FB). Since $v_A/d_A = 0.13$— a non-negligible perturbation— it is necessary to map the present (Eulerian) profile $u_A(r_A)$, to a Lagrangian comoving radius r_L and linear radial velocity $v_L(r_L)$ at some high redshift. Only then will comparisons with (linear) predictions be meaningful. For this purpose, BJ divide the flow into 15 shells, $2500\,\mathrm{km\,s^{-1}} \le r_A \le 6000\,\mathrm{km\,s^{-1}}$. The exact solution for spherical mass shells (see, e.g., Peebles 1980) is used to follow their evolution back in time. The resulting velocity profile is then evolved forward to the present as if the linear theory were exact. Figure 1 shows that the nonlinear correction is small for $\Omega = 1$ but significant for $\Omega \le 0.4$.

The velocity profile $v_L(r_L)$ is directly related to the overdensity, perturbing the Hubble flow: $3v_L(r_L) = -\Omega^{0.6}\Delta_L(r_L)r_L$. Here $\Delta_L(r_L)$ is the mass density contrast, averaged over a comoving spherical volume of radius r_L, drawn around the center of infall. Now we can answer the following question: *how many closed regions are there in a large volume, in which Δ_L exceeds a given value?* The mean number density of such regions is

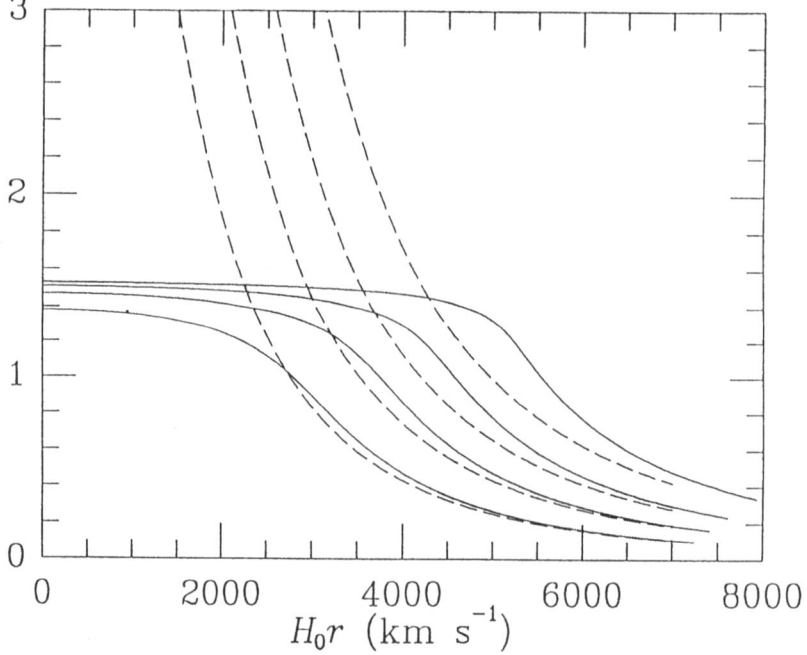

Fig.1. The mean density contrast in a sphere of radius r for the spherical Great Attractor model of FB. The dashed and solid curves give respectively: the profiles before ($\Delta_A = -3\Omega^{-0.6}u_A/r_A$ versus r_A) and after (Δ_L versus r_L) the nonlinear correction, for $\Omega = 1$ (lowest curve), 0.4, 0.2, 0.1 (highest curve). Our statistical analysis uses the range $2500\,\mathrm{km\,s^{-1}} \le H_0 r_A \le 6000\,\mathrm{km\,s^{-1}}$.

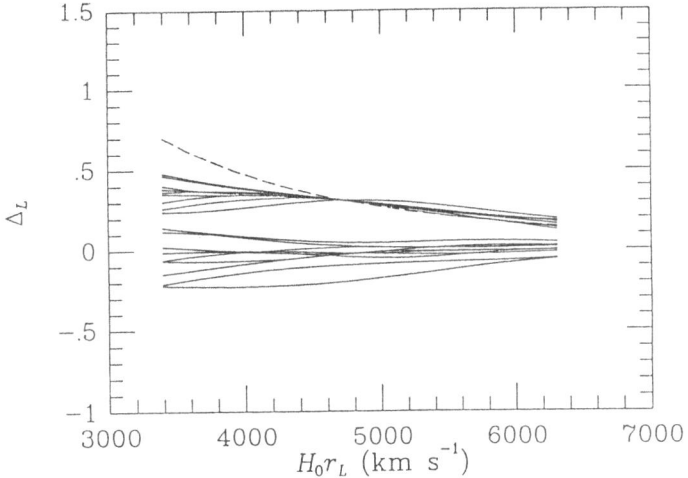

Fig.2. Random samples of the mean density profile for the $\Omega = 1$, $b = 1.5$ CDM model. The dashed curve gives the FB Great Attractor profile. The 10 lower curves give unconstrained profiles, while the 10 upper curves are constrained to fit the observations at the Local Group. The constraint probability corresponds to $\nu = 4.4$ standard deviations; even such a rare event still leads to profiles which are generally too shallow at small r_L. Of 10^6 unconstrained random samples in the $b = 1.5$ CDM model, none exceeded the model profile over the entire range shown.

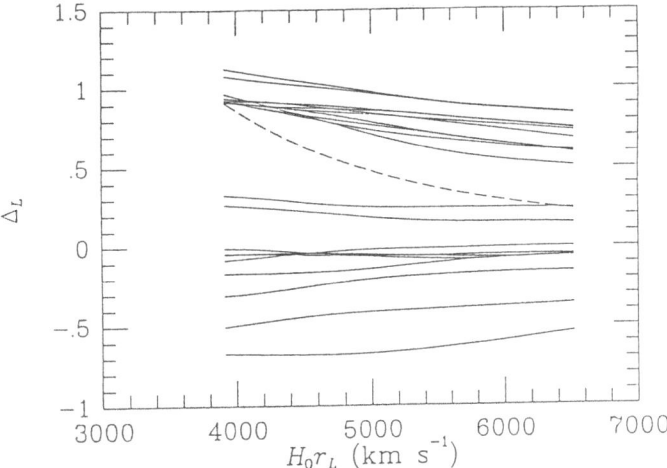

Fig.3. Samples of the mean density profile for the $\Omega = 0.4$, $n = -1$, $x_e = 1$ PIB model. The 10 lower curves are unconstrained samples, while the 10 upper curves have been selected to exceed the Great Attractor profile (dashed curve). Of 10^6 unconstrained samples, 827 passed this test. Note, however, the excessive coherence of the theoretical curves.

given approximately by (Doroshkevich 1970; Bardeen *et al.* 1986)

$$n_{up}(\nu) = (2\pi)^{-2}3^{-3/2}(\sigma_1/\sigma_0)^3(\nu^2 - 1)\exp(-\nu^2/2) , \qquad (3)$$

where $\nu = \Delta_L/\sigma_0$ and $\sigma_m^2(r_L) = (9/2\pi^2 r_L^2)\int_0^\infty P(k)k^{2m}j_1^2(kr_L)dk$, $m = 0,1$. The quantity $\sigma_0^2(r_L) = \langle\Delta_L^2\rangle$ is the variance of the field Δ_L, i.e., ν is the amplitude of the volume-averaged density contrast, expressed in standard deviations. Here and below $j_0(x) = \sin(x)/x$ and $j_1(x) = -dj_0(x)/dx$ are spherical Bessel functions. In Table 1 we give the mean number of such up-crossing regions within the Hubble volume $(c/H_0)^3$, for $\Delta_L(r_L)$ corresponding to $u_A(d_A)$. Similar results were recently obtained by Bond (1988).

One can also ask: how probable are *profile shapes* like that of the Great Attractor (GA) model? To answer this question, we need a large number of realizations of the density field averaged over N nested spheres drawn around a random point: $\Delta_l = \Delta(r_l)$, $l = 1,\dots,N$. The frequency with which $\Delta_l \geq \Delta_L(r_l)$ is given by

$$f(\Delta_L[r_l]) = \int_\omega p(\Delta_1,\dots,\Delta_N)\,d^N\Delta_l, \qquad (4)$$

where the domain of integration ω is $\Delta_l > \Delta_L(r_l)$ for $l = 1,\dots,N$. Here p is an N-dimensional Gaussian with covariance matrix

$$\mathbf{M}_{lm} = \langle\Delta_l\Delta_m\rangle = \frac{9}{2\pi^2}\frac{1}{r_l r_m}\int_0^\infty dk\,P(k)j_1(kr_l)j_1(kr_m) . \qquad (5)$$

A similar statistic using the dipole velocity with $N = 2$ was considered by Vittorio *et al.* (1986). In the limit $N \to \infty$, $r_m - r_{m+1} \to 0$, f is the probability that the curve $\Delta(r)$ lies above $\Delta_L(r)$ (Feynman and Hibbs 1965; Bertschinger 1987). BJ choose to approximate the continuum limit with $N = 15$; this is sufficient because the shells are thinner than the filter scale, $R_f = 400\,\mathrm{km\,s^{-1}}$. Then, a Monte Carlo integration (with the matrix $\mathbf{M}$ as an input) was used to calculate f. They obtained 10^6 independent samples of Δ and simply counted the number of samples lying above the model profile $\Delta_L(r)$ at each point r_l (see Table 1, Fig.2 and 3). These results have a standard error $\sim (10^6 f)^{1/2}$. Figures 2 and 3 also show random realizations of *constrained* profiles, generated from a conditional probability distribution, given that the velocity at the LG position fits the observations.

The elliptical sample does not include galaxies behind the GA. FB have recently reformulated the flow description for this sample, so that it is conservatively limited to the space thus far surveyed – the *cylindrical flow* model. Bertschinger and Juszkiewicz have included this model in their analysis. FB describe the flow within a cylinder of radius $R = 1750\,\mathrm{km\,s^{-1}}$ pointing toward the GA with the LG on its axis. At the far end of the cylinder, at a distance $r_A = 6700\,\mathrm{km\,s^{-1}}$ from the GA, the infall velocity is $400\,\mathrm{km\,s^{-1}}$. The flow velocity rises to $1000\,\mathrm{km\,s^{-1}}$ at the end near the Attractor at $r_A = 1200\,\mathrm{km\,s^{-1}}$. The velocity at LG position agrees with the spherical flow model. The FB cylindrical flow model can be represented by a quadratic fit to the one-dimensional velocity v_z as a function of distance along the long axis, defined as the z-axis. To map the nonlinear flow into an unevolved linear flow, BJ used the Zel'dovich (1970) solution. After this conversion, the cylinder was cut into 15 slices of equal thickness. Now, following BJ, let us consider the z-component of the peculiar velocity field, averaged over each slice, $V(z)$, and ask: how frequently the curve $V(z)$, generated in a random realization of the velocity field lies above the FB velocity profile along the cylinder? To calculate the frequency f of such events, BJ used a 10^6-point Monte Carlo integration, similar to that discussed above. Their results are shown in Table 1.

2.3. Velocity correlation tensor

In this section we describe some preliminary results obtained by Groth and Juszkiewicz (1987, GJ), who made an attempt to estimate the correlation tensor for the velocity field, $\xi_{\alpha\beta}(r) = \langle v_\alpha(\mathbf{r})v_\beta(0)\rangle$, where $\alpha, \beta = x, y, z$ denote different components of the peculiar velocity field $\mathbf{v}(\mathbf{r})$. This method directly uses *"raw data"*, i.e. positions and velocities of pairs of points at a given separation r in the sample. This provides an interesting alternative to tests, based on highly reduced representations of observations, described above. For a statistically homogeneous and isotropic velocity field, $\xi_{\alpha\beta}$ can be expressed in terms of two * functions (Monin and Yaglom 1975, Górski 1988):

$$\xi_{\alpha\beta} = (\Pi - \Sigma)\hat{r}_\alpha \hat{r}_\beta + \Sigma\delta_{\alpha\beta}, \tag{6}$$

where $\delta_{\alpha\beta}$ is the Kronecker delta. Here and below the hat denotes unit vectors, $\hat{\mathbf{r}} = \mathbf{r}/r$. $\Pi(r)$ and $\Sigma(r)$ are usually called the *radial* and the *transverse* correlation function, respectively. In a coordinate frame, in which the z- axis is parallel to $\mathbf{r}$, $\xi_{zz} = \Pi$, $\xi_{xx} = \xi_{yy} = \Sigma$, and $\xi_{\alpha\beta}$ is diagonal. The "dot product" correlation function, recently introduced by Peebles (1987) and considered by Kaiser in these proceedings, is given by

$$\xi_v(r) = \langle \mathbf{v}(\mathbf{r}) \cdot \mathbf{v}(0)\rangle = \Pi + 2\Sigma . \tag{7}$$

The observations provide only the line of sight components of $\mathbf{v}$. Under the assumption of statistical isotropy and homogeneity, the expectation value of the product of two radial velocities is given by

$$\langle (\mathbf{v}(\mathbf{R}) \cdot \hat{\mathbf{R}})(\mathbf{v}(\mathbf{S}) \cdot \hat{\mathbf{S}})\rangle = \kappa_\parallel \Pi(r) + \kappa_\perp \Sigma(r); \tag{8}$$

$$\kappa_\parallel = (\hat{\mathbf{R}} \cdot \hat{\mathbf{r}})(\hat{\mathbf{S}} \cdot \hat{\mathbf{r}}); \quad \kappa_\perp = \hat{\mathbf{R}} \cdot \hat{\mathbf{S}} - \kappa_\parallel; \quad \mathbf{r} = \mathbf{R} - \mathbf{S} . \tag{9}$$

To estimate the radial and transverse correlation functions, GJ binned all pairs of objects in the catalogue. The size of each bin in separation was $500\,\mathrm{km\,s^{-1}}$. Then, all radial velocity products within a bin were used to estimate Π and Σ via a least squares fit, using Eq.(8). The product of radial velocities for each pair was weighted inversely proportionally to the product of variances of these velocities. Propagation of errors, based on a (questionable) assumption that each velocity product is independent, was used to estimate formal errors in Π and Σ. The results for the FB elliptical sample are plotted in Fig.4. The corresponding ξ_v is shown in Fig.6. This is in good agreement with Kaiser's (1988b) estimate, based on a different weighting scheme. Our ξ_v is also consistent with the results, obtained recently by Górski and Davis (1988). To test the *GA* model, GJ replaced the "raw" velocities in the

---

* For vorticity free fields, Π and Σ are not independent. The condition $\nabla \times \mathbf{v} = 0$ gives $r\Sigma(r) = \int_0^r \Pi(x)dx$ (Monin and Yaglom 1975). In principle this constraint may be used as a test of gravitatational instability scenarios, all of which predict $\nabla \times \mathbf{v} = 0$ in the linear regime.

248

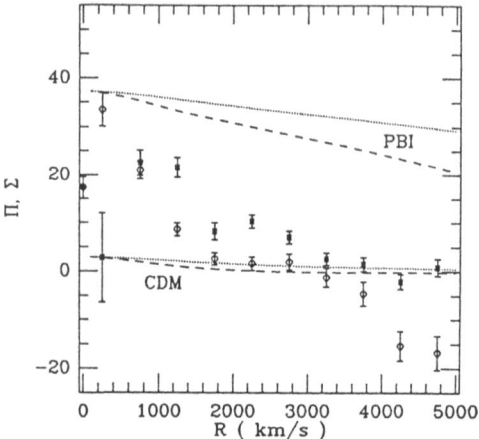

Fig.4. Parallel (Π) and transverse (Σ) velocity correlation functions in units of $(100\,\mathrm{km\,s^{-1}})^2$ *vs.* separation *r* in $\mathrm{km\,s^{-1}}$. The observational points have been calculated from the FB elliptical galaxy sample. Filled and open squares refer to Π and Σ, respectively. Error bars correspond to formal errors only. The theoretical predictions for CDM and PIB models are also shown. The dashed and dotted lines refer respectively, to Π and Σ.

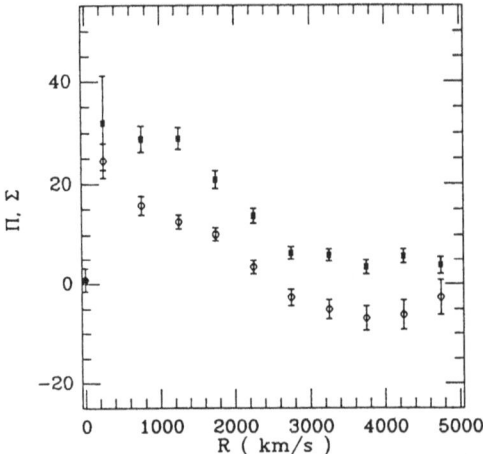

Fig.5. Π and Σ, computed from the set of positions of elliptical galaxy catalogue members and Great Attractor model velocities (Eq.[2]). Units, as in Fig.4. Filled and open squares refer to Π and Σ, respectively.

elliptical sample with values, given by Eq.(2). The resulting correlation functions, shown in Fig.5, appear to be in good agreement with the "raw" Π and Σ. For comparison, in Fig.4 we also show correlation functions,

$$\Sigma(r) = \frac{\Omega^{1.2}}{2\pi^2} \int_0^\infty P(k)(kr)^{-1} j_1(kr) dk \; ; \tag{10}$$

$$\Pi(r) = \frac{\Omega^{1.2}}{2\pi^2} \int_0^\infty j_0(kr) P(k) dk - 2\Sigma(r) \, , \tag{11}$$

predicted by the canonical CDM model ($b = 2.5$) and the PIB model ($b = 1$) with parameters $x_e = 1$; $\Omega = 0.4$; $n = -1$. Both spectra were smoothed over $R_f = 400 \, \mathrm{km \, s^{-1}}$. In Figure 6 we plot $\xi_v = 2\Sigma + \Pi$, derived for the CDM model from Eq.(11).

3. PECULIAR GRAVITY

The peculiar acceleration $\mathbf{a}$, exerted on the Local Group of galaxies is given by

$$\mathbf{a} = G\rho \int \delta(\mathbf{r})(\mathbf{r}/r^3) \, d^3r \, , \tag{12}$$

where G is the gravitational constant, $\delta(\mathbf{r})$ is the fractional density contrast at location $\mathbf{r}$ (with the LG at $\mathbf{r} = 0$), and $\rho = 3H_0^2\Omega/8\pi G$ is the mean mass density of the Universe. The induced peculiar velocity in linear gravitational instability theory is related to a by (Peebles 1980): $\mathbf{a}/H_0 = 3\Omega^{0.4} \mathbf{v}/2$. Since the LG velocity relative to the cosmic microwave background is known (Wilkinson 1988, and references therein), an independent estimate of the "dipole moment", or the integral $\mathbf{g} = (3/4\pi) \int \delta(\mathbf{r})(\mathbf{r}/r^3) \, d^3r$, can in principle

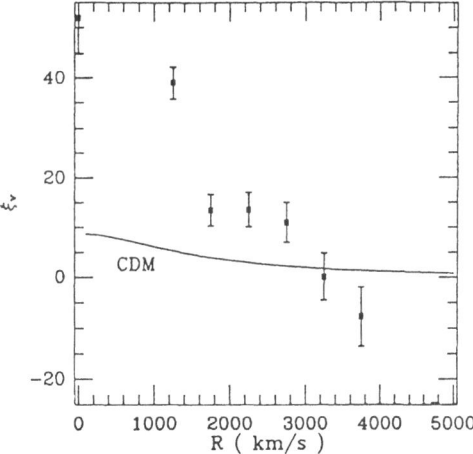

Fig.6. The "dot product" correlation function, $\xi_v(r)$. Filled squares correspond to the elliptical sample, while the solid curve represents the CDM prediction. For the PIB($n = -1$; $x_e = 1$; $\Omega = 0.4$) model, $\xi_v(r) \sim 100 \times (100 \, \mathrm{km \, s^{-1}})^2$, and is essentially constant in the shown range of separations.

lead to the determination of Ω. Several groups (Meiksin and Davis 1986; Yahil *et al.* 1986; Villumsen and Strauss 1987; Strauss and Davis 1987) have used the IRAS galaxies as tracers of the local desity field and attempted to estimate g. A similar approach has also been applied to optical galaxy catalogues (Davis and Huchra 1982; Lahav 1987). All such calculations assume that: (I) linear theory is valid; (II) galaxies trace the mass; (III) inhomogeneities on scales larger than the sample depth are negligible. The measurement of g can also be used to constrain the primeval power spectrum $P(k)$ on scales $k^{-1} \lesssim$ sample depth (Vittorio and Juszkiewicz 1987; Kaiser 1988b; Juszkiewicz, Vittorio and Wyse, hereafter JVW). Below we present some of the results, obtained recently by JVW. Our calculations allow for biasing – we relax the assumption (II).

3.1. Convergence in amplitude

The standard method used to justify the assumption (III) involves dividing the sample into a set of nested spherical subsamples. Then, a sequence of dipole moments, g_R, is calculated for the material, contained in spheres of increasing radii R. If the variations in $g_R = |\mathbf{g}_R|$ for $R > R_c$ remain within the noise, it is usually concluded that the sequence *converges* and the contribution from the material at $r > R_c$ can be neglected. It has been claimed in the literature that the IRAS data do support such "convergence" of the peculiar acceleration.

Unfortunately, the above criterion is very unreliable. Consider an estimate of g, derived from a spherically symmetric sample with an effective radius R:

$$\mathbf{g}_R = \int \delta(\mathbf{r}) w(r/R)(\mathbf{r}/r^3) \, d^3r \; . \tag{13}$$

Here w describes the spatial distribution of the galaxies in the sample ($w(r/R) \to 0$ for $r/R \to \infty$). The quantity g_R is expected to be a random variable, with Maxwellian distribution and a variance

$$\sigma_g^2(R) = \frac{9}{2\pi^2} \int_0^\infty dk P(k) \tilde{w}^2(kR) \; ; \tag{14}$$

$$\tilde{w}(x) = x \int_0^\infty dy \, w(y) \, j_1(xy) \; . \tag{15}$$

It is easy to show that unlike $W(k)$, which is a low pass filter (Eq.1), the function $\tilde{w}(kR)$ samples the *shortwave tail* of $P(k)$ (wavenumbers $k^{-1} \lesssim R$). Indeed, for a "top hat" $w(r/R)$, $\tilde{w}(x) = 1 - j_0(x)$, and $\tilde{w}(0) = 0$, while $\tilde{w}(\infty) = 1$. This is not surprising, since g_R is induced only by fluctuations within the sample volume. Now, let us consider an illustrative model which fits the observed galaxy autocorrelation function, $\xi(r) \propto r^{-1.8}$. The power spectrum in this case is $P(k) \propto k^{-1.2}$, and implies (Vittorio and Juszkiewicz, 1987)

$$\sigma_g(R) \propto R^{0.1} \; . \tag{16}$$

The low value of the exponent in this expression can mimic the apparent convergence of the peculiar acceleration. At the same time, the large scale power is so enormous, that the rms bulk velocity on the sample scale is divergent: $\sigma_v(R) = \infty$. JVW have investigated a more realistic example with large scale power: the PIB spectrum with parameters $\Omega = 0.4$; $n = -1$; $x_e = 1$. A sequence of rms accelerations for samples of

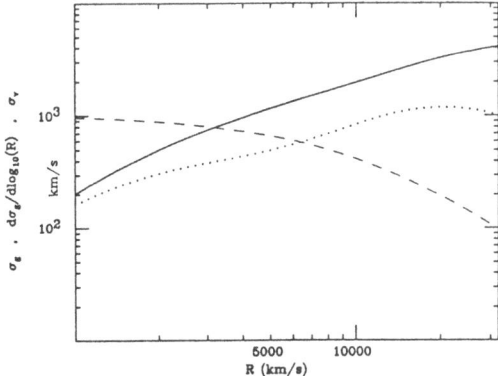

Fig.7. PIB($n = -1; x_e = 1; \Omega = 0.4$) model predictions for the rms "dipole moment" σ_g of the distribution of galaxies in the survey (solid line) and for the rms bulk velocity of the sample, σ_v, (dashed line) plotted as a function of depth of the catalogue R. The sample is defined by the IRAS window function of Yahil *et al.* (Eq. 17). The differential quantity $d\sigma_g(R)/dlog_{10}R$ (dotted line) determines the scale of inhomogeneities which provide the dominant contribution to the acceleration. The small-scale smoothing radius, adopted here, is $R_f = 500\,\mathrm{km\,s^{-1}}$.

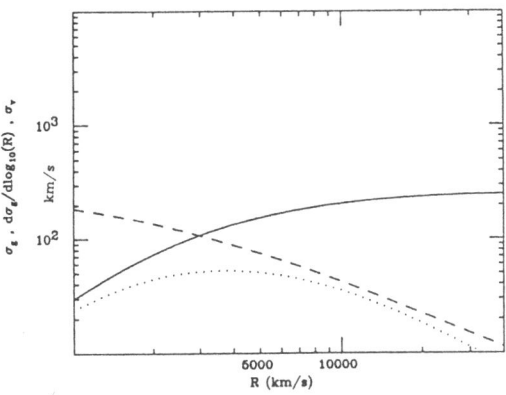

Fig.8. As in Fig.7, but for the CDM model, with $b = 2.5$.

increasing depth, $R \lesssim 2 \times 10^4 \, \mathrm{km\,s^{-1}}$, was calculated, using the window function of Yahil *et al.* (1986),

$$w(r/R) = [1 + (r^2/2.4R^2)]^{-2.4} \; , \tag{17}$$

to simulate the distribution of the IRAS sample. Our results are shown in Fig.7, where $\sigma_g(R)$ is plotted vs. R. At $R \sim 10^4 \, \mathrm{km\,s^{-1}}$, there is a low amplitude bump due to the large scale feature in the power spectrum. This feature is more prominent in the plot of the logarithmic derivative of $\sigma_g(R)$, also shown in Fig.7. In addition, the predicted rms bulk velocity of every subsample, $\sigma_v(R)$, is plotted as a function of its depth R. This quantity provides a measure of power at $k^{-1} > R$. A proper determination of Ω from g_R and the LG velocity, V_{LG}, is unlikely, unless $\sigma_v(R) \ll V_{LG} = 600 \, \mathrm{km\,s^{-1}}$. If there is substantial power at $k^{-1} \gtrsim$ sample depth, the the dipole moment g is likely to be underestimated, and this will result in an overestimate of Ω. The rms sample velocity, plotted in Fig.7, was calculated from Eq.(1), with

$$W(k) = 4\pi \int_0^\infty dx \, w(x/R) \, x^2 j_0(kx) \; . \tag{18}$$

Similar calculations were made for the CDM model (Fig.8). As expected, due to the lack of large scale power in this model, the acceleration converges for $R \sim 10^4 \, \mathrm{km\,s^{-1}}$.

3.2. Convergence in direction

As the acceleration is a vector, we should also discuss its convergence in direction. In linear theory one expects perfect alignment between **v** and **a**, provided that the power on scales beyond those probed by the sample is negligible. We also expect g_R to become parallel to **v** when $R \to \infty$. The direction of g_R, evaluated from the IRAS sample, is within $\alpha = cos^{-1}(\hat{g}_R \cdot \hat{v}) = cos^{-1}(\hat{g}_R \cdot \hat{g}_\infty) \sim 10° - 30°$ of the apex of the microwave dipole (Meiksin and Davis 1986; Yahil, Walker and Rowan-Robinson 1986; Harmon, Lahav and Meurs 1987; Strauss and Davis 1988). This has generally been considered as a small misalignment and as a qualitative argument for the lack of large scale power. We have made an attempt to quantify this problem. It turned out, that although the width of the probability distribution for α does depend on the shape of $P(k)$, the observed values are not small enough to exclude models with substantial amount of large scale power. Using the joint probability distribution for two Gaussian random vectors (e.g. Vittorio *et al.* 1986), it is easy to construct $p(\alpha|g_R, g_\infty)$– the conditional probability distribution for given g_R and g_∞. Here we present results for the case, when the dipoles are equal to their predicted rms values. The case $g_R = \sigma_g(R)$, $g_\infty = \sigma_g(\infty)$ is particularly simple since $p(\alpha|...)$ becomes independent of b. In Fig.9 and 10 we show the 5% and 95% conditional probability contours for $p(\alpha|g_R, g_\infty)$ for the CDM and one of the PIB models. As expected, the spread in α is wider when there is more power on large scales. The misalignment $\alpha \sim 10° - 30°$, observed for a sample of depth $R \sim 10^4 \, \mathrm{km\,s^{-1}}$ agrees reasonably well with the predictions of both models. The observed α can not be used to discriminate between PIB and CDM at a significant statistical level. An alignment as good as 8° (Harmon *et al.* 1987) is rather unlikely even in the CDM model. Shot noise at large R and non-linear effects at small R will tend to make the spread in α larger than predicted by linear theory (Villumsen and Davis 1986).

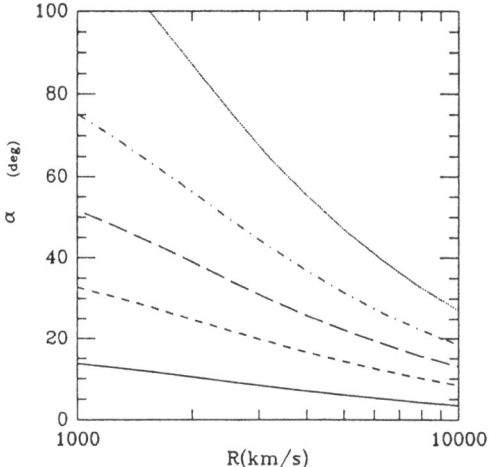

Fig.9. Confidence levels on the maximum misalignment angle expected in the CDM model, under the condition that both the acceleration and the CMB dipole are the rms values (see text): 5% (continuous line), 25%,50%,75%, 95% (dotted lines).

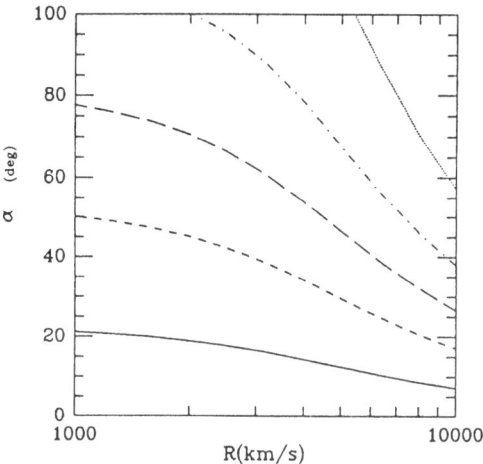

Fig.10. As in Fig.9, but for the PIB model, with $\Omega_0 = 0.4, n = -1$ and $x_e = 1$.

4. DISCUSSION

Each of the statistical tests we used has its own technical limitations and drawbacks. In particular, our conclusions based on the bulk flow and GA maximum likelihood fits to the data are only as strong as the studied *velocity models*. The effective value of R_* for the bulk flow solution is uncertain and the spherically symmetric GA infall model may be an overly coherent representation of the data. We note that in all cases the theoretical predictions fare much better against the FB cylindrical flow model than they do against the spherical GA (see Table 1). The estimate of ξ_v, which has the advantage of being derived *directly from the observations*, has potential problems of its own: it is vulnerable to Malmquist bias, to which the bulk flow solutions are insensitive (to linear order in δ, *cf.* Kaiser 1988b). Moreover, if there indeed is significant power on scales $\gtrsim 2000\,\mathrm{km\,s^{-1}}$, then much deeper surveys are needed to deal with sampling errors in the ξ_v estimate. Despite all this, we believe that the consistency among the different analyses presented here allows us to reach some robust general conclusions.

4.1. Cold Dark Matter

The $\Omega = 1$, $b = 2.5$ CDM model appears to be ruled out if the observations of large scale flows are taken at face value. The rms bulk flow velocities miss the value $V = 600\,\mathrm{km\,s^{-1}}$, reported by Dressler *et al.* by a large factor (Table 1). The density contrast, averaged over the Great Attractor volume, interior to the LG, represents a $7.3\sigma_0$ fluctuation. The mean number of such objects in a CDM universe is less than one per 10^5 Hubble volumes. No velocity profiles lying above the FB cylindrical flow model were found, out of 10^6 random samples examined. The estimate of the velocity correlation function, derived from the FB elliptical sample differs from the CDM prediction for $\xi_v(r)$, both in shape and amplitude. The observed velocity field appears to be much more correlated, than expected in the CDM model, in particular at large separations, $2000\,\mathrm{km\,s^{-1}} \lesssim r \lesssim 6000\,\mathrm{km\,s^{-1}}$. (Figures 4 and 6).

In summary, all new results, presented here confirm the conclusions reached earlier by Bond (1986) and Vittorio *et al.* (1986): large scale streaming motions are a serious embarassment for the CDM cosmogony.

It has been argued (Kaiser 1988a), that changing the normalization to $b = 1.5$ may salvage the CDM model. Switching to $b = 1.5$ does help but probably not sufficiently. With $b = 1.5$, only one of 10^6 samples of $V(r)$ was successful (as opposed to *none* with $b = 2.5$). Reducing the effective depth of the "Seven Samurai" sample also increases the predicted velocity, but, again, not significantly enough. Even for $R_* = 1500\,\mathrm{km\,s^{-1}}$, a very conservative lower limit, *and* $b = 1.5$ the observed bulk flow would still be a $2.4\sigma_v$ fluctuation, a rare event with Gaussian statistics.

Moreover, the biasing factor is not a parameter one may vary freely to maximize the probability of finding flows, coherent over large scales. N-body simulations of the motions of self-gravitating dust in a flat CDM model require $b = 2.5$. Lowering b would make the small scale peculiar velocities and galaxy clustering inconsistent with observations (Davis *et al.* 1985). A somewhat smaller b may be allowed when dissipation is included in the numerical experiments (Carlberg and Couchman 1988), although it would be probably premature to draw any quantitative conclusions from gas+dust simulations. These very interesting and worthwhile experiments are still in their early stages of development. Besides, there is another reason, which makes lowering b very difficult. All dynamical estimates give $\Omega \simeq 0.3 \pm 0.1$ (Peebles 1986 and references therein). To make these measurements consistent with an $\Omega = 1$ CDM universe, one needs $b \gtrsim 2$ to 3 (e.g. Dekel and Rees 1987).

Our discussion would be incomplete without mentioning the possibility that "the cosmic drift is is a cosmic myth". It may well be, that we are being deceived by systematic spatial variations in the Tully-Fisher and Faber-Jackson relations, used as distance estimators, while the Hubble flow is unperturbed (Silk 1988, Kaiser 1988b). However, as Silk points out, this explanation requires long range correlations, absent in models like CDM: the "missing power" problem reappears in a different guise.

4.2. Isocurvature models

The PIB model with $\Omega = 0.4$ naturally leads to large scale coherent flows and inhomogeneities, comparable to the GA. However, too much coherence in the velocity field is not good. The $n = -1$; $\Omega = 0.4$; $x_e = 1$ model produces profiles which decline too slowly with distance (Figure 3). The $n = 0$ case is better and we expect that reasonable agreement could be obtained by fine-tuning n. It remains to be seen if the value of n, derived from such phenomenological arguments can be justified by early universe microphysics (*cf.* Peebles 1988). The velocity correlation functions, Π, Σ and ξ_v, predicted by the fully ionized $n = -1$ model, also appear to have more large scale power than required by the observations. However, this is not necessarily a problem. The predicted half-power width of ξ_v is greater than the radius of the Seven Samurai survey. If the $n = -1$ model is correct, the catalogue is too shallow to constitute a fair sample. The estimate of ξ_v would then be dominated by sampling errors, making direct comparisons with the ensemble-averaged ξ_v difficult, if not impossible (however, a Monte Carlo simulation might do the job).

4.3. Local peculiar gravity and Ω

Estimates of Ω based on a comparison of the LG peculiar velocity with the peculiar acceleration, computed from galaxy catalogues are only valid provided the Universe is uniform on scales, comparable to the size of the survey. Two consistency tests have been discussed in the literature: convergence of the magnitude of the observed acceleration, and alignment of the direction of the observed acceleration with that of the cosmic microwave background dipole. The JVW analysis reported here has shown, that each of these two tests can yield misleading results. The dipole anisotropy results from the IRAS galaxies cannot be used to argue against cosmological models with substantial large-scale power. The estimates of Ω derived from these data are uncertain and sensitively depend on the implicitly assumed power spectrum.

Acknowledgments

The original research reported in this article was carried out in collaboration with Ed Bertschinger, Ed Groth and Rosie Wyse, and we are grateful for their permission to reproduce results from our joint papers. We thank Dave Burstein for providing us with observational data on streaming motions, and Jim Peebles for providing transfer functions for isocurvature models. This research was supported in part by the U.S. National Science Foundation.

REFERENCES

Bahcall, N. A. 1988, *Ann.Rev.Astron.Astrophys.*, **26**, in press.

Bardeen, J. M., Bond, J. R., Kaiser, N., and Szalay, A. S. 1986, *Ap.J.*, **304**, 15.

Bardeen, J. M., Bond, J. R., and Efstathiou, G. 1987, *Ap.J.*, **321**, 28.

Bertschinger, E. 1987, *Ap.J. (Letters)*, **323**, L103.

Bertschinger, E., and Juszkiewicz, R. 1988, *Ap.J. (Letters)* , **xxx**, in press, (BJ).

Blumenthal, G. R., Dekel, A., and Primack, J. 1988, *Ap.J.*, **326**, 539.

Bond, J. R. 1986, in *Galaxy Distances and Deviations from Universal Hubble Expansion,* ed. B. F. Madore and R. B. Tully (Boston: Reidel), p. 255.

Bond, J. R. 1988, in *Large Scale Motions in the Universe, Proceedings of the Vatican Conference,* ed. Rubin, V. N. (in press).

Burstein, D. 1988, this conference .

Carlberg, R., and Couchman, H. 1988, preprint .

Collins, C. A., Joseph, R. D., and Robertson, N. A. 1986, *Nature*, **320**, 506.

Davies, R. D., Lasenby, A. L., Watson, R. A., Daintree, E. J., Hopkins, J., Beckman, J., Sanchez-Almeida, J., and Rebolo, R. 1987, *Nature*, **326**, 462.

Davis, M., Efstathiou, G., Frenk, C. S., and White, S. D. M. 1985, *Ap.J.*, **292**, 371.

Davis, M., and Huchra, J. 1982, *Ap.J.*, **254**, 437.

Dekel, A., and Rees, M. J. 1987, *Nature*, **326**, 455.

Doroshkevich, A. G. 1970, *Astrofizika*, **6**, 320.

Dressler, A., Faber, S. M., Burstein, D., Davies, R. L., Lynden-Bell, D., Terlevich, R. J., and Wegner, G. 1987, *Ap. J. (Letters)*, **313**, L37.

Faber, S. M., and Burstein, D. 1988, in *Large Scale Motions in the Universe, Proceedings of the Vatican Conference,* ed. Rubin, V. N. (in press).

Feynman, R., and Hibbs, A. 1965, *Quantum Mechanics and Path Integrals* (New York: McGraw-Hill).

Górski, K. 1988, *Ap.J. (Letters)*, **xxx**, in press.

Górski, K., and Davis, M. 1988, in preparation .

Górski, K., and Hoffman, Y. 1988, preprint .

Groth, E., and Juszkiewicz, R. 1988, in preparation (GJ) .

Juszkiewicz, R., Vittorio, N., and Wyse, R. 1988, preprint (JVW) .

Harmon, R. T., Lahav, O., and Meurs, E. J. A. 1987, *M.N.R.A.S,*, **228**, 5P.

Kaiser, N. 1988a, *M.N.R.A.S.*, **231**, 149.

Kaiser, N. 1988b, this conference .

Lahav, O. 1987, *M.N.R.A.S.*, **225**, 213.

Lynden-Bell, D., Faber, S. M., Burstein, D., Davies, R. L., Dressler, A., Terlevich, R. J., and Wegner, G. 1988, *Ap.J.*, **326**, 19.

Meiksin, A. and Davis, M. 1986, *A.J.*, **91**, 191.

Melott, A. 1987, *M.N.R.A.S.*, **228**, 1001.

Monin, A. S., and Yaglom, A. M. 1975, *Statistical Fluid Mechanichs, vol.2* (Cambridge: MIT Press).

Peebles, P. J. E. 1980, *The Large-Scale Structure of the Universe* (Princeton: Princeton University Press).

Peebles, P. J. E. 1986, *Nature*, **321**, 27.

Peebles, P. J. E. 1987, *Nature*, **327**, 210.

Peebles, P. J. E. 1988, this conference .

Rubin, V. C., Ford, W. K., Thonnard, N., Roberts, M. S., and Graham, J. A. 1976, *A.J.*, **81**, 687.

Silk, J. 1988, preprint .

Strauss, M., and Davis, M. 1987, in *The Large Scale Structure of the Universe, IAU Symposioum N⁰130,* ed. Audouze, J., Pelletan, M. C., and Szalay, A. (Kluwer Academic Publishers).

Villumsen, J., and Strauss, M. 1987 , *Ap.J.*, **322**, 37.

Vittorio, N., Juszkiewicz, R. 1987, in *Nearly Normal Galaxies,* ed. S. M. Faber (New York: Springer-Verlag).

Vittorio, N., Juszkiewicz, R., and Davis, M. 1986, *Nature*, **323**, 132.

Vittorio, N., and Silk, J. 1985, *Ap.J. (Letters)*, **293**, L1.

Vittorio, N., Matarrese, S., and Lucchin, F. 1987, *Ap.J.*, **328**, 69.

Wilkinson, D. T. 1988, in *The Large Scale Structure of the Universe, IAU Symposium N⁰130,* ed. Audouze, J., Pelletan, M. C., and Szalay, A. (Kluwer Academic Publishers).

Yahil, A., Walker, D., and Rowan-Robinson, M. 1986, *Ap.J. (Letters)*, **301**, L1.

Zel'dovich, Ya. B. 1970, *Astron.Ap.*, **5**, 84.

DISCUSSION

PARTRIDGE: A word of caution on the direction of the CBR dipole. While, as I showed, there is good agreement on the amplitude of the dipole at different wavelengths, the directions are very susceptibile to systematic errors because of emission from the plane of the Galaxy (which of course has a dipole component). It wouldn't surprise me terribly if the CBR dipole was in error by 5 to 7 degrees and surely the same must be true for the IRAS dipole direction.

VITTORIO: Our theoretical predictions for the misalignement angle are of course independent of any observational uncertainties for the CBR and IRAS dipole. In any case, these uncertainties should only strengthen our conclusions: we can not use the observed misalignement angle to exclude the existence of density fluctuations on scales larger than the IRAS sample.

small b would also cause severe difficulties in reconciling the flatness of the CDM model with the dynamical estimates of based on the virgocentric infall and virial studies of clusters.

LUKASH: Your results on the large scale velocities in CDM models are somehow in disagreement with the previous statement by Kaiser. Could you comment on it?

VITTORIO: We have shown that the high value of b ($\sim$2 to 3) necessary for reconciling the CDM model predictions (Davis et al. 1985) with the observations of small scale velocities and galaxy clustering, as well as dynamical estimates, is inconsistent with the Seven Samurai results, at least at 95% confidence level. This conclusion is fairly independent on the details of the window function we have used to model the data. Kaiser used a different approach. He found that in a set of spectra with identical (CDM) shapes and with variable b, the "best" fit to the data is provided by the model with b=1.6. Among other things, this implies that the standard CDM model with b=2.5 does not work, and we agree with this particular implication.

COUCHMAN: I would like to make a comment, which I admit is somewhat an advertisement. Ray Carlberg and I have made some CDM N-body experiments which incorporate a simple gas dynamical scheme. We find a low bias at about 1.3. The pairwise velocities of galaxies on small scales are a factor of about two lower than those of the dark matter, thus allowing the observed cosmic virial theorem estimates to be satisfied. The low bias enables the observed streaming velocities on large scales to be consistent with the CDM model.

VITTORIO: The predicted velocities scale like 1/b, and lowering the biasing factor could help. However, choosing a small b would also cause severe difficulties in reconciling the flatness of the CDM model with the dynamical estimates of Ω based on the virgocentric infall and virial studies of clusters.

THE FRACTAL RANGE OF THE DISTRIBUTION OF GALAXIES; CROSSOVER TO HOMOGENEITY, AND MULTIFRACTALS

B.B. MANDELBROT
Physics Department, IBM T.J. Watson Research Center,
Yorktown Heights NY 10598, USA, and
Mathematics Department, Yale University, New Haven CT 06520

ABSTRACT. This paper takes as established that the large scale distribution of galaxies includes a self-similar fractal range. The astonishing power of simple fractal algorithms to generate rich form is recalled and illustrated, then two issues are tackled. A) Does the fractal range stop at 5 Mpc, as asserted by Peebles et al.? Or does it continue beyond? Does it stop before the limits of observation? That is, should one believe the conventional statistical arguments in favor of 5 Mpc? B) The simplest fractal distributions are "fractally homogenous," that is, homogeneous over a fractal set, and zero outside of this set. However, the distribution of galactic and inter-galactic mass is non homogenous to the extreme. It can be self-similar, in which case it follows a "multifractal measure," as discussed by the author in 1974. This paper is concerned with questions of method, and analysis of data is not included.

1. Introduction and Summary

1.1. THE FRACTAL RANGE

The existence of a fractal range in the distribution of galaxies does not appear to be questioned, though appearances sometimes suggest the contrary. Thus, some writers prefer to make use of terms such as "self similar range" or "scaling range," which have the same meaning, but sound less affirmative. Other writers say that they have proof that *the* fractal model fails, but they willfully restrict the scope of "fractal model" to the "pure mathematical" case of fractals without cross-over.

The evidence in favor of fractals is basically as follows.

On the one hand, the author's models (as restated in Section 2) are easily fitted to have not only the observed correlation properties, but also the observed visual appearance. That is, the prevalence of voids and filaments was not known when these fractal models were developed, yet these features do not have to be put in separately, but appear to be an unavoidable consequence of the fractal character of the distribution. This last fact is an empirical observation that is not fully understood, and deserves further examination, but cannot be dismissed.

On the other hand, suppose that a geometric model of distribution fails to be a fractal. Could it fit the observed power law correlation functions, and have the correct

259

M. Mezzetti et al. (eds.), Large Scale Structure and Motions in the Universe, 259–279.
© *1989 by Kluwer Academic Publishers.*

appearance, while invoking only a small number of parameters. There may be no mathematical proof this cannot be done, but there is no known example, either.

Therefore, a prudent student of the large-scale structure will find it worthwhile to keep in mind the properties of the fractals. He will attempt to know more about the unexpectedly rich variety of pattern allowed by simple fractal constructions, and to know better the techniques required for their study. Some references to fractals could be said to imply that fractal geometry reduces to a few simple arguments about how different quantities scale with respect to each other. Today, this is very far from being the case. The substantial body of knowledge that exists cries out to be tapped further in the study of the Universe.

1.2. A LIST OF POSSIBLE GEOMETRIC MODELS OF DISTRIBUTION

In the following list by increasing complication, the last possibility is phrased to cover every possible views of the relevance of fractals to the description of the Universe, a description that should be the first step towards understanding.

   - **1.** *Homogeneity.* The Universe is homogeneous, save for very local effects. This would be the simplest Possibility, but of course it fails very grossly to fit the data.

   - **2.** *Pure fractal.* Save for very local effects, the Universe is a self-similar fractal up to infinite scales. More precisely, the galaxies are of equal masses, and the Universe is fractally homogenous. This would be the next simplest Possibility. In this case, the appropriate statistical substitutes for the correlations follow power laws throughout, the principal parameter being one number called fractal dimension.

Comment. Possibility 2 predicts a vanishing overall density of matter, a property that is known to elicit fierce a *priori* opposition. Let us, therefore, comment immediately that other fractal models are available, as will be seen momentarily. Therefore, to disprove Possibility 2 is NOT to disprove the relevance of fractals.

   - **3.** *Pure multifractal.* The Universe is a self-similar multifractal throughout. The precise meaning of this term will be explained in Sections 4 and 5, but implies unequal galaxy masses and an interglacting medium. In this case, the principal parameter is not one number, but a probability distribution. The appropriate substitutes for the correlations follow power laws throughout, but with different exponents.

   - **4.** *One crossover.* The Universe is a self-similar fractal only up to a crossover scale that satisfies $0 < R_{cross} < \infty$, and the crossover between the fractal and the homogeneous ranges occurs very sharply. That is, there is no transient range of significant width, whose structure would require additional parameters in order to be specified.

Comment. To be the distance beyond which the large-scale structure of the Universe becomes homogeneous, R_{cross} should be the size of the largest significant structures (such as the voids). To understand the contrast between Possibilities 2 and 4, the telling background example is that of a structure very familiar in statistical physics: percolation clusters. Percolation clusters at criticality fall under Possibility 2. In percolation clusters above criticality, R_{cross} is finite, namely is the size of the largest observed patterns. These percolation patterns at criticality are "voids," whose shape would not surprise the astronomer of 1988.

   - **5.** *One crossover.* There is a crossover R_{cross} like in Possibility 4, but the Universe is multifractal.

– **6**. *Two crossovers*. The Universe involves a fractal range for "small" scales, an ultimate homogenous range for very large scales, and, in addition, involves a broad intermediate range characterized by significant structures. In this case, the single measure of scale R_{cross} of Possibility 2 is replaced by *at least two* measures of scale.

Comment. Insofar as we can tell, this Possibility is embraced by J.P.E. Peebles who invokes an intermediate range with negative correlations to explain the size of the observed voids. This is why the smaller measure of scale will be denoted by $R_{peebles}$, and the larger one will be denoted by R_{upper}. Clearly, $R_{peebles}$ is a lower bound to R_{upper}, but the Universe is *not* homogenous beyond $R_{peebles}$; in fact $R_{peebles}$ is far below the size of confirmed voids.

– **7**. *Two crossovers*. The Universe is like in Possibility 6, but with *fractal* replaced by *multifractal*.

– **8**. *Anything that does not fit in 1 to 7*. One possibility is that there are successive "rings" of different fractal dimension. An extreme possibility is that the distribution is ruled by a chaos without order.

Comment. We subscribe to the notion that the best models are the simplest. Therefore, in cases of doubt, we advocate the Possibility with the lower number.

1.3. THE ISSUE OF CROSSOVER AND THE RELIANCE UPON STATISTICS

Let us now elaborate. We first rephrase the above list of Possibilities by extracting two important issues. We assume that the fractal range, is not questioned for small distances. The horizon of observation will be denoted by R_{max}.

The first issue is whether this fractal zone crosses over sharply to a homogenous zone for some $R_{cross} < R_{max}$, or ends for some $R_{peebles} < R_{max}$ to be followed by a transient zone, or is a least as deep as R_{max}. In the latter case, we shall say that $R_{cross} = \infty$.

Let us recall that our book *The Fractal Geometry of Nature (FGN)* was clearly partial to Possibility 2, yet open-minded on this issue, primarily for lack of actual empirical checks. The recent analysis of the data in Pietronero 1987 and in Coleman et al. 1988, together with the rejoinder in Davis et al. 1988, have convinced us that, *either* Davis et al. is wrong in its criticism, *or* this issue *cannot* be resolved on purely statistical grounds. On the other hand, the visual analysis of the observation of increasingly large filaments and voids continues to encourage us to be open-minded, with clear partiality towards Possibilities 2 or 4.

The main fact is that to perform a statistical test between the alternatives $R_{cross} < R_{max}$ and $R_{cross} = \infty$ turns out to be a very delicate matter. Most important, it is necessary to limit statistics exclusively to tests that avoid prejudging a priori against the possibility $R_{mex} = \infty$. This happens to require fresh statistical thinking, instead of blind reliance on previously "proven" techniques. The reason is that all the existing methods of statistics make explicit or implicit assumptions that happen to become invalid when $R_{cross} = \infty$. In particular, customary and usually innocuous normalizations, like those involved in the definition of the usual correlation and of the pseudo correlation used by Peebles 1980, yield reported results that are processed to excess, and have been made practically impossible to interpret by the reader.

Section 3 will analyze some statistical tools, step by step. One neutral summary of the evidence is the mass radius function $M(R)$. Its derivative, divided by $4\pi R^2$, is a conditional occupation probability; it is better than $M(R)$ in some ways, but less acceptable in other ways. Arguments will be given against replacing it by the normalized pseudo correlation function of Peebles et al.. Normalization is proper in more conventional statistics, but in the case when the fractal range is significant normalization is risky and unacceptable.

Comment. The seasoned statistician knows that, when a theory is submitted to a sufficiently wide battery of tests, it often happens that every hypothesis fails at least one test. This is the case even when the tests themselves have proven their applicability. When the statistical techniques are new and unproven, one individual negative test *cannot* suffice to eliminate an otherwise attractive possibility.

Statistics is little used in the hard sciences, and we must confess puzzlement at the importance it has achieved in the present context. The facts that filaments and voids of increasing size continue to be observed when one reaches the deepest levels of observation is a very clear-cut symptom of an underlying fractal distribution; indecisive statistics involving arguable corrections to the data carry little weight in comparison.

1.4. THE ISSUE OF MULTIFRACTALITY

Granting that there is a fractal range, the second basic issue is whether (in this range) the distribution of mass is homogeneous on a fractal set, or whether instead it involves galaxies of varying mass, as well as interstellar matter. If either or both is the case, and the distribution is self-similar, it must be multifractal. Examples of fractally homogeneous measure are obtained by placing a uniform measure on either of the fractal dusts defined by the tales in Section 2. The resulting models can now be called *unifractal*. But it is nearly as easy to study fractal but non uniform measures. As a matter of fact, it is easy to construct a distribution of mass that combines self similarity, very high peaks (to be interpreted as galaxies of variable mass), and a low background (to be interpreted as a very variable interstallar mass). To fulfill this aim (in the analogous case of intermittent turbulence), we have introduced and developed the notion of *multifractal measure*, in 1968 then mostly in 1972-1976, but it has not acquired a large following until recently. The most widely known approach to multifractals is, unfortunately, quite unnecessarily complicated and artificial. Our original approach, in a recently completed form, is much more straightforward, and a survey is included as Section 5, in order to make it readily available to the student of large scale structure of the Universe.

1.5. THE ISSUES IN SECTION 1.3 AND 1.4 INTERACT

It will very soon become necessary to face them simultaneously.

2. Two Fancy Tales of How the World Began. Demonstration by Examples of the Power of Simple Fractal Models to Generate Unexpectedly Rich Structures.

This section begins by restating in fanciful style our two basic models of galaxy clustering using random fractals.

2.1. "THE SEEDING OF THE HEAVEN"

"In the beginning, the heaven was a void. And the Master of Matter, Light and Life proclaimed, Let there be matter: and matter was. It was one point. And the Master proclaimed, Let matter be seeded over the heaven, and Let every small part of the heaven be just like every other small part and like every large part. And two archangels set forth hopping; wherever they alighted, they left a pinch of matter and then resumed their journey as in its beginning. And the parts of the heaven were all made just alike. And the Master was everywhere, dwelling in every pinch of matter; and the heaven looked the same from every point where the Master dwelt."

2.2. "THE PARTING OF THE HEAVEN"

"In the beginning, the heaven was filled with matter. And the Master of Matter, Light and Life proclaimed, Let matter part away. Let it remove itself to form voids without number, and Let every small part of the heaven be just like every other small part and like every large part. And matter removed itself, and the Master was everywhere, dwelling in every place that was not in a void: and the heaven looked the same from every point where the Master dwelt."

2.3. COMMENT ON THE GENERATIVE POWER OF THE FRACTALS

In Chapters 33 to 39 of my book, *The Fractal Geometry of Nature (FGN)*, the above fancy tales are translated into sober fractal models, involving very simple statistical algorithms one can easily simulate on the currently available computers. A priori, one expects simple algorithms to generate nothing much of interest. In the specific case of galaxy modeling, this low expectation may be related to the natural but quite incorrect identification of all fractal models with the very early but extraordinary crude ones. For example, in the model of Fournier d'Albe, little is put in, and nothing more is obtained as output. Therefore, there is a strong tendency to use highly specified algorithms, in which every one of the features one wishes to see in the output (e.g. large voids) has been knowingly introduced in the input. An example of a needlessly over-specified fractal model is the one advanced by Soneira and Peebles. Similarly, Peebles 1980 proposes to correct in advance for presumed inadequacies of our "Seeding" model, by adopting multiple "Seeding centers," distributed uniformly. This introduces a finite R_{cross}, but is far too hasty a solution.

Given the a priori fear that simple models *must* be inadequate, it is a surprise to see that the simulations of our "Seeding" and "Parting" models look far more realistic than anticipated by anyone . . . even by us. Since "to see is to believe," many examples are shown in my book, *FGN*.

Yet another fractal construction, recently put forward by Szalay and Vicsek, has become famous as the cover of Audouze et al. 1988. It may or may not be physically realistic, but it strengthens further our belief that almost any sufficiently "natural" fractal dust would be reminiscent of the actual distribution of the galaxies.

These examples are meant to underline the power of simple fractal models to generate rich form. This power is the geometric facet of a fact that is growing in public awareness, namely of the power of simple dynamical systems to generate rich and seemingly chaotic orbits. Our hope is that the fractals' power will cease to be underrated, and that they will no longer be dismissed casually when it happens that some single statistical test appears to encounter difficulties.

3. On Diverse Statistical Summaries of the Data, and on the Pitfalls of "Normalizing" them

3.1. NON-TRUNCATED FRACTALS AND THE MASS-RADIUS FUNCTION

In order to simplify, this Section makes the "unifractal" assumption that all galaxies carry the same mass. This makes it possible to define the *mass-radius function*

$M(R)$ = mass within a radius R around a fixed (randomly selected), galaxy.

Under Possibility 2 of Section 1.2, on has

$$M(R) = F(R)R^{D}.$$

This is reminiscent of the formula valid when the distribution is completely homogeneous, but there is a *fundamental* difference. In the homogeneous case, the mass in a sphere is $M(R) = (4\pi/3)\delta R^{3}$, which is the power R^{3} with a *numerical* multiplicative prefactor. To the contrary, non-truncated fractals involve the very important fractal prefactor $F(R)$, which is *not* a constant. It is a stationary random function of $\log R$, and it can vary greatly.

The variability of $F(R)$ or of $\log F(R)$ around their expectations is an interesting characteristic of a fractal, independent of its D and recommended for further empirical and theoretical study. It is one aspect of a fractal's "lacunarity." For example, as variability measured by the usual variance increases, the fractal's lacunarity also increases (FGN, Chapters 34 and 35). We have investigated diverse families of fractals, in which different members differ solely by the value of the fractal dimension D. As D is varied from its highest possible value down to its lowest possible value (which is usually 0), the variability of F increases very sharply. The works of de Vaucouleurs and Peebles 1980 suggest $D \sim 1.2$, which is a small value for a dust in 3 dimensional space. Hence, our impression is that we should expect $F(R)$ to have a large fluctuation. This impression deserves to be subjected to a hard critical study.

In order to appreciate what happens as we move from $M(R)$ to the covariance and the pseudo-correlation, it is best to make the move in several distinct steps, and to introduce notation gradually.

3.2. NON-TRUNCATED FRACTALS AND THE LOCAL CONDITIONAL DENSITY, WITHIN THE VOLUME BOUNDED BY THE SPHERES OF RADII R AND $R + \Delta R$

Now consider the local conditional density at distance R from the point P, that is the mean density within the volume bounded by the spheres of radii R and $R + \Delta R$ centered at the point P. It is

$$\Delta M/4\pi R^{2}\Delta R = (1/4\pi)R^{D-3}[DF(R) + R\Delta F(R)/\Delta R].$$

The expectation $\langle F(R)\rangle$ is positive and finite, and is independent of R. And the expectation $\langle \Delta F(R)\rangle R/\Delta R = \langle \Delta F(R)\rangle/\Delta \log R$ vanishes, because we saw that the prefactor $F(R)$ is a stationary random function of $\log R$. Therefore, the plot of $\log \Delta M$ versus $\log R$ would run around a straight "trend line" of slope $D - 3$, with superposed fluctuations throughout. Several plots corresponding to different samples would have the same trend line, but entirely distinct fluctuations.

3.3. NON-TRUNCATED FRACTALS. THE USUAL NOTION OF "REPRESENTATIVE SAMPLE" IS NOT APPLICABLE. CONDITIONAL DENSITIES MUST NOT BE RENORMALIZED.

Statisticians have almost always dealt with situations where samples of sufficiently large size can be viewed as "representative" of the whole population, but this is not the case for non-truncated fractals. In particular, evaluate

$$\frac{\Delta M / 4\pi R^2 \Delta R}{M(R_{sample}) / 4\pi R_{sample}^3}$$

This is the ratio between the density at distance R and the overall density in a sample of size R_{sample}. For a standard distribution, the denominator is hardly random at all if the sample is sufficiently large, and the above renormalization is useful. But the fractal case is totally different. The ratio of the expectations would simply be

$$\left(\frac{R}{R_{sample}} \right)^{D-3},$$

and would provide marvelous material for the estimation of D. However, the ratio of non-averaged quantities is

$$\frac{DF(R) + R\Delta F(R)/\Delta R}{F(R_{sample})} \left(\frac{R}{R_{sample}} \right)^{D-3}.$$

Its overall trend is $(R/R_{sample})^{D-3}$, as expected. However, there is a random prefactor. Its numerator duly depends upon the randomness at the distance R, but its denominator mixes in the randomness at the distance R_{sample}. Again, plot log (ratio) against log R. There is still a straight trend line of slope $D - 3$. But the prefactor is reflected in a translation log (denominator of the prefactor), which depends on the overall sample. Thus, different samples have parallel but distinct trend lines. Any sensible scientist will push these line together in order to estimate their common slope, but the affect of renormalization upon the estimation of D is not helpful at all.

3.4. NON-TRUNCATED FRACTALS. THE COVARIANCE (OR "CORRECT CORRELATION FUNCTION") I.E., THE CONDITIONAL DENSITY AVERAGED OVER ALL ORIGINS.

Let us return to the "raw" prefactor $DF(R) + R(\Delta)F(R)/\Delta R$ of section 3.2. In order to average out its wild fluctuations, the best is to average for fixed R over all the spheres centers P. The resulting quantity is called "correct correlation function" by Pietronero 1987, but I prefer to continue to use the standard probabilistic term, which is *covariance*. Write $n(P)$ and $n(P + \bar{R})$ for the number of galaxies in a small sphere around P and around a point $P + \bar{R}$ whose distance to P is of length R. One has

$$\Gamma(R) = \frac{\langle n(P)n(P + \bar{R}) \rangle}{\langle n \rangle}.$$

Comment. It is an unavoidable feature that any given portion of space of radius R_{max} (e.g., the volume covered by a catalogue carried to a uniform depth) includes a small number of spheres of large radius, but a large number of spheres of small radius. When the sample average is carried over the origins of all the spheres within the catalog, the sample values for different R's fail to be independent. Hence, the fluctuations in F and ΔF average out even more poorly than if the samples had been independent. One deals with a reduced "effective sample size," whose value depends upon R. More precisely, when R is well below R_{max}, the effective sample size is near the actual sample size, and the conditional density is reliable. But for R close to R_{max}, the effective sample size becomes small, and the conditional density becomes greatly affected by the fluctuation of $F(R_{max})$. For example, while the expected conditional density decreases up to $R = R_{cross}$, the sample density may well actually increase in the range below R_{cross}.

3.5. NON-TRUNCATED FRACTALS. THE (PSEUDO) CORRELATION.

Let us now turn to the renormalized ratio of Section 3.3. By averaging over all the sphere centers P, one obtains the quantity

$$\frac{\Gamma(R)}{\langle n \rangle} = \xi(R) + 1.$$

Again, the renormalization via the division by $\langle n \rangle$ introduces entirely spurious effects.

The quantity $\xi(R)$ is called *correlation* by astronomers, but the statisticians' correlation is a different expression that ≤ 1 in absolute value, while $\xi(R)$ certainly can exceed 1, and the value where it reaches 1 is given a special standing.

3.6. THE DISTRIBUTIONS OF GALAXIES AND OF CLUSTERS, COMPARED

This section brings two remarks together. First, it was predicted in Section 3.3 and 3.5 that renormalization should be expected to introduce a spurious prefactor, hence a spurious translation in doubly logarithmic coordinates. Second, it can be shown, under a wide range of definitions of the notion of "galaxy clusters," that the functions Γ should be identical for galaxies and for their clusters.

The two remarks in the preceding paragraph make us expect that the $\xi(R) + 1$ function of galaxies and of galaxy clusters should differ by a factor, i.e., by a translation on doubly logarithmic coordinates. This is indeed the situation that is observed, as seen, e.g., in Szalay and Schramm 1985. This discrepancy has been viewed as a real empirical fact to be explained, and also as a counter argument to "the" (that is, the simplest) fractal description. To the contrary, we have always expected that this discrepancy will turn out to be almost certainly spurious. The data analysis in Pietronero 1987 confirms our expectation.

3.7. FRACTALS WITH ONE CROSSOVER

A wide class of random fractals with sharp crossover at R_{cross}, are covered by Possibility 4 of Section 1.2. They include variants of the Parting model of Section 2. For

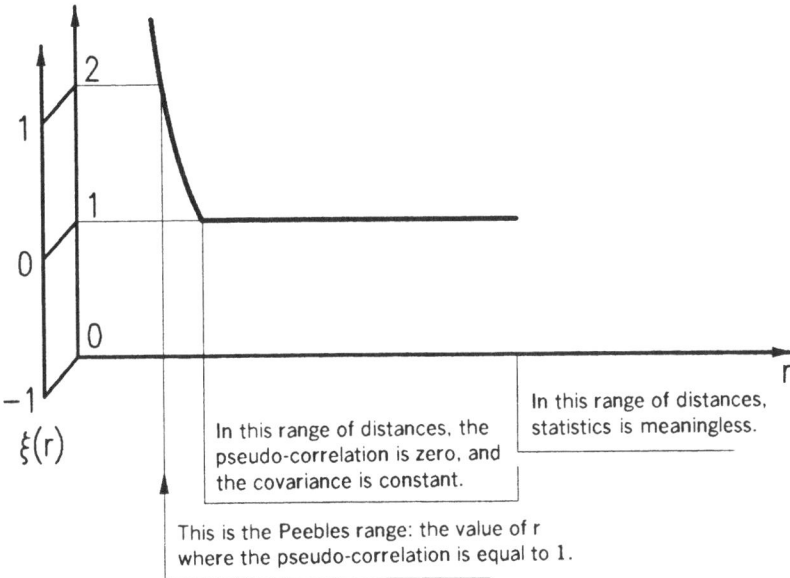

Figure 1. The relation between the covariance and the normalized "correlation" for a fractal universe with a sharp crossover to uniformity.

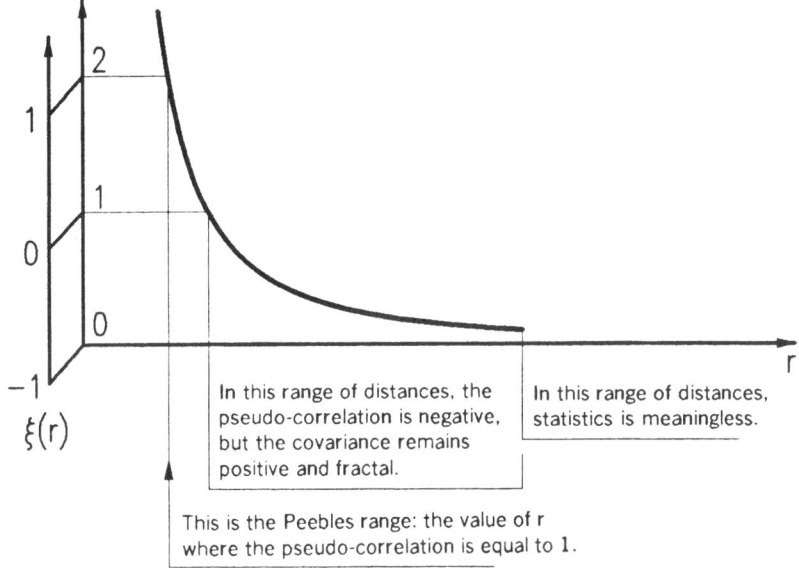

Figure 2. A typical renormalized correlation. The normalization does not use the value of $\langle n \rangle$ that is given by the last point of the sample covariance of the sample under investigation, but rather a value obtained by a larger but different sample.

these distributions, the fractal non homogeneity is described by the behavior of the mass-radius function $M(R)$. One has

$$M(R) = F(R)R^D \qquad\qquad \text{for } R < R_{cross}$$

$$M(R) = (4\pi/3)\delta R^3 + \text{ (a fluctuation } \sim \sqrt{(4\pi/3)\delta R^3}) \qquad \text{for } R > R_{cross}$$

In the homogenous range $R > R_{cross}$, things are classical: there is a non random factor involving a density $\delta,$ and a Poisson fluctuation factor, which is *additive,* and whose form is familiar to everyone.

To the contrary, the fluctuation factor in the fractal range $R < R_{cross}$ is multiplicative, like in Section 3.1. The local density described in the title of Section 3.2 is

$$\Delta M/4\pi R^2 \Delta R = (1/4\pi)R^{D-3}[DF(R) + \Delta F(R)R] \qquad \text{for } R < R_{cross}.$$

$$\Delta M/4\pi R^3 \Delta R = \delta + \text{ (a fluctuation term)} \qquad \text{for } R > R_{cross}.$$

On a plot of log (density) as function of log R, the fractal and the homogenous regimes give straight trend lines of respective slopes $3 - D$ and 0. The crossover occurs where these straight lines intersect each other.

In truncated fractal models I know well (and also for percolation clusters a little above criticality), the crossover between these two regimes is quite sudden, which is a great asset. (The corresponding expected mass-radius plot is the plot of an integral, hence the crossover is far more gradual, which is a drawback.)

What about the effect of renormalization on log (density) where there is a single crossover $R_{cross} < R_{max}$ we are in the standard statistical world, in which renormalization is perfectly legitimate, as seen on Figure 1.

For $R < R_{cross}$, the function $\xi(R) + 1$ stabilizes at 1, and the function $\xi(R)$ stabilizes at 0.

The radius where $\xi(R) = 1$ and $\xi(R) + 1 = 2$ has been singled out by P.J.E. Peebles in the analogous context to be tackled in Section 3.8. Let this value be denoted by $R_{peebles}$. Clearly, when $\xi(R) + 1 \sim R^{D-3}$, and $R < R_{cross}$, with $D = 1.2$, one has $R_{peebles} = 2^{-1/1.8} R_{cross} = .68 R_{cross}$.

3.8. FRACTALS WITH AT LEAST TWO CROSSOVERS. THE (PSEUDO) CORRELATION AND THE DEFINITION OF $R_{peebles}$.

Section 3.7. shows that $\xi(R)$ has a useful meaning when there is a single sharp crossover $R_{cross} < R_{max}$. Could $\xi(R)$ continue to be of use in other cases? This cautious writer expects little from statistics, and would not trust any expression "in the wild" before it has been "tamed" on a well-understood explicit example. However, something very much like $\xi(R)$, with one essential modification to be distributed shortly, has been extensively used in many works by P.J.E. Peebles, both those summarized in Peebles 1980 and more recent ones. This use is bold, in fact, it is reckless in our opinion, because the underlying model is never fully described — to our knowledge. Implicitly, Peebles does not believe in Possibility 4, since he describes a function he calls "correlation" behaves like on Fig. 2: it is $= 1$ for $R = 5$ Mpc; it is > 0 for $R < 1.46 \times 5$ Mpc and it is $= 0$ over some ultimate homogeneous range, but it is < 0 over an unspecified intermediate range beyond 1.46×5 Mpc. In this framework, the existence of the intermediate range, where $\xi(R)$ is most often negative, is indispensable to account for the existence of large voids and of all the other interesting structures observation keeps revealing.

Now to the essential modification Peebles brings to $\xi(R)$. In his view, in order to evaluate the global density of galaxies, it is not only permissible, but is desirable, to use counts that are broader than the counts that lead to $\Gamma(R)$.

Granted this modification, let us go beyond formal manipulation, and ponder the implications of Fig. 2, without questioning the validity of the essential modification. Given the list of Possibilities drawn in Section 1.2, it is clear that the simplest interpretation of Figure 2 involves Possibility 6, with $R_{peebles}$ being (within some factor of the order of 1) the radius where one leaves the fractal range. However, this is *not* the way $R_{peebles}$ is ordinarily presented. There is a widespread perception that $R_{peebles}$ (within some factor of the order of 1) is a measure of R_{upper}. We see that this perception is not warranted. As a matter of fact, there is invariably a gap in the graph of $\xi(R)$, between the largest R for which a correlation is reported, and the presumed asymptotics. The result is that given Figure 2, no numerical value can conceivably be inferred for R_{upper}. In words, the value of $R_{peebles}$ gives no hint of where the asymptotic homogeneity begins to prevail.

Let us now dig deeper, by questioning the assumption that one can plug into $\xi(R)$ a global density estimated on the basis of independent very deep data. This equality assumes that on the scale of the largest existing catalogue, the Universe is already homogeneous. Since this is what our task is to either confirm or contradict, assuming it in advance is *not permissible*. We also observe that deep surveys involve drastic corrections of uncheckable validity. The cosmologists feel that the density of visible matter is smaller than it "should" be, so that their corrections cannot inspire full confidence. This writer has attended the Seminars that followed the 1987 Balantonfüred Symposium of the IAU. While several speakers asserted that the global density is a well-defined number, its value did not lead to any consensus.

3.9. CONCLUSION OF SECTION 3

Many statistically dubious steps enter in the procedures customarily used to reach the widely accepted conclusion that the Universe becomes homogenous at comparatively short distances. Therefore, this conclusion is not persuasive. By its definition, $R_{peebles}$ is at best concerned with the range of significant fractal correlation, and is likely to underestimate it. More important, the value of $R_{peebles}$ gives no inkling whatsoever on the distances beyond which the universe is homogeneous.

Our feeling, therefore, is that, in the present state of knowledge, it is imperative *not* to renormalize the observed covariance $\Gamma(R)$. On Figure 2, a bold line represents the typical non normalized covariance $\Gamma(R)$. This line's overall shape suggests that these data are entirely compatible with Possibilities 2 and 4 of Section 1.2.

Analogous criticisms, combined with a fresh analysis of the data, are made in Pietronero 1987 and in Coleman et al. 1988. The counter analysis by Davis et al. 1988 involves unconvincing corrections. The strongest conclusion one may draw from this counter analysis is that the unquestioned data do not suffice to decide between the thesis of Peebles and Possibility 2 in the list in Section 1.2.

4. Spatial Variability Beyond the Fractally Homogenous Model

More detailed versions of Sections 4 and 5 are found in Mandelbrot 1989 and in our forthcoming *Selecta*.

In the models described in Section 2, mass is concentrated on a definite portion of space. Using the appropriate technical terms, mass is "supported by a closed fractal set." Furthermore, the mass distribution can be called "fractally homogenous." The rough meaning is that all galaxies are given the same mass, and the more precise meaning is that when two portions of the supporting set are identical except for translation, they support the same mass. These assumptions contradict two unquestionable facts.

First is the great inequality that prevails between galaxy masses.

Second (to quote from *FGN*, p. 376) is "our knowledge of the existence of interstallar matter. Its distribution is doubtless *at least* as irregular as that of the stars. In fact, the notion that it is impossible to define a density is stronger and more widely accepted for interstellar than stellar matter. To quote deVaucouleurs, 'it seems difficult to believe that, whereas visible matter is conspicuously clumpy and clustered on all scales, the invisible intergalactic gas is uniform and homogenous . . . [its] distribution must be closely related to . . . the distribution of galaxies. . . .

"[Thus, in the models of Section 2, parts of space] of less immediate interest were artificially emptied to make it possible to use *closed* fractal sets, but eventually these areas must be filled. This can be done using a fresh hybrid [now called *multifractals*. A multifractal] mass distribution in the cosmos will be such that no portion of space is empty, but, [given two] small thresholds θ and λ, a proportion of mass at least $1 - \lambda$ is concentrated on a portion of space of relative volume at most θ."

5. The Principal Ideas Underlying Multifractal Measures

5.1. AN OLD BUT GOOD ILLUSTRATION

Multifractals are old, insofar as we had first developed the basis of this technique in the years 1968 to 1976, in order to study different aspects of the intermittency of turbulence. Since "to see is to believe," Fig. 3 reproduces the earliest illustration of a multifractal, as it first appeared in our earliest full paper on this topic, Mandelbrot 1972. We expect it to inspire astronomers to many applications.

The horizontal axis shows "time" divided into small boxes of width Δt, and the vertical axis shows the sequence of the masses within these boxes. If the total integral measure over the total time span $[0, T]$ is set to 1, one can think of *the measure in a box* as *the probability of hitting this box*. If an analogous diagram were drawn for a measure having a density, it would be an approximation to this density — and a first characterization of our measure would be provided by the distribution of this approximate density along the horizontal.

In the present instance, however, the situation is extremely different. By design, the measure is approximately self-similar, in the sense to be discussed in Section 5.3. It follows that this measure grossly fails to have a density, nor is it discrete. For example, if the Δt is halved, the sharing of the measure in an original Δt between the two halves is usually very unequal. There is no such thing as a notion of "distribution" for the values of this measure. Fortunately, there is a very useful substitute.

5.2. THE NOTIONS OF LIMIT PROBABILITY DISTRIBUTION $\rho(\alpha)$, AND OF $f(\alpha)$.

Take different values of Δt, and, for each value of Δt, plot the corresponding measure distributions on doubly logarithmic coordinates. The measures we want to call multifractal have the following property. When both logarithmic coordinates of the plots drawn for different Δt's are reduced by the same factor $\log \Delta t$, the reduced plots of the distribution converge to a limit as $\Delta t \to 0$. This property can be turned around, and used to *define* the notion of multifractal. (But one must realize that the convergence to the limit may be slow.)

The reduced horizontal logarithmic coordinate is denoted by α, and will be seen to be a quantity called Hölder exponent. The reduced vertical logarithmic coordinate corresponding to the limit will be denoted by $\rho(\alpha)$. It will be seen that it is negative for all α, except where $\rho(\alpha)$ reaches its maximum, which is 0.

As has been first pointed out by Frisch and Parisi 1985, it is convenient to also introduce a quantity denoted by $f(\alpha) = \rho(\alpha) + 1$. When $f(\alpha) \geq 0$, one can interpret $f(\alpha)$ as being one fractal dimension of a suitable set. This is the only case Frisch and Parisi consider, and this is also the only case relevant to the application to astronomy. The replacement of $\rho(\alpha)$ by $f(\alpha)$ has virtues in some cases, but our feeling is that, fundamentally, it hides the nature of the multifractals.

Until the preprint of Frisch and Parisi was distributed in 1983, multifractals had continued to develop only in the sense that the mathematics was very much extended (see Kahane & Peyrière 1976). But they did not receive new applications, nor were they mentioned in *Astrophysical Journal Letters*. Their spread is a recent phenomenon, and most readers who have heard of them are likely to know presentations that follow the approach common to Frisch and Parisi and to Halsey et al.. Unfortunately, the algebra of these presentations is needlessly complicated, artificial, and of limited applicability and the terminology of Halsey et al. hides the extremely simple and almost familiar nature of the underlying structure. We shall, therefore, adopt the nota-

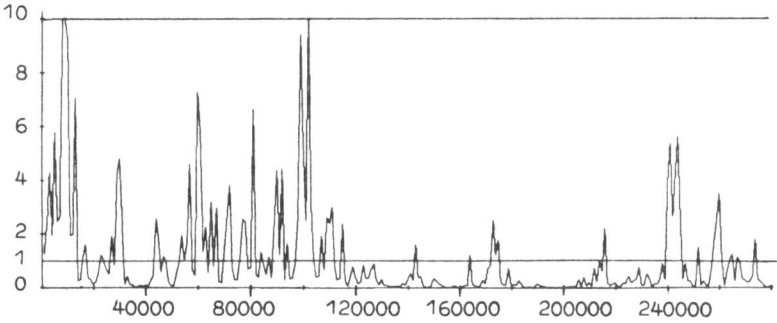

Figure 3. The earliest simulation of a sample from a multifractal measure. This measure is called limit lognormal in Mandelbrot 1972 (others call it "M's 1972 measure").

tion of Halsey et al., but follow our original approach in the form into which it has lately developed.

The multifractal formalism centers around the functions $\rho(\alpha)$ or $f(\alpha)$. In the simplest case, this paper obtains $f(\alpha)$ via the Lagrange multipliers procedure of statistical thermodynamics, which has long been familiar to every physicist. Later on, a full mathematical justification of the formalism, valid in a broader context in which $f(\alpha)$ can very well be negative, is provided by reference to existing (but little-known) limit theorems of probability due to Harald Cramèr, and concerned with "large deviations".

Mandelbrot 1974 considers two distinct kinds of random self-similar multifractals, respectively called *conservative* (or *microcanonical*) and *canonical*. This distinction is crucial to the study of low-dimensional cuts of multifractals embedded in a high dimensional space. But in astronomy this issue is not important.

5.3. SELF-SIMILAR MEASURES AND BEYOND

We need some definitions concerning measures $\mu(S)$. These $\mu(S)$ will be positive; therefore again, those not familiar with measure can think of μ as being the probability of hitting the set S. The multifractal measures obtained as a result of multiplicative cascades are the closest analog among measures of the exact self-similar fractal sets. Recall that a fractal set is exactly self-similar, if it can be decomposed into parts, each of which is obtained from the whole by a contracting similitude, $\mathscr{C}$. Such a set if fully determined by a collection of contractions. For example, each third of a basic fractal called Sierpinski gasket is obtained from the whole by a contracting similitude of ratio $r = 1/2$. Starting with any triangle in a "prefractal collection of triangles," the interpolation of the shape itself continues without regard to the "past" construction steps.

Now suppose that a (positive) measure $\mu(P)$ is defined for each third of the gasket, for each third of a third etc.. When the part P' is obtained from the part P by the contracting transformation $\mathscr{C}$, so that $P' = \mathscr{C}(P)$, the conditional measure of P' in P is defined just like a conditional probability, that is, by the ratio $\mu(P')/\mu(P)$ of the measure $\mu(P')$ to the conditioning measure $\mu(P)$. Now the idea strict of self-similarity for a measure is that the interpolation of the measure carried by a triangle in a prefractal collection of triangles also continues without regard to the "past" steps. That is as the parts contract, the measures they carry contract proportionately. To express this idea, take a second contracting transformation $\mathscr{L}$, and compare $\mu(P')/\mu(P)$ with $\mu[\mathscr{L}(P')]/\mu[\mathscr{L}(P)]$. If these conditional measures are identical, the measure μ will be called a *strictly self-similar multifractal*.

A random measure is called *statistically* self-similar if, given one or a finite collection of non overlapping parts $P_\gamma = \mathscr{C}_\gamma(P)$, the distribution or the joint distribution of the quantities $\mu(P_\gamma)/\mu(P)$ depends only on the contractions $\mathscr{C}_\gamma$.

Side remark. In a more general mathematical fractal set, the parts are obtained from the whole by transformations that are *non linear*. Examples where the contractions are in some sense *near linear* include the Julia sets of polynomial maps. The corresponding multifractals include the harmonic measures on these sets. Other examples of multifractal measures concern the limit sets of groups based upon inversions in circles (*FGN*, Chapters 18 and 20). A case when the limit set itself is a straight line is examined by Gutzwiller & Mandelbrot 1988. Finally, the *"fat fractals"* (new term for the fractals in *FGN*, Chapter 15) and the *Mandelbrot set* involve essentially non-linear transformations.

5.4. THE BASIC NON RANDOM SELF-SIMILAR MULTIFRACTALS

5.4.1. Basic background: the binomial multifractal measure. To construct this measure, given m_0 satisfying $\frac{1}{2} < m_0 < 1$ and $m_1 = 1 - m_0$, we spread mass over the halves of every dyadic interval, with the relative proportions m_0 and m_1. If $t = 0.\eta_1\eta_2 \ldots \eta_k$ is the development of t in the binary base 2, and φ_0 are φ_1 the relative frequencies of 0's and 1's in the binary development of t, the binomial measure assigns to the dyadic interval $[dt] = [t, t + 2^{-k}]$ of length $dt = 2^{-k}$ the mass

$$\mu(dt) = m_0^{k\varphi_0} m_1^{k\varphi_1}.$$

Adapting the classical notion of Hölder exponent to apply to the interval $[dt]$, we write

$$\alpha = \log[\mu(dt)]/\log(dt) = -\varphi_0 \log_2 m_0 - \varphi_1 \log_2 m_1,$$

and $0 < \alpha_{min} = -\log_2 m_0 \le \alpha \le \alpha_{max} = -\log_2 m_1 < \infty$. (The Hölder exponent has been given many new names. For example, it has been relabeled as "dimension" by Hentschel and Procaccia 1983, or as "pointwise dimension," but the term "dimension" *must* be reserved to sets.)

The number of intervals leading to φ_0 and φ_1 is $(k\varphi_0)!(k\varphi_1)!/k!$, giving the box fractal dimension

$$\delta = \log[(k\varphi_0)!(k\varphi_1)!/k!]/\log(dt).$$

For large k, the Stirling approximation yields

$$\delta = -\varphi_0 \log_2 \varphi_0 - \varphi_1 \log_2 \varphi_1.$$

Thus, α determines φ_0, hence $\delta = f(\alpha)$.

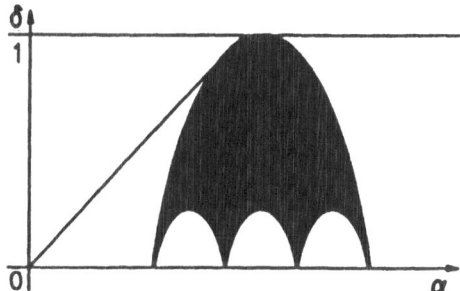

Figure 4. Rough idea of the domain of (α, δ) for a multinomial multifractal with $b = 4$, when the m_β differ from one another. The upper boundary defines the function $f(\alpha)$. Here, $\alpha_{min} = \min(-\log_b m_\beta) > 0$, and $\alpha_{max} = \max(-\log_b m_\beta) < \infty$.

A theorem by Eggleston relates δ to the Hausdorff dimension (Billingsley 1967).

5.4.2. *The multinomial measure.* To construct a multinomial measure of base $b > 2$, we require b masses m_β $(0 \le \beta \le b - 1)$. The b-adic intervals characterized by the frequencies φ_β of the digits β in the base-b development $0.\eta_1\eta_2 \ldots \eta_k$ yield

$$\alpha = -\sum \varphi_\beta \log_b m_\beta \text{ and } \delta = -\sum \varphi_\beta \log_b \varphi_\beta.$$

Now, the points (α, δ) cover a domain shown in black on Figure 4.

5.4.3. *The Lagrange multipliers argument, and the Legendre and inverse Legendre relations of the Gibbs theory.* The sets of φ_β's yielding the same α are dominated by the highest dimension term. This term maximizes $-\sum \varphi_\beta \log_b \varphi_\beta$, given $-\sum \varphi_\beta \log_b m_\beta = \alpha$, and $\sum \varphi_\beta = 1$. The classical method of Lagrange multipliers introduces a multiplier q, with $-\infty < q < \infty$, and yields

$$\varphi_\beta = \frac{b^{q \log_b m_\beta}}{\sum b^{q \log_b m_\beta}} = \frac{m_\beta^q}{\sum m_\beta^q}.$$

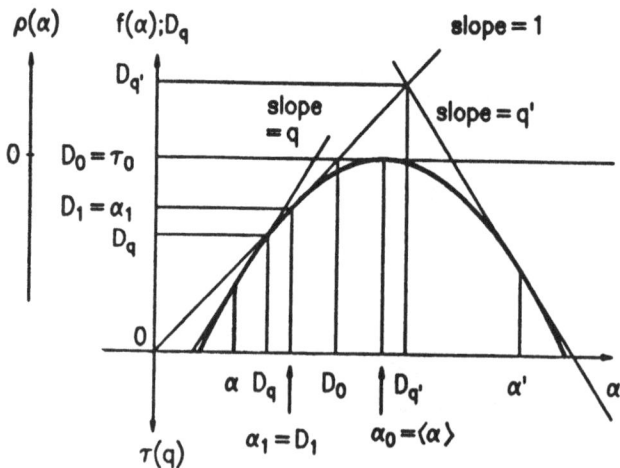

Figure 5. A multifractal diagram. The ordinate scale to the left shows $\rho(\alpha) = \lim_{k \to \infty}(1/k) \log_B$ (probability) versus the Hölder $\alpha = -(1/k) \log_b$ (measure). The ordinate scale to the right shows the function $f(\alpha) = \rho(\alpha) + D_0$. It is well-known that $q = f'(\alpha)$ and $-\tau(q)$ is the ordinate of the intercept of the tangent of slope q by the vertical axis. Let us add the observation that D_q is coordinate of the intercept of this tangent by the main bisector of the axes.

Define the quantity $\tau(q) = -\log_b \sum m_\beta^q$, which the mathematical statisticians call "cumulant generating function." In terms of τ, the Lagrange multipliers determine q and $f(\alpha)$ from α by

$$\alpha = -\sum \varphi_\beta \log_b m_\beta = -\frac{\partial}{\partial q} \log_b \sum m_\beta^q;$$

$$\max \delta = f(\alpha) = -\frac{\sum (q \log_b m_\beta - \log_b \sum m_\beta^q) m_\beta^q}{\sum m_\beta^q}.$$

That is,

$$\alpha = \frac{\partial \tau(q)}{\partial q} \text{ and } f(\alpha) = q \frac{\partial \tau}{\partial q} - \tau = q\alpha - \tau.$$

Figure 4 is now replaced by its upper boundary, which is the graph of a function $f(\alpha)$. Clearly, $\alpha > 0$ and $\delta \geq 0$, hence $f(\alpha) \geq 0$, $\alpha_{min} > 0$, $f(\alpha_{min}) \geq 0$ and $f'(\alpha_{min}) = \infty$, and $\alpha_{max} < \infty$, $f(\alpha_{max}) \geq 0$ and $f'(\alpha_{max}) = -\infty$. Multifractals that are not multinomial, yet posses these properties we call "pseudo-multinomial."

Formally, q = inverse temperature, τ = Gibbs free energy, and f = entropy.

5.4.4. The term "multifractal formalism" and the question of actual computation. The equations $\alpha = \tau'$ and $f = q\alpha - \tau$ are the "multifractal formalism." Frisch and Parisi 1985 and Halsey et al. 1986 obtain the same result via a steepest-descent argument, which experts will recognize as identical to the Darwin-Fowler justification of the Lagrange multipliers procedure. This is *not* the right way to proceed, just as *no one* will think of teaching thermodynamics by describing the Darwin-Fowler method directly, without first presenting the Lagrange multipliers. Therefore, Section 5.4.3 has taken the path towards the same formalism that involves the least effort and the fullest understanding. Section 5.5 describes the next simplest generalization.

A considerable literature has developed around ways of using this formalism. This literature is, obviously, unaffected by a change in the foundations.

5.5. THE RANDOM 1974 MULTIFRACTAL MEASURES

5.5.1. Generalization of the multifractal formalism by an application of Harald Cramèr's theorem on large deviations. Now we proceed to the exactly renormalized "1974 multiplicative multifractals" introduced in Mandelbrot 1974. First observe that the $f(\alpha)$ of a multinomial measure is unchanged if the indexes of the masses m_β are shuffled at random before each stage of the cascade that distributes mass. Next, suppose that $b = B^E$, with positive integers B and E. With no change in the algebra, the multinomial measure with random weight assignment can be interpreted as spread on cells in a E-dimensional cube of base B. The weights in the cells inject a random multiplier M that can take the values m'_β, with the probability $1/b$ for each value. In a cost-free generalization, consider a random multiplier satisfying only $M \geq 0$ and $\langle M \rangle = 1/b$.

Clearly, the mass $\mu(d\mathbf{x})$ in the b-adic cube of side B^{-k}, starting at $\mathbf{x} = 0.\eta_1\eta_2 \ldots \eta_k$, is

$$\mu(d\mathbf{x}) = M(\eta_1)M(\eta_1, \eta_2) \ldots M(\eta_1, \ldots \eta_k) \ldots .$$

Here, the successive M are identically distributed and independent. Hence

$$\alpha = -(1/k)[\log_b M(\eta_1) + \log_b M(\eta_1, \eta_2) \ldots]$$

is the average of k independent random variables. To tackle the distribution of α, "large deviations theorems" of H. Cramèr are available "off-the-shelf" (a pleasant surprise); see Book 1984, Chernoff 1952, Daniels 1954, 1987. These theorems establish that, as $k \to \infty$,

$(1/k)\log_b$ (probability density of α_L) converges to a limit, to be denoted as $\rho(\alpha)$.

The quantities $(1/k)\log_B$ (probability of values $> \alpha > \langle\alpha\rangle$) (resp., of values $< \alpha < \langle\alpha\rangle$) converge to the same limit. It is easily shown that

$$f(\alpha) = \rho(\alpha) + E = \rho(\alpha) + \text{dimension of the measure's support.}$$

It is a noteworthy fact that in the generalized Gibbs formalism resulting from the Cramèr theorem, different M's yield different $f(\alpha)$'s, and conversely.

Obviously, Cramèr-type theorems extend to the case when the factors M are weakly dependent or weakly non-identical.

5.5.2. *Comments concerning lognormality.* Section 5.5.1 may surprise those many readers who know the literature to the effect that $\log_b M(\eta_1) + \log_b M(\eta_1, \eta_2) + \ldots$ is asymptotically Gaussian, so that α is asymptotically lognormal. These assertions result from the application of a different renormalization, one that leads to the classical central limit theorem. It is indeed correct that the central limit theorem yields some information about the multiplicative multifractals. Also, this information is universal, but it is of very limited scope. It only implies that $\rho(\alpha)$ and $f(\alpha)$ are parabolic near their maximum. Away from the maximum, the behavior of $\rho(\alpha)$ and $f(\alpha)$ is *not* universal.

There is a seeming paradox here. On the one had, the probability outside of the central bell tends to 0 as $k \to \infty$, meaning that the tails become thoroughly insignificant. In the limit $k \to \infty$, the most probable value, the expectation and the other usual parameter of location all converge to each other. On the other hand, those "negligibly" few values in the tails are so huge that their contributions to all the moments of order $q \neq 0$, and to $\tau(q)$, are predominant. Moreover, the moments and $\tau(q)$ depend on the exact $f(\alpha)$, that is, are *non universal*.

In any event, the functions $(\rho)\alpha$ and $f(\alpha)$ are *not* like those from the lognormal M, except if the M are lognormal. Furthermore, lognormal M's require very special precautions. (Mandelbrot 1974 shows that the "principle of lognormality" claimed in 1962 by Kolmogorov is logically untenable. Frisch and Parisi 1985 have noted that, while their approach stemmed from Mandelbrot 1974, it was less general because it did not accommodate the lognormal.)

5.5.3. *The meaning of negative dimensions.* The special "pseudo-multinomial" situation requires $Pr\{M_{max}\} \geq b^{-1}$ and $Pr\{M_{min}\} \geq b^{-1}$. But a first feature of our 1974 multiplicative measures is that they allow $\alpha_{min} = 0$ and $\alpha_{max} = \infty$ and $f(\alpha_{min}) = \log_b[Pr\{M_{max}\}] + 1 < 0$ and $f(\alpha_{max}) < 0$.

When $f(\alpha)$ is viewed as a fractal dimension, $f(\alpha) < 0$ is impossible. However, in numerous applications, the restriction to $f(\alpha) \geq 0$ leads to self contradiction, or is otherwise not acceptable. However these applications involve either of the following two possibilities: a) the possibility of investigating the measure within the intersection a multifractal by a line (i.e., a thin cylinder) or by a plane (i.e., a flat pancake), and b) the possibility of taking successive independent samples from a population. Neither of these possibilities occurs in astronomy, hence it is safe to keep to $f(\alpha) \geq 0$

5.5.4. *The quantities* $D_q = \tau(q)/(q-1)$ *are "critical dimensions."* For 1974 measures, D_q is a *critical dimension* for the moments of order q, as shown in Mandelbrot 1974. However, the task of proving this property exceeds the space available here. It suffices to point out that the proof relies upon on of the most interesting aspects of the 1974 theory, namely on the study of low dimensional cuts of higher dimensional multifractals; see Mandelbrot 1989.

References

Audouze, J., Pelletan, M.-C. and Szalay, A., eds. *Large Scale Structures of the Universe* (IAU Symposium No. 130, Balantonfüred, Hungary), Dordrecht-Boston, Kluwer, 1988.

Billingsley, P. 1967. *Ergodic Theory and Information*, J. Wiley, New York, p.139.

Book, S.A. 1984. in *Encyclopedia of Statistical Sciences* Vol. 4, p. 476.

Chernoff, H. 1952. *Ann. Math. Stat* **23**, 493.

Coleman, P.H., Pietronero, L. and Sanders, R.H. 1988. *Absence of any characteristic correlation length in the DfA Galaxy Catalogue* (preprint).

Daniels, H.E. 1954. *Ann. Math. Stat.* **25**, 631.

Davis, M., Meiksin, A., Strauss, M.A., da Costa, L.N., and Yahil, A., 1988. *On the Universality of the Two-Point Galaxy Correlation Function* (preprint).

Daniels, H.E. 1987. *International Statistical Review* **55**, 37.

Frisch, U. and Parisi, G. 1985 in *Turbulence and Predictability in Geophysical Fluid Dynamics and Climate Dynamics*, International School of Physics "Enrico Fermi," Course 88, edited by M. Ghil et. al. North-Holland, Amsterdam, p 84.

Grassberger, P. 1983. *Phys. Lett.* **97A**, 227.

Gutzwiller, M.C. and Mandelbrot, B.B. 1988. *Phys. Rev. Lett.* **60**, 673.

Halsey, T.C., Jensen, M.H., Kadanoff, L.P., Procaccia, I. and Shraiman, B.I. 1986. *Phys. Rev.* **A33**, 1141.

Hentschel, H.G.E. and Procaccia, I. 1983. *Physica (Utrecht)* **8D**, 435.

Jones, B.J.T., Martinez, V.J., Saar E. and Einasto, J. 1988. *Multifractal description of the large scale structure of the Universe* (preprint).

Mandelbrot, B.B. 1972 in *Statistical Models and Turbulence* (Lecture Note in Physics, Vol. 12), Proc. Symp., La Jolla, Calif., M. Rosenblatt and C. Van Atta, eds., Springer-Verlag, New York, p. 333.

Mandelbrot, B.B. 1974. *J. Fluid Mech.* **62**, 331; also *Comptes Rendus* **278A**, 289, 355.

Mandelbrot, B.B. 1982. *The Fractal Geometry of Nature*. New York: W.H. Freeman.

Mandelbrot, B.B. 1988, in Audouze et al. 1988, 482-484.

Note. Mandelbrot 1988 is superseded by Section 3 of the present text. The old text is far too brief to make its point, and there are typographical errors.

278

Mandelbrot, B.B. 1989, in *Fluctuations and Pattern Formation*. H.E. Stanley and N. Ostrowsky eds., Boston: Kluwer.

Mandelbrot, B.B. (forthcoming Selecta, Vol. 1). *Fractals and Multifractals: Noise, Turbulence and Galaxies* New York: Springer.

Meakin, P. 1988, in *Phase Transitions and Critical Phenomena*, edited by C. Domb and J.L. Lebowitz, Academic Press, London, **12**, 335.

Peebles, P.J.E. 1980. *The Large Scale Structure of the Universe*. Princeton University Press.

Pietronero, L. 1987. 'The fractal structure of the Universe: correlations of galaxies and clusters and the average mass density.' *Physica* **144A**, 257.

Szalay, A.S. and Schramm, D.N. 1985, *Nature* **314**, 718.

Wiedenmann, C. and Atamspacher, H. 1988. *Integral pair correlations in the angular distribution of galaxies* (preprint).

Zeldovich, Ya.B., and Szalay, A.S. 1988. *Acta Physical Hungarica* (in the press).

DISCUSSION

Question by Valérie de Lapparent: Analysis of the recent CFA redshift survey shows that the galaxy distribution is characterized by well defined scale lengths: for example the thickness of the sheets and the mean separation of the galaxies with sheets are nearly constant throughout the observed region. This result seems difficult to reconcile with a self-invariant description of the galaxy distribution.

Reply by Benoit Mandelbrot: Thank you for the information you report, and allow me to respond by making two points.

The first point concerns method. Nowhere do my papers or books claim that a fractal description of a phenomenon in real space applies in *pure* mathematical rigor. To the contrary, I have always stressed that every real space model is an approximation that requires corrective terms, for example, cross-overs or cut-offs. (Let me add also that "pure fractals" are indeed often found in phase space.)

I have also stressed that the model one should prefer as a best first-order approximation should be the model that requires the smallest amount of second-order corrections. Therefore, even if the discrepancy you describe were to be confirmed, it would not *by itself* suffice to discard the fractal model as a first approximation valid within range of distances that is sufficiently broad to matter. Your earlier data suggest that its limit may not have been reached.

The second point concerns data analysis. The prowess of the observational astronomers never cease to amaze the mere theoreticians. But I am on the record as being less uniformly impressed by the methods used to analyze the resulting data. In fact, successive analyses of the same body of observations have often yielded contradictory results. New analyses will surely be made of the most remarkable empirical findings that you have reported a few years ago. Let us wait for the results before we rush to conclusions.

Question by P.J.E. Peebles: As you know, I argued in my book, *Large-Scale Structure of the Universe*, that the observations contradict the assumption that the galaxy distribution in the visible part of the universe is a pure scale invariant fractal. Nothing in the observational developments since then has caused me to change my mind.

Reply by Benoit Mandelbrot: Thank you for this opportunity to state my side in the friendly but inconclusive dispute we have been carrying on for a long time.

As I say in response to Valérie de Lapparent, "pure fractals" can be found in phase space, but I would not be completely surprised if they can *never* be found in any phenomenon in real space. Therefore, a lawyer may argue that we agree.

In any event, I am pleased that you *do not* specifically disagree with my minimal thesis, that an "impure fractal" model applies to galaxies, at least up to some crossover.

Now to the statement of some of our disagreements. In order to be used to contradict a theoretical model, observations must be subjected to careful analysis. This includes statistical analysis, and we certainly disagree on, a) which statistical tools are appropriate for the galaxy data and, b) what to do when one finds oneself beyond the range of applicability of statistics. Statistics tends to be a boring subject, but we agree that it does matter (though I have less confidence in it than you may have). By using the standard methods of statistical analysis, as you do, it may be true that one must indeed reach your conclusions, but I believe that your statistical methods grossly prejudge the issue, and are *not* applicable here. You are doubtless aware of the dispute that has recently pitted Pietronero et al. against Davis et al.. I tend to side with Pietronero, or − at worst − from our viewpoint to conclude that if the data are so poor as to demand brutal correction before analysis, one cannot conclude much on the basis of these data.

Second point. My criticism of your measure of the crossover has been countered by friends of yours, who assert that it does not matter whether the figure is 5 or 50, as long as the same definition is used throughout. I might perhaps have agreed, if the model being tested were characterized by a single crossover. (Possibility 4 in the terminology of Section 1.2.) But it appears that your account of the existence of the voids must accept the existence of an intermediate range of distances in which the correlations are negative. Therefore, you make an assumption that requires at least two crossovers. (Possibility 6 in the terminology of Section 1.2.) If so, the lower crossover, which I call $R_{peebles}$ to avoid ambiguity, might conceivably describe the point when one moves beyond the fractal range. But homogeneity would only be established after a much longer distance R_{upper}, which need not bear any simple relation to $R_{peebles}$. This R_{upper} is by far the more relevant notion, but it remains to be estimated. The evidence you provide does not exclude that $R_{upper} = \infty$, which would mean that the distribution beyond $R_{peebles}$ is neither fractal nor uniform, but a mess. Maybe this is the case, but we should do our best to avoid this conclusion, and find order in the large scale structure.

THE X-RAY BACKGROUND : DISCRETE SOURCES OR DIFFUSE PROCESSES?

G. Zamorani
Osservatorio Astronomico, Via G.B. Tiepolo 11
34131 Trieste, Italy
and
Istituto di Radioastronomia, CNR, Via Irnerio 46
40126 Bologna, Italy

ABSTRACT. In this paper we review the current status of the debate about the origin of the X-ray background. The existing observational data, when considered in their totality, seem to indicate that no simple scenario can easily explain in a direct way all the observed features of the X-ray background. After a brief discussion of the most important observational facts which have to be accounted for, we discuss the main problems, difficulties and astrophysical implications associated with each of the two main classes of models for the production of the X-ray background, i.e. X-ray emission from diffuse processes or from faint discrete X-ray sources.

1. Introduction

The existence of the diffuse X-ray background (XRB) has been known for more than 25 years (Giacconi et al. 1962). Since then a rapidly increasing amount of data has been acquired through rocket and satellite observations. The first important step with respect to our knowledge of the XRB has been made with the first all-sky survey (UHURU). The high degree of isotropy of the XRB revealed by this survey led almost immediately to realize that its origin has to be extragalactic. Later, with the launch of HEAO-1, it became possible to obtain detailed spectral information for both the XRB and a few bright extragalactic sources (clusters of galaxies and Active Galactic Nuclei), at a flux level of the order of a few x 10^{-11} erg cm^{-2} s^{-1}. Finally, the *Einstein* data allowed for the first time the study of the X-ray counts at faint fluxes (down to a limit of about 2.5×10^{-14} erg cm^{-2} s^{-1}) and produced enough high quality data to study the X-ray properties of various classes of objects (quasars, Seyfert galaxies, normal galaxies, clusters...) on both a single object basis and a statistical basis.

Despite this large amount of data the origin of the XRB is still controversial. Many different models have been proposed (see review by Setti and Woltjer 1982), but none of them appears to be able to explain in a simple way all the observed features of the XRB from a few keV to hundreds of keV. Broadly speaking, all the various proposed explanations can be arranged in two main classes. According to the first one, most of the energy contained in the XRB derives from a diffuse process. The most attractive for such a diffuse mechanism is thermal bremsstrahlung from a hot intergalactic gas (Field and Perrenod 1977; Guilbert and Fabian 1986).

M. Mezzetti et al. (eds.), Large Scale Structure and Motions in the Universe, 281–289.
© *1989 by Kluwer Academic Publishers.*

According to the second one, the XRB can be explained in terms of superposition of discrete X-ray sources. There are, of course, also intermediate scenarios in which diffuse and discrete source contributions are both dominant, but at different energies.

2. Observational Constraints

In order to decide in favor of one of the various proposed alternatives for the production of the XRB, we have to be able to reproduce, with reasonable physical assumptions, both the intensity and the spectral shape of the observed XRB. In doing so, we need to make use of all the relevant observational constraints which have been accumulated in the past 25 years. The most relevant observational facts in the field can be summarized as follows.

a) In the energy range 3-50 keV the shape of the XRB is well fitted by an isothermal bremsstrahlung model corresponding to an optically thin, hot plasma with kT of the order of 40 keV (Marshall et al. 1980). At low energy (up to 10-20 keV) the shape of the XRB can be fitted with a power law, with a best fit slope of the order of 0.4.

b) When viewed over large angular scales at high galactic latitudes the XRB is essentially isotropic. In the framework of discrete sources, lower limits of the order of 100 objects/$sq.deg.$ have been derived from isotropy and fluctuation analyses on UHURU, ARIEL V and HEAO-1 data (see Shafer 1983 and references therein). More recently, from an analysis of the arcminute scale fluctuations of the Deep Survey data gathered by the *Einstein Observatory*, Hamilton and Helfand (1987) have shown that, still in the context of discrete sources, the observed granularity of the *Einstein* images requires a number density of at least 5000 objects/$sq.deg.$ (see the end of Section 3).

c) Early results of the *Einstein Observatory* Deep Surveys in Draco and Eridanus showed that in the energy range 1-3 keV the surface density of extragalactic X-ray sources is (19 ± 8) objects/$sq.deg.$ at a limiting flux of about 2.6×10^{-14} erg cm^{-2} s^{-1}. The corresponding contribution to the XRB in this energy range is $(26\pm11)\%$ (Giacconi *et al.* 1979). More recently, a joint deep optical and X-ray survey in the Pavo region (Griffiths *et al.* 1988) has shown that X-ray selected quasars, with a surface density of about 32 per square degree, account directly for about 30% of the XRB in the energy range of the *Einstein Observatory*. The presently available optical identifications of the *Einstein* Deep Sensitivity Survey suggest that about 70% of these faint X-ray sources are Active Galactic Nuclei (AGNs), with the rest being normal galaxies and clusters of galaxies. The percentage of AGNs is of the same order also in the *Einstein* Medium Sensitivity Survey (Gioia *et al.* 1984). The X-ray sources of this survey are fully identified and have an average X-ray flux 3-10 times higher than those in the Deep Sensitivity Survey.

d) As for the X-ray spectra of the various classes of X-ray sources, a brief summary of our present knowledge is as follows:
i) The X-ray emission from clusters of galaxies is well fitted by optically thin bremsstrahlung models with characteristic temperatures ranging from 2 to 7 keV as a function of the dynamical evolutionary stage of the cluster itself (Forman and Jones 1982).
ii) For normal galaxies there is some indication that the average spectra of spiral galaxies are harder ($kT \geq 2$ keV; Fabbiano and Trinchieri 1987) than those of

elliptical galaxies (kT of the order of 0.5-2.0 keV; Forman, Jones and Tucker 1985). iii) About 30 active galaxies, mainly Seyfert 1 galaxies, have reliable 2-20 keV spectral information (Mushotzky 1984), mostly from HEAO-1 data. They show a tight uniformity in their spectra, which are well fitted by a single power law with an average energy index of the order of 0.65 and a small dispersion ($\sigma = 0.15$) around the average value. All these objects have an X-ray flux greater than 10^{-11} erg cm^{-2} s^{-1}.

iv) At lower energy (0.2-3.5 keV) and down to fluxes of the order of 10^{-12} erg cm^{-2}s^{-1}, the *Einstein Observatory* data on a sample of 33 quasars look quite different (Wilkes and Elvis 1987). The best fit power law slopes have a wide range ($-0.2 \leq \alpha \leq 1.8$), with an average value of 0.5 for radio loud quasars and 1.0 for radio quiet quasars.

v) Recently, Maccacaro et al. (1988) have studied the spectral properties of about 600 sources extracted from the *Einstein Observatory* Extended Medium Sensitivity Survey. Because of the faint average flux of these sources (reaching 10^{-13} erg cm^{-2}s^{-1}), the uncertainty on the measurement for each single object is quite large. Exploiting, however, the large number of objects at their disposal, they find through a statistical analysis that the X-ray selected active galactic nuclei are characterized by a variety of spectral indices with an average slope 1.03 ± 0.05 and an intrinsic dispersion $\sigma = 0.36$. Since most of these AGNs are radio quiet, this result is in very good agreement with the result obtained by Wilkes and Elvis (1987) at higher fluxes.

vi) A similar analysis applied to 55 X-ray selected extragalactic sources extracted from the EXOSAT survey (Giommi, Tagliaferri and Angelini 1988) yields an even steeper average slope ($\alpha = 1.4$, with an associated 1 σ error of about 0.4). In comparing this result with those derived from the *Einstein* data one has to keep in mind that the EXOSAT energy band (0.05-2.0 keV) is softer than the *Einstein* energy band.

This, probably incomplete, list of observational results immediately explains why no firm conclusion on the nature of the XRB has been reached so far. Point a) seems to support the diffuse scenario, in particular the thermal bremsstrahlung model from a hot intergalactic gas; point c), however, tells us that, at least in the *Einstein* energy band, the contribution of discrete sources is substantial (greater than 25%); the X-ray spectra of the bright counterparts of the classes of objects which have been detected in the X-ray surveys, however, appear to be significantly different from the observed spectrum of the XRB (Point c)); this discrepancy remains true also at the average fluxes of the *Einstein* Medium Sensitivity Survey (10^{-13} - 10^{-12} erg cm^{-2} s^{-1}).

Before proceeding any further, it is important to remind here that at faint fluxes the only available spectral information (from *Einstein* and EXOSAT) refers to the soft X-ray energy band, where the spectrum of the XRB is not well measured. Vice versa, the available spectral information at higher energy refers only to bright and relatively near-by sources, whose contribution to the XBR is of the order of a few percent.

3. Minimum Contribution of Known Sources to the Soft XRB

The first and crucial point for a discussion of the origin of XRB is to start from an estimate of the contribution of 'known' sources. Since the statistical properties

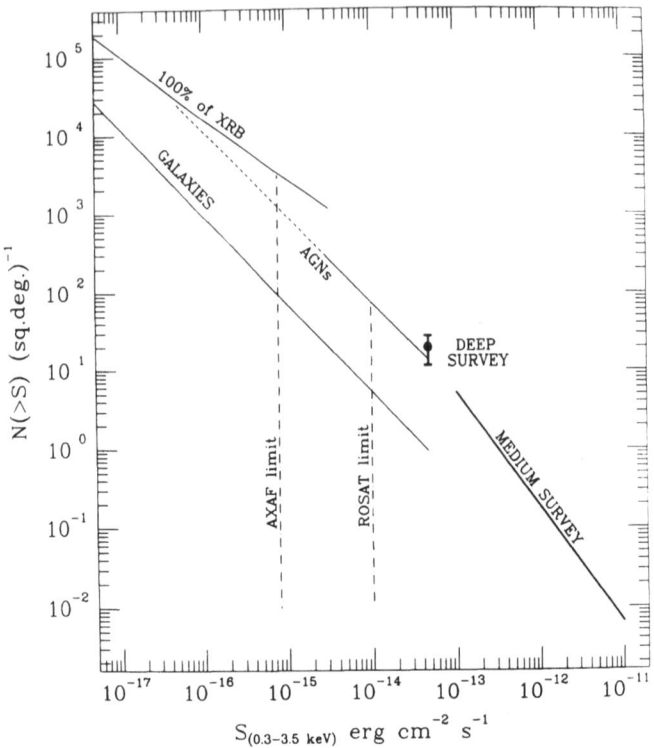

Figure 1. X-ray log N - log S at 0.3-3.5 keV. Thick solid line and solid circle with error bar represent the results from the *Einstein* Medium and Deep Sensitivity Surveys. Thin solid line represents our estimates for the minimum expected contribution from galaxies and AGNs. (Adapted from Giacconi and Zamorani 1987).

of the X-ray sources and the X-ray counts at faint fluxes are better known at soft energies, we will now concentrate on the *Einstein* energy range (0.3-3.5 keV). To overcome the problem that at this energy range the XRB is not well measured, we compute the percentage contribution of discrete sources assuming that the XRB in this energy range can be smoothly extrapolated, without change in slope, from what is observed above 3 keV. Then the results obtained in this way are converted to higher energies with educated assumptions on the spectra of the sources.

The contribution of discrete sources to the XRB has been computed by various authors, with results not always in agreement with each other (see Giacconi and Zamorani 1987 and references therein). Here, following Giacconi and Zamorani, we present an estimate which is based as much as possible on observed quantities, rather than on model dependent quantities such as luminosity and evolution functions. The observational inputs for this computation are:

i) The X-ray extragalactic log N - log S from the *Einstein* Medium and Deep Sensitivity Surveys (Gioia *et al.* 1984; Giacconi *et al.* 1979; Griffiths *et al.* 1988);

ii) The optical identification content of these surveys;

iii) The optical counts at faint magnitudes of quasars (Koo, Kron and Cudworth 1986; Marano, Zamorani and Zitelli 1988 and references therein) and galaxies

TABLE I

Percentage of the 2-10 keV Background due to Known Sources

Slope	Motivation	Percentage
0.4	XRB slope	50%
0.7	Seyferts (HEAO-1)	38%
1.0	QSOs + *Einstein* Medium Survey	30%

(Tyson 1988).

Figure 1 shows our estimate of the minimum extrapolation of the expected log N - log S at fluxes below the *Einstein* limit for galaxies and AGNs. These extrapolations have been computed with the following prescriptions.

1. As for galaxies, we have used an average value for the ratio of X-ray to optical fluxes for spiral and elliptical galaxies (derived from Fabbiano and Trinchieri 1985, and Trinchieri and Fabbiano 1985, respectively), to convert the optical log N - m, reaching $m_B = 27$ (Tyson 1988) into an expected X-ray log N - log S.

2. As for AGNs we used the information that they represent about 70% of the extragalactic X-ray sources at the *Einstein* limit. This corresponds to the number density of AGNs found in optical surveys at $m_B \simeq 20.0$. Since the optical surveys show a flattening of the log N - m relationship at about this magnitude, we assume a similar flattening in the X-ray log N - log S at fluxes immediately below the *Einstein* Deep Survey point. We have then extrapolated, for the equivalent of 3 magnitudes, the X-ray log N - log S with the same slope (1.1) as measured at optical wavelengths down to $m_B \simeq 23.0$ (Koo, Kron and Cudworth 1986).

The integral of the two expected log N - log S curves drawn in Figure 1 (thin solid lines) corresponds to a contribution to the soft XRB of $\sim 40\%$ for AGNs and $\sim 13\%$ for galaxies. It is likely that both these values should be considered as lower limits rather than best estimates (Giacconi and Zamorani 1987), so that we can conclude that it is very unlikely that the contribution of known sources to the 2 keV background is smaller than 50%. To convert this percentage to the 2-10 keV energy range, where the shape of the spectrum of the X-ray background is better known, we have to make an assumption on the average spectrum of the sources. For pure power law spectra we obtain the results shown in Table I.

The extrapolation of the X-ray log N - log S showed in Figure 1 is consistent with the recent results by Hamilton and Helfand (1987). From a detailed analysis of the arcminute scale fluctuations in deep *Einstein* observations they have shown that, if the log N - log S continues with a slope 1.5 below the *Einstein* limit, a cutoff has to exist at a flux which is about one third of the Deep Survey limit, corresponding to a contribution of about 50% from discrete sources. If, on the other hand, the slope flattens just at the Deep Survey limit, they find that for any slope flatter than 1.2 both the observed flux and the observed granularity of the XRB may be understood as coming entirely from point sources. Equivalently, the majority of a point-like contribution to the XRB must come from sources with a number density of at least 5000 objects/*sq.deg.* (see Figure 1). They also conclude that there is no way to use fluctuation analyses on the *Einstein* data to distinguish between a large population of faint point-like sources and a truly diffuse component.

4. The Spectrum of the Residual Background

Since fluctuation analyses do not allow to distinguish between the contribution from faint point-like sources and a truly diffuse component, we have to find an other way to distinguish between these two possibilities. One possibility is to use the available information on the spectrum of the XRB, coupled with the previously derived estimate of the contribution from known sources and some educated guesses on the average spectrum of these sources. With these inputs we can try to estimate the spectrum of the residual spectrum. By 'residual' we mean here the X-ray flux which remains after subtracting the estimated contribution of known sources from the observed background. Such a decomposition of the spectrum of the XRB (known sources plus residual) has been computed by various authors (Leiter and Boldt 1982; Setti 1985; Giacconi and Zamorani 1987). The common feature of all these decompositions is a very flat residual background up to at least 10-20 keV (see, for example, Figure 2 in Giacconi and Zamorani 1987).

If this residual background is due to discrete sources, what can we say about their properties? The first possibility is that this residual background is made of the same classes of sources which have been detected in the *Einstein* surveys. On the basis of the current estimates these sources should mainly be AGNs. For this to be true, the simplest interpretation of the available data would require a strong spectral evolution for these sources: in the energy range 3-20 keV their mean spectral index at faint fluxes should be significantly flatter than what is observed at high fluxes (see De Zotti *et al.* 1982). An alternative model, which does not require any evolution of the X-ray spectra of active galactic nuclei, has recently been presented by Schwartz and Tucker (1988). Abandoning the assumption that the X-ray spectra of AGNs are described by a single power law, they find that spectra which tend to flatten with increasing energy (consistently with the results from EXOSAT, *Einstein*, and HEAO-1; see previous Section) provide a satisfactory explanation of both the spectrum and the intensity of the XRB in terms of AGNs. No real observational test of these two alternatives can be obtained until accurate spectra for many AGNs, at both low and high redshifts, are measured over the entire range of 0.1 to 10 keV by the AXAF Observatory. The second possibility is that a completely new class of objects is responsible for the bulk of the X-ray background. Precursor active galaxies and young quasars have been suggested in a series of papers (Leiter and Boldt 1982; Boldt and Leiter 1984 and 1987) as potential important contributors to the XRB. In these objects a flat X-ray spectrum, consistent with the spectrum of the residual background can result from either thermal Comptonization of cyclotron photons from an equipartition magnetic field (Zdziarski 1988) or a nonthermal Comptonization (Fabian, Done and Ghisellini 1988).

Alternatively, can the residual background be explained as the result of a purely diffuse emission? Giacconi and Zamorani (1987) concluded that a pure thermal bremsstrahlung emission from hot gas can **not** reproduce the residual, essentially because of the very flat slope in the energy range 3-20 keV obtained by subtracting the contribution from known sources. This conclusion, however, being based on a number of extrapolations and assumptions, can still be challenged. Guilbert and Fabian (1986), for example, by reinvestigating the possibility of a hot intergalactic medium as the origin of the XRB, conclude that an acceptable fit to the residual background can be achieved. They assume that the gas is heated at an epoch z_h by some mechanism which leaves the electrons and the ions in thermal equilibrium. The electrons and the ions are then assumed to interact only through

Coulomb scattering. Using semi-relativistic calculations for the electron thermo-dynamics and bremsstrahlung, and including cooling by adiabatic expansion and Compton scattering with the microwave background radiation, they find that the X-ray spectrum of the residual background can be fitted within the observational errors provided that:

i) $z_h \leq 6$; for larger z_h, in fact, the Compton cooling of the electrons on the microwave background causes the resulting X-ray spectrum to steepen unacceptably.

ii) the contribution at 5 keV of Seyfert-like objects ($\alpha \sim 0.65$) is lower than 20%. This value is lower than our estimate for the contribution from this kind of sources (see Table I), and represents, therefore, a possible problem for this model.

Other difficulties with this model are:

a) The required Ω_B, in order to produce enough X-ray flux, is of the order of 0.25-0.30, which is higher than what is allowed by the standard big bang nucleosynthesis models. One possible way to overcome this difficulty is to clump the intergalactic medium. As the emitted power is proportional to the square of the density of the gas, some degree of clumpiness would reduce the effective number of baryons needed. Barcons and Fabian (1988) have recently studied a two phase model, in which dense 'cold' blobs at a temperature of about 5×10^8 K are surrounded and confined by a hotter (T $\geq 6 \times 10^9$ K) tenuous medium. This hot phase would not contribute substantially to the 3-300 keV XRB, but could provide a significant fraction of the 'MeV bump'. Pressure equilibrium and $\Omega_B \sim 0.1$ fix the filling factor of the cold blobs at about 0.2. A clumpy intergalactic medium produces, however, fluctuations in both the XRB and the temperature of microwave background. The present upper limits to $\Delta T/T$ at sub-arcminute scale require blobs not much larger than about 20 kpc, i.e. a galaxy size!

b) A very large amount of energy, corresponding to about 40% of the energy in the microwave background radiation, has to be injected into a uniform intergalactic medium to heat it. This corresponds to about 10^{64} erg per each large galaxy and could be supplied if each galaxy contains a black hole of mass $\geq 10^{10}$ $M_\odot$, whose binding energy was efficiently converted to thermal energy during formation. The energy required to heat the two-phase intergalactic medium discussed by Barcons and Fabian (1988) is at least twice that needed to heat a uniform medium.

c) The isotropy of the XRB requires a very uniform heat source, implying a high space density of heating centers, comparable to that of bright galaxies. If this is the case, why not to think that the XRB is due to these heating centers themselves?

5. Conclusions

The main conclusions of this paper can be summarized as follows:

1. The contribution to the soft extragalactic XRB due to known classes of sources (galaxies, clusters of galaxies, AGNs) is likely to be at least of the order of 50%.

2. The X-ray spectra of these sources, known from HEAO-1 data to be different from that of the XRB at high fluxes, appear to maintain this difference also at the fluxes typical of the *Einstein* Medium Sensitivity Survey, i.e. about two orders of magnitude below the typical HEAO-1 fluxes.

3. If discrete sources provide the rest of the background, their surface density has to be ≥ 5000 objects/$sq.deg.$, with typical fluxes of about 10^{-16} erg cm^{-2} s^{-1}, and their spectra have to be substantially different from those detected so far (proto-AGNs with partially comptonized spectra?).

4. Emission from a diffuse gas is still a viable possibility only if the contribution from known sources (with typical Seyfert-like spectra) at 2-10 keV is $\leq 20\%$. This implies a very sharp flattening of the X-ray log N - log S immediately below the *Einstein* Deep Survey limit, which may be soon tested by ROSAT surveys. The main difficulty for this model is the large required Ω_B.

5. Clumping of the gas would alleviate the problem with Ω_B, but the existing constraints from fluctuations on the temperature of the microwave background require clumping on sizes of the order of 20 kpc, bringing us back to discrete sources!

References

Barcons, X., and Fabian, A.C. 1988, M.N.R.A.S., **230**, 189.

Boldt, E., and Leiter, D. 1984, Ap.J., **276**, 427.

Boldt, E., and Leiter, D. 1987, Ap.J.(Letters), **322**, L1.

De Zotti, G., Boldt, E., Cavaliere, A., Danese, L., Franceschini, A., Marshall, F.E., Swank, H.J., and Szymkowiak, A.E. 1982, Ap.J., **253**, 47.

Fabbiano, G., and Trinchieri, G. 1985, Ap.J., **296**, 430.

Fabbiano, G., and Trinchieri, G. 1987, Ap.J., **315**, 46.

Fabian, A.C., Done, C., and Ghisellini, G. 1988, preprint.

Field, G.B., and Perrenod, S.C. 1977, Ap.J., **215**, 717.

Forman, W., and Jones, C. 1982, Ann. Rev. Astr. Ap., **20**, 547.

Forman, W., Jones, C., and Tucker, W. 1985, Ap.J., **293**, 102.

Giacconi, R., Gursky, H., Paolini, F.R., and Rossi, B.B. 1962, Phys. Rev. Letters, **9**, 439.

Giacconi, R., *et al.* 1979, Ap.J.(Letters), **234**, L1.

Giacconi, R., and Zamorani, G. 1988, Ap.J., **313**, 20.

Gioia, I.M., Maccacaro, T., Schild, R.E., Stocke, J.T., Liebert, J.W., Danziger, I.J., Kunth, D., and Lub, J. 1984, Ap.J., **283**, 495.

Giommi, P., Tagliaferri, G., and Angelini, L. 1988, in *X-Ray Astronomy with EX-OSAT*, ed. N.E. White and R. Pallavicini, Mem.S.A.It., no. 58, in press.

Griffiths, R.E., Tuohy, I.R., Brissenden, R.J.V., Ward, M., Murray, S.S., and Burg, R. 1988, P.A.S.P., in press.

Guilbert, P.W., and Fabian, A.C. 1986, M.N.R.A.S., **220**, 439.

Hamilton, T.T., and Helfand, D.J. 1987, Ap.J., **318**, 93.

Koo, D.C., Kron, R.G., and Cudworth, K.M. 1986, P.A.S.P., **98**, 285.

Leiter, D., and Boldt, E. 1982, Ap.J., **260**, 1.

Maccacaro, T., Gioia, I.M., Wolter, A., Zamorani, G., and Stocke, J.T. 1988, Ap.J, **326**, 680.

Marano, B., Zamorani, G., and Zitelli, V. 1988, M.N.R.A.S., **232**, 111.

Marshall, F.E., Boldt, E.A., Holt, S.S., Miller, R.B., Mushotzky, R.F., Rose, L.A., Rothschild, R.E., and Serlemitsos, P.J. 1980, Ap.J., **235**, 4.

Mushotzky, R.F. 1984, *Advances in Space Research*, Vol. **3**, no. 10-12, *High Energy Astrophysics and Cosmology*, ed. G.F.Bignami and R.A. Sunyaev (New York:Pergamon Press), p. 157.

Schwartz, D.A., and Tucker, W.H. 1988, Ap.J., in press.

Shafer, R.A. 1983, Ph.D. dissertation, University of Maryland, NASA Technical Memorandum 85029.

Setti, G., and Woltjer, L. 1982, in *Astrophysical Cosmology, Proc. Vatican Study Week on Cosmology and Fundamental Physics*, ed. H.A.Bruck, G.V.Coyne, and M.S.Longair (Vatican City: Pontificia Academia Scientiarum), p. 315.

Setti, G. 1985, in *Nonthermal and Very High Temperature Phenomena in X-Ray Astronomy*, ed. G.C.Perola and M.Salvati (Rome: Istituto Astronomico, Universitá 'La Sapienza'), p. 159.

Trinchieri, G., and Fabbiano, G. 1985, Ap.J., **296**, 447.

Tyson, J.A. 1988, in *Towards Understanding Galaxies at High Redshift*, ed. R.Kron and A.Renzini (Dordrecht: Reidel), in press.

Wilkes, B.J., and Elvis, M. 1987, Ap.J., **323**, 243.

Zdziarski, A.A., 1988, M.N.R.A.S., in press.

THE X-RAY LOG N - LOG S RELATION

Elihu Boldt
Laboratory for High Energy Astrophysics
NASA / Goddard Space Flight Center
Greenbelt, Maryland 20771
USA

ABSTRACT.   Results from various surveys are reviewed as regards X-ray source counts at high galactic latitudes and the luminosity functions determined for extragalactic sources. Constraints on the associated log N-log S relation provided by the extragalactic X-ray background are emphasized in terms of its spatial fluctuations and spectrum as well as absolute flux level. The large number of sources required for this background suggests that there is not a sharp boundary in the redshift distribution of visible matter.

1. INTRODUCTION

As the sky in the microwave band is dominated by cosmic radiation, so too is the X-ray sky (>3keV) mainly indicative of an isotropic extragalactic background. Relative to a possible preferred axis, the large-scale asymmetry of this XRB (X-ray background) is $\leq 1\%$ (Shafer 1983).   If most of the XRB arises from sources at the highest redshifts, then this limit on anisotropy provides an upper bound of about $0.1(c/H_0)$ for our offset distance from the center of the possible "baryonic island" discussed by Novikov (1988). However, the portion of this XRB (>3keV) accounted for by actually resolved (surveyed) sources is still less than 20% (Boldt 1987). This review gives the status of our current knowledge of sources already resolved and describes constraints on those yet to be directly resolved.

2. EXPERIMENTS

The most definitive information presently available on X-ray source populations comes from surveys carried out with the HEAO program (i.e., the all-sky HEAO-1 mission and the HEAO-2 Einstein Observatory) and EXOSAT. Table 1 summarizes the principal characteristics of these missions that have already been completed as well as future experiments that should be particularly relevant in this regard. They are classified here as "large sky" and "deep sky" in order to exhibit the distinction between angular coverage and faint source sensitivity, respectively.   The large-sky/low-sensitivity survey (LSS) at high galactic latitudes performed with the HEAO-1 A2 experiment (Piccinotti et al. 1982) is complete and optically identified down to the level of about a milliCrab; the HEAO-1 A4 modulation collimator (MC) experiment enabled several of these

291

M. Mezzetti et al. (eds.), Large Scale Structure and Motions in the Universe, 291–302.
© 1989 by Kluwer Academic Publishers.

Table 1

PRINCIPAL SURVEYS

EXISTING   PLANNED

Large Sky

HEAO-1   ROSAT
$E = 0.1keV \rightarrow 1MeV$ $E < 2keV$
LSS $(2\rightarrow 30keV)$: $S > 5 \times 10^2 \, S_o$ $S > 4S_o$
PC/MC: 1'   IPC: 1'
Map XRB $(\sim 9 \, deg^2)$ Map XRB (galactic)

Deep Sky

HEAO-2 "Einstein"   AXAF
HSS: 1"   CCD: 1"
$E < 3.5keV$ $E < 10keV$
$S > S_o$ $S > 2 \times 10^{-2} \, S_o$

MSS: 1' (IPC)   ASTRO-D: 1'
$E = 0.5keV \rightarrow 3.5keV$ $E < 10keV$
$S > 4 \, S_o$ $S > S_o$ (spectra)
Map XRB (1')   Map XRB (1')

EXOSAT: 10" (CMA)   MMX: <1'
$E = 0.05keV \rightarrow 1.5keV$ $E < 10keV$
$S > 4 \, S_o$

$S_o \equiv 2.6 \times 10^{-14}$ ergs $cm^{-2} \, s^{-1}$ (1-3keV)

identifications. The LSS overlaps in spectral coverage (2-60keV) with observations of the extragalactic X-ray background (XRB) obtained with the same detectors (Boldt 1987). Observed XRB surface brightness fluctuations have been used to set contraints on the log N - log S relation for sources an order of magnitude fainter than those resolved in the LSS (Shafer 1983). The all-sky survey to be provided by ROSAT covers X-ray energies below 2keV; this will be particularly useful for investigating the population of low temperature clusters previously inaccessible to large-sky surveys and in the identification of AGN (active galactic nuclei) with soft X-ray spectral components. However, XRB mapping with ROSAT is likely to be dominated by galactic effects.

Deep sky surveys over limited angular regions of the sky with imaging detectors at the focus of grazing incidence X-ray telescopes have resolved sources substantially fainter than a milliCrab. In particular, the high sensitivity survey (HSS) with HEAO-2 (Giacconi et al. 1979; Griffiths et al. 1983) has a source detection threshold S_0(1 - 3keV) which, when extrapolated to the band 3 - 10keV, corresponds to about 2 microCrabs. The medium sensitivity survey (MSS) with HEAO-2 (Gioia et al, 1984) and the high latitude survey with EXOSAT (Giommi and Tagliaferri 1987) have used serendipitous source detections to obtain relatively large samples of objects at a somewhat reduced senstivity from that of the HSS. The effective response of the EXOSAT CMA (channel multiplier array) emphasizes lower energies than HEAO-2 and is only sensitive to "point-like" (<10") sources (e.g., not clusters). The HEAO-2 imaging proportional counter (IPC) is sensitive to surface brightness and has been used to set limits on XRB fluctuations over arc-minute pixels devoid of resolved sources (Hamilton and Helfand 1987), albeit at energies (<3keV) where our knowledge of the extragalactic XRB is relatively uncertain. The extended energy response and "fast optics" anticipated for ASTRO-D should be particularly powerful in extending such analysis of arc-minute XRB fluctuations into the crucial 3-10keV band. A significant advance in directly resolving faint sources up to 10keV, however, must await the AXAF mission, where the sensitivity is expected to be about 50 times better than that of the HSS (Giacconi and Zamorani 1987).

3. SPECTRAL CONSIDERATIONS

Relating source counts obtained in various different spectral bands involves using a standard underlying source spectrum. Apart from BL Lac type objects, AGN tend to exhibit power-law spectra characterized by an energy spectral index $\alpha \approx 0.7$ (Mushotzky 1982 ; Petre et al. 1984 ; Pounds and Turner 1988); for the brightest of these observed with the HEAO-1 A4 experiment this power-law extends to ≥ 100keV (Rothschild et al. 1983). However, it is becoming apparent from EXOSAT data (Pounds and Turner 1988) that many AGN exhibit significant soft spectral components over and above the canonical power-law. For example, although the 0.2-50keV spectrum of Mrk-509 obtained with HEAO-1 A2 is very well described by the canonical AGN power-law down to 1keV, a soft excess emerges at lower energies which (at 0.2keV) becomes an order of magnitude greater than the extrapolation of the power-law fit obtained at higher energies (Singh, Garmire and Nousek 1985). Another spectral complication is that relatively low luminosity AGN tend to exhibit pronounced absorption effects below a few keV (Lawrence and Elvis 1982 ; Danese et al.1986 ; Reichert et al 1985). For example, the beautiful canonical power-law observed with HEAO-1 for Cen A over the remarkably broad band 5keV $\rightarrow$ 1MeV, when extrapolated down to lower energies, is found to reside well above the observed spectrum at ≤ 3keV by a factor ≥ 3 (Baity et al.

1981). This sort of effect is obviously very important for the interpretation of surveys carried out with the HEAO-2 Einstein Observatory and EXOSAT as well as with future soft X-ray survey missions (e.g., ROSAT). For such missions even the less evident effects of absorption due to our galaxy must also be considered (Zamorani et al. 1988).

The x-ray spectrum from clusters of galaxies is not affected by self-absorption. Furthermore, the average spectrum for Abell clusters observed with the HEAO-1 A2 experiment (>2keV) is adequately described by that of an optically thin isothermal plasma characterized by $kT \approx 7keV$ (Stottlemyer and Boldt 1984) and thereby serves as a valid standard. The X-rays arising from the cooling cores of centrally condensed clusters would appear at lower energies as spatially spiked soft emission (Mushotzky 1984); this could complicate the interpretation of soft X-ray surveys.

In general, clusters characterized by kT<2keV would tend to be undersampled in the LSS of HEAO-1 A2 relative to softer X-ray surveys. To evaluate a limit on the magnitude of this effect we note that luminosity is well correlated with temperature (Mushotzky 1987) and that the luminosity function flattens below about 5×10^{43} ergs s^{-1} (Kowalski et al. 1984). In the Euclidean approximation suitable to the LSS

$$N(>S) \propto S^{-1.5} \int L^{1.5} (dn/dL)\, dL \qquad (1)$$

where the luminosity function $dn/dL \propto L^{-2.2}$ (Mushotzky 1987) for $L = 5 \times 10^{43} \to 2 \times 10^{45}$ ergs s^{-1} and flattens at lower L. Using the aforementioned correlation between L and T we infer that $L > 5 \times 10^{43}$ ergs s^{-1} for kT >2keV . Evaluating the integral in equation (1) for this lower-limit to L and comparing it with the maximum total integral (down to L=0) we obtain

$$N(S;\ kT>2keV) > 0.7\ N(S;\ all\ T) \qquad \text{for clusters.} \qquad (2)$$

Hence, the underestimation in cluster counts for the LSS due to a spectral bias against X-ray sources with kT <2keV is not expected to be extremely significant.

Finally, we note that there is a rather severe spectral bias associated with the detection of stars. In particular, for most quiet stellar coronae kT<1keV. Hence, surveys at energies ≥1keV are expected to undersample stellar X-ray sources.

4. SOURCE COUNTS

The count of sources detected at high galactic latitudes with HEAO-1, HEAO-2 and EXOSAT are exhibited in Table 2, classified according to optical identifications. The sources listed under HEAO-1 are those detected at a level of statistical significance >5σ in the LSS (Piccinotti et al. 1982; Shafer 1983); bright sources associated with our galaxy (i.e., Sco X-1, Her X-1) and "local" extragalactic objects (associated with the LMC, SMC and M31) have been excluded. Sources detected over a comparable region of the sky with the HEA0-1 A1 experiment provide a larger flux-limited sample (337 objects), one that is currently being investigated with MC data (from HEAO-1 A4) for obtaining the refined positions needed for making identifications (Bradt et al. 1988). The survey of a restricted region (314 deg^2) of maximum HEAO-1 exposure, associated with the north ecliptic pole, has yielded a sample of 21 sources of which 14 have proposed extragalactic identifications (Shrader, Wood and Matilsky 1986) .

Table 2

Results of "Complete" Surveys

| | HEAO-1 | HEAO-2 | | EXOSAT | HEAO-2 |
	LSS	MSS	EMSS		HSS
Coverage (deg.^{2})	27×10^3	90	780	570	2.3
Sources		112	836	143	27
Clusters	30	18	(74) <z>=.18	0	0
AGN	29	57	(246) <z>=.40	(25)	8→19 <z>=.58
BL Lac	4	4	(22)	(8)	
Galaxies	1	5	(14)	(1)	2→13
Stars	11	28	(200)	(83)	6

Sources detected (at the 5σ level) in the HEAO-2 MSS have been completely identified. With substantially increased angular coverage the Extended Medium Sensitivity Survey (EMSS) is an on-going project that has yielded an order of magnitude more sources (detected at the 4σ level) 67% of which have already been identified (Gioia et al. 1987a,b). The results from the EMSS confirm those from the MSS. In particular, the sample of AGN is about three times that of clusters; this is in sharp contrast to the LSS where they are comparable. The results listed in Table 2 for the HEAO-2 HSS correspond to recently extended angular coverage involving detections at 4.5σ (Primini 1988). These results are consistent with those of the MSS as regards the population of stars relative to the total sample ($\approx 1/4$) and the absolute number of sources (i.e., an extrapolation of the MSS predicts 21 HSS objects, only 2 of which are expected to be clusters). Of the 27 HSS sources, 16 are clearly identified; the 11 remaining objects are likely to be extragalactic (AGN and galaxies).

The results of the EXOSAT high latitude survey (Giommi and Tagliaferri 1987) presented in Table 2 correspond to an update of an on-going analysis effort (Giommi 1988). Since the EXOSAT CMA is relatively insensitive to extended sources (>10") there are no clusters in this sample, as expected. It is interesting to note, however, that the sensitivity for detection of stars is remarkably high. Of the sources already identified (82% of the total sample) 71% are stars. This is to be compared with the much smaller relative population of stellar sources in the HEAO-2 surveys (see Table 2).

5. LOG N - LOG S RESULTS

The number of sources (N) as a function of flux (S) is usually expressed in the following form:

$$dN/dS = K\ S^{-\Gamma}. \tag{3}$$

This is a particularly useful form for regimes where K is essentially constant. In any event, however, we can define an apparent effective Γ as

$$\Gamma \equiv -d[\log(dN/dS)]\ /d[\log(S)]. \tag{4}$$

For a homogeneous distribution of sources in Euclidean space K is constant and $\Gamma = 2.5$. In fact, assuming $\Gamma = 2.5$ for the sources of the LSS gives an acceptable fit to the data in determining a best value for K (Piccinotti et al. 1982). However, simultaneous fitting of both K and Γ yields $\Gamma = 2.72(-0.10, +.15)$, where the errors are 1σ. The MSS (Gioia et al. 1984; Maccacaro et al. 1984) yields $\Gamma = 2.45(-.12, +.12)$ for the total sample. However, this apparent agreement with the Euclidean approximation appears to be a coincidence resulting from compensating effects since the individual source populations are characterized by different values for Γ. For AGN the best fit is $\Gamma = 2.71(-.15, +.15)$ while for clusters $\Gamma = 2.04(+.23, -.23)$. That the cluster fraction for the MSS is smaller than for the LSS (see Table 2) is a reflection of this difference in Γ.

The effects of relativistic geometry on equation (3) are quite pronounced, even at the modest redshifts characterizing the source populations of the MSS/EMSS of HEAO-2 (see Table 2). Taking the ideal case of a constant comoving density of like luminosity sources and assigning a fixed $\Gamma=2.5$ in equation (3) leads to a variable value of K that depends strongly on S; for S corresponding to $z\approx0.3$, K is only about 40% of its value for the brightest ($z<<0.3$)) sources. This effect manifests itself as a flattening of the LogN-logS relation at decreasing values for S. Considering the definition given by equation (4) this flattening may be expressed as a decrease in the apparent value of Γ which, for S corresponding to z >0.3, amounts to $\Delta\Gamma>0.5$ (Schwartz 1978). In light of this, $\Gamma\geq2.5$ obtained in the MSS for AGN implies strong evolution with redshift for this source population, whereas $\Gamma<2.5$ obtained for clusters implies relatively weak evolution or possibly even "anti" evolution (Cavaliere and Colafrancesco 1987).

The logN-logS results obtained from the LSS and MSS have been expressed in terms of S(0.3-3.5keV) and S(2-10keV), respectively. A comparison of these results for extragalactic sources requires a knowledge of the correct underlying spectrum for these objects over the composite band 0.3-10keV. For the canonical AGN power-law spectrum we note that

$$S(0.3\text{-}3.5\text{keV}) \approx S(2\text{-}10\text{keV}) \qquad \text{for } \alpha=0.7. \qquad (5)$$

Comparing values of K determined for Euclidean fits (i.e., $\Gamma=2.5$) to the MSS and LSS data we obtain

$$K_{LSS} / K_{MSS} \approx 3 \qquad (6)$$

where K_{LSS} refers to the value determined with the LSS over the range $S_{LSS}=(2.8\rightarrow16)\times10^{-11}\text{ergs}/(\text{cm}^2\text{s})$ and K_{MSS} comes from the MSS for $S_{MSS}\approx 10^{-13}\rightarrow10^{-11}\text{ergs}/(\text{cm}^2\text{s})$. Using the MSS logN-logS result for AGN alone (corresponding to $\Gamma=2.71$) to predict the number to be expected in the LSS we obtain N=8 as compared with the 29 actually observed (see Table 2).

To investigate the origin of the pronounced discrepancy between the LSS and MSS results it is instructive to examine surface brightness fluctuations in the unresolved X-ray background (XRB) observed with the HEAO-1 A2 experiment in the course of carrying out the LSS over the same spectral band (Shafer 1983). Referred to 9 degree^2 pixels, these fluctuations have an rms value $\approx6\%$ of the extragalactic XRB. The greatest constraint on dN/dS placed by these fluctuations is on sources an order of magnitude in flux below those resolved in the LSS; as such, this sort of analysis addresses the sensitivity regime of the MSS, albeit referred to the spectral band of the LSS.

Assuming the Euclidean approximation (i.e., constant K, $\Gamma=2.5$) Shafer (1983) finds an acceptable fit for the fluctuations with

$$K(\text{fluctuations}) /(K_{LSS})^* =1.1\rightarrow1.4 \qquad \text{at 90\% confidence} \qquad (7)$$

where $(K_{LSS})^*$ refers to the LSS with stars included (see Table 2) and the limits are on K(fluctuations) at $S(2-10keV) \approx 3 \times 10^{-12}$ ergs cm^{-2} s^{-1} (i.e., the sensitivity regime of the MSS). Considering the confidence limits on $(K_{LSS})^*$ as well, the ratio expressed by equation (7) is consistent with unity (Shafer 1983). Equations (6) and (7) imply that the MSS determination of K is an underestimate reflecting some systematic bias (e.g., such as could arise from a significant population of sources exhibiting highly absorbed spectra).

To evaluate the case for unabsorbed spectra we compare the LSS and MSS as regards clusters of galaxies. For the optically thin thermal bremsstrahlung spectrum characteristic of clusters (with kT=6keV) we note

$$S_{LSS} > 3.1 \times 10^{-11} \text{ ergs(cm}^2\text{s)}^{-1} \qquad \text{for kT=6keV} \qquad (8)$$

$$S(0.3-3.5keV) \approx 1.4 \times S(2-10keV) \qquad \text{for kT=6keV.} \qquad (9)$$

Using equations (8) and (9) for extrapolating the MSS result for clusters (corresponding to Γ=2.04) to predict the number to be expected in the LSS we obtain N=34 as compared with the 30 actually observed. Combining all LSS and EMSS data on clusters indicates that N(>S) is probably continuous over a range in S that spans the two surveys (Gioia et al. 1987c). These aspects of agreement suggest that sources with unabsorbed spectra are not subject to the same bias that appears to be involved in the MSS sample of AGN. Even though the logN-logS relation for clusters appears to be continuous between the MSS and LSS, however, these two studies are inconsistent as regards luminosity function (Schmidt and Green 1986); this could be due to evolution (Cavaliere and Colafrancesco 1987) and/or a systematic observational bias (e.g., arising from effects of surface brightness and angular size as well as spectral coverage).

6. AGN LUMINOSITY FUNCTION

For the AGN observed in the HEAO-1 LSS the luminosity function over the range $L=4 \times 10^{42} \rightarrow 10^{45}$ergs/s is well described by a power-law, viz:

$$dn/dL \propto L^{-\gamma} \qquad (10)$$

where $\gamma = 2.75 \pm 0.15$ (Piccinotti et al. 1982). However, we already know that the luminosity function must eventually flatten since an extension of the power law somewhat below 4×10^{41}ergs/s would infer that the composite flux from AGN exceeds the XRB. On the other hand, if the luminosity function power-law flattens below about 4×10^{42}ergs/s then the AGN contribution would amount to about a third of the XRB at 3keV (Rothschild et al. 1983).

To examine the spectral consequences of subtracting various estimated amounts of AGN foreground from the XRB measured with HEAO-1, we have fit the residual energy spectrum (3-60keV) with the functional form

$$I_E \equiv dI/dE \propto E^{-\alpha} \exp(-E/B) \qquad (11)$$

where α and B are parameters determined from the spectral fit and E is photon energy. In doing so we assume that AGN have canonical spectra (i.e., α=0.7, B>60keV). For zero foreground, we recover α=0.29 and B=40keV (Boldt 1987) corresponding to the XRB thermal bremmstrahlung spectrum characterized by kT=40keV (Marshall et al. 1980). We note that for an AGN contribution exceeding about 30% (at 3keV) α<0.2 and B<30keV; this would imply that the candidate "thermal" sources of the residual XRB have kT>200keV in their proper frame and that they are located at z>6 (Boldt 1987), well beyond the highest redshift quasars. If the AGN foreground exceeds 50% (Giacconi and Zamorani 1987) then the residual XRB spectrum would exhibit an observable peak at E>3keV. Since the contribution of AGN to the XRB is proportional to $(L_{min})^{-0.75}$ [where L_{min} is the lower limit luminosity of the regime for the luminosity function characterized by γ =2.75] this puzzling case is avoided for L_{min} >10^{42}erg/s. Therefore, we conclude that HEAO-1 LSS and XRB data, when considered together, allow us to bracket the acceptable range for L_{min}, as follows:

$$4x10^{42} > L_{min} > 10^{42} ergs/s. \tag{12}$$

A recent study by Persic et al.(1988) has provided us with direct information about the AGN luminosity function for $L(2-10keV)<10^{42}$ergs/s as well as higher luminosities. In contrast to the luminosity function determination based on the X-ray selected HEAO-1 LSS, they have used an optically selected AGN sample in examining HEAO-1 data for obtaining associated X-ray luminosities, including upper limits. Although in agreement with the results of Piccinotti et al. (1982) in the regime of luminosity overlap, the analysis of Persic et al.(1988) demonstrates that the luminosity function is remarkably flat at $L \leq 2x10^{42}$ergs/s, corresponding to $|\Delta\gamma|$ >1 relative to the higher luminosity portion.

7. SOURCES OF THE RESIDUAL BACKGROUND

Under the assumption that the residual XRB (i.e., that unresolved with HEAO-2) is not diffuse and that it ultimately arises from discrete sources, severe integral constraints on the responsible population have been obtained by evaluating upper limits to the apparent surface brightness fluctuations of the XRB observed on the scale of arc-minute pixels. In this way Hamilton and Helfand (1987) have used HEAO-2 IPC data to conclude that the number of point sources needed would have to exceed $5x10^3$degree^{-2}, much larger than the estimated total number of quasars (Schmidt and Green 1986). Extending the arc-minute fluctuation analysis to the known XRB spectral regime (>3keV) is clearly needed and should be carried out with ASTRO-D. Directly resolving the individual sources of the residual XRB is a major challenge for future experiments such as AXAF. Using the source confusion criterion of one resolved source per 40 beams (Murdoch, Crawford and Jauncey 1973) we conclude that the required spatial resolution must be better than 8 arc seconds. The average flux from these sources is obtained by considering the residual XRB spectrum (Boldt 1987), yielding

$$<S(3-10keV)> = 2.6x10^{-15}(dN/d\Omega)^{-1} ergs (cm^2 s)^{-1} \tag{13}$$

where $dN/d\Omega$ is the number of objects per square arc minute. Using the lower limit $dN/d\Omega > 1.4$ sources per square arc minute in the evaluation of equation (13) yields that $<S>$ is less than 1.9×10^{-15} ergs cm^{-2} s^{-1} (3-10keV). With an effective energy spectral index $\alpha \leq 0.3$ over the band 1 - 10keV (Worrall and Marshall 1984) this implies that $<S>$ corresponds to less than $0.03 S_0$ (where $S_0 = 2.6 \times 10^{-14}$ ergs cm^{-2} s^{-1} is the threshold for the HSS in the 1-3keV band).

Using a recent estimate of the present-epoch density of AGN (Persic et al. 1988) as the invariant comoving density at earlier epochs as well, we find that $N > 5 \times 10^3$ degree^{-2} (as required for the residual XRB) provided that the integration extends to lookback times greater than 80% of the Hubble time (i.e., to z >2). However, canonical AGN do not exhibit the requisite X-ray spectra for the residual XRB. On the other hand, if the residual XRB is due to spectrally suitable precursor AGN at z >3 (with the same comoving density as local AGN) then these X-ray sources would be sufficiently numerous only for the case where they exist over a lookback time interval that is not less than about a tenth of the Hubble time (i.e., corresponding to $\Delta z > 3$ for $q_0 = 1/2$). As such, this scenario (Boldt and Leiter 1987) would preclude the need for invoking a "baryonic island" (Novikov 1988) characterized by a sharp redshift boundary for visible matter.

8. ACKNOWLEDGEMENTS

It is a pleasure for me to express my thanks to Isabella Gioia, Paolo Giommi, Richard Mushotzky and Francis Primini for valuable information about observed sources and to James Peebles for pointing out the relevance of the XRB isotropy to the question of a possible baryonic island.

REFERENCES

Baity, W. et al, Ap J. **244**, 429 (1981)
Boldt, E., Physics Reports **146**, 215 (1987)
Boldt, E. and Leiter, D., Ap. J. 322, L1 (1987)
Bradt, H. et al., ISAS Symposium, Tokyo (1988)
Cavaliere, A. and Colafrancesco, S., in NATO Advanced School *Hot Thin Plasmas in Astrophysics* (ed., R. Pallavicini) Cargese, France (1987)
Danese, L. et al., Astron. Astrophys. **161**, 1 (1986)
Giacconi, R., AS&E Report **1567** (1966)
Giacconi, R., et al., Ap. J. **234**, L1 (1979)
Giacconi, R. and Zamorani, G. Ap. J. **313**, 20 (1987)
Gioia, I., et al., Ap. J. **283**, 495 (1984)
Gioia, I., et al., in Observational Cosmology (A. Hewitt et al.,eds) p593 IAU (1987a)
Gioia, I., et al., in Greenbank Workshop on Large Scale Surveys of the Sky, Sept. (1987b)
Gioia, I., et al., CfA #2529 (1987c)
Giommi, P. and Tagliaferri, G., in Observational Cosmology (A. Hewitt et al., eds) p601 IAU (1987)
Giommi, P. personal communication (1988)

Griffiths, R., et al., Ap. J. **269**, 375 (1983)

Hamilton, T. and Helfand, D., Ap. J. **318**, 93 (1987)

Kowalski, M. et al., Ap. J. Suppl. **56**, 401 (1984)

Lawrence, A. and Elvis, M., Ap. J. **256**, 410 (1982)

Maccacaro, T. et al., Ap. J. **284**, L23 (1984)

Marshall, F. et al., Ap. J. **235**, 4 (1980)

Murdoch, H., Crawford, D. and Jauncey, D. Ap. J. **183**, 1 (1973)

Mushotzky, R., Ap. J. **256,** 92 (1982)

Mushotzky, R., Physica Scripta **T7**, 157 (1984)

Mushotzky, R., in NATO Advanced School *Hot Thin Plasmas in Astrophysics* (ed.,
    R. Pallavicini) Cargese, France (1987)

Novikov, I., this symposium (1988)

Persic, M. et al., Ap. J., submitted (1988)

Petre, R. et al., Ap. J. **280**, 499 (1984)

Piccinotti, G. et al., Ap. J. **253**, 485 (1982)

Pounds, K. and Turner, T., ISAS Symposium, Tokyo (1988)

Primini, F., personal communication (1988)

Reichert, G. et al. Ap. J. **296**, 69 (1985)

Rothschild, R. et al. Ap. J. **269**, 423 (1983)

Schmidt, M. and Green, R., Ap. J. **305**, 68 (1986)

Schwartz, D., in Symp. on Non-solar X-ray Astronomy (L. Peterson and W. Baity,eds)
    p453 COSPAR (1978)

Shafer, R., PhD dissertation, University of Maryland, NASA TM **85029** (1983)

Shrader, C., Wood, K. and Matilsky T., Ap. J. Suppl. **61**, 353 (1986)

Singh, K., Garmire, G. and Nousek, J., Ap. J. **293**, 633 (1985)

Stottlemyer, A. and Boldt, E., Ap. J. **279**, 511 (1984)

Worrall, D. and Marshall, F., Ap. J. **276**, 434 (1984)

Zamorani, G. et al., Astron. Astrophys. (submitted) CfA #**2555** (1988)

DISCUSSION

DE ZOTTI: Danese et al. (1986) found that photoelectric absorption is unlikely to be the only reason for the apparent discrepancy between A2 and MSS counts. In addition to that there may be a genuine difference in the absolute calibration. The fact that there is no discrepancy in the case of clusters of galaxies may be somewhat fortuitous: the A2 experiment misses a good fraction of richness 1 clusters and essentially all richness 0 clusters, which, on the contrary, may be detected by the Einstein observatory.

BOLDT: Photoelectric absorption may not be the only complication in comparing HEAO-1 and HEAO-2 results but is a major factor for obtaining agreement.

RICHTER: Is the apparently smaller amplitude of the XRB dipole anisotropy consistent with the CMB anisotropy in terms of the standard kinematical interpretations?

BOLDT: They both give about the same value for our velocity relative to the proper form of the background.

ELLIS: What are the limits on the X-ray quadrupole anisotropy?

BOLDT: The dipole dominates, but is itself of limited statistical significance. Variations in the galactic background with longitude might well mask additional extragalactic anisotropies.

X–RAY CONSTRAINTS ON LARGE SCALE CLUSTERING

G. DE ZOTTI[1], M. PERSIC[2,3], E.A. BOLDT[3], F.E. MARSHALL[3],
L. DANESE[4], A. FRANCESCHINI[4], and G.G.C. PALUMBO[5,6]

[1] *Osservatorio Astronomico, Padova, Italy*
[2] *Osservatorio Astronomico, Trieste, Italy*
[3] *Laboratory for High Energy Astrophysics, NASA/GSFC, Greenbelt*
[4] *Dipartimento di Astronomia, Padova, Italy*
[5] *Dipartimento di Astronomia, Bologna, Italy*
[6] *I.Te.S.R.E./CNR, Bologna, Italy*

ABSTRACT. We have analyzed the HEAO–1 A2 all-sky survey data to investigate the autocorrelation function of the surface brightness fluctuations of the extra-galactic 2–10 keV X-ray background on angular scales ranging from 3° to 27°. No significant signal has been detected on any scale. The derived upper limits set important bounds on correlated emission on scales ≥ 10 Mpc. X–ray data are found to be compatible with optical estimates of the two–point spatial correlation function of rich clusters of galaxies and with recent predictions for their cosmological evolution. Constraints on clustering of Active Galactic Nuclei (AGNs) and on its dependence on cosmic time are analyzed in the light of results of recent optical studies. We also discuss bounds on very large scale structures, on correlations between AGNs and clusters and between clumps of hot gas.

1. Introduction

It has long been pointed out that anisotropies of the X-ray background (hereafter XRB) can provide an interesting probe of the lumpiness in the universe over a large range of scales, not easily accessible with other techniques. A large scale distribution pattern involving X–ray sources would produce intensity variations in excess of the Poisson fluctuations due to the finite number of sources in the beam (Fabian 1972, 1981; Shafer 1983; Shafer and Fabian 1983; Meszaros and Meszaros 1988; Barcons and Fabian 1988a), and, of course, would translate into an autocorrelation of the XRB fluctuations (Dautcourt 1977; Peebles 1980; Danese and De Zotti 1986).

The A2 experiment on HEAO–1 (Rothschild *et al.* 1979) had several unique features for determining the surface brightness distribution of the X–ray sky: very

M. Mezzetti et al. (eds.), Large Scale Structure and Motions in the Universe, 303–316.
© 1989 by Kluwer Academic Publishers.

stable performance of the detectors, good statistics, continuous monitoring of the relatively low and very stable internal background, and a large effective detector area. A study of the XRB autocorrelation function, based on these data, allows to derive constraints on correlated emission on scales exceeding ~ 10 Mpc, due either to clustering of discrete sources or to large clumps of hot plasma (impulsively heated and then adiabatically cooled, see Guilbert and Fabian 1986, or more continuously heated, see Ostriker, Thompson and Witten 1986).

The spatial resolution of the A2 experiment allow us to deal with angular scales $\geq 3°$. An analysis of the correlation function of the 1–3 keV X-ray background on scales from 1' to 15' has been performed by Barcons and Fabian (1988b), from 5 EINSTEIN IPC fields.

A Hubble constant $H_o = 50$ km $s^{-1} Mpc^{-1}$ is used throughout this paper.

2. Autocorrelation Analysis of Intensity Fluctuations

Each scanning circle of the HEAO–1 A2 experiment is at approximately fixed ecliptic longitude λ, and the detector's field of view has FWZI of 3° and 6° along the scan direction and perpendicular to it, respectively. We have selected 1800 non overlapping fields in the ecliptic–latitude range $-60° \leq \beta \leq +60°$, according to $\Delta\beta = 3°$, $\Delta\lambda = 6°$. This criterion ensures that no long-term variation in the detector's internal-background level affects appreciably the computation of the correlation function.

Of the total number of fields, 940 were at high galactic latitude ($|b| \geq 20°$) and were not contaminated by (i.e., were $\geq 6°$ away from) known X–ray sources (Piccinotti et al. 1982). The fluctuation intensity, δI at each field's central position was evaluated by fitting the signal (both positive and negative values allowed) at the target position with a point source plus a constant background upon which as many as required off-axis point sources (with intensity and position as free parameters) were superposed in order to bring the reduced χ^2 as close as possible to unity. The mean offset from zero, $\langle \delta I \rangle = 0.048 \pm 0.009$ $R15$ $counts$ s^{-1} (for a definition of $R15$ $counts$ s^{-1}, see Marshall et al. 1979), was subtracted out from the individual values of δI.

The products $(\delta I \cdot \delta I')_\theta$ were subsequently computed for angular separations of 3°, 6°, ..., 27° between the centers of each field pair. Since the maximum resolution attainable along the scan direction is 3°, any other angular separation would result in a loss of either data or statistical independence.

The sample averages $C(\theta) = \langle (\delta I \cdot \delta I')_\theta \rangle$ are a measure of the angular correlation function of the XRB on the selected angular scales. In Table I we list the values of $C(\theta)$, obtained weighting the individual values with their statistical errors, and the associated uncertainties $[\sigma = (sample\ variance/N)^{1/2}]$.

No significant signal (either positive or negative) has been detected on any an-

Table I. Results of the autocorrelation analysis

θ	$10^3 \times C(\theta)$	N [a]
$3°$	$+1.1 \pm 2.8$	734
$6°$	-5.2 ± 3.2	563
$9°$	$+6.4 \pm 3.5$	433
$12°$	$+0.98 \pm 3.9$	324
$15°$	-0.01 ± 4.5	233
$18°$	-0.2 ± 5.2	156
$21°$	-0.98 ± 8.1	97
$24°$	$+0.49 \pm 9.2$	55
$27°$	$-5.1 \pm 11.$	21

[a] Number of pairs of fields.

gular scale. The the formal 1σ upper limits to

$$\Gamma(\theta) = [C(\theta)]^{1/2} / \langle I \rangle \tag{1}$$

range from $\simeq 1.4 \times 10^{-2}$ for $\theta = 3°$ to $2 \div 3 \times 10^{-2}$ for larger angular scales. Further details can be found in Persic $et\ al.$ (1988).

Note that the quoted limits must be regarded as conservative, since several "local" effects can broaden the distribution of $(\delta I \cdot \delta I')_\theta$. We expect that a more refined analysis, currently underway, fitting the expected probability distribution of $(\delta I \cdot \delta I')_\theta$, will yield appreciably tighter limits.

3. The Intensity Correlation Function

In most cases of interest, the maximum scale of appreciable clustering is much smaller than the Hubble radius and the angular separation θ is much less than one radian. If we further assume that the luminosities of sources are statistically independent of their positions relative to the other sources, the observed angular correlation of the intensity fluctuations for beam axes separated by an angle $\theta_\star$ writes (for a more general treatment and further details, see De Zotti *et al.* 1988):

$$
C(\theta_\star) = \left(\frac{c}{4\pi H_o}\right)^2 \int d\omega f(\vartheta, \varphi) \int d\overline{\omega} f(\overline{\vartheta}, \overline{\varphi}).
$$
$$
\cdot \int_{z_m(L_{min},S)}^{z_{max}} dz \frac{j_{eff}^2(z)}{(1+z)^4(1+\Omega z)} \int_{Max[z_m-z,-\Delta(r_{max})]}^{\Delta(r_{max})} d(\delta z)\xi(r,z), \tag{2}
$$

where $f(\vartheta, \varphi)$ is the response function of the detector, $\Delta(r_{max})$ is the value of δz corresponding to the maximum scale of clustering,

$$
j_{eff}(z) = \int_{L_{min}}^{min[L_{max}, L(S,z)]} d\log L \, L \, n_c(L,z) K(L,z) \tag{3}
$$

is the effective volume emissivity, $n_c(L,z)$ is the *comoving* number density of sources, and $L(S,z)$ is the luminosity that, at the redshift z, yields a flux equal to the detection limit S $[S(2-10) \simeq 3.1 \times 10^{-11} erg \, cm^{-2}s^{-1}$ for the A2 survey]. Note that, unlike Poisson fluctuations, intensity correlations do not diverge for $S \to \infty$, so that, in principle, there is no need for a cutoff at high fluxes; in practice, however, such cutoff is usually expedient.

A convenient model for the epoch–dependent spatial correlation function, that fits the available data is (Peebles 1980; Bahcall and Soneira 1983; Sebok 1986)

$$
\xi(r,z) = D^2(z)\xi_o(r)
$$
$$
\xi_o(r) = (r_o/r)^\gamma, \tag{4}
$$

with $\gamma \simeq 1.8$. In numerical calculations we have adopted a slightly more refined model for $\xi_o(r)$:

$$
\xi_o(r) = \begin{cases} [r_o/(r + r_c)]^\gamma & \text{if } r \leq r_{max} \\ 0 & \text{if } r > r_{max}, \end{cases} \tag{5}
$$

making the density flat within regions of size r_c, that may be of the order of the physical size of sources, and explicitly incorporating a cutoff radius r_{max}.

The factor $D^2(z)$ allows for possible evolution of the clustering. A frequently used parametrization is

$$
D^2(z) = (1 + z)^{-(3+\epsilon)}. \tag{6}
$$

$\epsilon < 0 \, (> 0)$ means that clustering is decaying (growing) with cosmic time. If $\epsilon = -3$ or $\epsilon = \gamma - 3$, ξ is constant in physical or in comoving coordinates, respectively. A self–similar evolution of the correlation function (requiring $\Omega = 1$ and a power–law spectrum of initial density perturbations) yields $\epsilon = 0$. The same value applies (barring the case of small–scale structure being systematically disrupted as larger condensations form) for $\xi \gg 1$, whereby the the number of objects $n(t)\xi(t)$ at fixed proper separation r is constant (statistically stable clustering).

On the other hand, some scenarios for galaxy formation would rather predict correlations at early epochs comparable to, or even larger than, those observed today (i.e. $\epsilon \leq -3$). In a "pancake" picture (hot dark matter) high-z QSOs should be found in a small number of high density regions, with angular size $\approx 2° \div 3°$ (Rees 1986). Subsequent formation of more numerous sources would dilute their clustering. In cold dark matter scenario, luminous high redshift QSOs and rich clusters are associated with rare high density peaks of the density perturbation field and are thus expected to have enhanced clustering (Kaiser 1986; Efstathiou and Rees 1988; Cole and Kaiser 1988).

In the power–law model [eq. (4)] with $\gamma \simeq 1.8$ and r_o not too small, the integral over δz is only weakly dependent on r_{max}, so that, to a first order approximation, the integration can be extended from $-\infty$ to ∞, to obtain (Totsuji and Kihara 1969):

$$
C(\theta_\star) = H_\gamma \left(\frac{c\omega_{eff}}{4\pi H_o} \right)^2 \left(\frac{H_o}{c} r_o \right)^\gamma \theta^{1-\gamma}
$$
$$
\cdot \int_{z_m(L_{min},S)}^{z_{max}} dz \, j_{eff}^2(z) \frac{D^2(z) \left(\frac{H_o}{c} d_A(z) \right)^{1-\gamma}}{(1+z)^2 (1+\Omega z)^{1/2}}. \tag{7}
$$

where H_γ is a product of Gamma functions (for $\gamma = 1.8$, $H_\gamma \simeq 3.7$) and ω_{eff} is the effective solid angle, taking into account the response function of the detector, and d_A is the angular diameter distance.

In several interesting cases (see below) $dC(\theta_\star)/d\log z$ peaks at $z_p \ll 1$. If so, $H_o d_A/c \simeq z$, and the integral in eq. (7) can be easily evaluated. We get:

$$
\Gamma(\theta_\star) \simeq \left(\frac{H_o}{c} r_o \right)^{\gamma/2} \left(\frac{z_p^{2-\gamma}}{2-\gamma} \right)^{1/2} \left[1 - \left(\frac{z_{m(L_{min},S)}}{z_p} \right)^{2-\gamma} \right]^{1/2}
$$
$$
\cdot \frac{j_{eff}}{j_{XRB}} \theta_\star^{(1-\gamma)/2}, \tag{8}
$$

where j_{XRB} is the volume emissivity of the XRB.

4. Constraints on Correlation Functions

4.1 UNEVOLVING CLUSTERS OF GALAXIES

The local 2–10 keV luminosity function of rich clusters of galaxies in the range $3 \times 10^{43} \leq L(2\text{–}10\ keV)\ erg\ s^{-1} \leq 3 \times 10^{45}$ has been derived by Piccinotti et al. (1982). From the first pass data of the HEAO-1 A2 survey, they found

$$n_o(L) \simeq 8.1 \times 10^{-7} \left(\frac{L}{L_o}\right)^{-1.15} Mpc^{-3}(d\log L)^{-1}, \tag{9}$$

or

$$n_o(L) \simeq 4.6 \times 10^{-7} \left(\frac{L}{L_o}\right)^{-1.03} Mpc^{-3}(d\log L)^{-1}, \tag{10}$$

including the Virgo cluster in the sample, or excluding it, respectively. In the above equations, and in the following, $L_o = 10^{44} erg\ s^{-1}$ in the 2–10 keV band.

In computing the K–correction we must take into account that the temperature of the X–ray emitting gas appears to decrease with decreasing luminosity, although with a considerable scatter (Mushotzky 1984). A fit to the HEAO-1 data discussed by Mushotzky (1984) yields

$$T \simeq 2.25 \left(\frac{L}{L_o}\right)^{0.425} keV. \tag{11}$$

A consequence of eq. (11), coupled with the correlation between luminosity and Abell richness class (Johnson et al. 1983), is that the pure hard X–ray selection leading to the luminosity function of Piccinotti et al. (1982) strongly favours clusters of higher richness classes. There are 19 Abell cluster in Piccinotti's sample; only one (A2052) is of richness class $R = 0$, 8 are of class $R = 1$ and 10 are of class $R = 2$. For comparison, in the "redshift sample" (Abell's distance class $D \leq 4$) of Bahcall and Soneira (1983) there are 82 objects of richness class $R = 1$ and 22 with $R \geq 2$; in their comparison sample ($D = 5 + 6$) there 1125 $R = 1$ and 422 $R = 2$ clusters.

According to Bahcall and Soneira (1983) and Postman, Geller, and Huchra (1986) the amplitude of the cluster correlation function strongly increases with richness, clusters with $R \geq 2$ showing stronger correlations by a factor ≈ 3 as compared with $R = 1$ clusters, whose correlation scale–length was found to be $r_o \simeq 45\ Mpc$. For the Piccinotti et al. (1982) sample, the intermediate value $r_o \simeq 60\ Mpc$ may be appropriate. Note that the relatively flat X–ray luminosity function of clusters implies that all logarithmic luminosity intervals contribute almost equally to Γ if $\xi(r)$ is luminosity–independent; if the correlations are increasingly stronger for richer systems, the latter will determine the effective value of r_o.

The predicted cluster contribution to $\Gamma(\theta_\star)$ for $\gamma = 1.8$, $-3 \leq \epsilon \leq 0$, $r_c = 5\ Mpc$, $r_{max} = 300\ Mpc$ are

$$\Gamma(\theta_\star) \simeq (0.56 \div 1.2) \times 10^{-2} \left(\frac{r_o}{60Mpc}\right)^{0.9} \left(\frac{\theta_\star}{3^\circ}\right)^{-0.4}. \tag{12}$$

Most of the spread in the results comes from uncertainties in the local luminosity function [eq. (9) yields values $\simeq 60\%$ higher than eq. (10)]. Varying ϵ from 0 to -3 increases $\Gamma(\theta_\star)$ only by $\simeq 35\%$, due to the steep K–correction that effectively suppresses the contributions from z larger than a few tenths. The results are essentially independent of the assumed density parameter, again because most of the integral comes from $z \ll 1$, and depend very weakly on r_c and r_{max}. Note the good agreement with the estimates that can be simply obtained from eq. (8) with $j_{cluster}/j_{XRB} \simeq 0.04$ (Piccinotti *et al.* 1982), $z_p \simeq 0.2$ and $z_m \simeq 0$.

The X-ray upper limits are then compatible with the optical estimates of the spatial correlation function of rich clusters by Bahcall and Soneira (1983) and Klypin and Kopylov (1983).

4.2 CONSTRAINTS ON CLUSTER EVOLUTION

Kaiser (1986) has recently pointed out that, in the framework of hierarchical theories with approximately scale–free primordial density fluctuations (such as the "cold dark matter" or "cosmic string" pictures), X-ray properties of rich clusters are expected to evolve strongly, and has worked out detailed predictions.

For his fiducial model (post–recombination spectral index of the perturbation field $n = -1$), the *total* comoving emissivity grows with z by a factor $(1+z)^{2.5}$. At the same time, however, the characteristic temperature of the X-ray emitting gas decreases as $(1+z)^{-1}$. Hence, as z increases, a rapidly increasing fraction of the emitted power drops out of the spectral window of the A2 experiment. Correspondingly, $\Gamma(\theta_\star)$ increases by only $\simeq 40\%$. Hence the model is compatible with the X-ray upper limits. Note, however, that the quoted evolution of the gas temperature was derived ignoring any energy input from galaxies. If the latter was important, the predicted $\Gamma(\theta_\star)$ could be substantially larger.

On the other hand, for a "white noise" density perturbation spectrum ($n = 0$), the caracteristic temperature varies quite slowly $[T \propto (1+z)^{-1/3}]$, and the volume emissivity evolves somewhat faster $[j \propto (1+z)^{17/6}]$ than for $n = -1$. As a consequence, the predicted $\Gamma(\theta_\star)$ jumps to $\simeq 4.9 \times 10^{-2}$, for $r_o = 60\ Mpc$ and $\epsilon = -3$, making the model inconsistent with the data.

4.3 LOCAL AGNs

The A2 data are much more sensitive to clustering of AGNs than of rich clusters because the former objects: i) have a substantially larger volume emissivity: ii) have "hard" power–law spectra, so that the contributions to $\Gamma(\theta_\star)$ from high z are not suppressed by the K–correction factor; iii) have a strongly evolving luminosity function.

A firm lower limit to the contribution of AGNs to $\Gamma(\theta_\star)$ can be obtained ignoring cosmological evolution. Assuming a power–law spectrum with energy index $\alpha_x = 0.7$ and $\epsilon = 0$ (stable clustering), we easily estimate z_p to be $\simeq 0.07$. The local volume emissivity of AGNs brighter than $L_x(2\text{–}10\ keV) = 3 \times 10^{42}\ erg\ s^{-1}$ amounts

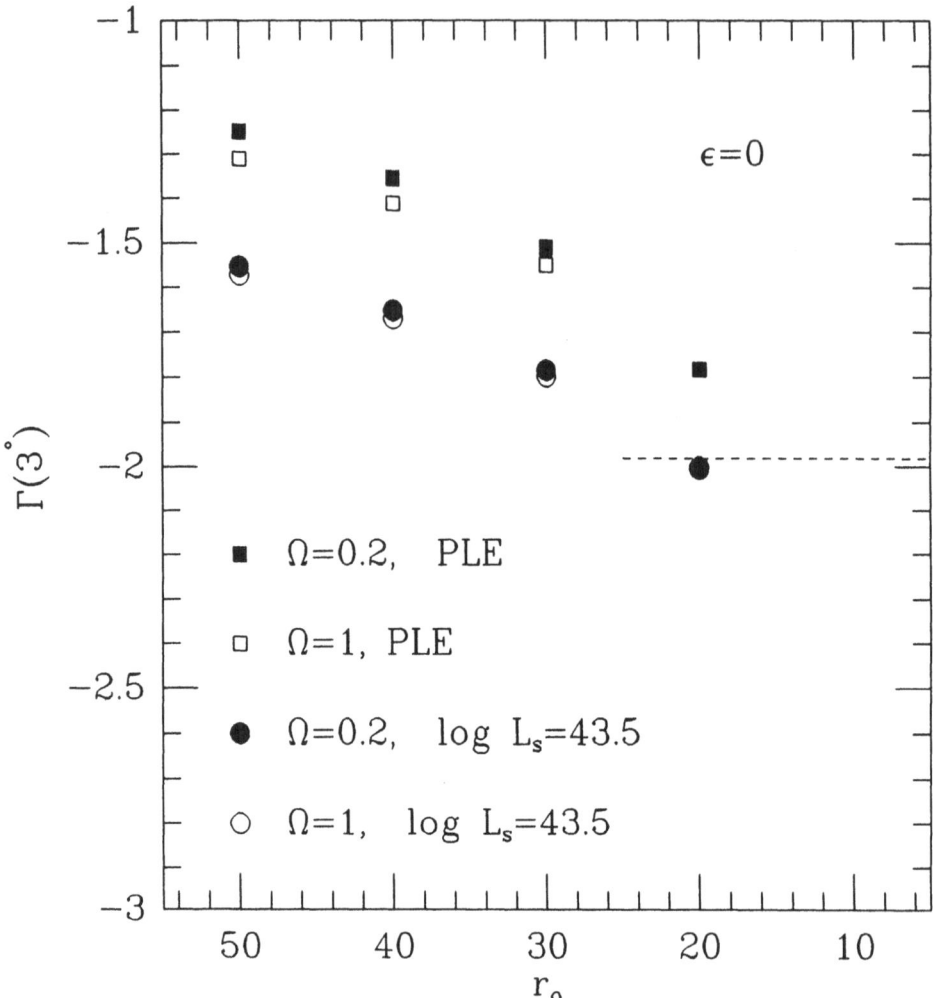

Figure 1. Predicted $\Gamma(3°)$ for AGNs brighter than 3×10^{42} erg $s^{-1}(2\text{–}10$ $keV)$, in the case of stable clustering [$\epsilon = 0$, see eqs. (4) and (6)], as a function of the clustering scale r_o. Squares correspond to pure luminosity evolution (PLE); circles to models allowing for luminosity evolution only above $\log L = 43.5$ [eq. (14)]. We have set $r_{max} = 6r_o$, $r_c = 0.1$ Mpc [see eq. (5)]. The dashed line shows the observational 1σ upper limit.

to $\simeq 0.2 \times j_{XRB}$ (Piccinotti et $al.$ 1982). Then eq. (8) yields

$$\Gamma(\theta_\star) \simeq (1.1 \div 1.6) \times 10^{-2} \left(\frac{r_o}{30Mpc} \right)^{0.9} \left(\frac{\theta_\star}{3°} \right)^{-0.4} \frac{j_{AGN}}{j_{XRB}}, \qquad (13)$$

implying that their spatial correlation scale, r_o, cannot be as large as that found by Bahcall and Soneira (1983) for rich clusters.

On the other hand, as soon as some information on the $\xi(r)$ of AGNs will become available, the observational upper limits on $\Gamma(\theta_*)$ will translate, trough eq. (13), into constraints on their global contribution to the XRB, without evolution. As is well known, such contribution depends strongly on the poorly known space densities of intrinsically weak sources. For example, if AGNs cluster at least as optically selected galaxies ($r_o \simeq 10\ Mpc$, Davis and Peebles 1983), the limit on $\Gamma(3°)$ implies $j_{AGN}/j_{XRB} \leq 0.7$.

4.4 EVOLVING AGNs

The epoch–dependent X–ray luminosity function of AGNs have been recently discussed by many authors (see Setti 1987 and Danese and Franceschini 1988 for recent reviews). Although the problem is still far from settled, luminosity evolution models turn out to provide remarkably successful representations of the data.

Danese *et al.* (1986) have explored a full set of models of this general type, whereby the luminosity varies with cosmic time according to

$$L(t) = \begin{cases} L_s + (L_o - L_s)\exp(\kappa\tau) & \text{if } L_o > L_s \\ L_o & \text{if } L_0 \leq L_s, \end{cases} \tag{14}$$

where $L_o \equiv L(t_o)$ is the present luminosity, $\tau = (1 - t/t_o)$, and κ is a constant. For $L_s = 0$ we have pure luminosity evolution (PLE); $L_s > 0$ corresponds to the case that the emission of intrinsically weaker sources is steadily fuelled (e.g. by stellar mass loss) for times $\approx H_o^{-1}$ (Cavaliere, Giallongo, and Vagnetti 1985). Models consistent with the available X–ray survey data range from PLE to $L_s \leq 10^{43.5}\ erg\ s^{-1}$.

The two extreme possibilities, to be considered in the following, imply substantially different predictions for contributions to the XRB (Danese *et al.* 1986). In the case of PLE, AGNs provide $\simeq$ 60% of the XRB intensity in an Einstein–de Sitter universe ($\Omega = 1$), and up to 100% in an open universe; if $L_s \geq 10^{43.5}\ erg\ s^{-1}$ the expected contributions drop to $\simeq$ 25% for $\Omega = 1$ and to $\simeq$ 30% for $\Omega = 0.2$.

The predicted $\Gamma(3°)$ for several values of r_o are summarized in Figs. 1 (for $\epsilon = 0$) and 2 ($\epsilon = -3$), where the observational 1σ upper limit is represented by a dashed line.

5. Conclusions

5.1 STRUCTURES ON VERY LARGE SCALES

To the extent that clusters and AGNs reflect the distribution of galaxies, the present limits on $\Gamma(\theta)$ imply that any very large scale structure (yet much smaller than the

Hubble radius) must have a small density contrast. Suppose, for example, that the space correlation function ξ has a value Ξ on a scale R. Then we get:

$$\Gamma(\theta_\star) \approx 0.5 \, \Xi^{1/2} \left(\frac{H_o}{c} R \right)^{1/2} \frac{j_{sources}}{j_{XRB}}, \qquad (15)$$

where $\theta_\star$ is the effective angular size of the structure. Taking into account that clusters and AGNs contribute at least 30% of the XRB,

$$\Xi \leq 1.8 \times 10^{-2} \left(\frac{\Gamma(\theta_\star)}{2 \times 10^{-2}} \right)^2 \left(\frac{H_o}{c} R \right)^{-1} \left(\frac{j_{sources}/j_{XRB}}{0.3} \right)^{-2}. \qquad (16)$$

5.2 RICH CLUSTERS OF GALAXIES

The present constraints on $\Gamma(\theta)$ are compatible with Bahcall and Soneira (1983) estimate of the correlation length scale for rich clusters. Evolution of the kind predicted by Kaiser (1986) is also allowed, provided that the gas temperature decreases significantly with increasing z.

5.3 AGNs

Optical evidence for clustering of QSOs with an average redshift $z \simeq 1$ has been found by Shanks *et al.* (1988), Shaver (1988), Kruszewski (1988), Iovino and Shaver (1988). The estimated amplitude of the quasar correlation function is $\xi \approx 1$ on a *proper* scale $\simeq 10 \, Mpc$. This would correspond to a present clustering length $r_o \approx 30 \, Mpc$ for stable clustering ($\epsilon = 0$), or to $r_o \approx 10 \, Mpc$ if the spatial correlation function $\xi(r)$ is constant in physical coordinates ($\epsilon = -3$).

As shown by Fig. 2, the latter possibility is fully compatible with the X–ray upper limits on $\Gamma(3°)$, independently of the assumed value of the density parameter Ω and of the evolution model for the luminosity function.

On the other hand some indications have been reported (Kruszewski 1988; Shaver 1988; Iovino and Shaver 1988) that the amplitude of the quasar–quasar correlation function is decreasing relatively rapidly with increasing z, perhaps as fast as implied by the stable clustering model [$\xi \propto (1 + z)^{-3}$]. In that case, however, the predicted $\Gamma(3°)$ would be somewhat in excess of the observational limits: we find $\Gamma(3°) \simeq 1.6 \times 10^{-2}$ if only sources with $\log L > 43.5$ are allowed to evolve, $\Gamma(3°) \simeq 2.8 \times 10^{-2}$ or $\simeq 3.1 \times 10^{-2}$ for pure luminosity evolution models and $\Omega = 1$ or 0.2, respectively. Although the present uncertainties are too large to allow firm conclusions, if the indications of strong evolution of quasar clustering will be confirmed, differential luminosity evolution models (implying relatively low contributions of AGNs to the XRB) would be favoured.

A word of caution is in order, however. While optical results refer to QSOs, the X–ray predictions include the contribution of lower luminosity AGNs (Seyfert

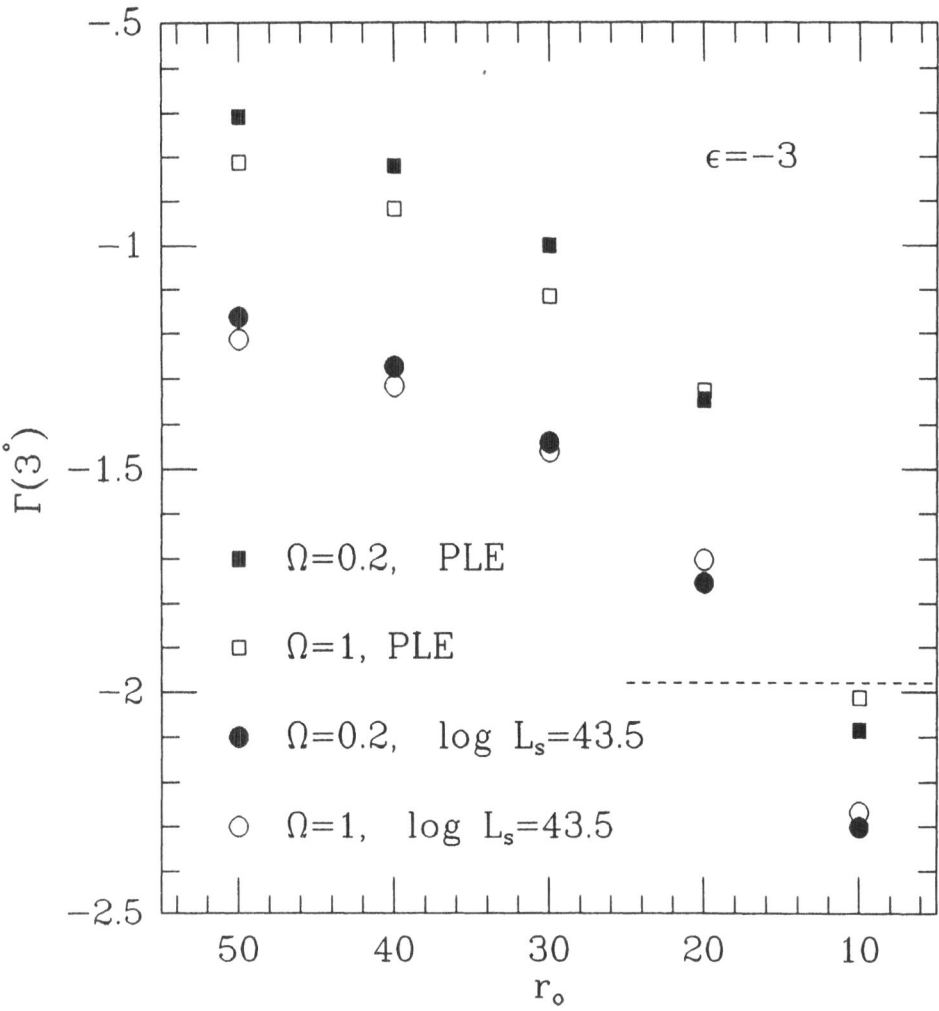

Figure 2. Same as Fig. 1, but for constant correlation amplitude ($\epsilon = -3$).

nuclei). In the framework of luminosity evolution models it seems natural to extend the continuity between the two classes of sources also to their clustering properties. But if Seyfert galaxies are much more weakly clustered than QSOs, the values of $\Gamma(3°)$ reported in Figs. 1 and 2 must be revised downwards. For example, if we take into account only objects whose X–ray luminosity (2–10 keV) is larger than that of $Fairall\ 9\ (2.6 \times 10^{44} erg\ s^{-1})$, sometimes assumed as defining the boundary between Seyfert nuclei and QSOs, we find $\Gamma(3°) \simeq 6 \times 10^{-3}$ for models with $\log L_s = 43.5$ and $\Gamma(3°) \simeq (0.8 \div 1) \times 10^{-2}$ for pure luminosity evolution models.

5.4 AGN–CLUSTER CORRELATIONS

The now abundant evidence for quasar–galaxy associations (cf. Shaver 1988, and references therein) indicates that AGNs are preferentially located in groups or clusters of galaxies. A significant positive contribution to $\Gamma(\theta)$ may then come from AGN–cluster correlations. For a power–law two–point correlation function such contribution can be obtained from eq. (8) replacing j_{eff} with $[j_{AGN} \cdot j_{cluster}]^{1/2}$. Using the values of 0.2 and 0.04 for the background–normalized volume emissivities, we find $r_o(AGN-cluster) \leq 50\ Mpc$.

5.5 CLUMPS OF HOT PLASMA

The present limits on $\Gamma(\theta)$ do not significantly constrain hot-bubble models. In the scenario involving decaying superconducting cosmic strings proposed by Ostriker, Thompson and Witten (1986), the loop-decay redshift is $z_{decay} \leq 0.5$ for bubbles to have an angular size of $3°$, for a suitable range of electromagnetic-to-gravitational energy ratios. For a filling factor of about unity, an estimate of the 2–10 keV volume emissivity of such bubbles is $j_o(bubbles) \simeq 10^{38}(\Omega_b/0.03)^2(\frac{\delta\rho}{\rho}/3)^2(T/30\ keV)^{1/2}\ erg\ s^{-1} Mpc^{-3}$, i.e. $\approx 4\%$ of the XRB. In connection with eq. (8) and with the upper limit on $\Gamma(3°)$, this figure is compatible with a broad range of values for the correlation length, in particular with the ones suggested for optically selected galaxy clusters (e.g., Bahcall and Soneira 1983) and superclusters (Bahcall and Burgett 1986).

On the larger angular scales, the estimated 3σ upper limit to the 2–10 keV volume emissivity from hot intrasupercluster gas (Persic, Rephaeli and Boldt 1988), $j_o(SCs) \leq 10^{36}\ erg\ s^{-1} Mpc^{-3}$, coupled with the suggested value $R_o(SCs) \simeq 110\ Mpc$ (Bahcall and Burgett 1986), would generate a correlation strength for fluctuations well below the current limits.

Additional bounds on the two-point correlation function of X-ray sources come from the limits on the XRB fluctuations in excess of the Poisson contributions from known classes of objects (Shafer and Fabian 1983; Barcons and Fabian 1988; Meszaros and Meszaros 1988).

Acknowledgements. We are indebted to J. Swank for her constant help. L.D., G.D.Z., and G.G.C.P. are grateful to S. Holt and to the GSFC/LHEA for their warm hospitality. This research was started while M.P. and G.G.C.P. were NRC/NAS Research Associates at NASA/GSFC. Work supported in part by MPI and CNR (through GNA and PSN).

References

Bahcall, N.A., and Burgett, W. 1986, *Ap. J. (Letters)*, **300**, L15.

Bahcall, N.A., and Soneira, R.M. 1983, *Ap. J.*, **270**, 20.

Barcons, X., and Fabian, A.C. 1988a, *M.N.R.A.S.*, **230**, 189.

Barcons, X., and Fabian, A.C. 1988b, these proceedings.

Cavaliere, A., Giallongo, E., and Vagnetti, F. 1985, *Ap. J.*, **296**, 402.

Cole, S., and Kaiser, N. 1988, preprint

Danese, L., and De Zotti, G. 1986, in *Galaxy Distances and Deviations from Universal Expansion*, ed. B.F. Madore and R.B. Tully (Dordrecht:Reidel), p. 215.

Danese, L., De Zotti, G., Fasano, G., and Franceschini, A. 1986, *Astr. Ap.*, **161**, 1.

Danese, L., and Franceschini, A. 1988, in *The Post-Recombination Universe*, ed. N. Kaiser and A.N. Lasenby , in press

Dautcourt, G. 1977, *Astr. Nachr.*, **298**, 141.

Davis, M., and Peebles, P.J.E. 1983, *Ap. J.*, **267**, 465.

De Zotti, G., Persic, M., Boldt, E.A., Marshall, F.E., Danese, L., Franceschini, A., and Palumbo, G.G.C. 1988, in preparation.

Efstathiou, G., and Rees, M.J. 1988, *M.N.R.A.S.*, **230**, 5P.

Fabian, A.C. 1972, *Nature Phys. Sci.*, **237**, 19.

Fabian, A.C 1981, in *Proc. Tenth Texas Symposium on Relativistic Astrophysics*, *Ann. N.Y. Acad. Sci.*, **375**, 235.

Guilbert, P.W., and Fabian, A.C. 1986, *M.N.R.A.S.*, **220**, 439.

Iovino, A., and Shaver, P.A. 1988, *Ap. J. (Letters)*, submitted.

Johnson, M.W., Cruddace, R.G., Ulmer, M.P., Kowalski, M.P., and Wood, K.S. 1983, *Ap. J.*, **266**, 425.

Kaiser, N. 1986, *M.N.R.A.S.*, **222**, 323.

Klypin, A.A., and Kopylov, A.I. 1983, *Pis'ma A. Zh.*, **9**, 75 [*Sov. Astr. Letters*, **9**, 41].

Kruszewski, A. 1988, preprint

Marshall, F.E., Boldt, E.A., Holt, S.S., Mushotzky, R.F., Pravdo, S.H., P.J. Rothschild, and Serlemitsos, P.J. 1979, *Ap. J. Suppl.*, **40**, 657.

Meszaros, A., and Meszaros, P. 1988, *Ap. J.*, **327**, 25.

Mushotzky, R.F. 1984, *Physica Scripta*, **17**, 157.

Ostriker, J.P., Thompson, C., and Witten, E. 1986, *Phys. Lett. B*, **180**, 231.

Peebles, P.J.E. 1980, *The Large Scale Structure of the Universe* (Princeton: Princeton University Press).

Persic, M., De Zotti, G., Boldt, E.A., Marshall, F.E., Danese, L., Franceschini, A., and Palumbo, G.G.C. 1988, *Ap. J. (Letters)*, submitted.

Persic, M., Rephaeli, Y., and Boldt, E.A. 1988, *Ap. J. Lett.*, **327**, L1.

Piccinotti, G., Mushotzky, R.F., Boldt, E.A., Holt, S.S., R.E.,Marshall, F.E., Serlemitsos, P.J., and Shafer, R.A. 1980, *Ap. J.*, **235**, 4.

Postman, M., Geller, M.J., and Huchra, J.P. 1986, *A. J.*, **91**, 1267.

Rees, M.J. 1986, in it The Structure and Evolution of Active Galactic Nuclei, ed. G. Giuricin, F. Mardirossian, M. Mezzetti, and M. Ramella (Dordrecht: Reidel), p. 447.

Rothschild, R.E., *et al.* 1979, *Space Sci. Instr.*, **4**, 269.

Sebok, W.L. 1986, *Ap. J. Suppl.*, *62*, 301.

Setti, G. 1987, in *IAU Symposium 124, Observational Cosmology*, ed. A. Hewitt, G. Burbidge, and L.Z. Fang (Dordrecht:Reidel), p. 579.

Shafer, R.A. 1983, Ph. D. thesis, University of Maryland.

Shafer, R.A., and Fabian, A.C. 1983, in *IAU Symposium 104, Early Evolution of the Universe and its Present Structure*, ed. G.O. Abell and G. Chincarini (Dordrecht:Reidel), p. 333.

Shanks, T., Boyle, B.J., and Peterson, B.A. 1988, in *Proc. Tucson Workshop on QSO Surveys*, in press.

Shaver, P.A. 1988, in IAU Symposium 130, The Evolution of Large Scale Structure in the Universe, ed. J. Audouze and A. Szalay (Dordrecht:Reidel), in press.

Totsuji, H., and Kihara, T. 1969, *Publ. Astr. Soc. Japan*, **21**, 221.

DISCUSSION

KRUSZEWSKI: How do you think your results are influenced by not using the known X-ray sources. At angular separation equal to 3 degrees any correlations would be contributed mainly by object with redshifts smaller than 0.1 and many of these objects must have been catalogued.

DE ZOTTI: While most of the Poisson contribution to cell-to-cell fluctuations comes from the brightest sources in the sample (and, in fact, the appropriate integral diverges as the flux S approaches infinity), the mean effect for autocorrelations comes from weak sources [see, e.g. eq. (7)], and there is no need for a cutoff at high fluxes. Indeed we have directly verified that the results are not significantly affected by including catalogued sources in the analysis.

POSTER PAPERS

Angular Momentum Transport in a two-component Protogalaxy.

V. Antonuccio-Delogu

International School for Advanced Studies
Strada Costiera 11
34014 - Trieste
ITALY

1.) - **Introduction**. - The dynamics of the collapse and formation of cosmological perturbations leading to spiral galaxies is still far from being completely understood. One important issue is the dynamical role of a Dark Halo on the collapse process, and especially on the final mass and angular momentum distribution of the configuration. Recently, **Barnes aand Efstathiou (1987)**, **Quinn et al. (1987)**, **Zurek et al. (1987)** have analyzed numerically this problem. Their conclusions cannot be taken as definitive, but suggestive of some trends that can be verified by inventing and studying exact models of the collapsing configuration.
We have recently worked on a model of a two-component **Dark Halo -Stellar Disk** system, with the aim of calculating the amount of angular momentum exchanged between these two components during the accretion of the disk in the equatorial plane of a rotating, axisymmetric dark matter halo. At variance with preceding work by **Fall and Efstathiou (1980)**, **Ryden and Gunn (1987)** and **Ryden (1987)**, we have a rotating, flattened halo described by an anisotropic distribution function, as it should be expected if Dark matter formed by collapse of initially slowly ($\lambda \sim .08$) rotating halos.

2) - **Model** - We model the Halo as a spheroidal, stationary and collisionless system having a density profile:

$$\rho(r) = \rho_c \left[1 - \frac{R^2}{a^2} - \frac{z^2}{c^2} \right] \tag{1}$$

This density distribution generates a quartic potential (**Chandrasekhar, 1969**):

$$\Phi = \frac{1}{2} \pi G \rho_c \left[4A_1 a^2 + 2A_3 c^2 - 2A_1 R^2 - 2A_3 z^2 + A_{11} R^4 + 2A_{13} R^2 z^2 + \right.$$
$$\left. + A_{33} z^4 \right] \tag{2}$$

319

M. Mezzetti et al. (eds.), Large Scale Structure and Motions in the Universe, 319–320.
© 1989 by Kluwer Academic Publishers.

The distribution function to the first order in the eccentricity e is:

$$f_0 = \frac{\rho_c}{\sqrt{2}} \cdot \frac{1}{6\pi^2} \cdot \frac{\left[1 - 5\chi J/4\psi\right]}{\left[\chi J - \psi - E\right]} \cdot \frac{H\left[-E - 6\psi/5 + 3\chi J/2\right]}{\left[-E - 6\psi/5 + 3\chi J/2\right]} + O(e) \qquad (3)$$

(Hunter, 1974), where: J is the total angular momentum, E the energy, and χ and ψ are some constants.
For the disk we take a density profile:

$$\rho = \rho_* \left[a_k^2 - R^2\right]^n \qquad (4)$$

for which equilibrium distribution functions are known also for the not self-gravitating case (Zweibel, 1978).
We then solve **exactly** the linearized system of Vlasov-Poisson equations for this two-component system (for the details see Antonuccio-Delogu, 1988), and calculate the total angular momentum transferred outside. **A typical result for**: $n = 3/2$, $a_k = 20$ Kpc, ρ_* = .03 $M_\odot/Kpc^3$, ρ_c = 15 $M_\odot/$ Kpc^3, a = 60 Kpc, e = .08, is: **35%** of angular momentum within the radius at which initially half of the total angular momentum of the disk is contained is transferred outside. We refer to Antonuccio-Delogu (1988) and to forthcoming work for more detailed results.

Bibliography

Antonuccio-Delogu, V. *Sissa Preprint*   (1988)
Barnes, J and Efstathiou, G., Astrophys. J., 319, 575 (1987)
Chandrasekhar, S.: *Ellipsoidal Figures of Equilibrium*, Dover (1969)
Fall, S.M and Efstathiou, G.: M.N.R.A.S., 193, 189(1980)
Hunter, C: in *Iau Symp. n. 69*, A Hayli ed. , p.200   (1974)
Quinn, P.J and Zurek, W.H.: Astrophys. J., in press   (1988)
Ryden, B and Gunn, J.E.: Astrophys. J., 318, 15   (1987)
Ryden, B.: *Ias Preprint*   (1988)
Zweibel, E.: Astrophys. J., 222, 103   (1978)
Zurek, W.H., Quinn, P.J., and Salmon, J.K.: Los Alamos *Preprint*,
                                                    |(1988).

ANISOTROPIES OF THE MICROWAVE BACKGROUND RADIATION PRODUCED BY SPHEROIDS

F. Argüeso and E. Martínez-González
Dpto. de Física Moderna
Universidad de Cantabria
39005-SANTANDER (SPAIN)

ABSTRACT. We analyze the effect of large scale inhomogeneities on the MBR. These structures are treated as linear perturbations in an otherwise homogeneous and isotropic Einstein-De-Sitter universe. In this universe, the MBR large angular scale anisotropies are produced by the Sachs-Wolfe and Doppler effects. We consider these effects as produced by an isolated homogeneous spheroidal structure, since this is a natural extension of the spherical case widely treated in the literature.

INTRODUCTION

Observations of the microwave background radiation show a dipole component which amounts to a Local Group velocity $V_{LG}=550\pm 40Kms^{-1}$ towards l=267 b=31, relative to the cosmic barckground rest frame. The fit to the dipolar and quadrupolar spherical harmonics is consistent with a quadrupole bound of $7x10^{-5}$.

Spheroidal objects can be used to represent large filamentary as well as pancake-like structures, whose existence is very likely according to recent observations. By comparing our calculations with the dipole and quadrupole measurements, we draw some conclusions about the existence and properties of these elongated or flattened lumps (or voids) around us.

CONCLUSIONS

We can express the dipole a_1 and the quadrupole a_2 as functions of five parameters: a) the density contrast Δ which we will take constant all over the structure; 2) the volume of the spheroid, that we will write in terms of the "equivalent radius" R (R is the radius of the sphere with the same volume than our spheroid); 3) the "aspect ratio" x, x means the ratio between the major and minor axis, this quantity is related to the eccentricity $x^2= (1-e^2)^{-1}$ and is more intuitive and easier to handle; 4) the distance λ from the observer to the spheroid center; 5) the orientation of the structure, given by the angle θ , formed by our direction of observation and the

321

M. Mezzetti et al. (eds.), Large Scale Structure and Motions in the Universe, 321–322.

spheroid different axis. Therefore, we will be able to calculate a_1 and a_2 for any homogeneous spheroid, wherever it is situated as a_1, a_2 (Δ^2, R, x, λ, θ).

We can conclude that spheroids centered very close to our position produce a much higher quadrupole than that of spheres with the same density and volume. On the other hand, beyond a distance R x $\%\sqrt{3}$, γ:2/3,1/3 (obl) spheroids and spheres behave in a similar way reaching the maximum value for a_2 on the photosphere, except for very elongated filaments, which cross the horizon, whose quadrupole remains nearly constant at long distances. In the case of oblate spheroids with θ=0, it is obtained a null quadrupole at certain distances (Figure 1).About the dipole, this is maximum when the observer is on the structure surface decreasing with the eccentricity; we need higher densities to get the same effect than for spheres with the same volume. The dipole decreases very steeply outside the structure (Figure 2).

Density constraints are imposed for very large structures due to the quadrupole limit $a_2 \leqslant 7 \times 10^{-5}$. So, for a filament centered inside the horizon with x=50 and R=1000h^{-1}Mpc, the density contrast must be $\Delta \leqslant 10\%$. For a equivalent (same volume and eccentricity) disk-like object we have $\Delta \leqslant 40\%$ for λ=0 but $\Delta \leqslant 4\%$ for λ=1. Finally, we want to remark that we need at least a 3 σ peak to generate the observed dipole a_1=2x10^{-3} with a large structure R $\geqslant$ 50h^{-1}Mpc in a scale invariant CDM scenario. The height of the peak grows with the eccentricity and the size of the spheroid.

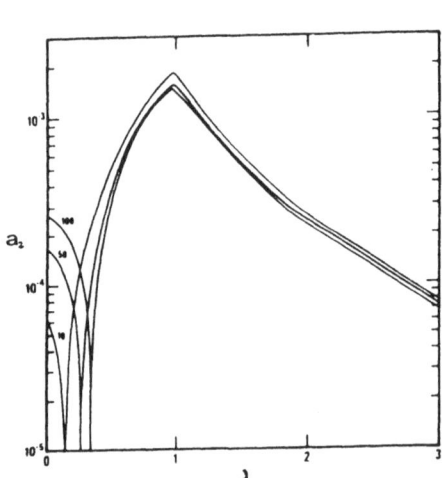
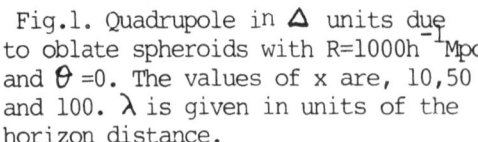
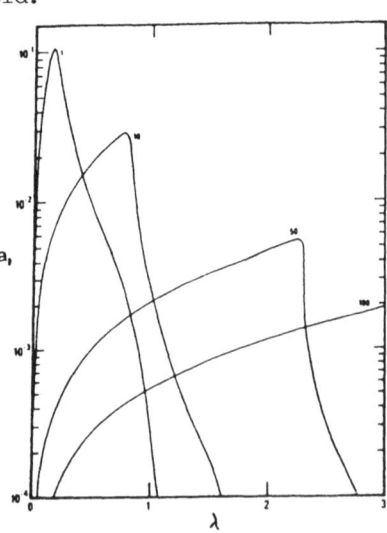

Fig.1. Quadrupole in Δ units due to oblate spheroids with R=1000h^{-1}Mpc and θ=0. The values of x are, 10,50 and 100. λ is given in units of the horizon distance.

Fig.2. Dipole in Δ units due to prolate spheroids with θ=0, R= =1000h^{-1}Mpc and x takes the values 1,10,50 and 100.

FEATURES OF THE COSMIC BACKGROUND RADIATION MAPS

F. Argüeso, E. Martínez-González & J.L. Sanz
Dpto. de Física Moderna
Universidad de Cantabria
39005- SANTANDER (Spain)

ABSTRACT. We study the influence of a large spherical structure on the microwave and X-ray backgrounds. If we substract the monopole and dipole components, we observe rings and spots in the sky patterns caused by such structure. We can then use the available sky maps (Strukov & Skulachev 1987; Lubin et al. 1985) to identify a dominant mass concentration (or void) in the Universe.

INTRODUCTION

The microwave background radiation MBR has propagated since $Z \approx 10^3$, carrying information about the geometry and the distribution of matter in the universe. A dipole anisotropy is observed in this radiation, usually explained as result of the Local Group peculiar motion with respect to the general expansion. Although the cause of this motion is not clear (a great Attractor?), it seems essentially due to local structures at distances $r_0 \leqslant 50h^{-1}$ Mpc.

In order to analyze the influence of a very big spherical structure of radius $R \gtrsim 200h^{-1}$ Mpc, situated far away from our position, on these two backgrounds, we propose the following strategy: we calculate the anisotropies created by this object, and after removing the monopole and dipole components, we compare our results with a sky map from which these components have been substracted. With this method, we are able to eliminate the contamination of local structures.

RESULTS

In the theoretical calculation of the anisotropies, we assume an isolated linear structure and we find that the basic contributions are due to the Sachs-Wolfe effect (potential on the photosphere) and the emitters velocity. We also find rings and spots in the corresponding sky patterns, of different nature than those arising for instance in the Bianchi models. To illustrate the kind of sky maps that spherical structures generate, we show below two figures which

323

M. Mezzetti et al. (eds.), Large Scale Structure and Motions in the Universe, 323–324.
© *1989 by Kluwer Academic Publishers.*

324

give us information about the influence of different structures (a complete calculation is given elsewhere, Argüeso et al. 1988).

Figure 1 shows the MBR sky patterns after substracting the dipole created by an object situated at a) $r_o = 5700h^{-1}$ Mpc, $z=400$, b) $r_o = 3000h^{-1}$ Mpc, $z=3$. $\delta T/T$ is given in quadrupole units and in both cases we can appreciate a diffuse hot ring at 65° and 75° from the structure center respectively and a coldspot in the direction of the center. This coldspot is quite remarkable when the structure is very close to the photosphere.

Figure 2 shows the intensity fluctuations (in quadrupole units) in the XRB, after substracting the dipole produced by a structure at a) $r_o = 2700h^{-1}$ Mpc with a radius $R=900h^{-1}$ Mpc; in this case there appears a picked ring at $\simeq 17°$ surrounding a spot, both of them hot, b) $r_o = 3300h^{-1}$ Mpc and $R=1500h^{-1}$ Mpc. The hot ring is now diffuse and surrounds a weak coldspot.

REFERENCES

F. Argüeso, E. Martínez-González and J.L. Sanz, preprint 1988.
A.A. Klypin, M.V. Sazhin, I.A. Strukov and D.P. Skulachev. Sov. Astron. Lett. **13**, 104 (1987).
P. Lubin, T. Villela, G. Epstein, G. Smoot. Astrophys. J. **298** L1. 1985.

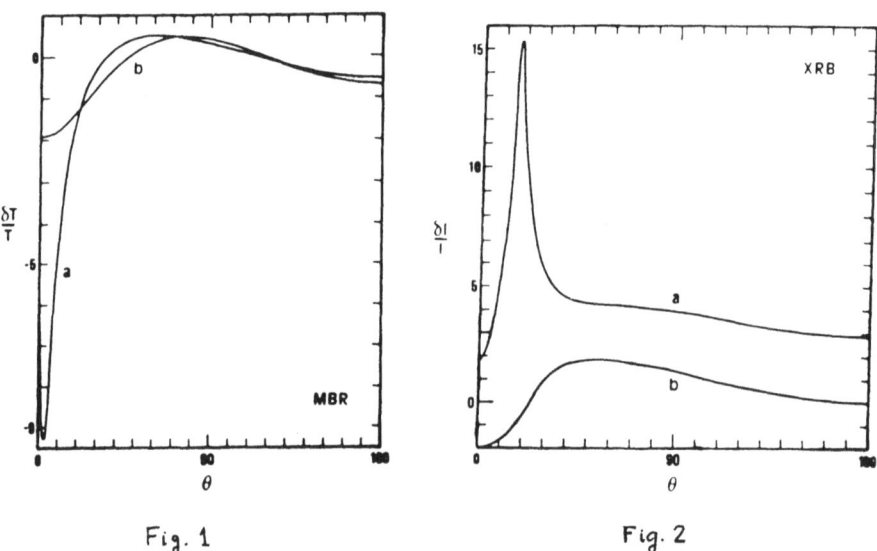

Fig. 1   Fig. 2

THE AUTOCORRELATION FUNCTION OF THE X-RAY BACKGROUND ON SMALL ANGULAR SCALES

X. Barcons
Departamento de Física Moderna, Universidad de
Cantabria, 39005 Santander, Spain
and
A.C. Fabian
Institute of Astronomy, Madingley Road
Cambridge CB3 OHA, United Kingdom

ABSTRACT. We have searched for small scale anisotropies
in the 1-3 keV X-ray background, using data from 5 Einstein
IPC deep fields. Although we find a signal on <5 arcmin,
its significance and origin are not clear. An upper limit
for the correlation function at 5 arcmin of $\sim 9\%$ (95%
confidence) is found, which poses some interesting constraints on the clustering of the X-ray Universe.

1. INTRODUCTION

The spatial distribution of the X-ray background (XRB) is,
at present, one of the few available pieces of direct
information on the "galaxy formation" Universe ($1 < z \ll 10^3$).
Fluctuation studies of the 2-10 keV XRB on scales of 5°
(Shafer & Fabian 1983) have shown to provide constraints
on the clustering of the X-ray Universe (Barcons & Fabian
1988). At the same energies, and for angular separations
of 3 to 12 degrees, there are no significant anisotropies,
which is just compatible with the cluster correlation
function (De Zotti 1988).
   At lower energies (1-3 keV), and angular scales of ~ 1
arcmin, fluctuation analyses of the XRB have given very
severe constraints on the log N - log S relation for X-ray
sources (Hamilton & Helfand 1987). In this contribution,
we present the results on the Auto-Correlation Function
(ACF) of the XRB at these energies, which should be able
to constrain the clustering scale of X-ray sources.

2. THE AUTOCORRELATION FUNCTION OF THE X-RAY BACKGROUND

Subtraction of $> 3\sigma$ sources has been carried out in the
0.25 square degrees central parts of 5 high-galactic

325

M. Mezzetti et al. (eds.), Large Scale Structure and Motions in the Universe, 325–326.
© *1989 by Kluwer Academic Publishers.*

latitude Einstein IPC deep fields. The ACF of the XRB was
then calculated from 1 to 15 arcmin (with a count smoothing
pixel of about 1 arcmin). A gaussian fit to the ACF, by
minimization of the χ^2 function, yields a correlation
angle of about 4 arcmin and a correlation amplitude at 5
arcmin of about 5 % (detector vignetting and other effects
could contribute to this amplitude). As it is very unlikely
that any effect decreases the ACF, we can establish, at
95 % confidence, an upper limit for the ACF at 5 arcmin
of 8.8 %.

3. DISCUSSION

As the ACF of the XRB maps the correlation function of the
background-contributing X-ray sources, we can constrain
the clustering of these by using the upper limit discussed
in last Section. Basically, if the XRB is produced in a
narrow redshift range ($\Delta z \simeq 1$) around a redshift from 2
to 5 by a population of point-like clustered sources, their
correlation length cannot be greater than $\sim 10h^{-1}$Mpc. This
limit is, however, model dependent, and the maximum clustering scale can change from model to model.

4. REFERENCES

Barcons, X. & Fabian, A.C., 1988. M.N.R.A.S., **320**, 189
De Zotti, G., 1988. These Proceedings
Hamilton, T.T. & Helfand, D.J., 1987. Ap. J., **318**, 93
Shafer, R.A. & Fabian, A.C., 1983. In: IAU Symp. 104, Early
     Evolution of the Universe and its Present Structure,
     eds. G.O. Abell & G. Chincarini, Dordrecht:Reidel

CORRELATION BETWEEN FLATTENING AND RADIO EMISSION AMONG E AND S0 GALAXIES

MASSIMO CALVANI
International School for Advanced Studies
Strada Costiera 11, 34014 Trieste, Italy
GIOVANNI FASANO
Padova Astronomical Observatory
vicolo Osservatorio, 35122 Padova, Italy
ALBERTO FRANCESCHINI
Department of Astronomy, University of Padova
vicolo Osservatorio, 35122 Padova, Italy

We collected a complete sample of 779 early type galaxies (313 E's and 466 S0's) to test the dependence between their radio emission and flattening.

The sample is based on three independent surveys of optically selected galaxies. Radio data were taken from Disney & Wall (Mon. Not. R. astr. Soc., **179**, 235 (1977)), Dressel & Condon (Ap. J. Suppl., **36**, 53 (1978)) and Sadler (Astron. J., **89**, 53 (1984)). By means of "survival analysis" techniques, we have fully taken into account both radio detections (140 objects) and upper limits. To include very powerful radio objects, we have also considered some radio selected samples. All available catalogues have been exploited in order to obtain reliable morphological types and axial ratios.

We confirm, to a high degree of significance, the previous finding by Disney et al. (Mon. Not. R. astr. Soc., **206**, 899 (1984)) and Sparks et al. (Mon. Not. R. astr. Soc., **207**, 445 (1984)) — based on a quite smaller sample — that radio–loud ellipticals are rounder on average than the radio-quiet ones. A possible interpretation could be that high radio luminosity objects are either cD galaxies or the effects of galaxy collisions and mergers (Heckman et al., Ap. J., **311**, 526 (1986)) It is plausible that in both cases gravitational interactions could finally give rise to intrinsically rounder configurations.

As for the S0 galaxies in our sample, we surprisingly find a similar correlation of both optical and radio luminosities with flattening. Unless we assume that dramatic absorption effects are operating in the optical selection, we are led to conclude that some high-luminosity S0's in the sample are misclassified ellipticals or cD galaxies. Indeed, we have verified that this is the case for the S0's with $P_5 > 10^{23}$ W/Hz/sr.

M. Mezzetti et al. (eds.), Large Scale Structure and Motions in the Universe, 327.
© *1989 by Kluwer Academic Publishers.*

VLA OBSERVATIONS OF 24 SPIRAL GALAXIES IN THE VIRGO CLUSTER

V. Cayatte[1], J. Van Gorkom[2], C. Balkowski[1],
C. Kotanyi[3], P. Guhathakurta[4]

[1] Paris Observatory, DAEC – University
F92195 Meudon Principal Cedex – FRANCE

[2] NRAO and Columbia University, Department of Astronomy
538W 120th Street, New York, NY 10027 – USA

[3] Toulouse Observatory, Université de Toulouse III
14 Av. Belin, F31400 Toulouse – FRANCE

[4] Princeton University, NJ 08540 – USA

Being the nearest cluster, the Virgo cluster can be studied in great detail, especially concerning the effects of the environment on the galaxies. HI observations have shown that the deficiency of neutral gas in Virgo cluster spirals is more important in galaxies lying in the central region (Chamaraux et al. 1980).

To better understand how the hydrogen is removed, we have obtained twenty–four HI maps of the brightest Virgo cluster spiral galaxies. Preliminary results have already shown that for ten of the brightest spiral galaxies, the size of the hydrogen disks compared with the optical size is strongly dependent of the projected center distance (van Gorkom and Kotanyi, 1985).

We present here the whole survey on a map summarizing the integrated HI distribution of spiral galaxies (Fig. 1). The galaxies which are at smaller projected distances than $3°$ from M87 present HI disks within the optical image, indicating that the rampressure sweeping by the interstellar gas is responsible of the external hydrogen removal. Several galaxies seem just begin a pre–sweeping stage showing an asymmetric distribution pointing the sharp edge toward M87 both optically and in HI. The Virgo cluster central zone contains a number of anemic spiral galaxies with a ring shaped characteristic HI distribution leading the question of the evolutionary role of the intergalactic gas like in NGC 4579. Figure 2 gives an example of HI maps of some observed galaxies at different projected distances of M87, superimposed on optical picture. The whole survey will be soon submitted to Astron. J. (Cayatte et al., 1988).

The HI rotation curves of 21 galaxies in the Virgo cluster will soon appear in Astron. J. (Guhathakurta et al.). The curves look similar to those of spirals in the field in that the great majority of them is flat or still rising at the Holmberg radius. Our main conclusion is that the mass distribution of spirals is not affected in an obvious way by their environment. A few galaxies very close to the center have lost their HI to within almost the turnover point of the asymmetries in their velocity field as would be expected if rampressure sweeping were taking place.

M. Mezzetti et al. (eds.), Large Scale Structure and Motions in the Universe, 329–330.
© 1989 by Kluwer Academic Publishers.

330

REFERENCES

Cayatte, V., van Gorkom, J., Balkowski, C., Kotanyi, C. : 1988, in preparation.
Chamaraux, P., Balkowski, C., Gérard, E. : 1980 *Astron. Astrophys.* **83**, 38.
van Gorkom, J., Kotanyi, C. : 1985, Proceedings of the ESO Workshop on the Virgo Cluster, p. 61.
Guhathakurta, P., van Gorkom, J., Kotanyi, C., Balkowski, C. : 1988, *Astron. J.*, in press.

CAPTIONS FOR FIGURES

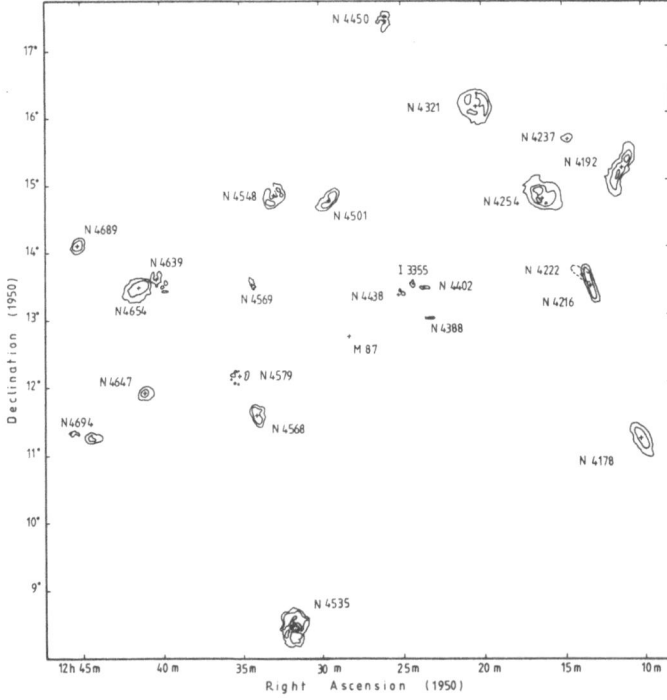

Figure 1 : Integrated neutral hydrogen in the brightest spiral galaxies of the Virgo Cluster center.

Figure 2 : Example of HI maps of some observed galaxies at different projected distances of M87 : NGC 4438 (at 1°), NGC 4579 (at 1.9°) and NGC 4254 (at 3.7°), superimposed on optical pictures.

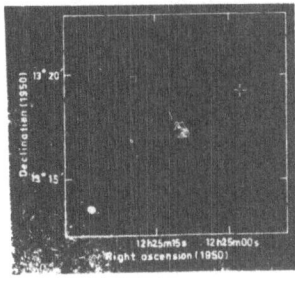
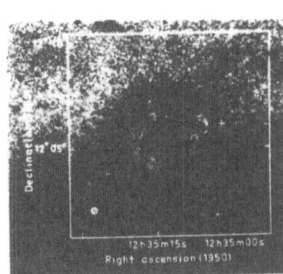
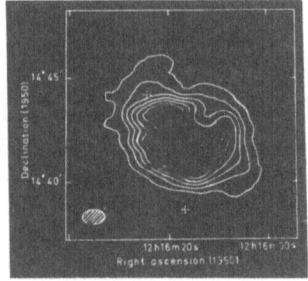

MEASUREMENT OF THE RADIO QUASAR CLUSTERING

Yaoquan Chu and Xingfen Zhu
Center for Astrophysics
University of Science and Technology of China
Hefei, China

The knowledge about quasar clustering is rapidly improving. Statistical results show strong evidence for quasar clustering at small redshifts and some evidence for its possible evolution (Chu and Fang, 1987; Zhu and Chu, 1988).

Recently, using the new Hewitt-Burbidge Quasar Catalog (1987), we have measured the correlation function of quasar clustering for different types of subsamples (i.e. radio, optical, color and x-ray selected quasars). We compared the numbers of quasar pairs with small projected and radial separations and those with large separations, and found the excess of quasar pairs with small separations and redshift differences. A very interesting result first discovered by our study is that an apparent difference of clustering exists between radio-loud and radio-quiet quasars (Fig. 1).

The radio quasars show stronger clustering than other types of quasars. The significant level for the difference between radio and optical quasar clustering is about 2 sigma. The correlation amplitude for radio quasars seems to be quite closed to that for clusters of galaxies, when we extrapolate it to small redshift. A possible interpretation is that the activity of radio quasars may occur preferentially in clusters of galaxies.

References:
Chu, Y. and Fang, L. Z., 1987, in "Observational Cosmology", IAU Symp. No. 124 (ed. A. Hewitt and G. Burbidge and L. Z. Fang), p. 627.
Chu, Y. and Zhu, X., 1988, A. Ap. in Press.
Hewitt, A. and Burbidge, G., 1987, Ap. J. Suppl., 63, No. 1.
Zhu, X. and Chu, Y., 1988, in "Proceeding of 2nd Ringsberg Workshop on High Energy Astrophysics", Tegernsee, (ed. G. Boerner).

M. Mezzetti et al. (eds.), Large Scale Structure and Motions in the Universe, 331–332.
© *1989 by Kluwer Academic Publishers.*

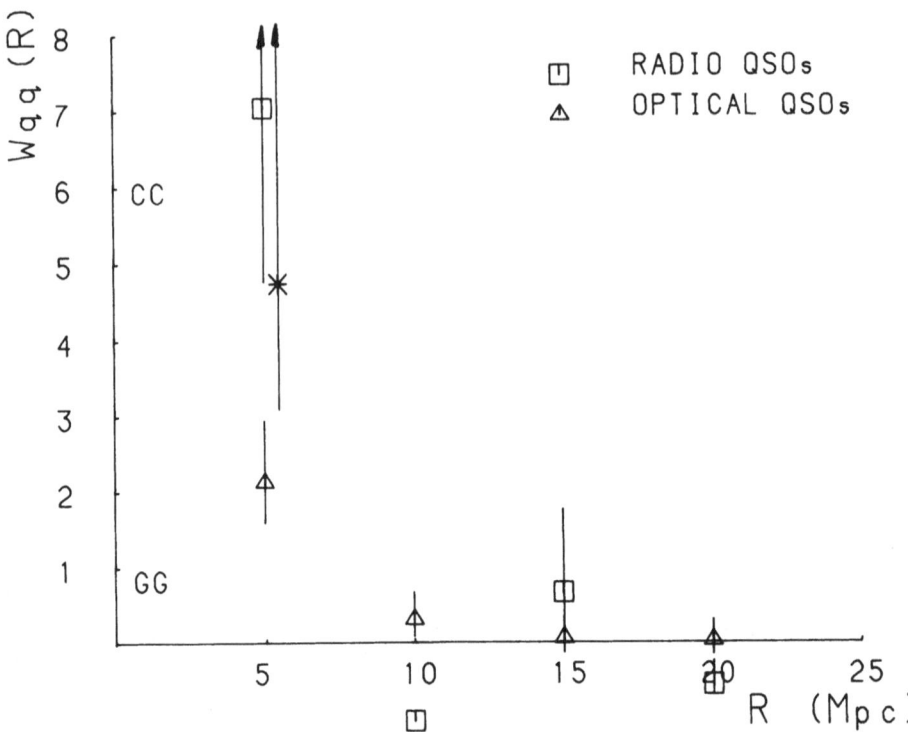

Fig. 1: The correlation functions for radio and optically selected quasars. The symbol * : radio quasars excluding 2 gravitational lens candidates.

BIASED GALAXY FORMATION IN THE COLD-DARK-MATTER SCENARIO

Shaun Cole
Institute of Astronomy,
Madingley Road,
Cambridge.
CB3. 0HA.

ABSTRACT. We consider a simple 'theory' of galaxy formation in which the luminosity L of a galaxy is determined by a universal Faber-Jackson law $L/L_* = (V/V_*)^4$, regardless of formation redshift. One consequence of such a model is that long wavelength density perturbations present in the CDM Gaussian initial conditions naturally induce a positive bias in the galaxy distribution. On large scales we find the bias factor b, the factor by which fluctuations in the galaxy number density are amplified relative to the underlying mass distribution, to be $b = 1.4 - 1.5$. While identifying rich clusters as recently virialized systems we obtain a factor 3 enhancement for their L/M relative to the global value. Thus the bias we find using this simple model is substantial, though somewhat smaller than seems to be required to reproduce the virial analysis.

Introduction

In the standard CDM scenario, non-linear dark matter condensations have been found to grow by hierarchical clustering, such that at any epoch one finds virialized structures with a wide range of masses, with a progression to higher masses with increasing time.

If galaxies form in the potential wells of these CDM condensations then galaxies are in some way fossils of certain members of this evolving hierarchy of condensations. Thus before the present day distribution of galaxies can be compared with the CDM predictions, a "theory" of galaxy formation is required to relate specific members of the time series of CDM condensations to galaxies of given luminosities.(Cole and Kaiser 1988 and Kaiser and Lahav 1988)

Any viable theory of galaxy formation in this scenario must be biased in that the distribution of galaxies that it predicts must be more clustered than the underlying mass distribution. This is necessary in order to reconcile the dynamically determined density parameter $\Omega_{dyn} = 0.2 \pm 0.1$ with the $\Omega = 1$ required by inflation.

One way in which biasing can occur naturally in the CDM scenario is through the affect of long wavelength modes present in the CDM Gaussian initial conditions. If one accepts that to a useful approximation one can associate the halos which have just formed at any epoch with overdense regions in the initial state (e.g. Press and Schechter, 1974; Bardeen *et al.*, 1986) then the effect of long wavelength modes is simply to modulate the collapse times of the dark matter condensations - perturbations will turn around earlier in a region subject to a positive long wavelength perturbation.

We shall consider a "theory" of galaxy formation in which the depth of the potential well determines the luminosity and not the amount of gas present, and assume that there exists some mechanism which causes galaxy luminosities to obey the Faber-Jackson law $L \propto V^4$ regardless of their formation redshift, and that $V_{stars} \simeq V_{halo}$.

M. Mezzetti et al. (eds.), Large Scale Structure and Motions in the Universe, 333–334.
© 1989 by Kluwer Academic Publishers.

Formalism and Results

To enable us to make quantitative estimates of the evolution of the dark matter condensations we have adopted the formalism of Press and Schechter (1974), which has been found be in reasonable agreement with N-body computations (Davis *et al.*).

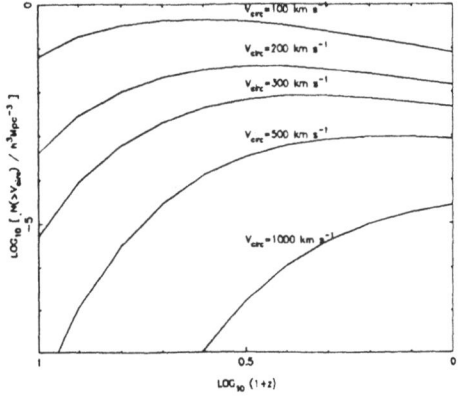

Figure 1. We plot for the Press-Schechter distribution the number density of halos $N(\geq V)$ with circular velocities greater than some value V_{circ}, against collapse redshift, for a normalization, defined by the rms in spheres of $8h^{-1}Mpc$ of $\sigma_8 = 0.6$

We see that there is a well defined era of galaxy formation, where the number density of halos with velocity dispersions $\sim V_\star = 300\ kms^{-1}$, reach their peak number density of $8 \times 10^{-3}\ h^3 Mpc^{-3}$ at a redshift of $z \simeq 1$, which agrees with the corresponding present number density of galaxies. Furthermore the mass and radius of such a condensation at redshift $z = 1$ are $M = 2.2 \times 10^{12}\ h^{-1}M_\odot$ and $r = 100\ h^{-1}kpc$, which are also reasonable for the formation of a typical $L_\star$ galaxy, after the baryons have dissipated their binding energy and collapsed by a factor ~ 10.

Having made this tentative identification of the galaxy distribution and the distribution of the CDM halos we are in a position to address the question of biasing. If we consider adding a small linearly growing background density perturbation to a given region, this will have the effect of reducing the threshold for collapse in that region, and cause any given perturbation to collapse earlier. Thus in this region the distribution of halo masses will be modified from the standard Press-Schechter distribution. We find that the resultant peak number density has a strong non-linear dependence on the amplitude Δ_0 of this background fluctuation, with the number density of galaxies with luminosity L, having a roughly exponential dependence on Δ_0, $n_{gal}(L, \Delta_0) \propto exp^{\beta(L)\Delta_0}$. Furthermore the parameter $\beta(L)$ increases with increasing luminosity, indicating that in this model the rarer more luminous galaxies are more strongly clustered.

For instance on large scales where Δ_0 is small the amplification factor by which the fluctuations in the galaxy distribution are enhanced relative to the mass distribution is $b = 1 + \beta$, and for typical $L_\star$ galaxies $\beta = 0.6$ and the bias parameter $b = 1.6$.

Another interesting parameter that we can calculate is the amount by which the L/M for clusters will be enhanced over the global value. If we identify clusters as recently virialized systems, it is then simple to calculate the modification to their luminosity functions and integrating over this new luminosity function, we find the enhancement to L/M in the clusters is about a factor 3.

References

Bardeen, J.M., Bond, J.R., Kaiser, N. & Szalay, A., 1986. Astrophys. J., **304**, 15.

Cole, S., & Kaiser, N., 1988 in preparation

Davis, M., Efstathiou, G., Frenk, C.S., & White, S.D.M., 1988. Preprint

Kaiser, N., & Lahav, O. 1988. Proceedings of the Vatican study week 'Large scale motions in the universe'

Press, W.H. & Schechter, P., 1974. Astrophys. J., **187**, 425.

TESTING COSMIC FLUCTUATION SPECTRA

H.M.P. Couchman and J.R. Bond
Canadian Institute for Theoretical Astrophysics,
University of Toronto,
Toronto, ON M5S 1A1,
Canada

ABSTRACT. We show that the angular correlation function of galaxies, $w_{gg}(\theta)$, may be used as a direct test of large scale power in cosmic density fluctuation spectra. Groth and Peebles (1986) found a break in $w(\theta)$ from a power-law at $\theta \sim 3°$ in an analysis of the Shane–Wirtanen galaxy counts. Maddox and Efstathiou (1987) find evidence for a similar break in $w(\theta)$ in the Southern Sky Survey. Testing for a break is a powerful probe of the *linear* regime of fluctuation spectra. The CDM model has such a break; almost all candidate Gaussian theories with enough large scale power to explain the clustering of clusters do not.

Calculation of $w(\theta)$

We have calculated the galaxy angular correlation function directly from various fluctuation power spectra using only linear dynamical corrections. Provided that we probe wavenumbers, k, that are large enough, $k^{-1} \gtrsim 5\,h^{-1}$ Mpc, non-linear dynamical evolution on these scales is small in most candidate theories (Bond and Couchman 1987). This is the case for the $w(\theta)$ break.

For this paper we have calculated the $w(\theta)$ using the small-angle, non- relativistic Limber equation. We have

$$w_{gg}(\theta) = \int_0^\infty b_g^2 \, \mathcal{P}_\rho(k) \, \mathcal{F}_{eff}(k; \theta, D) \, d\ln k,$$

where $\mathcal{P}_\rho(k) \equiv (k^3/2\pi^2) \langle |\delta(k)|^2 \rangle$ and b_g is the bias factor. Since an integral ($J_3(r)$) of $b_g^2 \mathcal{P}_\rho$ is typically used to normalize spectra, w_{gg} calculated in this way will be independent of the bias factor. Here J_3 has been compared with the data at $r = 10\,h^{-1}$ Mpc. The filter $\mathcal{F}_{eff}(k)$, is shown in Figure 1 for various values of θ. It can be seen that it strongly filters out short wavelengths, demonstrating that for $\theta \gtrsim 1°$ we are only probing the linear regime.

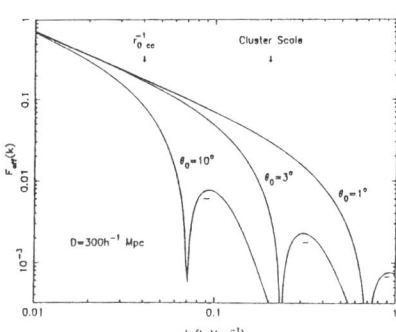

Figure 1. The effective power spectrum filter shown for three different angles. The curves show the rapid cutoff at large wavenumbers supporting the assertion that we are probing the linear regime. Also shown are the wavenumbers corresponding to cluster scales and to the cluster correlation length. We have adopted a Schechter luminosity function with $\alpha = -1$ (Efstathiou 1987) and an effective depth of $D = 300\,h^{-1}$ Mpc. (This choice is compared, for standard CDM, with a Schechter function using $\alpha = -1.25$ and $D = 360\,h^{-1}$ Mpc in Figure 2.)

To be more precise we should use the relativistic Limber equation; this refinement will have

335

M. Mezzetti et al. (eds.), Large Scale Structure and Motions in the Universe, 335–336.
© *1989 by Kluwer Academic Publishers.*

little influence on the results.

The qualitative features of our calculated angular correlations, such as the presence or absence of a break, are a reflection of the underlying spectrum. Only the position and amplitudes of the features can vary somewhat with different choices for the luminosity function and limiting magnitude.

Results

Although the standard $\Omega = 1$ CDM model does quite well in reproducing the break (Figure 2 gives plausible variations in the parameters defining it), open isocurvature baryon models and open CDM models (or equivalently similar $\Omega = 1$ models with large cosmological constants $\Omega_{vac} = 1 - \Omega_X - \Omega_B$) which have ample large scale power do not.

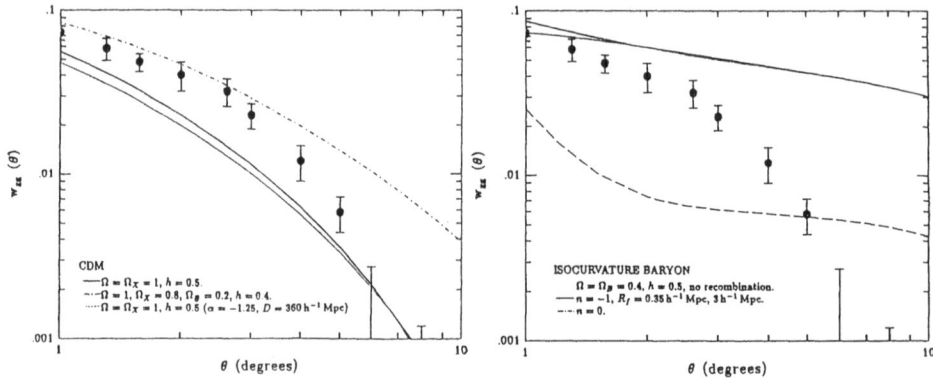

Figure 2. The figures show the calculated angular correlations compared with the data of Groth and Peebles. The individual cases are detailed for each figure. The selection function used was derived from a Schechter function with $\alpha = -1$. The depth was $300\,h^{-1}$ Mpc.

The critical wavenumber range for the break is $5\,h^{-1}\,\text{Mpc} \lesssim k^{-1} \lesssim 20\,h^{-1}\,\text{Mpc}$, over which $\mathcal{P}_\rho$ must have a significant downturn. In order to have large scale power as well, the power would have to upturn beyond this regime. The large scale power models shown in the figures do not have a sufficient feature of this kind. One possibility would be to add a peak of extra power, centred at wavenumber $k_P^{-1} \sim 100\,h^{-1}\,\text{Mpc}$ with density fluctuation amplitude $\sigma_{\rho P} \sim 0.1$, to the usual scale invariant initial conditions for the CDM model. Although such features are difficult to realize in inflation models, they would substantially increase the cluster correlation function on large scales. The penalty is the addition of a term $\sim b_g^2 \sigma_{\rho P}^2 \mathcal{F}_{eff}(k_P; \theta, D)$ to w_{gg}, which would be approximately constant for $\theta \lesssim 40°$. Since large scale gradients in the Lick catalogue were subtracted by Groth and Peebles, they may have missed such a term. Although Maddox and Efstathiou do not suffer from this subtraction difficulty over intermediate angular scales, they cannot determine whether there is an offset on $\sim 40°$ scales. (Their depth is $D \sim 500\,h^{-1}\,\text{Mpc}$). It is too early to say whether their data can warrant rejection of this possibility.

References

Bond, J.R. and Couchman, H.M.P., 1987. In: *Proceedings of the Second Conference on General Relativity and Relativistic Astrophysics*; eds. C. Dyer, A.Coley. World Scientific (1987).

Efstathiou, G.E., 1987. Private communication

Groth, E.J. and Peebles, P.J.E., 1986. *Ap. J.*, **310**, 499.

Maddox, S.M. and Efstathiou, G.E., 1987. Preprint.

DISSIPATIVE N-BODY SIMULATIONS OF CDM COSMOLOGIES

H.M.P. Couchman
Canadian Institute for Theoretical Astrophysics,

R.G. Carlberg
Department of Astronomy,
University of Toronto, Toronto, ON M5S 1A1, Canada

ABSTRACT. A dissipative N-body code is used to follow the evolution of an $\Omega = 1$, $h = 0.5$ Cold Dark Matter spectrum in cubes of 40, 80 and 200 Mpc on a side. Half of the 524288 particles, comprising one eleventh of the total mass, are treated as a gas that can cool and turn into stars. We find that the the bias present in the model, at the time identified as the present, is relatively small. The pairwise velocities on small scales are a factor two smaller in the galaxies than in the dark matter. Mergers are found to play an important role in distinguishing the histories of galaxies and dark matter haloes.

Introduction

To reconcile the standard $\Omega = 1$ Cold Dark Matter model with the $\Omega \simeq 0.2$ inferred from the motions and clustering of galaxies requires that galaxies form in such a way that they reflect only a fraction of the total matter. There have been a variety of estimates of the amount of bias present in the galaxy distribution. Davis *et al.* (1985) labelled particles in a N-body model as galaxies only if they exceeded an overdensity threshold 2.5 times the RMS fluctuation amplitude on galaxy scales as measured in the initial fluctuations. This gives an effective bias of around 2.05 (Lilje and Efstathiou 1988). A somewhat lower bias, b=1.7, was predicted using the statistics of peaks (Bardeen *et al.* 1986).

An important constraint that requires a fairly small bias is the large-scale streaming velocities of galaxies observed by Lynden-Bell *et al.* who conclude that $b = 1$ is compatible with the data but that $b = 2$ or higher is improbable.

Numerical Technique

We generate a realization of the CDM spectrum in the following way. Half of the particles are initially randomly distributed within the box. The white noise is then reduced on long wavelengths by Fourier transforming the random positions and calculating particle perturbations that cancel these waves. The expected amplitudes of the CDM spectrum are then added to this distribution using the standard Fourier technique. Gas particles are then superimposed on each dark matter particle.

The forces on the particles are calculated using a P^3M code. Gas particle collisions result in three possible outcomes; no cooling, instantaneous cooling, or star formation. The outcomes are regulated by the density and temperature of the gas as set by a simplified version of the cooling curve. If the density exceeds a critical value $d_{cool} \simeq 10^{-2} \mathrm{cm}^{-3}$ and the equivalent collision velocity temperature exceeds 10^6 K, the gas will instantaneously cool to 10^4 K. Otherwise the collision is completely elastic. If the density exceeds $d_* \simeq 5 \times 10^3 \mathrm{cm}^{-3}$ then the particles turn into stars which are subsequently collisionless.

Several tests were carried out in the smallest box to assess the sensitivity of these models to

M. Mezzetti et al. (eds.), Large Scale Structure and Motions in the Universe, 337–338.

the choice of gas dynamical parameters. The number of particles that turn into stars is sensitive to d_{cool} and not strongly influenced by d_*, implying that star formation is mostly regulated by the density of the gas that can cool. Galaxies are identified with a link length of 0.1 grid units (a 64^3 grid was used) and contain a minimum of 5 particles. If the softening length was small, fewer, but more massive, galaxies formed. This suggests that softening may affect merger dynamics in these experiments. The correlation length of galaxies is relatively insensitive to most of the parameters, a 10% variability, however the small softening experiment does show an enhanced clustering of the galaxies. It is to be expected that these few heavier galaxies will show an enhanced clustering. The relative constancy of these properties reflects the dominant influence of dark matter on the clustering and merger dynamics. We can deduce that the properties of the experiments are more strongly compromised by resolution than by the gas dynamics.

Conclusions

The significant features of these simulations are summarised below. The similar results given by experiments on three different scales lend credence to the overall success of this model.

The bias at the present epoch is found to be relatively small. Estimated values range from 1.0 to 1.6 with a preferred value of 1.3 ± 0.2. Since bias decreases as the model evolves this particular value is dependent on the normalization to the current epoch. We find that the dark haloes have $b < 1$ on small scales as the result of mergers. The $\Omega = 1$ model and the low bias factor for the galaxies remain consistent with an effective $\Omega \simeq 0.2$ because the galaxies have pairwise velocities that are typically a factor two less than those of the dark matter. On scales much larger than the correlation length galaxies have streaming velocities essentially equal to those of the dark matter, and agree well with linear theory. Because of the low bias value these velocities are compatible with the observations of large-scale streaming motions. The velocity ellipsoid of the galaxies is found to be more radial than that for the dark matter.

Differences between galaxies and dark haloes are largely the result of different merger cross sections. The merger rate of galaxies is roughly three times less than that of haloes. For galaxies the merger rate per galaxy ($M_B \leq -18.5$) is 0.01–0.02 /Gyr. This is 4 times Toomre's value, but similar to the values deduced by Bahcall and Tremaine (1988) from the data of Malin and Carter (1983). Star formation peaks near $z \simeq 5$ with galaxies being assembled at $z \sim 1.5$: Massive galaxies are almost certain to be the result of mergers. At the present epoch 42 out of 65 luminous galaxies formed by merging. We find that the peaks theory works remarkably well for galaxies assuming a peak threshold $\nu_t = 1.3(1 + z)/\sigma_0(M)$. Haloes are better described with the Press–Schecter model using an identical ν_t. For the standard measured galaxies the threshold is relatively low; $\nu_t = 0.65$.

References

Bahcall, S. and Tremaine, S.D. 1988. *Ap, J. (Letters)*, **326**, L1.

Bardeen, J.M., Bond, J.R., Kaiser, N. and Szalay, A.S., 1986. *Ap. J.*, **321**, 28.

Carlberg, R.G. and Couchman, H.M.P., 1988. Preprint.

Davis, M., Efstathiou, G.E., Frenk, C.S. and White, S.D.M., 1985. *Ap. J.*, **292**, 371.

Lilje, P. and Efstathiou, G.E., 1988. *Mon. Not. R. astr. Soc.*, submitted.

Lynden-Bell, D., Faber, S.M., Burstein, D., Davies, R.L., Dressler, A., Terlevich, R.J. and Wegner, G., 1988. *Ap. J.*, **326**, 19.

Malin, D.F. and Carter, D., 1983. *Ap. J.*, **274**, 534.

IUE OBSERVATIONS OF QUASARS IN GALAXY CLUSTER ENVIRONMENTS

E. Ellingson
University of Arizona

Richard F. Green
Kitt Peak National Observatory

H. K. C. Yee
Universite de Montreal

In previous direct imaging surveys of fields around quasars, it was found that radio-loud quasars with redshifts > 0.6 are often situated in galaxy clusters as rich as Abell class 1, while their lower redshift counterparts rarely are. We are currently engaged in a detailed study of several of these rich clusters which involves ultraviolet spectroscopy, multicolor photometry and optical multislit spectroscopy.

To quantify the richness of each individual quasar environment, we calculate the quasar-galaxy spatial covariance amplitude, B_{qq} , a measure of the enhancement of the probability of finding galaxies in nearby space. Abell richness class 1 corresponds to a B_{qq} of about 800, and richness class 0 to 300.

It has been suggested (Foltz et al., 1986) that the existance of C IV absorption features at $z_{abs} \sim z_{em}$ may be correlated with quasar environment. They claim that these systems are commonly found in the spectra of steep-spectrum radio sources, and that the absorption may represent either absorption in the host galaxy of the quasar, or in halos of galaxies in a rich cluster about the quasar.

We are conduction a detailed survey of IUE low resolution spectra of quasars at $z \sim 0.5$ for which we have deep optical images to determine if there are a correlation between the incidence of C IV absorption and the richness of the quasar environment. Such a correlation would enable us to verify the proposed nature of the absorbers as well as provide us with a valuable tool for spectrally determining the environment of higher redshift quasars.

M. Mezzetti et al. (eds.), Large Scale Structure and Motions in the Universe, 339–340.

340

The quasars in our preliminary sample are listed in Table I.
They were chosen to have a wide range of B_{qq} and were constrained by their visual magnitudes and the availability of IUE Arch-
ival data. Figure 1 shows C IV emission in the spectrum of the
quasar PKS 0405-12. None of our spectra show strong absorptions
similar to those detected by Foltz et al. Simulations indicate
that absorptions with rest EW greater than about 4 A would have
been detectable in a spectrum with a signal-to-noise ratio equal
to that of our worst spectrum.

The results from this small preliminary sample suggest that
the existance of strong absorption systems with $z_{abs} \sim z_{em}$ is
not correlated with rich galaxy cluster environment. This may
imply that the absorptions arise within the quasar host galaxy,
perhaps from material which has been accelerated by the quasar.
An alternate scenario is that halos of galaxies in rich clusters
have evolved between z=1.5 and z=0.5, causing the lack of strong
absorptions in our sample

REFERENCES
Foltz, C.B. et al. 1986, Ap. J. <u>307</u>, 504.
Yee, H.K.C., R.F. Green, H.S. Stockman 1986, Ap.J.Suppl. <u>62</u>, 681.

TABLE I

Object	m_v	z	log P (Watts/Hz/Str)	B_{qq}	total exposure time (min)
PKS 0405-12	14.8	.58	26.374	1260	622
4C 41.21	17.0	.61	26.001	-210	355
4C 31.30	15.8	.46	25.851	145	770
3C 263	16.2	.65	26.283	993	130
3C 345	15.5	.59	27.199	773	220

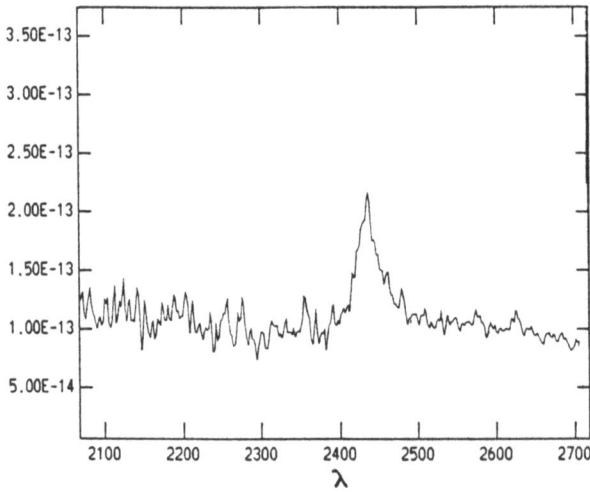

Figure 1   IUE spectrum of the quasar PKS 0405-12.

SIMULATED X-RAY CLUSTERS IN THE COLD DARK MATTER MODEL

August E. Evrard
Institute of Astronomy, Madingley Road, Cambridge CB3 0HA, UK

ABSTRACT. A "catalogue" of x-ray clusters in the cold dark matter (CDM) model is constructed numerically. The x-ray properties of these rare objects are used to constrain the normalization of the CDM spectrum, resulting in a preferred "bias parameter" $b = 1.7$. The simulated clusters also provide a means of testing the accuracy of observational x-ray mass estimates — the simplest hydrostatic isothermal model is found to underestimate the true mass within Abell radii by a systematic factor of 1.5.

341

M. Mezzetti et al. (eds.), Large Scale Structure and Motions in the Universe, 341.

SOUTHERN REDSHIFT PLOTS

A. P. Fairall
Department of Astronomy
University of Cape Town
Rondebosch
7700 South Africa

The plots below are samples from a much larger set of Declination and Right Ascension "slices", available in the catalogue by Fairall and Jones (1988) which lists optical redshifts for over 6200 galaxies south of Declination $-17\frac{1}{2}°$ (as available to late 1987). Although the data are not controlled by any uniform magnitude or diameter limit, past experience and comparison show the plots can be successfully used for the identification and delineation of superclusters and voids. In all, six superclusters and twelve voids are identified. A number of the slices reveal a distribution with a foamlike texture. Discussion, references and cross identifications are available in the catalogue.

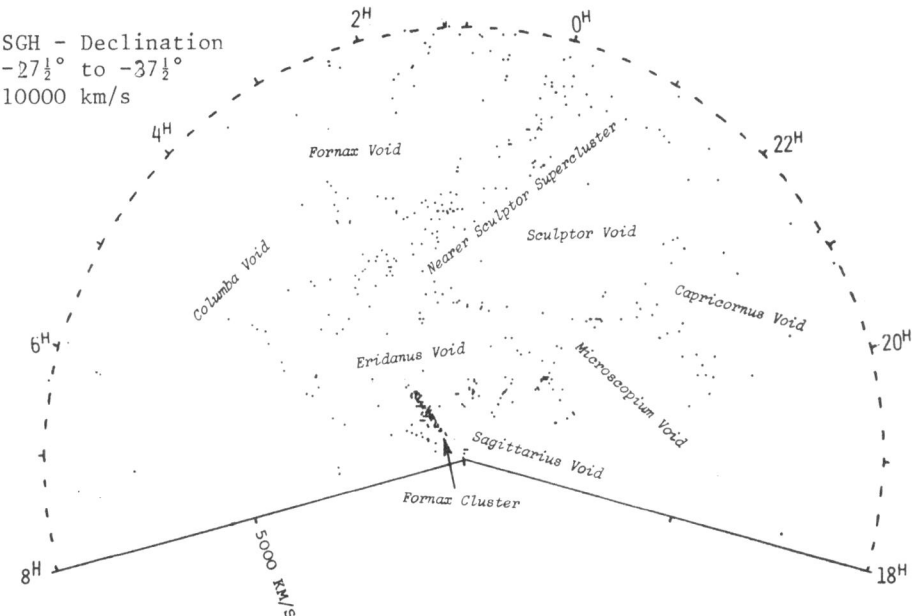

343

M. Mezzetti et al. (eds.), Large Scale Structure and Motions in the Universe, 343–344.
© 1989 by Kluwer Academic Publishers.

344

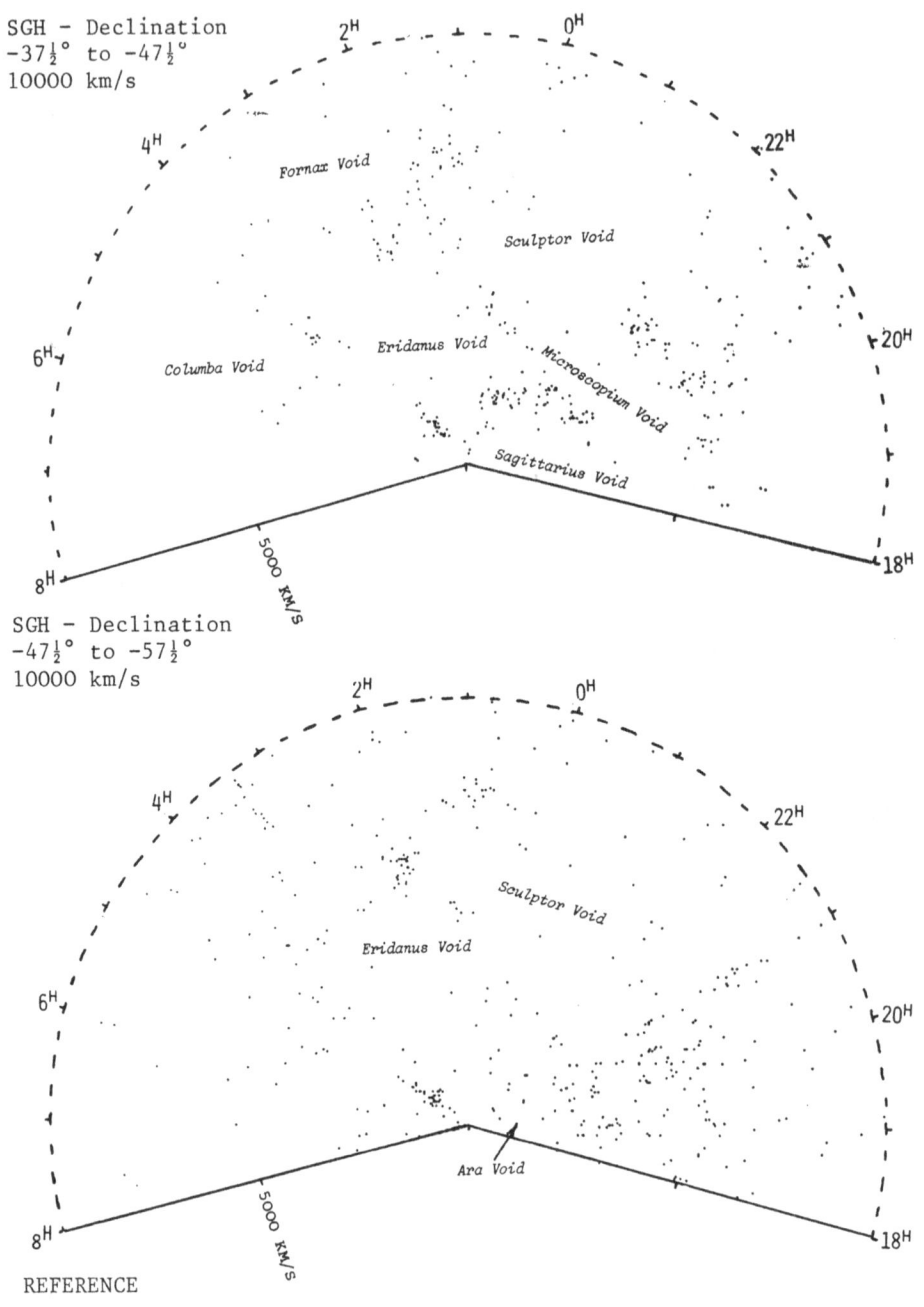

SGH – Declination
$-37\frac{1}{2}°$ to $-47\frac{1}{2}°$
10000 km/s

Fornax Void

Sculptor Void

22^H

4^H

2^H

0^H

6^H

Columba Void

Eridanus Void

Microscopium Void

20^H

Sagittarius Void

5000 KM/S

8^H

18^H

SGH – Declination
$-47\frac{1}{2}°$ to $-57\frac{1}{2}°$
10000 km/s

2^H

0^H

4^H

22^H

Sculptor Void

Eridanus Void

6^H

20^H

Ara Void

5000 KM/S

8^H

18^H

REFERENCE

Fairall, A.P. and Jones, A., 1988. Southern Redshifts – Catalogue and
Plots, Publ. Dept. Astr. Univ. Cape Town, No. 10 (available in printed
or floppy-diskette form from the author).

A filament of galaxies in the Camelopardalis region.

P. Focardi (1), B.Marano (2), R.Merighi (3), G. Vettolani(4)
(1) Dipart.to di Astronomia, Universita' di Bologna
(2) Osservatorio Astronomico di Catania
(3) Osservatorio Astronomico di Bologna
(4) Istituto di Radioastronomia, CNR, Bologna

New redshifts obtained by means of photographic spectra at the 152 cm telescope in Loiano have allowed us to derive a sample of 189 UGC [1] galaxies with $m_{ph} \leq 14.5$ and lying at RA $\in$ [6^h, 14^h], DEC $\in$ [$63°$,$78°$]. The region had been only partially covered by the Cfa Survey [2]. The sample is 84% complete.

Fifty of the galaxies constitute an extensive narrow filament having $\langle Vr \rangle$ = 2680 Km/s and RMS = 320 km/s. There are no nearby Abell clusters [3] within the limits of our region, although 4 of the 11 nearby Zwicky clusters [4] lie within the filament velocity range [5].

The structure of the filament is defined by early-type galaxies.

The filament is probably connected with the huge "bubble like" structure centered at RA = 13^h DEC = $45°$ (see Fig 1).

There is a huge void beyond the filament, between Vr $\in$ [4600, 6500] Km/s (see Fig 2) where due to our imposed sample limit in apparent magnitude, the maximum probability distribution could have been expected [6]. The existence of another void is suspected, in the range Vr $\in$ [7000,8500] km/s (see Fig 2), although measurements at these "distances" suffer from a lack of completeness effect. The velocity distribution of the near Zwicky clusters present in this region suggests the existence of the two voids.

References

[1] Nilson, P.: 1973, Uppsala General Catalogue of Galaxies, Nova Acta R. Soc. Scient. Uppsaliensis, Ser. V: A., vol. 1, Uppsala.
[2] Huchra, J., Davis, M., Latham, D., Tonry, J.: 1983, Astroph. J. Suppl. Ser., 52, 39.
[3] Abell, G.O.: 1958, Astrophys. J. Suppl. Ser., 3, 211.
[4] Zwicky F., Herzog, E., Wild, P., Karpowicz, M., Kowal, C.T.: 1961-1968, Catalogue of Galaxies and Clusters of Galaxies, 6 Volumes (Caltech, Pasadena)
[5] Baiesi-Pillastrini, G.C., Palumbo, G.G.C., Vettolani, G.: 1984, Astron. Astrophys. Suppl., 56, 363.
[6] Chincarini, G.L.: 1978, Nature, 272, 515.

M. Mezzetti et al. (eds.), Large Scale Structure and Motions in the Universe, 345-346.
© 1989 by Kluwer Academic Publishers.

346

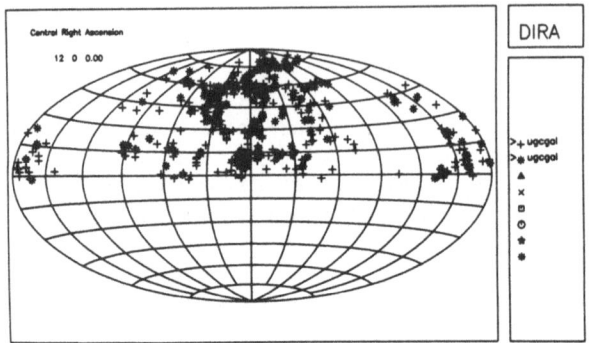

Fig 1   Hammer–Aitoff equiarea projection of UGC galaxies having
          mph ≤ 14.5 and Vr ∈ [2000,3300] Km/s. Clearly visible is the
          Camaleopardalis filament. The symbol * denotes early–type
          galaxies.

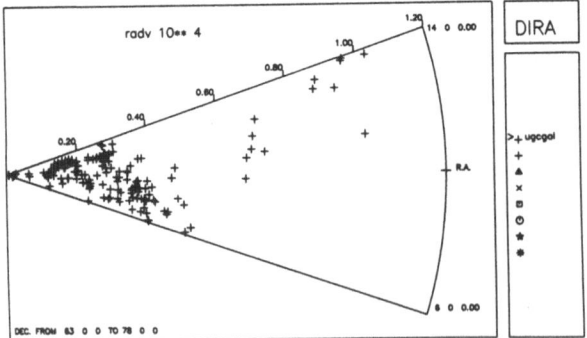

Fig 2   Wedge diagram of the surveyed region.

E and S0 Galaxies in the Perseus Supercluster

P. Focardi (1), B.Marano (2), R.Merighi (3), G. Vettolani(4)
(1) Dipartimento di Astronomia, Universita' di Bologna
(2) Osservatorio Astronomico di Catania
(3) Osservatorio Astronomico di Bologna
(4) Istituto di Radioastronomia, CNR, Bologna

The Perseus Supercluster is a huge structure belonging to the Southern Galactic Hemisphere and probably extending to the Northern one [1]. It has been the subject of many different studies.

The existence of the Perseus Supercluster was pointed out [2] on the basis of the two-dimensional distribution of the galaxies. Several redshift surveys have been performed in the Perseus region, including those described in [3], [4], [5], [6], [7].

The bulk of available data has been provided by [5] and [6].

Morphological segregation ,i.e. a higher concentration of early-type galaxies in the higher density regions, has been evidenced [8] in the Perseus Supercluster structure.

This morphological-effect seems to be quite common and can put constraints on the structure formation models.

In this paper we present results obtained from our redshift survey of the E-S0 galaxies belonging to the central main filament of the supercluster between RA [0^h-4^h] DEC [26°-44°].

There was in fact a lack of redshift data on early-type galaxies, due to the failure in detecting low HI content objects suffered by radio-redshifts (which at present constitute the largest bulk of data on the Perseus Supercluster).

In the region of interest 312 UGC [9] galaxies with $m_{ph} \leq 14.5$ are found, of which 94 are E or S0s; our new data allow us to reach 99% completeness for the E-S0 sample and 92% for the total sample.

We are extending our early-type galaxies redshift survey down to m_{ph} = 15.0, our deeper survey is presently 62% complete.

Our spectra are photographic and have been obtained with a Boller and Chivens spectrograph equipped with an EMI intensified tube, mounted at the 152 cm Loiano telescope.

Our data show that early type galaxies have a narrower spatial distribution than spirals and do define the filamentary structure; the same trend is present in the deeper sample. From the velocity histogram (fig 1) an overpopulation can be seen at higher radial velocity values for the deeper early-type sample. This is indicative of both little "virialization" in the supercluster structure and of the

347

M. Mezzetti et al. (eds.), Large Scale Structure and Motions in the Universe, 347–348.
© 1989 by Kluwer Academic Publishers.

348

presence of a further structure beyond the "PISCES VOID".

References:

[1] Focardi P.,Marano B.,Vettolani G. 1984 Astron. Astrophys.,136,178
[2] Einasto, J.,Joeveer,M.,Saar,E.:1980,Monthly Notices
     Roy.Astron.Soc. 193,353
[3] Gregory,S.A.,Thompson,L.A.,Tifft,W.G.:1981,Astrophys.J,243,411
[4] Focardi, P., Marano, B., Vettolani, G.: 1982, Astron. Astrophis.,
     113, 15.
[5] Giovanelli, R., Haynes, M.P.: 1985, Astron. J., 90, 2445.
[6] Giovanelli, R., Haynes, M.P., Myers, S.T., Roth, J.: 1986, Astron.
     J., 92, 250
[7] Huchra, J., Davis, M., Latham, D., Tonry, J.: 1983, Astroph. J.
     Suppl. Ser., 52, 39.
[8] Giovanelli, R., Haynes, M.P., Chincarini, G.L.: 1986, Astophys.
     J., 300, 77.
[9] Nilson, P.: 1973, Uppsala General Catalogue of Galaxies, Nova Acta
     R. Soc. Scient. Uppsaliensis, Ser. V: A., vol. 1, Uppsala.

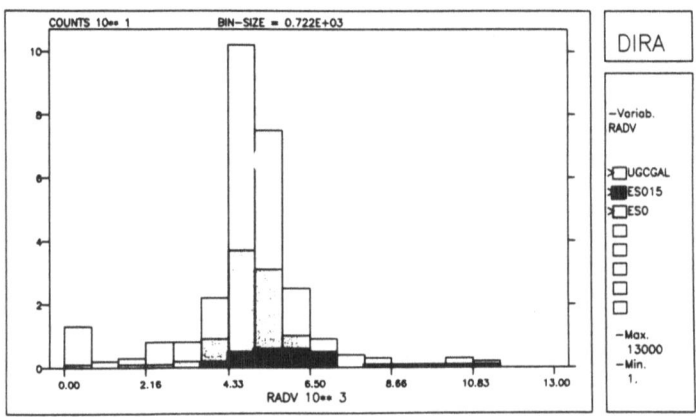

Fig 1   Velocity Histogram of our surveyed region.
          White represents the late-type UGC mph ≤ 14.5 galaxies.
          Grey represents the E-S0 UGC mph ≤ 14.5 galaxies.
          Black represents the E-S0 UGC mph ∈ [14.5,15] galaxies.

FAR-INFRARED PROPERTIES OF ACTIVE GALAXIES

K. J. Fricke, H.-H. Loose, W. Kollatschny
Universitäts-Sternwarte Göttingen
Geismarlandstrasse 11
D-3400 Göttingen
Federal Republic of Germany

1. SAMPLE OF GALAXIES

Our sample includes all active galaxies from the catalog of Véron-Cetty
& Véron (1987), which have been classified as S1 (Seyfert 1), S2 (Seyfert 2), S3 (Liner), BL (BL Lac object),or H2 (nuclear HII-region), and
in addition, which have been detected by the IRAS satellite in all four
wavelength bands at 12, 25, 60, and 100μm. The infrared fluxes have
been taken from "Cataloged Galaxies and Quasars Observed in the IRAS
Survey" (Jet Propulsion Laboratory, 1985).
 The criteria are fulfilled by 96 galaxies: 31 of type S1, 44 of
type S2, 18 of type H2, two of type S3, and one BL object.

2. IR-LUMINOSITY FUNCTION

The sample is flux-limited by the sensitivity of the IRAS detectors at
12μm. This allows an estimate of the local space density of Seyfert
galaxies, characteristic for our sample. Fig.1 shows the differential
infrared luminosity function, defined as $dN = \Phi(\log L) \, dV \, d(\log L)$,
where the luminosity L is given by $4\pi D^2 \nu f_\nu$ at 12μm.

3. MODELS FOR THE IR-FLUX

The infrared continua of Seyfert galaxies, as measured by IRAS, may in
principle include contributions from thermal and nonthermal radiation.
From an analysis of the IRAS-colors of our sample galaxies we find that
the nonthermal contribution, if present, cannot be represented by a
power law. One prominent exception is the quasar 3C273 (S1), for which
a nonthermal emission of $f_\nu = 1.85 \text{Jy} \, (\lambda/100 \text{ m})^{0.6}$ between 12 and 100μm
is deduced. This power law, extrapolated to radio frequencies, yields
excellent agreement with measurements between 0.3 and 3cm.
 The infrared continua of a large fraction of galaxies of our
sample are dominated by thermal dust emission. Hence, we have analysed
the infrared fluxes f_ν of all galaxies in terms of a simple, purely

349

M. Mezzetti et al. (eds.), Large Scale Structure and Motions in the Universe, 349–350.
© 1989 by Kluwer Academic Publishers.

thermal two-component model, where $f_\nu = (K(\nu)/D^2)[M_1 B_\nu(T_1) + M_2 B_\nu(T_2)]$.
Here, $B_\nu(T)$ is the Planck function, $K(\nu) \propto \nu$ the dust opacity, D the
distance to the galaxy, M_1 and M_2 the mass of a hot and a cool dust
component, respectively. T_1 and T_2 are the associated temperatures.
In comparison to the H2-galaxies of our sample no unusual temperatures
nor dust masses must be assumed to reproduce the IRAS-colors of the
Seyfert galaxies. In this model the relatively flat infrared continua
of many Seyfert galaxies are due to a higher ratio of hot to cool dust
M_1/M_2.

The emission at 12 and 25μm shows a stronger correlation to measured nuclear Hα-line luminosities than the far-infrared emission.
Hence, the hot dust component, which contributes mainly at 12 and 25μm,
must be more concentrated to the active nucleus than the cool dust.

Fig.2 shows the correlation of the derived masses of the hot dust
M_1 to the Hα-line luminosities for Seyfert 2 galaxies. The correlation
between M_1 and $L(H\alpha)$ is much strongerthan between M_2 and $L(H\alpha)$. We,
therefore, associate the hot dust component to the narrow line regions
of the Seyfert galaxies. Assuming a dust opacity of $K(\lambda) = 2500\text{cm}^2\text{g}^{-1}$
$(\lambda/\mu m)^{-1}$ (Hildebrand, 1983), an electron temperature of 10^4K, and an
electron density of 10^3cm^{-3}, the derived gas to dust ratios $M(\text{ion})/M_1$
lie mainly between 100 and 1000, which is not very different from
'normal' gas to dust ratios.

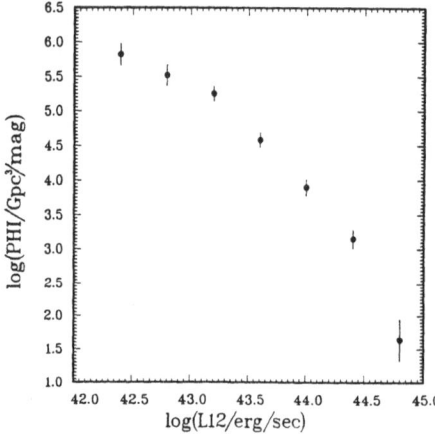
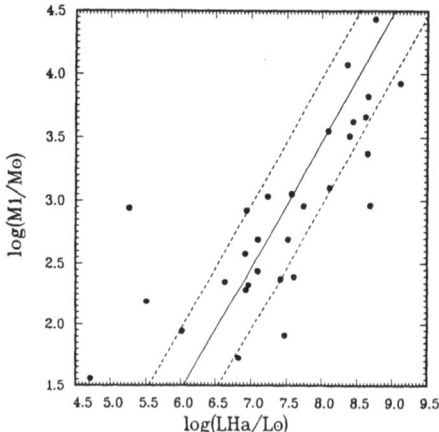

Figure 1. Combined local space
density of Seyfert 1 and 2 galaxies as a function of luminosity
L_{12} at 12μm. Error bars are 1σ
Poisson statistical errors.

Figure 2. Seyfert 2 galaxies: Mass
of hot dust $\log(M_1/M_\odot)$ versus nuclear
Hα-luminosity $\log(L(H\alpha)/L_\odot)$.The lines
are for constant gas to dust ratio;
from left to right $\log(\text{Mion}/M_1) = 2$,
2.5, and 3.

Hildebrand, R. 1983, Quaterly Journ. R. A. S. 24, 267
==
Véron-Cetty, M.-P. & Véron, P. 1987, ESO Scientific Report No. 5

CLUSTERS OF GALAXIES: CAN THE X-RAY EMITTING GAS MIMIC THE UNSEEN MATTER ?

D. GERBAL & R.SADAT
Department d'Astrophysique Extragalactique et de Cosmologie
Observatoire de Paris
92195 - Meudon Principal cedex
France.

1 - **WE SIMULATE A FAMILY** of hydrostatic two-component clusters of galaxies.

The gravitational potential is generated by both the galaxies and the X-ray emitting gas (ICM). When a dynamical analysis is performed, taking into account the luminous component only, the Dynamical unseen Mass (Mu dyn) depending on the relative concentration, is quite different from the actual or true unseen mass (Mu t).
Masses are given in unit of M_g

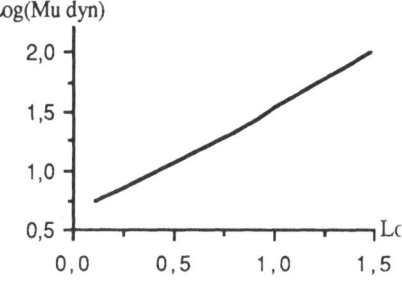

2 - **THEN, WE ADDRESS THE QUESTION:** can the x-ray emitting gas (ICM, indexed by $_x$) mimic a part or the whole amount of the unseen Mass ? If ICM is concentrated enough, does a dynamical analysis lead to an overestimate of the dynamical mass as it is suggested by the effect describe above (Limber effect)? To answer this question, we analyse two-component hydrostatic models.

Model	Theoretical unseen Mass	Dynamical unseen Mass
A	3.7	7
B	2.5	2

(table 1)

We have chosen models, such as:

- Mg = Mx, where Mg and Mx are respectively the total luminous mass and the total mass of ICM
- Temperatures are such as :

$(kTx/\mu m_p) = 0.7 \ V^2g$ (Model A),

$(kTx/\mu m_p) = 0.9 \ V^2g$ (Model B)

Tx is the temperature of the gas , V^2g is

M. Mezzetti et al. (eds.), Large Scale Structure and Motions in the Universe, 351–352.

the velocity dispersion of the galaxies (μ,k, m_p are respectively the mean molecular weight, the Boltzmann constant and the proton mass).

3 - COMMENTS: we have fitted galaxy density profile (GDP) by a King law and the x-ray surface brightness (XSB) by the modified Hubble law: $B(s) = B(0) (1+s^2/s_c^2)^{-3\beta + 1/2}$. The results of the dynamical analysis are given in table 1 (col. 3) and are compared to the theoretical results(col. 2).

* We find in the case of model A that the he Dynamical Unseen Mass is 7 times the luminous one. i.e. **the Dynamical total mass is 9 times the luminous mass.** But the GDP shows a bump in the central part while the XSB is well fitted ($\beta = 1.03$).

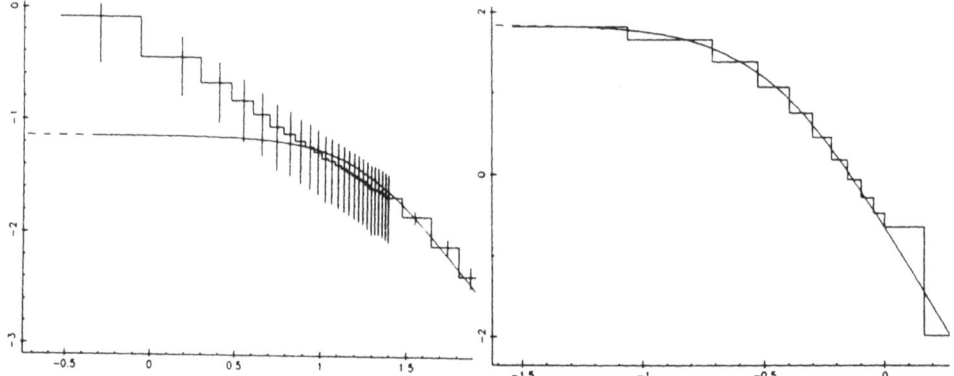

* For the model B, we find that he dynamical unseen Mass is about twice the luminous mass, **the Dynamical total mass is even 4 times the luminous mass.** The GDP and the XSB are well fitted ($\beta=0.85$).

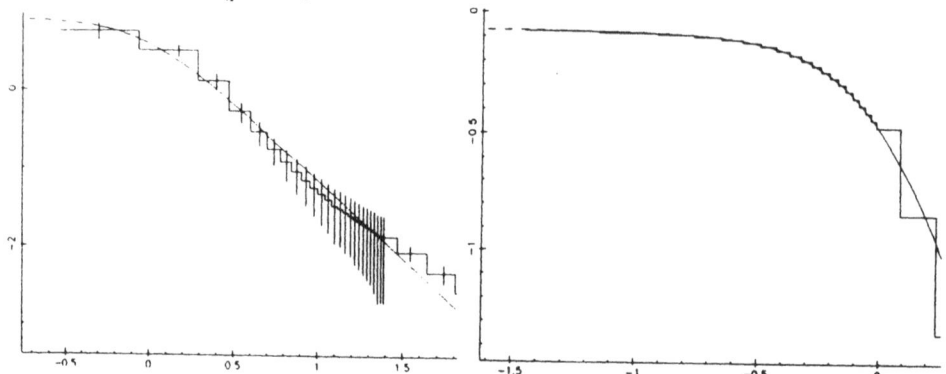

4 - CONCLUSION: a provisional conclusion arises from this analysis of two examples: the dynamical Mass is not a reliable estimate of the actual missing mass, but the x-ray emitting mass can only partly mimic (in the sense defined above) the true missing mass .

REFERENCES: Limber, D. N., 1959, Ap.J.,130,414.
                  Sadat & Gerbal (1988) Balatonfured, UAI symposium n° 130
                  Gerbal & Sadat (1988) preprint Observatoire de Paris.

THE EVOLUTIONARY STATUS OF GROUPS OF GALAXIES

G. Giuricin (1), P. Gondolo (1), F. Mardirossian (1)
M. Mezzetti (1), M. Ramella (2)
(1) Dipartimento di Astronomia
    Universita' degli Studi di Trieste
(2) Osservatorio Astronomico di Trieste

We present a method capable of determining the evolutionary stage of bound , isolated systems of galaxies. In particular, we shall apply it to a sample of groups, belonging to the Local Supercluster, taken from the Geller and Huchra's (1983) Catalogue. The knowledge of the evolutionary stage of a system of galaxies is interesting in the framework of theories concerned with the formation of large-scale structures in the universe. Moreover, if one knows the evolutionary stage, it is possible, at least in principle, to determine the mass of the system even when it is not in virial equilibrium, as well as to infer some properties of its initial condition.

In accordance with our method, we find that most of the groups in our sample are still in the phase of collapse and, thus, are not yet virialized. Their evolutionary stages are distributed around that of the Virgo cluster, which lies in a central position in the region considered. The median M/L ratio of the groups in our sample does not change significantly with respect to the usual "virial" determination, even if it proves to be slightly larger than 500 h Msun/Lsun(B) (Ho = h 100 km/sec/Mpc).

M. Mezzetti et al. (eds.), Large Scale Structure and Motions in the Universe, 353.
© *1989 by Kluwer Academic Publishers.*

VERY LARGE-SCALE STRUCTURE: OBSERVATIONAL TESTS

L.J. Goicoechea and J.M. Martín-Mirones
Departamento de Física Moderna
Universidad de Cantabria
Avda. de los Castros, s/n
39005-Santander
Spain

ABSTRACT: Assuming the existence of a very large-scale (> 100 Mpc) structure in our universe, we study the perturbation of the m-z relation by this inhomogeneity.

1. INTRODUCTION

Some recent papers have predicted the existence of a very large--scale structure in the universe (Collins et al. 1986, Lynden-Bell 1986, Melnick and Moles 1987, Shaver 1987). This inhomogeneity is the cause of the dipole anisotropy in the cosmic background radiation and deviations from the Hubble flow.

We will study the magnitude-redshift test for a universe with the above Great Attractor. We assume that this inhomogeneity is a spherically symmetric perfect fluid without pressure (Tolman 1934). Its centre will be placed at redshifts $0.1 \lesssim z \lesssim 0.5$ (in the general direction of Hydra--Centaurus).

Hubble diagrams constructed from cosmological objects at redshifts $z > 0.3$ lead to high values for q_0 (Lilly and Longair 1984, Wampler 1987). The evolutionary properties of galaxies and quasars can explain this overestimate. However, the evolutionary effects are isotropic and generate a systematic shift of points in the m-z test. The observed dispersion only can be interpreted as showing experimental problems or inhomogeneity and anisotropy effects.

The detailed anlysis of inhomogeneity in the Hubble diagram is a mathematically tedious work (MacCallum and Ellis 1970, Goicoechea and Martín-Mirones 1987). The formalism is based on the hypothesis that $1 + z$ can be expanded in powers of the luminosity distance around our present position. In this communication we wish to show the dispersion at $z \lesssim 0.1$ (we do not consider the contribution due to evolution).

2. RESULTS

Assuming the inhomogeneity as a perturbation with respect to the

355

M. Mezzetti et al. (eds.), Large Scale Structure and Motions in the Universe, 355–356.

standard models, we can obtain effective parameters H(eff) and q(eff) (where H(eff) and q(eff) are the respective standard parameters plus the anisotropic contributions ΔH and Δq due to the Great Attractor). In this corrected Hubble diagram the quadrupole term (related to ΔH) is comparable to the octopole one (related to Δq).

First, we consider a standard density contrast ($\delta \sim r^{-2}$). If the inhomogeneity is centred on z= 0.15 and our infall velocity toward the Great Attractor is v= 470 km/s (by subtracting the infall toward Virgo from the cosmic microwave background dipole amplitude), we obtain the maximum relative deviations from the standard deacceleration parameter

$$\Delta q_{max}/q_0 = 140\% \quad (q_0 = 0.1, \; M_{GA} = 4.6 \times 10^{17} h^{-1} \; M_\odot)$$

and

$$\Delta q_{max}/q_0 = 28\% \quad (q_0 = 0.5, \; M_{GA} = 2.3 \times 10^{18} h^{-1} \; M_\odot),$$

where $H_0 = 100h$ km/s Mpc is the Hubble constant and $M_\odot$ is the solar mass.

Finally, we have analyzed the Shaver's hypothesis about the possible existence of a very large-scale structure which is at distance of $\sim 800h^{-1}$ Mpc and has a diameter of $\sim 400h^{-1}$ Mpc, in a flat universe.

The relation of inhomogeneous models to observational data is discussed further. The angular dependence in ΔH and Δq allows the possibility to analyze how is the structure and where is its core (Goicoechea and Martín-Mirones 1988).

3. REFERENCES

Collins, Joseph and Robertson (1986), Nature, 320, 506.
Goicoechea and Martín-Mirones (1987), Astron. Astrophys., 186, 22.
Goicoechea and Martín-Mirones (1988), Astron. Astrophys. (submitted).
Lilly and Longair (1984), Mon. Not. R. Astr. Soc., 211, 833.
Lynden-Bell (1986), Q. Jl. R. Astr. Soc. 27, 319.
MacCallum and Ellis (1970), Commun. Math. Phys., 19, 31.
Melnick and Moles (1987), Rev. Mexicana Astron. Astrof., 14, 72.
Shaver (1987), Nature, 326, 773.
Tolman (1934), Proc. Natl. Acad. Sci., 20, 169.
Wampler (1987), Astron. Astrophys. 178, 1.

INTERGALACTIC DUST MATTER

T. Grabińska

Wrocław Technical University, Wrocław,
Department of Fundamental Research,
Institute of Physics, 50-370 Poland

ABSTRACT. An operational definition of the absorption in intergalactic space is presented. It is shown that in the case of the Central Void-discussed previously in literature- the parameter of extragalactic extinction is not zero i.e. is not neutral. Extragalactic extinction has been operationally defined.

Non-integer corrections ε_{IJ}, where I and J stand for different colours, disturb the operational sense of searches of dust matter (Rudnicki, 1983; Rudnicki et al., 1985; Zabierowski, 1983, 1985, 1988; Grabińska and Zabierowski, 1984, 1987). Galaxy counts have been searched for six possible subregions of the Central Void and the appropriate representation of Θ has been used (Zabierowski, 1985, 1988). For I=B and i=2,...,6 and J=Y and i=1, ...,5 we have obtained one value of $\gamma > 0$ and five values $\gamma < 0$; typically $\gamma = -0.03 \pm 0.3$, hence we have interpreted the results in terms of the limiting (critical) case characterized by the unique value of $\gamma =: 0$. The Central Void has been treated as a consequence of nearby interstellar cloud with $E = 0\overset{m}{.}3$ - in full agreement with the average (typical) values of physical parameters for interstellar clouds (although the problem appeared with the proper interpretation of the great z-distance of the cloud above the plane of the Milky Way).

The more sophisticated solution of the enigma of the Void is the superposition of galactic and extragalactic extinction. The extragalactic E-contribution ($E \ll 0\overset{m}{.}3$)

M. Mezzetti et al. (eds.), Large Scale Structure and Motions in the Universe, 357–358.
© 1989 by Kluwer Academic Publishers.

is empirically not excluded. The effect of possible Milky
Way and intergalactic superposition of extinction may be
in principle differentiated in this case only via comparis-
ion of $\gamma_1 (B_{i=2,\ldots,6}, Y_{i=1,\ldots,5})$ and $\gamma_2 (B_{i=3,\ldots,6}, Y_{i=2,\ldots,5})$. Negativeness of γ ought to be less distinguished
for γ_2 because dust intergalactic extinction requires
correlation with the limiting apparent magnitude μ. We find
that $\gamma_2 = +0.047$ ($\sigma_n = 0.060$) and $\gamma_1 = -0.028$ ($\sigma_n = 0.026$) hence
extragalactic extinction seems to be present and its
contribution is given by $E = 0\overset{m}{.}02$, $a = 0\overset{m}{.}09$. One can claim
that the Central Void is also a consequence of the obscuring
intergalactic matter; it has been demonstrated for five
subregions of the Central Void. This result supports the
early Rudnicki and Zabierowski interpretation of the Central
Void and in the light of the previous discussions and uncertainties the measured effect ($E = 0\overset{m}{.}02$) is extremely
important (Grabińska and Zabierowski, 1985; Rudnicki et al.,
1985).

Grabińska, T. and Zabierowski, M.: 1984, Nuovo Cimento 82B,
    235.
Grabińska, T. and Zabierowski, M.: 1985, in: Cosmic Back-
    ground Radiation and Fundamental Physics, F. Melchiorri
    (ed.), Soc. Italiana di Fisica, Bologna, 89-101.
Grabińska, T. and Zabierowski, M.: 1987, Astrophys. Space
    Science 129, 403.
Grabińska, T. and Zabierowski, M.: 1987, in: Evolution of
    Galaxies, J. Palous (ed.), IAU, Prague, 441.
Rudnicki, K.: 1983, private communication.
Rudnicki, K., Grabińska, T. and Zabierowski, M.: 1985,
    Nuovo Cimento 8C, 368.
Zabierowski, M.: 1983, unpublished, and communicated to K.
    Rudnicki.
Zabierowski, M.: 1985, Astrophys. Space Science 117, 179.
Zabierowski, M.: 1988, in: Stars and Quasars, S. Grudzińska
    (ed.), ('Is the Intergalactic Dust Matter Hypothesis
    Unfounded?'), University of M. Kopernik at Toruń, Toruń.

PECULIAR VELOCITIES OF CLUSTERS OF GALAXIES FROM m*

Donald H. Gudehus
Randall Laboratory
University of Michigan
Ann Arbor, MI 48109

Investigations over the past 15 years have gradually led to the belief that clusters of galaxies partake of large peculiar motions[1,2,3]. For example, Gudehus[1], using m^* for several clusters as a standard candle, and radius-parameter for 362 cluster galaxies as a standard yardstick, found that Virgo was moving faster than its Hubble expansion velocity by 315 and 441 km s^{-1}, respectively.

CCD photometry of A1689 and other clusters by means of a new method of galaxian photometry which takes account of overlapping images, plus revised photometry and redshifts, now allows extension of the m^* Hubble diagram to twice the redshift of the most distant cluster used in my previous study[2]. A nonlinear least squares solution yields $q_0 = 0.34 \pm 0.50$ with a mean error for an observation of average weight of $0\overset{m}{.}07$. A peculiar space velocity for each cluster is calculated by adding to the peculiar radial velocity, a tangential component which is equal to zero in a frame at rest in the Local Group. These vectors are plotted in Figure 1 as projected on the best fitting plane through the cluster coordinates on the sky. The average of the space velocity vectors points to $\alpha(1950.0) = 10\overset{h}{.}80 \pm 0\overset{h}{.}28, \delta(1950.0) = -15\overset{\circ}{.}7 \pm 5\overset{\circ}{.}2$ with a magnitude of 685$\pm$96 km s^{-1}. The null result for the peculiar velocity of Virgo found by Kormendy[4] is thought to be due to a newly discovered bias present in surface brightness vs. $\log(radius-parameter)$ plots.

To test for the effects of dynamical evolution on m^*, I present in Figure 2, Hubble radius-parameter plotted against absolute magnitude for 344 galaxies in 5 clusters (H_0=50 km s^{-1} Mpc^{-1}). These data are in agreement with an earlier study[1] and suggest that it is unlikely that $M_V^* = -21\overset{m}{.}15$ is affected by dynamical evolution since it is $1\overset{m}{.}53$ fainter than the change of slope at $M' = -22.68$ ($a' = 823$ pc).

The Virgocentric velocity is calculated in a model and Hubble constant independent way as $v_{Virgocentric} = v_{r,pec,Virgo} - v_{LG,pec}\cos\theta = 528 - 546\cos\theta = 70 \pm 68$kms^{-1}, where $\theta = 33\overset{\circ}{.}1$. There is no infall, but an outflow, once the Hubble expansion is removed. The Hubble constant is found to be $H_0 = (965 \pm 30)$x$10^{-0.2(m-M)_{Virgo}+5}$ km s^{-1} Mpc^{-1}. If $(m - M)_{Virgo}$=31.70[5], H_0=44.1$\pm$1.6 km s^{-1} Mpc^{-1}. If $\Omega = 1$, the age of the Universe is (14.8 ± 0.5)x10^9 yr, giving agreement with recent estimates of the ages of globular clusters[6]. Thus there is no necessary conflict between the global and galactic time scales. The required mass

359

M. Mezzetti et al. (eds.), Large Scale Structure and Motions in the Universe, 359–360.
© *1989 by Kluwer Academic Publishers.*

360

of the hot dark matter candidate, the neutrino, is decreased to 17 eV because of the smaller value of H_0.

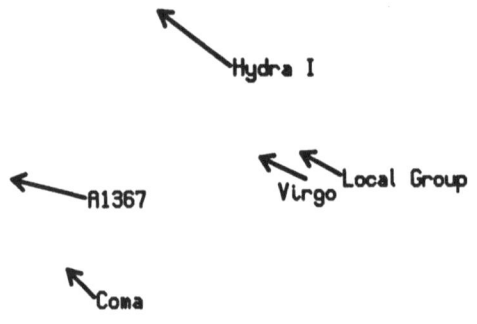

Figure 1. Peculiar space velocity vectors of several nearby clusters

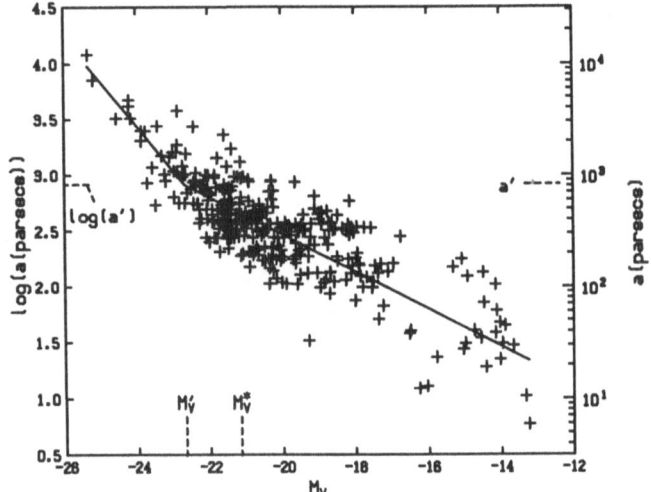

Figure 2. Radius-parameter vs. absolute magnitude for 344 galaxies in 5 clusters

REFERENCES

1. Gudehus, D. H. *Astron. J.* **78**, 583-593, (1973).
2. Gudehus, D. H. *Nature* **275**, 514-515 (1978).
3. Dressler, A., Faber, S. M., Burstein, D., Davies, R. L., Lynden-Bell, D., Ter-levich, R. J., & Wegner, G. *Astrophys. J. Lett.* **313**, L37-L42 (1987).
4. Kormendy, J. *Astrophys. J.* **218**, 333-346 (1977).
5. Sandage, A. & Tammann, G. A. *Astrophys. J.* **210**, 7-24 (1976).
6. Van den Berg, D. *Bull. A. A. S.* **19**, 675-675 (1987).

COULD THE GRAVITATIONAL LENSING ALTER THE INTERPRETATIONS OF THE OBSERVATIONAL DATA IN THE EXTRAGALACTIC FIELD ?

François HAMMER
Observatoire de Meudon-DAEC
92195 Meudon Principal Cedex
FRANCE

To deduce intrinsic source properties (absolute luminosity, radio flux, size...), we currently use z as a distance indicator; following the Friedmann (or R-W) model, the main relations are based on the **Cosmological Principle.** However, there are inhomogeneities of (luminous or dark) matter up to 100 Mpc which potentially modify light beams passing near or across them. This is describe by the gravitational optics theory. It leads to :

→ a change of the source location (/ of the null deflecting mass).

→ temporal effects (time-delay and redshift effects).

→ affect the light beam properties following the Optical Scalar Equations (Sachs, 1961). It depends on two main terms, both being proportional to the geometrical term $D_d D_{ds}/D_s$ (angular diameter distances, d : deflector, s:source) : (1) K (Ricci term) due to the matter contained into the beam, which provides an expansion of the beam without distorts it; (2) Γ (shear term) due to the matter contained into the beam, which distorts and may multiply the beam. Note that the source surface brightness is unchanged by the gravitational lensing, but its surface is amplified by $Amp = [(1 - K)^2 - \Gamma^2]^{-1}$ It does not depend on the wavelength (geom. optic approxim.), but may apparently change the source color by either a differential lensing of a source for which the spatial structure depends on wavelength, or a blend of the lens with the source images (case of a galaxy source).

WE HAVE TO TAKE INTO ACCOUNT GRAVITATIONAL LENSING EFFECTS FOR VERY DISTANT SOURCES AS DISTANT GALAXIES (3 CR) OR QSOS :

→ foreground galaxies (z > 0.05) gravit. amplify and could multiply background sources (because of a strong shear term).Imagery by Tyson (1987) of an apparently empty field, shows an 1/3 sky coverage by galaxies up to R=26 (i.e. dominated by z~2 objects).

→ foreground clusters of galaxies (z > 0.1) could amplify sources by several magnitudes but rarely multiply them (predominant Ricci term).Clusters of galaxies are covering the sky up to z = 1 to 2 .

→ **SELECTION EFFECTS**:

* Gravitationally amplified sources are preferentially selected in a flux limited sample (with respect to the non-amplified sources). If we build a sample of sources having a flux close to the flux limit (or detection limit), it may be filled by only gravit. amplified sources !.

* To deduce intrinsic source properties, we need to build an unbiased sample with respect to the foreground matter : strong gravitational amplification of few sources is balanced by very faint attenuation of a large number of sources.

CONSEQUENCES :

→ Overestimation of absolute luminosities and/or sizes of distant sources by several magnitudes, meaning an overestimation of the source evolution and/or the q0 value . Gravitational amplification of light by foreground matter affects the Hubble diagram of Brightest Cluster Galaxies (Hammer,

M. Mezzetti et al. (eds.), Large Scale Structure and Motions in the Universe, 361–362.

Nottale, 1986, Astron. Astrophys., 167, 1) as well as the 3CR high-z galaxies (Hammer et al, 1986, Astron. Astrophys., 169, L1 and the poster by O. Le Fevre and F. Hammer, here). The following text presents some new results on QSOs.

→ Explanation of certain association of galaxies as VV 172, for which one galaxy has a discordant redshift (see Hammer and Nottale, 1986, Astron. Astrophys., 155, 420).

→ Provides spectacular images for observers as the ring-like structures in the center of rich distant clusters of galaxies (see the following text).

SOME NEW RESULTS ON GRAVITATIONAL LENSING

1) Effect of the intervening matter associated with absorption lines on the QSO luminosities.

Presence of narrow metallic absorption lines (or damped Ly α) with $z_{abs} < z_{em}$ is currently interpreted as caused by intervening matter (galaxies, haloes) along the QSO line of sight. It may affect the QSO luminosity : using samples containing absorbing QSOs (Hewitt and Burbidge, 1986; Wampler et al, 1984; Young et al, 1982; Lanzetta et al, 1987), we find that *gravitational amplification of QSOs by absorbing foreground matter is present ($\approx$ 1-2 mag) at 4-5 σ significance level (Nottale and Hammer, in prepar.).* The correlation remains available if we deal with the 26 QSOs from the Young et al (1981) and Lanzetta et al (1987) samples, which verified $1.6 < z_{em} < 2.4$ (to avoid evolution effects) and had been spectroscopically investigated with FWHM ≤ 5 A from blue to red wavelengths (to avoid biases coming from spectroscopic instrumentation).

2) Gravitational lens model for the A 370 ring (Hammer, 1987, 3rd IAP meeting, P.467).

→ source is a z=0.72 galaxy (Soucail et al, 1988) lensed by matter in the A370 center (z=0.374).

→ dark matter is assumed being associated with luminous or X matter.

→ A 370 cluster and the (16) cluster or field galaxies are modelized by a de Vaucouleurs law.

The left figure shows the isophotal contours of the A 370 center to be compared with the model predictions shown in the right one (collaboration with F. Rigaut); the main results are:

* to predict a cluster mass compatible with the virial mass ($1.5 \ 10^{15}$ $M_\odot$, see Mellier et al, 1987).
* the 2nd image is strongly unstable to a secondary deflector; it is attenuated if $M_{\neq 20} \geq 5 \ 10^{12}$ $M_\odot$.
*the break of the ring eastern part is well explained if the two galaxies have $M_{\neq 37} \geq 1.5 \ 10^{12}$ $M_\odot$.
*to predict an elliptical shaped source (a:b$\approx$2) which account for the differential thickness from we to east of the ring, and which is compatible with the arc spectroscopy indicating a spiral galaxy.
*this model depends on *only 2 free parameters* and may *reproduce pixel to pixels* the observations.

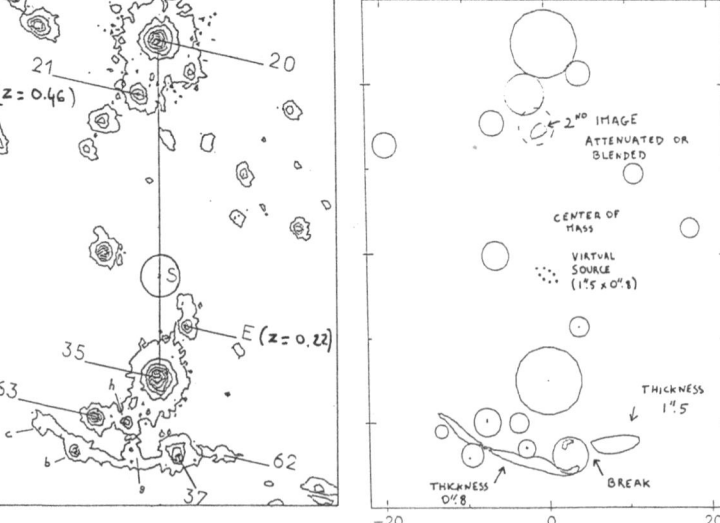

GALAXIES FOUND AT 21 CM IN A COSMIC VOID

P. A. Henning and F. J. Kerr
Astronomy Program
University of Maryland
College Park, MD 20742
U.S.A.

Recent redshift surveys indicate that the vast majority of galaxies included in magnitude-limited samples appear to lie on the surfaces of "bubbles" that surround regions of low galaxy density, known as "voids". There is no hard evidence, however, that these voids are not populated by low surface brightness (LSB) galaxies. One constraint on models of large-scale structure formation is the relative distributions of the LSB and high surface brightness (HSB) galaxies. For instance, biased cold dark matter models predict that LSB galaxies should be more uniformly distributed than HSB galaxies. In this scenario, cosmic voids could be filled with LSB objects while HSB objects would be found preferentially in the denser environment of the shells.

We have searched for neutral hydrogen in the cosmic void at $23^h \leq \alpha \leq 01^h$, $30° \leq \delta \leq 33°$ using the 300-ft telescope at Green Bank, WV. Our search differs from optical and radio redshift surveys in that we did not start with a catalog, but used a blind search method. We observed over 250 arbitrarily chosen lines of sight which penetrated the void and pierced its back wall. This corresponds to about 7% of the region. We found one, and possibly two sources of 21 cm emission inside the void. Our next step is to observe the galaxy with the VLA in order to pin down its position more accurately. Presently, we know its position only to within one 300-ft beam area (about 10' between half-power points). There are no catalogued galaxies within the beam at that position, but there is a small, uncatalogued, blue galaxy visible on the Palomar Observatory Sky Survey prints within a beam area of the HI detection. At this point all we can say is that this object may be the galaxy we detected but we cannot be sure until we determine its redshift and check its optical position against a more accurate HI position. Also, during the search we detected two catalogued galaxies which lie on the edge of the void.

M. Mezzetti et al. (eds.), Large Scale Structure and Motions in the Universe, 363–364.
© *1989 by Kluwer Academic Publishers.*

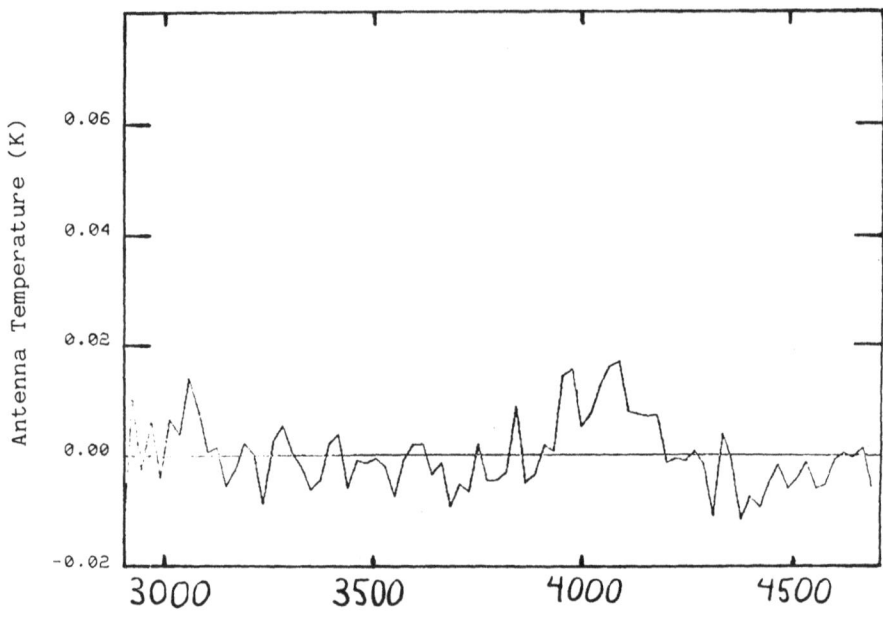

Heliocentric Velocity (km/s)

HI profiles of the galaxies found in the void. Because of the large beam size of the 300-ft. telescope at this wavelength, we do not yet know the precise position of our sources. Accurate position measurements will lead to definitive values for the redshifts and HI flux densities.

The Near Background of the Virgo Cluster

G. Lyle Hoffman
Department of Physics
Lafayette College
Easton, PA 18042 USA

Bright galaxies are sparse in the volume of the universe immediately behind the Virgo Cluster out to the Coma/A1367 supercluster at $V_\odot \sim 6000$ km s^{-1}. All *Virgo Cluster Catalog* (Binggeli *et al.* 1985) dwarf irregular galaxies, and all spiral galaxies labelled "member" or "possible member" of the cluster in that catalog, along with most of the "background" spirals, now have HI observations from Arecibo (Helou *et al.* 1984; Hoffman *et al.* 1987, 1988 and work in progress), and we can explore the large-scale structure of that void using a fainter sample than has been previously available. All but one of the galaxies in the *VCC* survey area with $3000 < V_\odot < 6000$ km s^{-1} lie in a well-defined filamentary structure as shown in Fig. 1, which plots the positions on the sky of all *VCC* galaxies with known velocities in the range $3000 < V_\odot < 6000$ km s^{-1}. There are *no* galaxies with $3000 < V_\odot < 3500$ km s^{-1}. The symbols indicate $V_\odot$: triangles indicate $3500 < V_\odot < 4000$ km s^{-1}, and the symbol acquires an additional side for each additional 500 km s^{-1}. The circle represents a 5° cluster core, and the irregular boundary shows the limits of the *VCC* survey. All galaxy types are represented, and the shape of the luminosity function is quite similar to that for the Virgo Cluster itself. (There is no preponderance of dwarf galaxies *unless* a substantial number of dwarf elliptical galaxies, few of which have redshifts measured as yet, should turn out to lie at these velocities.) A Tully-Fisher analysis suggests that on average the filament lies about 3 times further than the Virgo Cluster—*i.e.*, about twice as far away as the W group. Although the W group lies near the filament on the sky, there is an 800 km s^{-1} gap in velocity between the highest $V_\odot$ in the W group and the lowest $V_\odot$ in the filament. There is too much scatter in the Tully-Fisher relations for us to determine whether the structure is truly a filament (no deeper than it is wide), or a sheet seen on edge.

The W group has been recognized for some time as an aggregation of mainly early-type galaxies southwest of the Virgo Cluster, at a distance approximately twice that of the cluster itself. In Fig. 2, the "classical" W group is the dense knot around 12^h18^m, 6°, with a number of other early-type galaxies spread more diffusely up to 12^h11^m, 8°, and down to 12^h22^m, 4.5°. (In the map, all *VCC* galaxies with $1500 < V_\odot < 3000$ km s^{-1} are shown. The velocities are coded as in Fig. 1, but with triangles representing $1500 < V_\odot < 1800$ km s^{-1} and each additional side representing an additional 300 km s^{-1}.) We find that this knot is surrounded by a cloud of dwarf irregular galaxies stretching especially to the

M. Mezzetti et al. (eds.), Large Scale Structure and Motions in the Universe, 365–366.

south. In particular, almost half the BCDs in the *VCC* area with known velocity in the Virgo Cluster range lie within this cloud. The velocities of the spirals and dwarf irregulars tend to be smaller than those of the early-type galaxies, in the range 1500-2500 km s^{-1} while the latter occupy mainly 2000-3000 km s^{-1}. The blue Tully-Fisher relation applied to the spiral and dwarf irregular galaxies in the cloud gives a mean distance of 1.7 d_{Virgo}, consistent with previous determinations for the brighter spirals alone (Huchtmeier 1984).

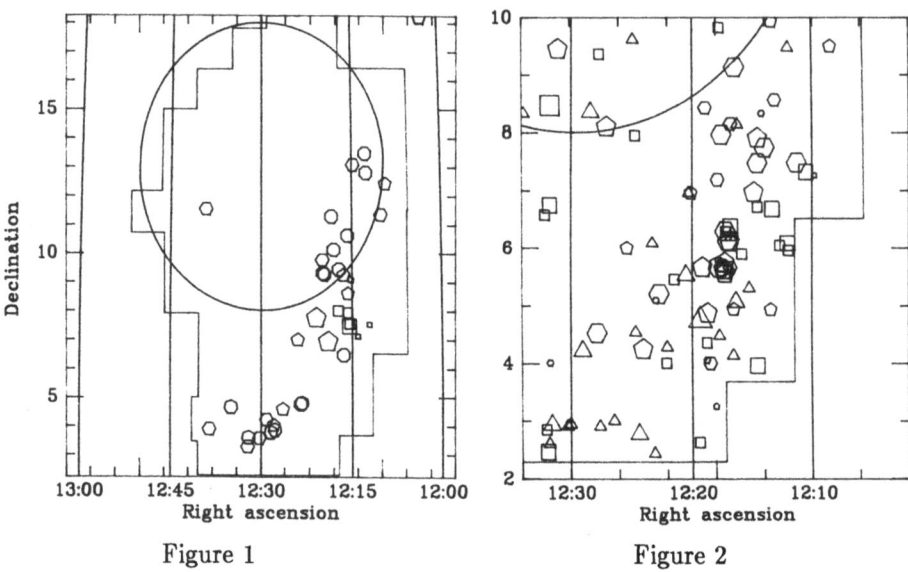

Figure 1   Figure 2

The assistance of B.M. Lewis and H.L. Williams in acquiring the data, and many discussions with E.E. Salpeter, are gratefully acknowledged.

References:

Binggeli, B., Sandage, A., and Tammann, G.A. 1985, *A. J.*, **90**, 1681.
Helou, G., Hoffman, G.L., and Salpeter, E.E. 1984, *Ap. J. Suppl. Ser.*, **55**, 433.
Hoffman, G.L., Helou, G., Salpeter, E.E., Glosson, J., and Sandage, A. 1987, *Ap. J. Suppl. Ser.*, **63**, 247.
Hoffman, G.L., Lewis, B.M., Helou, G., Salpeter, E.E., and Williams, H.L. 1988, *Ap. J. Suppl. Ser.*, submitted.
Huchtmeier, W.K. 1984, in *The Virgo Cluster*, ed. O.-G. Richter and B. Binggeli (Garching: European Southern Observatory), p. 23.

PROPERTIES OF GALAXIES AND THE DISTANCE OF THE VIRGO CLUSTER

W.K.Huchtmeier

Max-Planck-Institut fur Radioastronomie

Auf dem Hugel 69

5300 Bonn-1 ,F.R.G.

A nearly comlete sample of 252 galaxies of type Sa and later in the Virgo cluster area is detected in the 21cm line of neutral hydrogen (Huchtmeier & Richter, 1986a). Another 46 galaxies - from their (optical) radial velocities to be considered as members of the cluster - were not detected in the 21cm line. Global parameters of the galaxies of this sample have been compared with a sample of nearby galaxies (v < 500 km/s ; Huchtmeier & Richter, 1986b). Galaxies of similar luminosity in both samples have similar sizes and total masses. However, a definite deficiency in HI content is observed in the Virgo cluster. This deficiency is found to increase towards the centre of the cluster and towards early type galaxies.

The Tully-Fisher relation for the nearby sample and the cluster sample agrees assuming a Virgo distance of 21 Mpc (see Fig. 1) . The slope of the blue Tully-Fisher relation as derived in this

M. Mezzetti et al. (eds.), Large Scale Structure and Motions in the Universe, 367–368.

paper is -7.17 (+- 0.20) combining the Virgo sample with the sample of nearby galaxies, which is exactly the value of our calibration of the blue TF for nearby groups of galaxies (Richter & Huchtmeier, 1984). Now the IR Tully-Fisher relation seems to converge to a slope of -7, too (e.g. Geller, this conference).

References:

Huchtmeier, W.K., and Richter, O.-G.: 1986a Astron.& Astrophys.Suppl.Ser. 64, 111

Huchtmeier, W.K., and Richter, O.-G.: 1986b Astron.& Astrophys.Suppl.Ser. 63,323

Richter, O.-G., and Huchtmeier, W.K.: 1984 Astron.& Astrophys. 132, 253

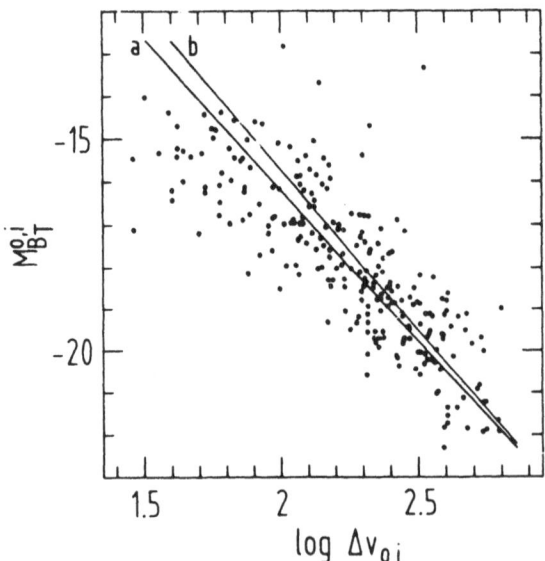

Fig. 1 : The correlation between blue absolute magnitude and the HI-line width (i.e. the Tully-Fisher relation) for Virgo galaxies (line b) is very similar to that describing the comparison sample of nearby galaxies (line a).The average slope of both lines corresponds to -7.17 .

THE CLUSTERING OF QUASARS

A. IOVINO[1], P.A. SHAVER[1],
P.S. OSMER[2], P. HEWITT[3],
D. CRAMPTON[4], A.P. COWLEY[5], F.D.A. HARTWICK[6],
C. BARBIERI[7], S. CRISTIANI[7]
[1]*European Southern Observatory,*
[2]*National Optical Astron. Obs.,* [3]*Univ. of Cambridge,*
[4]*Dominion Astrophys. Obs.,* [5]*Arizona State Univ.,* [6]*Univ. of Victoria,*
[7]*Univ. of Padova*

ABSTRACT. In an attempt to obtain the highest possible sensitivity to quasar clustering using homogeneous samples, we have combined four deep surveys totalling 579 quasars. This reveals a 5.5σ excess of pairs with separations $\leq 10h^{-1}$ Mpc at $z \leq 1.5$.

1. INTRODUCTION

Because of the low space density of quasars, it has proven difficult to measure their clustering at a high level of significance. One approach has been to apply novel analysis techniques to large but heterogeneous catalogues of quasars. Another is to assemble quasar surveys of sufficient depth and homogeneity that clustering can be detected using conventional methods. It has become clear that, at the present sensitivity of such surveys, several hundred quasars are required before any significant statements can be made regarding clustering. We here present the results for the largest sample of quasars from deep surveys so far assembled.

Four homogeneous samples have been used here: the CFH/MMT Grens survey (201 quasars, Crampton et al. (1987) and unpublished data), a UVX survey published by Boyle, 1986 (171 quasars), a grism survey by Osmer and Hewitt, in preparation (127 quasars), and a UVX survey by Barbieri et al., in preparation (80 quasars). The total number of quasars in these deep surveys is thus 579.

2. ANALYSIS AND RESULTS

The most conservative and conventional analysis technique has been used here (*cf.* Iovino & Shaver, 1988). For each sample the overall distributions of the objects in both redshift and position were reproduced in large randomly generated samples. Comparison of the real and random samples as a function of comoving linear separation of quasar pairs then yielded the correlation function in the standard way (*e.g.* Peebles, 1980).

The results are shown in Table 1 and in figure 1. There is a 5.5σ excess of pairs with separation $\leq 10h^{-1}$ Mpc at redshifts ≤ 1.5, and no significant excess at higher redshifts. The difference between the correlation amplitudes at low and high redshifts has a significance of 2.6σ.

369

M. Mezzetti et al. (eds.), Large Scale Structure and Motions in the Universe, 369–370.
© *1989 by Kluwer Academic Publishers.*

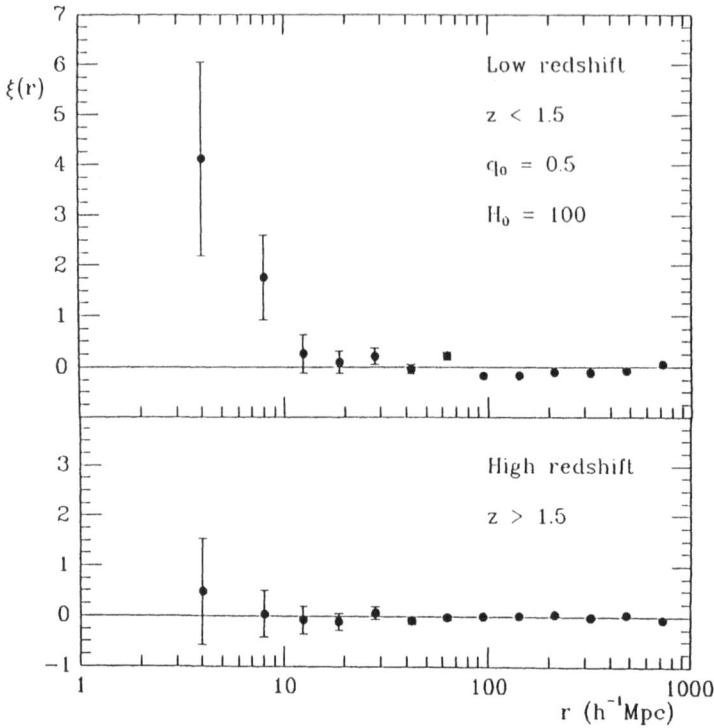

Fig. 1 Quasar two-point correlation function at low (z < 1.5) and high (z > 1.5) redshifts, from the samples used.

TABLE 1

Number of Quasar Pairs with Separation $\leq 10h^{-1}$ Mpc

	z < 1.5	z > 1.5	Data Set
observed (expected)	9 (2.3)	0 (1.4)	Crampton et al., 201 quasars
observed (expected)	6 (2.1)	4 (3.0)	Boyle (1986), 171 quasars
observed (expected)	0 (0.2)	2 (1.5)	Osmer & Hewett, 127 quasars
observed (expected)	3 (1.0)	1 (0.2)	Barbieri et al., 80 quasars
observed (expected)	18 (5.3)	7 (6.2)	Total Sample, 579 quasars

REFERENCES

Boyle, B.J., 1986, *Ph. D. thesis*, University of Durham.
Crampton, D., Cowley, A.P., Hartwick, F.D.A., 1987, *Ap.J.*, **314**, 129.
Iovino, A., Shaver, P.A., 1988, *Ap. J. Letts*, in press.
Peebles, P.J.E., 1980, *The Large Scale Structure of the Universe*, Princeton Series in Physics.

THE CLUSTERING OF HII GALAXIES

A. IOVINO, J. MELNICK, P.A. SHAVER
European Southern Observatory
Karl-Schwarzschild Str. 2
D-8046 Garching bei München, W. Germany

ABSTRACT. HII galaxies, a class of gas-rich low-mass galaxies, are three times less clustered than normal galaxies, with a clustering scale length of $2.7h^{-1}$ Mpc.

1. SAMPLE USED

We used a sample of 230 HII galaxies, from the Spectrophotometric Catalogue of HII Galaxies (Terlevich et al., 1987). These galaxies, obtained from southern objective prism surveys, have an estimated mass in the range between $10^{7.5} M_\odot$ and $10^{9.5} M_\odot$, with a median value of $10^{8.8} M_\odot$.

2. METHOD OF ANALYSIS

While the sky distribution of HII galaxies is rather haphazard, the redshift distribution is fairly uniform and can be used to define a random comparison sample by simply randomizing the redshifts in accordance with the observed distribution. $\xi(r)$ results then from comparison of the observed and random samples using the usual formula:

$$\xi(r) = N(observed)/N(random) - 1$$

In addition, our sample covers the same redshift range as the Durham/AAT galaxy survey (Peterson et al., 1986) which, therefore, provides a suitable sample of normal galaxies to which the same method can be applied. This permits a direct and unambiguous comparison between the clustering properties of HII galaxies and those of normal galaxies.

3. RESULTS

Fig. 1 shows that HII galaxies are significantly clustered on scales $\leq 10h^{-1}$ Mpc, but with an amplitude considerably lower than for AAT galaxies. The form of the HII galaxy correlation function is similar to that of the AAT sample. Fitting our data below $10h^{-1}$ Mpc with the usual power law:

$$\xi(r) = \left(\frac{r}{r_o}\right)^\gamma$$

we obtain $\gamma \sim 1.3$, and $r_o \sim 2.7h^{-1}$ Mpc. The ratio of the integrated correlation amplitudes of the HII galaxies relative to the Durham/AAT galaxies below $10h^{-1}$ Mpc is 0.32 ± 0.08.

M. Mezzetti et al. (eds.), Large Scale Structure and Motions in the Universe, 371–372.
© *1989 by Kluwer Academic Publishers.*

This result, pertaining to a class of low mass objects, represents an important observational constraint for cosmological theories of large scale structure.

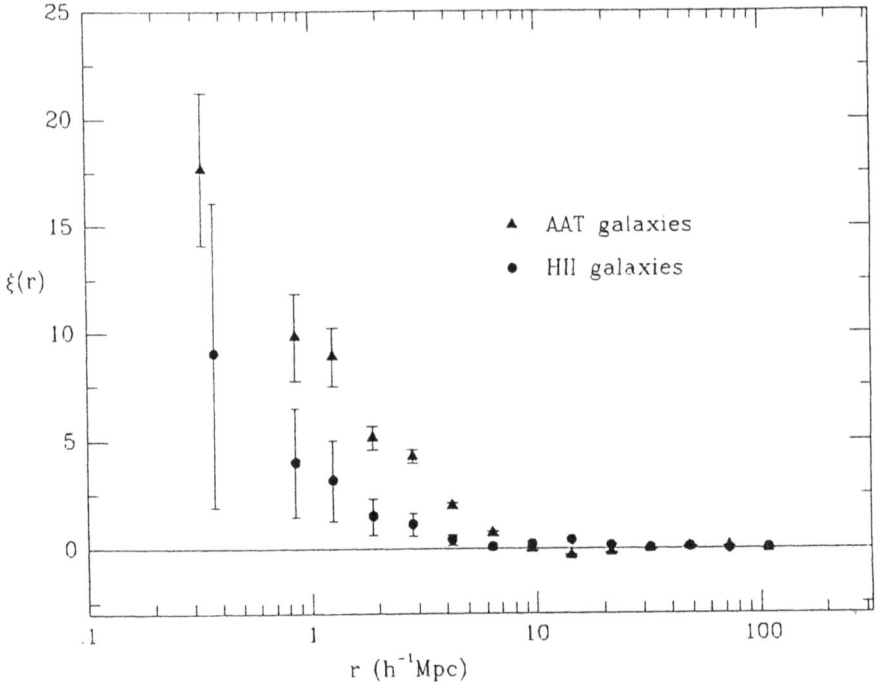

Fig. 1. Correlation functions for the HII galaxies and the AAT galaxies.

REFERENCES

Peterson, B.A., Ellis, R.S., Efstathiou, G., Shanks, T., Bean, A.J., Fong, R., Zen-Long, Z., 1986, *M.N.R.A.S.*, **221**, 233.

Terlevich, R., Melnick, J., Masegosa, J., Moles, M., 1987, *A Spectrophotometric Catalogue of HII Galaxies*, in preparation.

VOIDS IN THE CORONA BOREALIS REGION

M. Kalinkov, I. Kuneva
Dept. of Astronomy, Bulg. Acad. Sci., Lenin 72, Sofia 1784, and
A. Kopylov
Special Astrophys. Obs., Nizhnij Arkhys, Zelenchukskaya,
Stavropolskij kraj, 357 147 USSR

The distribution of the galaxies and the clusters of galaxies in the region of the CrB supercluster of galaxies is extremely complicated (de Lapparent, Geller, and Huchra, 1986, LGH; Postman, Huchra, and Geller, 1986, PHG; Kalinkov, Kuneva, and Kopylov, 1988b, KKK).

A void centered at 6500 km/s and a thin overdense shell at 9500 km/s were detected by LGH. The width of the shell is about 1000 km/s. Another void in the range 11000 - 17000 km/s is centered at 13500 km/s (PHG).

It was shown (Kalinkov, Kuneva, and Kopylov, 1988a; KKK) that the CrB supercluster is not a single supercluster. Two separate superclusters are identified with mean coordinates and velocities

	α (1950) δ		V,km/s	n
CrB N	$15^h23^m.7$	$+29°33'$	22218	7
st.dev.	5.6	1 29	1895	
CrB D	15 19.9	+30 31	35132	7
st.dev.	4.6	1 30	1546	

The cone diagram for the clusters (Table 1, KKK) in the region $14^h52^m \le \alpha \le 15^h48^m$, $25° \le \delta < 33°$ (1950) is shown in Fig. 1. Obviously there is a void between CrB D and CrB D superclusters. No A-clusters are to be found there.

Fig. 2 presents radial-velocity histograms for some A-clusters with more than a few redshifts.

Fig. 3 summarizes all velocities for the galaxies in the 14 clusters, members of the CrB superclusters. The dotted line shows the supplement of those galaxies, given in Field 1525+2805 and Table II from PHG.

Therefore, a third void is centered at 28700 km/s in the CrB region.

REFERENCES

de Lapparent, V., Geller, M., and Huchra, J. 1986, Ap.J., 302, L1 (LGH)
Kalinkov, M., Kuneva, I., and Kopylov, A. 1988a, in IAU Symp. No. 130,
    Evolution of Large Scale Structures in the Universe, eds. J.
    Audouze and A. Szalay (KKK)

M. Mezzetti et al. (eds.), Large Scale Structure and Motions in the Universe, 373–374.
© 1989 by Kluwer Academic Publishers.

374

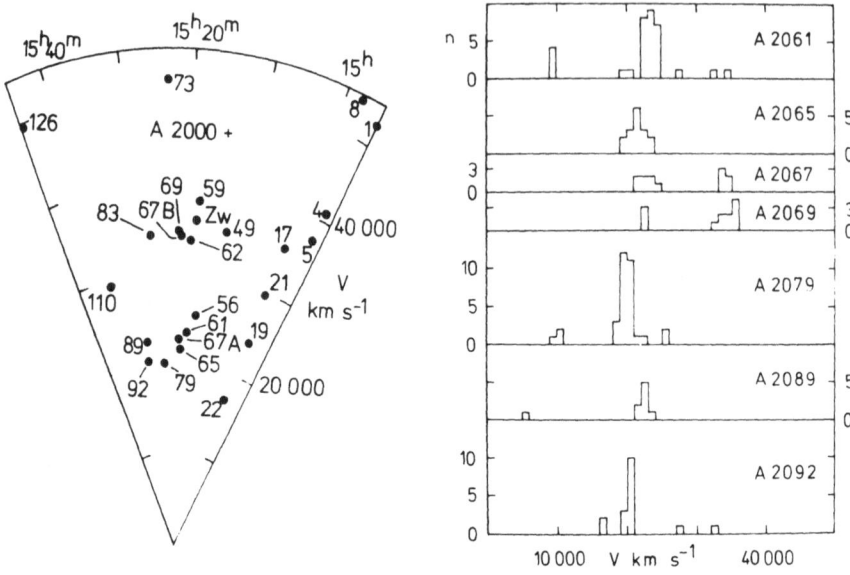

Fig. 1. Cone diagram for CrB region

Fig. 2. Radial-velocity histograms for some A-clusters

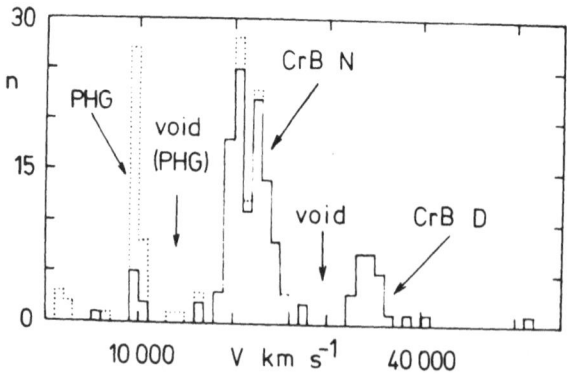

Fig. 3. Radial-velocity histogram for galaxies in the direction of CrB superclusters

Kalinkov, M., Kuneva, I., and Kopylov, A. 1988b, this volume
Postman, M., Huchra, J., and Geller, M. 1986, A.J., 92, 1238 (PHG)

REDSHIFTS FOR CLUSTERS OF GALAXIES IN THE CORONA BOREALIS REGION

M. Kalinkov, I. Kuneva
Dept. of Astronomy, Bulg. Acad. Sci., Lenin 72, Sofia 1784, and
A. Kopylov
Special Astrophys. Obs., Nizhnij Arkhys, Zelenchukskaya,
Stavropolskij kraj, 357 147 USSR

We have undertaken a redshift survey in the direction of the CrB super-cluster of galaxies. Our first communication (Kalinkov et al., 1988) reported the preliminary conclusion: two superclusters of galaxies in the direction of the CrB region were found, namely CrB N(ear) and CrB D(istant) which are superimposed.

The observations were carried out with the 6 meter telescope (Special Astrophysical Observatory, USSR, Academy of Sciences) during 1985, May 17-22. We have measured 50 velocities of galaxies in the clusters A2049, 2056, 2059, 2061, 2062, 2067, 2073, 2079, 2089, 2092 and in a Zwicky cluster, Z7412 (1518.5+2814, mc, pop 160, MD). The number 7412 refers to the machine-readable Catalog of Abell and Zwicky clusters (Kalinkov and Kuneva, 1987) where all 9134 Z-clusters are sorted by R.A.(1950). This cluster is located between A2056 and A2065. We assume that Z7412 is not very different from both A-clusters and its Abell richness group is 1. Besides we regard A2067 as two superimposed clusters

We gathered from the literature all velocities for galaxies in A-clusters in the region defined by $14^h52^m \leq \alpha \leq 15^h48^m$ and $25° \leq \delta \leq 33°$ (1950). The mean galactocentric velocities V are given in Table 1, where n is the number of galaxies.

Table 1. Galactocentric velocities for the clusters in the CrB region

A-cl	V km/s	n	Note	A-cl	V km/s	n	Note	A-cl	V km/s	n	Note
2001	52471	3	1	2056	25368	1	9	2069	34778	9	14
2004	40952	1	2	Z7412	36028	1	9	2073	51522	1	9
2005	37684	1	4	2059	39144	1	9	2079	19924	28	15
2008	54262	2	4	2061	23458	24	10	2083	34266	1	16
2017	35580	0	5	2062	33660	3	11	2089	22262	8	17
2019	23983	0	6	2065	21561	16	12	2092	20194	13	18
2021	29799	1	7	2067A	22760	7	13	2110	29400	3	19
2022	16938	1	8	2067B	33939	5		2126	49496	0	6
2049	35110	1	9								

M. Mezzetti et al. (eds.), Large Scale Structure and Motions in the Universe, 375–376.
© 1989 by Kluwer Academic Publishers.

Notes. (1) Ciardullo et al. (1983)-CFBH propose the existence of A2001A and A2001B. V = 33277 for A2001A, two measurments of one galaxy, and V = 52471 for A2001B - four measurments of three galaxies (Mason et al., 1981 & CFBH). The examination of the POSS does not confirm the existence of A2001A. (2) We assume that the galaxy in A2004A (CFBH) with V = 19187 is a foreground one. (3) Two redshifts for one galaxy (Hoessel et al., 1980 - HGT & CFBH). The group 33 in CFBH is A2005 (Struble and Rood, 1987). (4) CFBH. (5) For two galaxies Rhee and Katgert (1988) - RK give V = 53630, 87720, and 33160, 53693. Our estimate with multiple regression formulae (Kalinkov and Kuneva, 1988) is in the Table. (6) Our esimate. (7) CFBH. (8) HGT. (9) Our V. (10) Hintzen et al., 1980 - HSM & HGT & Postman et al. (1986)-PHG & our (n = 11) V. (11) Three velocities for two galaxies - RK & our V. (12) HMS & Spinrad (1977) & HGT. (13) HSM & HGT & our (n = 10) V. (14) Gioia et al. (1982). (15) Melnick and Sargent (1977) & HGT & Smith et al. (1985) & PHG & our (n = 9) V. (16) Sarazin et al. (1982). (17) PHG & our (n = 1) V. (18) HGT & PHG & our (n = 13) V. (19) Schneider et al. (1983) & RK.

Evidently, the front supercluster CrB N includes A-clusters 2056, 2061, 2065, 2067A, 2079, 2089 and 2092, and the back one, CrB D - A2049, Z7412, A2059, A2062, A2067B, A2069 and A2083.

REFERENCES

Ciardullo, R., Ford, H., Bartko, F., and Harms, R. 1983, Ap.J., 273, 24 (CFBH)

Gioia, I., Geller, M., Huchra, J., Maccacaro, T., and Steiner, J. 1982, Ap.J., 255, L17

Hintzen, P., Scott, J., and McKee, J. 1980, Ap.J., 242, 857 (HSM)

Hoessel, J., Gunn, J., and Thuan, T. 1980, Ap.J., 241, 486 (HGT)

Humason, M., Mayall, N., and Sandage, A. 1956, A.J., 61, 97 (HMS)

Kalinkov, M., and Kuneva, I. 1987, Catalog of Abell and Zwicky clusters of galaxies (on magnetic tape). An older version is distributed by Centre de Données Stellaires, Strasbourg

Kalinkov, M., and Kuneva, I. 1988. In preparation

Kalinkov, M., Kuneva, I., and Kopylov, A. 1988, in IAU Symp. No. 130 Evolution of Large Scale Structures in the Universe, eds. J. Audouze and A. Szalay

Mason, K., Spinrad, H., Bowyer, S., Reichert, G., and Stauffer, J. 1981, A.J., 86, 803

Melnick, J., and Sargent, W. 1977, Ap.J., 215, 401

Postman, M., Huchra, J., and Geller, M. 1986, A.J., 92, 1238 (PHG)

Rhee, G., and Katgert, P. 1988, A & A Suppl., 72, 243 (RK)

Sarazin, C., Rood, H., and Struble, M. 1988, A 5 A, 108, L7

Schneider, D., Gunn, J., and Hoessel, J. 1983, Ap.J., 264, 337

Smith, R., Efstathiou, G., Ellis, R., Frenk, C., and Valentijn, E. 1985, M.N.R.A.S., 216, 71p

Spinrad, H. 1977, P.A.S.P., 89, 116

Struble, M., and Rood, H. 1987, Ap.J. Suppl., 63, 543

A PHOTOMETRICAL STUDY OF A DISTANT CLUSTER
OF GALAXIES A665

A.T. Kalloghlian
Byurakan Astrophysical Observatory, Armenia, USSR

D.Nanni and A.Vignato
Rome Astronomical Observatory

A B and V photometry of nearby and distant clusters of galaxies has been under-taken by using different telescopes, namely, the 2.6 m telescope of the Byurakan Observatory for clusters of Abell's intermediate distance groups, 6 m telescope for clusters of higher redshifts and the Tautenburg 2 m Schmidt telescope for nearby clusters. Some results of this investigation are already published (Kalloghlian et al., 1983; Egikian et al., 1984; Kalloghlian et al., 1987).

In this report we present preliminary results of BVR photometry of A 665 (z=0.1832); it is of richness class 5 and BM type III. The plates were scanned with the PDS 1010G of the Rome Astronomical Observatory.

In Fig.1 the bidimensional galaxy distribution for the cluster, together with some isodensity contours are shown; the cluster has a clumpy structure with two evident subcondensations.

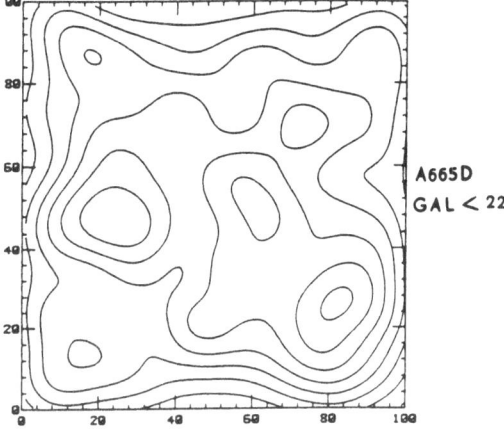

Fig.1 Galaxy isodensity contours for A665 (V < 22)

M. Mezzetti et al. (eds.), Large Scale Structure and Motions in the Universe, 377–378.
© *1989 by Kluwer Academic Publishers.*

378

A bimodal structure was observed also in clusters A777 ($z=0.224$) and A 910 ($z=0.205$) on the 6 m telescope plates (see Fig.7a and 7b in Iannicola et al., 1987).

The LF of A665 in BVR system (Fig.2) is not corrected for the cosmological counts. In all 3 colours the LF in the faint part is flatter than the corresponding one in the Coma cluster. In the V band the LF was approximated by the Schechter's function. The best fit is with $M_V^*=-20.5$ and $\alpha=-1.22$ in good agreement with standard values of these parameters.

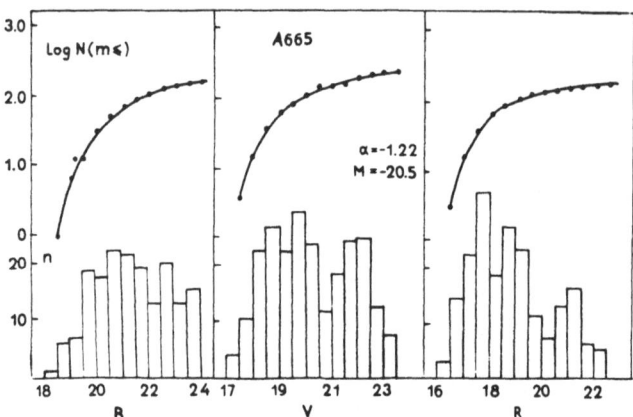

Fig.2 Differential (bottom) and integral (top) LF of A665 in B,V and R colour bands.

Because of the clumpy structure of A665 it is difficult to choose an adequate centre for the cluster; for this reason the Butcher and Oemler parameter f_B was computed for all galaxies brighter than $M_V=-20$. We obtain $f_B=0.17$; because of colour segregation effect and field contamination this value must be regarded as an upper limit.

References

Egikian A.G.,Kalloghlian A.T., Nanni D., Trevese D., Vignato A.: 1984, Astrophysics **27**, 417

Iannicola G.,Kalloghlian A.T., Nanni D., Vignato A.: 1987, Astron. Astrophys. **182**, 189

Kalloghlian A.T., Egikian A.G., Nanni D., Trevese D., Vignato A.: 1983, Astrophysics **19**, 183

Kalloghlian A.T., Richter G., Thanart W., Nikogossian E.H.: 1987, Astrophysics **27**, 417

EXTENDING THE STRUCTURAL PATTERN INTO THE ZONE OF AVOIDANCE

F. J. Kerr and P. A. Henning
Astronomy Program
University of Maryland
College Park, MD 20742
U.S.A.

The zone of avoidance has always presented difficulties for extragalactic studies, because at least a quarter of the sky is affected to a greater or lesser extent by Galactic obscuration and by confusion produced by the high stellar density. These problems are especially serious in large-scale structure studies, because almost no optical information is available over a substantial portion of the sky. There are several cases where "sheets" disappear into the Milky Way, and their whole structure cannot be followed.

We have carried out a pilot study using the 91-meter telescope at Green Bank, West Virginia, to test the feasibility of searching at 21 cm for galaxies hidden behind the Milky Way. We used a blind-search procedure and in 4000 arbitrarily-chosen points we found 23 galaxies, plus a number of other possibles which will need reobservation. Some of these are clearly spirals, judging from the shapes of their line profiles, while others are probably irregulars. We searched a velocity range out to about 7000 km s^{-1}. So far the galaxy positions are not accurate, as we only know that a galaxy is within one particular beam area of the 91-meter telescope, which has a half-power beamwidth of 10 arcminutes.

In addition to our study of the hidden region, we used the same method to look at 2000 points in the "clear" region away from the Milky Way, and showed that we could detect (known) spirals, blue compact dwarfs, etc., in this way. It is important that we should continue to study the clear region, to see how the sample collected from 21-cm observations compares with the population of galaxies known optically. In particular, there may be a greater percentage of gas-rich objects with low optical surface brightness.

For the immediate future, a study of the newly-discovered galaxies by other means is under way. We plan VLA observations first to improve the positions, and then infrared imaging, CO, and radio continuum observations. This will be the first time a sample of galaxies has been studied without benefit of optical images.

M. Mezzetti et al. (eds.), Large Scale Structure and Motions in the Universe, 379–380.
© *1989 by Kluwer Academic Publishers.*

Having shown that hidden galaxies can be found at 21 cm, we are now considering what we would need in order to carry out a full census of spiral and irregular galaxies in the zone of avoidance. Even with a multifeed and multibackend arrangement, this could be quite a long project, but it would make important contributions to the delineation of large-scale extragalactic structure.

Hidden galaxies can also be found by looking at IRAS sources whose spectral characteristics suggest that they may be galaxies. This is probably a faster method for locating a substantial number of galaxies, but it is less complete and systematic than the 21-cm approach, because it is essentially limited to spiral galaxies, without revealing objects of other types, and also the IRAS sky is confused at the lowest latitudes by the high density of sources.

The first list of new galaxies found at 21 cm is given by Kerr and Henning, 1987, Astrophys. J. Letters, 320, L94.

GROUPS AROUND QUASARS

W.Kollatschny, K.J.Fricke
Universitäts-Sternwarte Göttingen
Geismarlandstrasse 11
D-3400 Göttingen
Federal Republic of Germany

ABSTRACT. Deep images of fields around Quasars(radioloud and radioquiet) have been obtained at ESO with the 3.6m telescope for objects with red-shifts up to z = 0.6. In many cases there are clear signs of tidal interaction with neighbouring faint galaxies. Multiobject spectroscopy of galaxies on the CCD-frames confirmed quasars to be located in groups or clusters. The verified neighbouring galaxies often show emission lines.

1.MOTIVATION

A recent spectroscopic survey of groups around Seyfert galaxies has yielded the following results (Kollatschny and Fricke 1988):

- Close companions show significantly stronger nuclear emission-line activity than more distant($r \geq 250$ kpc) members.

- This phenomenon can be explained with a model in which tidal interactions trigger starburst activity lasting over a period of$\sim 10^8$years.

- The closer companions appear strongly reddened. Therefore, in those galaxies only the spectral region around Hαcan be measured reliably.

In this contribution we extend this investigation to QSO's and the surrounding faint galaxies.

2. OBSERVATIONS

We selected bright QSO's with m_V ~15.3 - 17.0 and z ~ 0.12 - 0.52. The galaxies in the field for which we obtained spectra have m_V~ 18 - 20. The projected CCD-field corresponds to a linear scale ~ 500 kpc to 2000 kpc.
Direct imaging and multi-object-spectroscopy was done with EFOSC at the ESO 3.6m telescope. The wavelength range needed to detect redshifted Hα was 5500 - 10000 Å. Beyond 7000 Å the spectra are heavily contaminated by night-sky emission OH - emission lines. Typical exposure time for

381

multi-object-spectroscopy was   1 hr.

3. RESULTS

TABLE 1: Galaxy redshifts found in seven QSO environments from a preliminary evaluation of EFOSC fields

QSO	Q1340	PG1416	PG1552	PK1725	PK2135	PK2300	PK2349
m_V(QSO)	16.	15.4	16.02	16.09	15.53	16.38	15.33
z(QSO)	.326	.133	.122	.301	.206	.517	.178
z(comp. gal.)	.330 .297	.135 .128	.117 .118	.300 .301 .302 .303	.191 .205 .206 .208	.519	.177 .177
z(field- gal.)		.088 .089 .242(A) .339	.19 .19(A)		.179(A)	.238 .070(A)	.245

(A) indicates that absorption lines only are present in the spectrum

This Table shows that the close companions of the quasars have emission line spectra showing at least Hα . the direct images demonstrate that interactions of the quasars with very close companions are frequent.

A Galaxy Search in the Zone of Avoidance near Hydra

Renée C. Kraan-Korteweg, Astronomisches Institut der Universität Basel, Venusstr. 7, Ch-4102 Binningen, Switzerland

Abstract. - A survey of 10 fields of the ESO/SRC survey revealed 830 galaxies in the galactic plane ($|b| \leqslant 10°$). Their distribution suggests an overdensity around $l = 278°$, $b = 0°$, but an extension of the searched area is required for confirmation.

After subtraction of the local Virgocentric infall, the MWB dipole requires a Local Group motion of $\sim$500 km/s toward $l = 274°$, $b = 12°$. This poses the question as to the accelerator(s) responsible for this peculiar motion. At present, neither their number nor their identity are agreed on.

As the apex region lies between the Hydra cluster ($l = 275°$, $b = 25°$) and the galactic plane, part of the cluster or a larger galaxy complex may be obscured. To test this possibility a galaxy search was begun in the zone of avoidance.

The IIIaJ films of the ESO/SRC sky survey are well suited for this search. The fields were systematically searched through visual inspection with a viewer magnifying 50 times. A lower diameter limit of D = 0!2 was imposed to avoid confusion with stars. On the ten fields searched so far, 830 galaxies were detected of which 800 were previously unknown. Diameters were measured, the surface brightness estimated and the morphological types were roughly described. Positions were accurately determined with the OPTRONICS at ESO in Garching.

The majority of the galaxies have magnitudes in the range $B_i = 17$-18 mag with the faintest galaxies at $B_i \sim 19$ mag. The diameters are generally in the order of 0!3-0!5. However, the observed diameters are substantially reduced because of the heavy front absorption, which ranges from $A_B = 1.0$-1.5 mag at $b = 5°$ to $A_B = 2.5$-5 mag at $b = 0°$ (Neckel and Klare, 1980). The diameter reduction factor is typically $f = 2.0$-3.5 for absorption values of $A_B = 1.5$-2.2 mag. This diameter loss causes an additional dimming of the magnitudes: depending on galaxy type the total absorption is about 1 mag higher for a diameter reduction of $f = 3$. Hence, many of the detected galaxies could lie at

383

M. Mezzetti et al. (eds.), Large Scale Structure and Motions in the Universe, 383–384.
© 1989 by Kluwer Academic Publishers.

the distance of the nearby Hydra cluster and therefore be part of a larger complex.

The distribution of the 830 galaxies with $D \geq 0!2$ is illustrated below. The outlined area encloses the investigated 10 fields (~ 300 sq degrees); solid symbols mark certain galaxies (739) and open symbols represent galaxy candidates (91). Two regions of higher density are apparent from the distribution of the galaxies. These seem connected across the galactic plane. It is remarkable that these connected regions lie close to the apex of the MWB dipole corrected for the infall into Virgo (denoted by the circle). The number counts of detected galaxies are, of course, no objective measure of densities since they are strongly influenced by the patchiness of the galactic dust clouds. However, neither the extinction determinations of Neckel and Klare (1980) nor the Dark Cloud Catalogue of Feitzinger and Stüwe (1984) give any indication that the higher densities or the bridge across the galactic plane are an artifact of low extinction values.

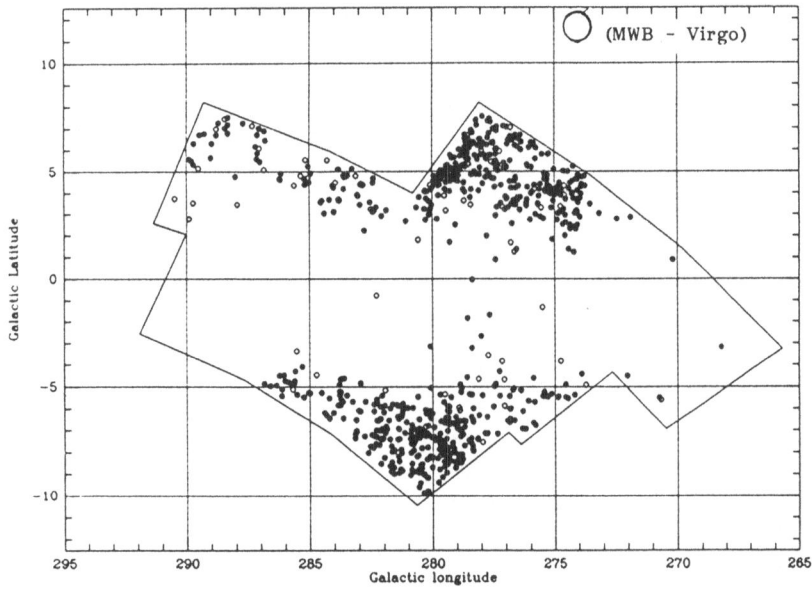

The search is presently extended to a larger area. This will allow a better differentiation between relatively transparent regions in the galactic plane and real density enhancements. Determinations of (radio) redshifts will eventually expose the three dimensional picture.

References:
Feitzinger, J.V., and Stüwe, J.A. 1984, Astr. Ap. Suppl. 58, 365.
Neckel, Th., and Klare, G. 1980, Astr. Ap. Suppl. 42, 251.

NEW RESULTS ON QUASAR CLUSTERING

A. KRUSZEWSKI
Warsaw University Observatory
Al. Ujazdowskie 4
00-478 Warsaw
Poland

Quasar clustering have been studied using both complete and incomplete samples of objects. The set of complete samples (Kruszewski 1988) have been enlarged by adding two new samples and contains now 1254 quasars that are brighter than −22.5 mag. The incomplete sample is formed by combining Hewitt and Burbidge (1987) catalogue with the third edition of Véron–Cetty and Véron (1987) catalogue and supplementing it with two recently published samples. There are 3741 quasars in the combined catalogue brighter than the same absolute magnitude limit. The resulting correlation functions show at small redshifts the correlations which are 3–6 times larger than the galactic correlations at the current epoch. The quasar correlations are not detectable for redshifts larger than 1.5. The data for all quasars irrespective of redshifts give correlations that are twice stronger than in the case of nearby galaxies. The results from two kinds of samples agree well. They are also consistent with the results of the earlier investigations (Shaver 1984, 1987, Kruszewski 1986, 1988, Shanks *et al.* 1987, Clowes, Iovino and Shaver 1987, Iovino and Shaver 1987, Anderson, Kundt and Sargent 1987). The evolutionary effect in the quasar clustering properties is thus confirmed with more accurate data.

References

Anderson, N., Kunth, D., and Sargent, W.L.W. 1987, preprint.
Clowes, R.G., Iovino, A., and Shaver, P.A. 1987, *M.N.R.A.S.* **227**, 921.
Hewitt, A. and Burbidge, G. 1987, *Ap. J. Suppl.* **63**, 1.
Iovino, A. and Shaver, P.A. 1987, In *Evolution of Large Scale Structures in the Universe*, IAU Symposium No. 130, Balatonfured.
Kruszewski, A. 1986, submitted to *Ap. J.*
Kruszewski, A. 1988, *Acta Astronomica*, in press.
Shanks, T., Fong, R., Boyle, B.J., and Peterson, B.A. 1987, *M.N.R.A.S.* **227**, 739.
Shaver, P.A. 1984, *Astron. Ap.* **136**, L9.
Shaver, P.A. 1987, In *Evolution of Large Scale Structures in the Universe*, IAU Symposium No. 130, Balatonfured.
Véron-Cetty, M.-P. and Véron, P. 1987, *ESO Scientific Report*, No. 5.

M. Mezzetti et al. (eds.), Large Scale Structure and Motions in the Universe, 385.

Clustering of Extragalactic Radio Sources in the MG 5GHz Survey

G. LANGSTON
Max Planck Institute für Radioastronomie
Auf dem Hügel 69
5300 Bonn-1
W. Germany

ABSTRACT. The Mit-Greenbank 6cm survey is presented and angular clustering of radio sources is observed. Significant clustering of similar sources in VLA observations is predicted.

M. Mezzetti et al. (eds.), Large Scale Structure and Motions in the Universe, 387.
© *1989 by Kluwer Academic Publishers.*

A HIGH SPATIAL RESOLUTION IMAGING SURVEY
OF VERY DISTANT 3CR GALAXIES

Olivier Le Fèvre[1,2] and Francois Hammer[2]
1. Canada-France-Hawaii Telescope Corporation, P.O.Box 1597, Kamuela, HI96743, USA.
2. DAEC, Observatoire de Paris-Meudon, 92195 Meudon, France.

ABSTRACT: We present here part of a survey of the very distant $0.8 \leq z \leq 1.841$ 3CR galaxies done at the prime focus of the Canada-France-Hawaii telescope with a CCD under excellent seeing conditions. The mean FWHM is 0"8 with values as good as 0"6. All the galaxies observed have been resolved into multiple components.

I The 3CR radio-galaxy sample.

The very distant 3CR galaxies constitute a sample of $\sim$30 galaxies for redshifts from 1 to 1.841, and are therefore expected to give information on the evolution of galaxies at early epochs. One can also hope that they will allow us to derive fundamental cosmological parameters (Spinrad and Djorgovski, 1987).

Deep photometric and spectroscopic observations have shown that this sample contains elongated objects (Djorgovski, 1988), most of the time resolved into multiple components (Djorgovski et al. 1987; Le Fèvre et al., 1987, 1988a,b). Strong line emission spatially as extended as the continuum (or even more) have also been observed. An interpretation in terms of merging of galaxies was first proposed (Djorgovski et al., 1987).

However there is ongoing evidence that this galaxy sample does not contain normal elliptical galaxies as was the hope when studies of these objects started, and that they can not be compared easily to low redshift brightest ellipticals.

First, two groups reported a striking alignment of the optical and radio morphologies for some of these objects, showing a significant difference between them and lower redshift radio-galaxies (McCarthy et al., 1987b; Chambers et al., 1987). A possible interpretation was given in terms of outflow of matter along the radio axis. At such redshifts one may also consider the possibility of observing galaxies in the process of formation. In this context, the galaxy 3C326.1 has recently been interpreted as a proto-galaxy (McCarthy et al., 1987a).

In another way, we have suggested that the sample may be highly contaminated by foreground objects like clusters of galaxies and galaxies leading to possible gravitational amplification/lensing (Hammer et al., 1986, Le Fèvre et al., 1987, 1988a, 1988b). A close object near the line of sight may induce a large flux amplification for the background 3CR galaxy. In the case of a foreground galaxy and with a typical angular separation of $\sim$5 arcsec, a $M_R \sim$-22, M/L$\sim$10, galaxy can easily produce an amplification of more than one magnitude on a 3CR object. For closer angular separations one can have the occurence of multiple gravitational imaging; we have suggested a possible gravitational lens mechanism to explain the observations of 3C324 (Le Fèvre et al., 1987). A background 3CR source at a distance up to a few arcminutes from the center of a moderately rich cluster with intermediate to high redshifts ($z \geq 0.1$) can be amplified by more than one magnitude (Hammer, 1986); for close

389

M. Mezzetti et al. (eds.), Large Scale Structure and Motions in the Universe, 389–390.
© *1989 by Kluwer Academic Publishers.*

alignments with the peak density of the cluster and the brightest central galaxies, multiple imaging may occur, a beautiful example being given by the recent interpretation of the giant luminous arcs found in distant clusters in terms of gravitational lensing (Soucail et al., 1988).

II. New observations

We summarize here the results of the observations done at the prime focus of the CFH 3.6m telescope with a double density RCA CCD.

• 2 galaxies have been observed near bright foreground galaxies (to be added to 3C13, Le Fèvre et al., 1988a).

- 3C194 (z=1.779) is ∼ 4" from a z=0.312, M_R=-23.5 galaxy, and an amplification of more than one magnitude of the background 3CR galaxy should therefore be present. Moreover, a bright component of the elongated optical object associated with the radio source, not detected in U, may be also foreground and in this case gravitational lensing effects are expected (see figure 1).

- 3C225A (z=1.565) is at less than 4" from a z=0.134, M_R=-21.5 galaxy and therefore its flux may also be amplified by more than one magnitude.

The probability to find 3 galaxies in a sample of 30, with foreground galaxies at less than 4" is 10^{-4}, showing a strong bias in the sample.

• Galaxies with aligned multiple components and total optical extension up to 160kpc like 3C68.2 (z=1.575) and 3C437 (z=1.48) have been observed. We do not have 2D faint spectroscopy available so 2 cases have to be considered: either all the components are at the same redshift and one may consider these objects as proto-galaxies, or we are observing a chance projection of galaxies at different redshifts and gravitational amplification effects may play an important role.

• Some objects like 3C266, 3C454.1, have observed properties compatible with a gravitational lensing hypothesis, including excellent optical/radio correspondence. Deep high spatial resolution 2D spectroscopy is needed to conclude.

• Galaxies with bright quasi-stellar components like 3C208.1 (z=1.02), 3C297 (z=1.406) or 3C325 (z=0.86). These objects have properties closer to AGNs or QSOs than to galaxies and may be misclassified in the radio-galaxy sample.

III. Conclusion

The peculiarity of the high redshift 3CR galaxy sample seems to prevent its use to test for galaxy evolution by means of standard magnitude/color-redshift relationships before having a better understanding of which physical processes are at play.

References

- Chambers, Miley, van Breugel, 1987, Nature, **329**, 604.
- Djorgovski, 1988, Proc. Moriond Astrophysics Workshop *'Starburst and Galaxy Evolution'*, Th. Montmerle (Ed.), Editions Frontières.
- Djorgovski, Spinrad, Pedelty, Rudnick, Stockton, 1987, AJ, **93**, 1307.
- Hammer, 1986, PhD, Université de Paris VII.
- Hammer, Nottale, Le Fèvre, A&A, **169**, L1.
- Le Fèvre, Hammer, Nottale, Mathez, 1987, Nature, **326**, 268.
- Le Fèvre, Hammer, Nottale, Mazure, Christian, 1988a, ApJ, **324**, L1.
- Le Fèvre, Hammer, Jones, 1988b, ApJ, in press.
- McCarthy, Spinrad, Djorgovski, Strauss, van Breugel, Liebert, 1987a. ApJ, **319**, L39.
- McCarthy, van Breugel, Spinrad, Djorgovski, 1987b, ApJ, **321**, L29.
- Soucail, Mellier, Fort, Mathez, Cailloux, 1988, A&A, **191**, L19.
- Spinrad, Djorgovski, 1987, IAU Symp.124, *'Observational Cosmology'*, Burbidge (Ed.), Reidel.

NUMERICAL SIMULATIONS OF SELF-GRAVITATING FLUIDS

J. LÉORAT and G. ALECIAN
DAEC, UA 173 (CNRS),
Observatoire de Paris-Meudon
F-92195 Meudon Principal Cedex
France

ABSTRACT. Up to now, the numerical approach of cosmological flows have been based mainly upon "particle" methods. We have investigated numerical simulations applied to continuous medium using the hydrodynamical equations and spectral methods with a 1D code in the framework of the newtonian cosmology.

A recent study by Alecian & Leorat (Astron. Astrophys., 1988, in press) has been extended here to high resolution computations (with 512 grid points). This resolution allows us to examine the evolution of density and velocity spectra in linear and non-linear regimes over two decades of wave numbers.

The non-linear phase of such cosmological flows do not build new spatial structures: the initial density fluctuations are the seeds of the future large-scale structures.

The basic equations of newtonian cosmology are:

$$a^2 \, \partial_t \rho + \nabla.(\rho \, \mathbf{u}) = 0$$
$$a^2 \, \partial_t \mathbf{u} + \mathbf{u}.\nabla \mathbf{u} = - M^{-2} \, a^{5-3\gamma} (\gamma\rho)^{-1} \, \nabla\rho^\gamma + (2/3) \, a \, \mathbf{g} + \mu \, a \, \nabla.\Sigma$$
$$\nabla.\mathbf{g} = - (\rho-1)$$

where

$$M^{-2} = 4 \, \pi^2 \, (s_i \, t_i / L_i)^2 = 4.10^{-8} \, \gamma \, (h_0 l_i) Mpc^{-2} \, (1+z_i)^{-1}$$

The scale factor $a(t)$ depends on the model of Universe (here Einstein-de Sitter); γ is the polytropic index, s_i is the initial sound velocity (the computations are made with $\gamma = 1$).

The dimensionless variables have been defined as follows: the numerical space variable extends over 2π; the characteristic length is L_i (l_i Mpc), the characteristic density is the mean density at initial time t_i (which is the Hubble time $2/(3H_i)$ and also the time unit), and the characteristic speed is $L_i/2\pi t_i$.

These equations are solved using a Fourier pseudo-spectral method and periodic boundary conditions.

The viscous dissipation term ($\mu \, a \, \nabla.\Sigma$) is needed for numerical reasons (its effects are negligible except in the smallest scales).

Because of lack of space, we present here only a computation of gravitational collapse in an expanding universe (512 grid points) in order to illustrate a typical numerical experiment.

M. Mezzetti et al. (eds.), Large Scale Structure and Motions in the Universe, 391–392.
© *1989 by Kluwer Academic Publishers.*

In the run shown on the figure, L_j = 2.5 kpc (10 Jeans lengths); z_i = 10; $a(t_i)$=1, $a(t_{final})$ = 3.3; t_{final} = 6 t_i ; the velocity unit is $3/(4\pi)$ times the Hubble velocity over L_j.

The initial density and velocity spectra are flat: the excited density and velocity modes lie between k = 20 and 100 (with random phases). The maximum density (resp. velocity) fluctuations is 50% of the mean density (resp. velocity unit).

The plot at left represents, at equally spaced interval of time, the density (continuous line) and the velocity (dashed line); the density spectrum is shown at right.

Although the initial velocities are supersonic in this run, expansion prevents shock formation and velocities decay rapidly to values small compared to the Hubble velocity of the largest scale.

The density fluctuations are first amplified due to the combined effect of gravity and intial velocities. Then the fluctuations remain almost frozen (see the saturation in the density spectra), there is no creation of new scales even through non-linear interactions.

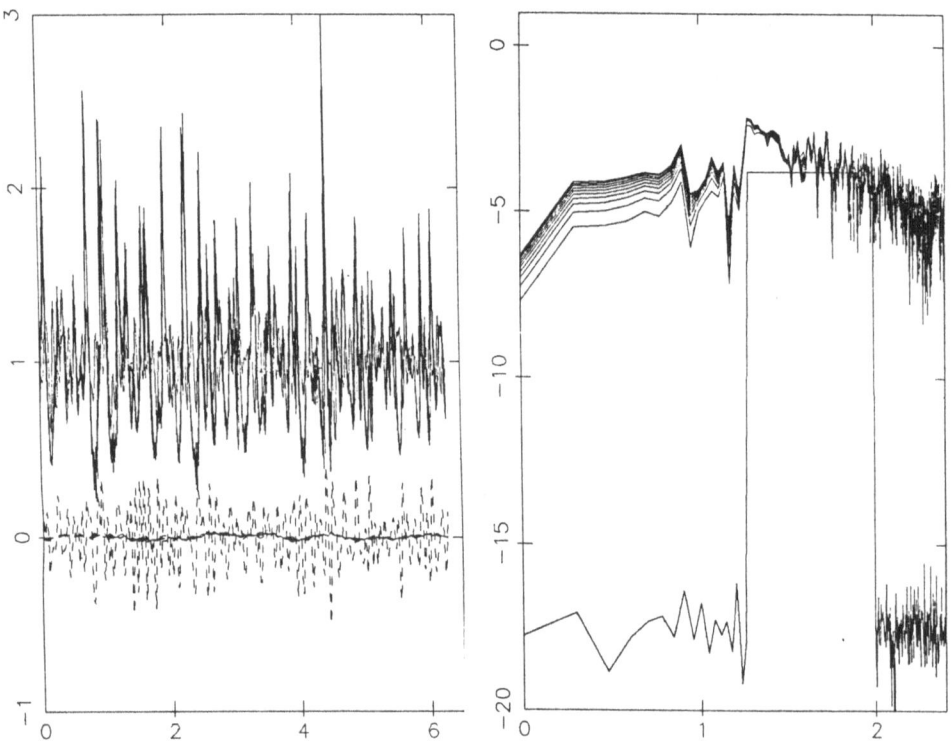

CORRELATION FUNCTIONS IN REDSHIFT SPACE

Per B. Lilje* and G. Efstathiou
Institute of Astronomy
Madingley Road
Cambridge CB3 0HA
England

ABSTRACT. We apply N-body simulations of a cold dark matter universe to test the effects of peculiar velocities on the shape of the galaxy-galaxy ($\xi_{gg}(\mathbf{s})$) and the cluster-galaxy ($\xi_{cg}(\mathbf{s})$) correlation functions measured in redshift space. We show that anisotropies in $\xi_{gg}(\mathbf{s})$ arising from radial infall between galaxy pairs should be observable in redshift surveys now in progress, but that radial infall into rich clusters is masked by random motions. Results for the the direction average of $\xi_{gg}(\mathbf{s})$ agree extremely well with the linear theory model of Kaiser, even on scales where the density perturbations are mildly non-linear ($\xi_{gg}(r) \sim 1$). It should therefore be possible to check whether any bias in the distribution of galaxies relative to the mass varies with scale and to check the predictions of the $\Omega = 1$ cold dark matter model.

1. Introduction

For galaxies at large separations it is assumed that that the velocity field is well described by linear theory. One then expects streaming of pairs of galaxies towards each other and a flattening of the galaxy two-point correlation function normal to the line of sight. This effect may be used as an important cosmological diagnostic. At small separations, non-linear effects are important and the correlation function is extended along the line of sight. It is not clear at which separation the non-linear effects become insignificant, and if linear theory is applicable at any separation where the amplitude of the correlation function is strong enough to be measured. In this study we therefore apply N-body simulations of a cold dark matter universe to test the effects of peculiar velocities on the shape of the redshift space galaxy-galaxy ($\xi_{gg}(\mathbf{s})$) and the cluster-galaxy ($\xi_{cg}(\mathbf{s})$) correlation functions at large separations ($> 5\,h^{-1}\mathrm{Mpc}$).

2. Results

Our main results are shown in figures 1) and 2). Anisotropies in $\xi_{gg}(\mathbf{s})$ arising from radial infall between galaxy pairs are large enough to be observable in redshift surveys now in

---

* Present address: CITA, University of Toronto, 60 St. George Street, Toronto, Ontario M5S 1A1, Canada

M. Mezzetti et al. (eds.), Large Scale Structure and Motions in the Universe, 393–394.

progress where a high signal to noise ratio for ξ_{gg} at large separations is secured by sparse sampling.

The detailed shape of $\xi_{gg}(\mathbf{s})$ (see figure 1), however, deviates quite strongly from the predictions of linear theory. But results for the direction average of $\xi_{gg}(\mathbf{s})$ agree extremely well with the linear theory model of Kaiser, where the ratio between the direction average of $\xi_{gg}(\mathbf{s})$ and the real space correlation function ($\xi_{gg}(r)$) equals $1 + \frac{2}{3}\Omega^{0.6} + \frac{1}{5}\Omega^{1.2}$. Here Ω denotes the effective density parameter, which in the biased galaxy formation scenario equals $b^{-5/3}\Omega_0$ (b being the bias parameter). The agreement (see figure 2) is good even on scales where the density perturbations are mildly non-linear ($\xi_{gg}(r) \sim 1$). By comparing the results from this method with estimates of Ω on smaller scales (by i.e., the cosmic virial theorem), it should therefore be possible to check whether any bias in the distribution of galaxies relative to the mass varies with scale and to check the predictions of the $\Omega_0 = 1$ cold dark matter model.

Anisotropies in the cluster-galaxy cross-correlation function are totally different from the predictions of linear theory. Radial infall into rich clusters is masked by random motions of galaxies on all scales where ξ_{cg} is measurable.

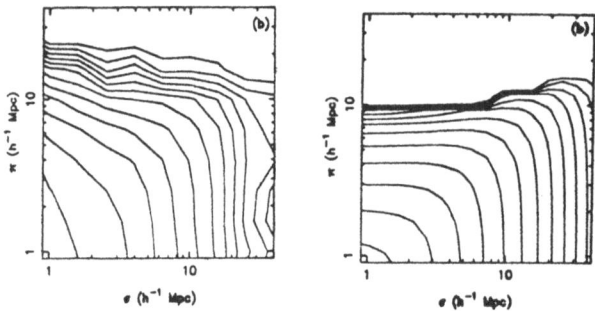

Figure 1. ξ_{gg} in redshift space. The direction along the line of sight is denoted π and the direction perpendicular to the line of sight σ. The lefthand figure shows the result from the simulations, while the righthand figure shows the prediction of linear theory.

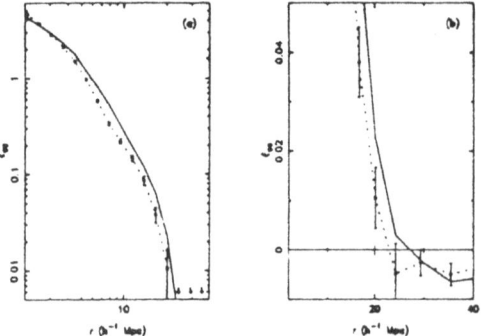

Figure 2. The direction average of the redshift space galaxy correlation function (continous line) and the real space correlation function (dotted line). At left in log-log and at right plotted linearly.

MICROWAVE BACKGROUND ANISOTROPY DILUTION BY LENSING

ERIC V. LINDER
Max-Planck-Institut für Astrophysik
8046 Garching bei München, FRG

Light propagation effects in an inhomogeneous universe impose difficulties on straightforward association of linear and angular scales. The existence of structure, *i.e.* clumpiness, leads to two effects: shrinking of angular scales (by up to a factor of 250) because of decreased Ricci focusing, and small scale smearing by gravitational lensing from either recent clumps or early almost smooth density perturbations. Conversely, observed absence of these effects could constrain the evolution of structure and the equation of state of the universe.

1. INTRODUCTION

In a less than perfectly symmetric cosmology the question of how to relate the properties of a source to the light received at the observer is nontrivial. Here we consider the association of a linear (metric) size, for example a Fourier wavelength or related mass scale at the last scattering surface, with an angular scale characteristic of anisotropies in the microwave background radiation. Within the geometric optics approximation (Sachs 1961) and a cosmology that is globally Friedmann-Lemaître but with local inhomogeneities (clumps) with global density equal to $1 - \alpha$ times the total density, the relation between proper source size d and observed angular size Θ is

$$d = \Theta\, r(z), \tag{1}$$

where the proportionality factor $r(z)$ is the angular diameter distance, dependent on redshift z.

2. SHRINKING OF ANGULAR SCALES

The presence of matter within the source beam induces Ricci focusing, causing the angular distance to decrease and hence the observed angular scale to increase, by equation (1). Therefore, either a low density universe or a clumpy one where the light misses clumps (low probability of gravitational lensing) has decreased focusing and hence smaller angular scales for a fixed linear scale. We can quantify this by comparing a cosmology with density parameter Ω and smoothness parameter α to a standard $\Omega = 1$, $\alpha = 1$ (flat and smooth) model.

Forming the ratio $R(\alpha, \Omega) = r(\alpha, \Omega)/r(1, 1)$ gives the shrinking factor of angular scales as a function of redshift. Numerical results are presented in Linder (1988b); simple analytic limits for $z \gg 1$ are:

$$
\begin{aligned}
R(1, \Omega) &\approx \Omega^{-1}, & \Omega &\gg z^{-1}, \\
R(\alpha_*, \Omega) &\approx 0.6\,\Omega^{-1} z_0, & \Omega &\gg z_0^{-1}; \quad 1 \ll z_0 \ll z, \\
R(0, \Omega) &\approx 0.2\, z,
\end{aligned}
\tag{2}
$$

M. Mezzetti et al. (eds.), Large Scale Structure and Motions in the Universe, 395–396.

where $\alpha_* = 0$ for $z < z_0$, $= 1$ for $z > z_0$. Note this implies that in a universe clumpy back to last scattering at $z \approx 10^3$ an anisotropy we expect to see at $1°$ would appear at $18''$! More realistically, a low density universe ($\Omega = 0.2$) clumpy since $z_0 = 5$ shrinks scales by a factor of 10.

3. BLURRING OF ANISOTROPIES

Although lensing conserves the temperature of the background radiation, it alters the apparent sky position. When observing a region of the sky we find that the temperature entering the correlation function has the form

$$T(\phi) = \int d\theta\, p(\theta)\, T(\Phi = \phi + \theta),\qquad(3)$$

where Φ is the true position of the source, ϕ the apparent (image) position, and θ the deflection angle with probability distribution $p(\theta)$. Thus, the observed temperature is effectively a weighted average of the true temperatures in a finite neighborhood, and so closer to the mean — *more isotropic.*

For a Gaussian deflection probability with zero mean and standard deviation θ_{rms} the correlation function $C(\psi) = \langle \delta_T(\phi)\, \delta_T(\phi') \rangle$, where $\psi = |\phi - \phi'|$ and $\delta_T = \Delta T/T$, behaves like

$$C(0) = C_{true}(0)/[1 + 2(\theta_{rms}/\Phi_c)^2],\qquad(4)$$

where Φ_c is the coherence angle of the anisotropies. This is totally analogous to beam smearing.

Of course Φ_c is also subject to scale shrinking so the total dilution is

$$(\Delta T/T)_{obs}/(\Delta T/T)_{true} \approx [\sqrt{2}R(\alpha,\, \Omega)\,\theta_{rms}/\Phi_c]^{-1}.\qquad(5)$$

That is, *large scales shrink to small ones and small scales are diluted.*

If we ignore lensing effects due to clumpiness in the recent universe we still have "smooth" lensing due to the linear density perturbations $\delta\rho$ present before structure forms. These give a deflection

$$\theta \approx \int_0^L dx\, \nabla(\delta\rho/\rho),\qquad(6a)$$

leading to an average blurring over scales

$$\theta_{rms} \approx 27'' \,[l/10\,h^{-1}Mpc]^{3/2}\,[z_0/10]^{1/2}\,[(\delta\rho/\rho)_{rec}/10^{-3}].\qquad(6b)$$

Comparing this to the coherence scale of the anisotropies, and integrating over an invariant spectrum of perturbations yields

$$\Phi_c \approx (kL)^{-1} \approx 10'\,[l/10\,h^{-1}Mpc][z_0/10],\qquad(6c)$$

$$\text{Dilution} \approx 1.2,\qquad(7)$$

where l is the proper size of a density fluctuation.

This calculation is discussed in more detail in Linder (1988b) but the conclusion remains that "smooth" lensing has negligible effect. Preliminary results (Linder and Schneider 1988) indicate that clumpy lensing in the recent universe could give rise to a $\theta_{rms} \approx 10'$ so this is a more promising avenue for isotropization of the microwave background. Coherence between the perturbations responsible for the anisotropies and those for the inhomogeneities also increase the dilution.

Finally, note that $r(z)$ and hence the shrinking factor also depend on the equation of state (EOS) of the universe. Lack of observed scale shrinking could put a limit

$$\Omega_\sigma < 10^{-9\sigma}\qquad(8)$$

on the present energy density in a component with EOS $p = \sigma\rho$ (Linder 1988a).

REFERENCES

Linder, E.V. 1988a, submitted to *Astr. Ap.*
Linder, E.V. 1988b, in preparation.
Linder, E.V. and Schneider, P. 1988, in preparation.
Sachs, R.K. 1961, *Proc. Roy. Soc. London A* **264**, 309.

COSMIC STRING AND THE DISTRIBUTION OF QUASAR ABSORPTION LINES

Li-Zhi Fang (1), Yaoquan Chu (2) and Xingfen Zhu (2)
(1) Beijing Astronomical Observatory,
    Chinese Academy of Sciences, Beijing, China.
(2) Center for Astrophysics
    University of Science and Technology of China, China.

We investigate the distribution of quasar absorption lines from the cosmic string scenario for the formation of large scale structure, in a sense of the fractal scheme. Using a quite homogeneous sample of Ly-α absorption lines with high resolution (0.8 - 1.5 Å FWHM) (Murdoch et al, 1986) we calculate the count of Ly-α absorption lines N(s) for each quasar and find that all N(<s) can be fitted for a power law as

$$\overline{N}(<s) \sim s^d$$

where s is the space distance between two absorption lines. The value of d is about 1, namely, for the primordial hydrogen cloud (PHC) the fractal dimension D is about 3. It means that the PHC is distributed completely different from galaxies, which has local fractal dimension $D_\ell \sim 1$ and Global dimension $D_g \sim 2.5$ (Fang, 1986). Thus, the PHC should be formed by different way from galaxies. The matter surrounding cosmic string loops became galaxies and the matter in the interloop space became PHC. These two kinds of objects should be anti-associated in space.

We use cross-correlation function (Chu, Zhu et al. 1984) to study the expected anti-association between Ly-α absorption lines (PHC) and heavy element absorption systems (galaxies). The results show that all the cross-correlation functions are less than 1 and most of them are negative near $s \sim 0$.

Our results may imply that there are two kinds of objects formed by different processes of clustering. This is favourable for the cosmic string theory on the formation of large scale structure in the universe. In the string modles the objects can be divided into two kinds according to their clustering with or without string loops as their initial density perturbations.

References:
Chu, Y., Zhu, X., Burbidge,G.,and Hewitt, A., 1984, A. Ap., 138, 408.
Fang, L.-Z., 1986, Mod. Phys. Lett., A1, 601.
Murdoch, H.S., Hunstead, R.W., et al., 1986, Ap. J., 309, 19.

M. Mezzetti et al. (eds.), Large Scale Structure and Motions in the Universe, 397.
© 1989 by Kluwer Academic Publishers.

MULTIDIMENSIONAL ANALYSIS OF THE DISTRIBUTION OF GALAXIES WITH DIFFERENT LUMINOSITY

V.J. Martínez

Departament de Matemàtica Aplicada i Astronomia
Universitat de Valencia
Burjassot,46100 Valencia, Spain
and
R. Domínguez-Tenreiro
Departamento de Física Teórica
Universidad Autónoma de Madrid
Cantoblanco,28049 Madrid, Spain

Abstract

We have used the multidimensional or multifractal formalism to study the large scale luminosity segregation of the CfA catalogue. In every sample we have analyzed, it has been found that the spectrum of scaling indices is scale invariant and that bright galaxies are more clustered than faint galaxies.

Some statistical indicators to describe the large scale structure of the universe have been recently introduced which surmount the deficiencies of the conventional methods (1). These last are mainly based in the use of two point correlation functions (2). The use of the multifractal formalism (3) to analyze the spatial distribution of galaxies has been proposed by Jones et al.(4).

A multifractal is a fractal characterized by an infinite hierarchy of scaling indices α. These are defined as follows: to each point i of a set of N points, a probability at scale ϵ, $p_i(\epsilon)$, is assigned in such a way that:

$$p_i(\epsilon) = An_i(\epsilon) = \epsilon^{\alpha_i} \tag{1}$$

where $n_i(\epsilon)$ is the number of set points inside a sphere of radius ϵ centered at the point i, A is a normalization factor and the α_i are known as the scaling indices. The number of times α takes a value in the interval $[\alpha', \alpha' + d\alpha']$ is expressed as:

$$n(\alpha')d\alpha' = \epsilon^{-f(\alpha')}d\alpha' \tag{2}$$

and $f(\alpha)$ is known as the α spectrum.

In the limit $\epsilon \to 0$, $N \to \infty$, the Legendre transformed of the pair (α, f), (q, τ), is related to the generalized dimensions D_q (5) by:

$$D_q = (q - 1)^{-1}\tau(q) \tag{3}$$

For finite N, $\epsilon \leq (V/N)^{(1/3)}$ is meaningless. If the scaling indices do not depend on the scale (or at least this holds for a finite interval I of ϵ values), an effective function $\tau_{eff}(q)$ is defined

399

M. Mezzetti et al. (eds.), Large Scale Structure and Motions in the Universe, 399–400.
© *1989 by Kluwer Academic Publishers.*

by:

$$\tau_{eff}(q) = \frac{\log \sum_{i=1}^{N} p_i(\epsilon)^q - \log(N\epsilon^{D_0})}{\log \epsilon} \tag{4}$$

and then an effective $f(\alpha)$ curve can be obtained as the Legendre transformed of $\tau_{eff}(q)$. It does not depend neither on N nor on ϵ, for $\epsilon \in I$ and, consequently, characterizes the structure of the point distribution.

This formalism has been applied to the CfA catalogue (Huchra's compilation of redshifts corrected as in Einasto et al.(6)). We have found that, for every analyzed sample, the α spectrum is scale invariant for a certain interval of ϵ values around the correlation length. This means that $f(\alpha)$ is a good tool to analyze the large scale luminosity segregation (7). We present the results of this analysis for the samples specified in Table I. The velocities are in km/s and $H_0 = 100 km/s/Mpc$. They are all conical complete samples, corresponding to $b \geq 40°$. The $f(\alpha)$ curves for these samples are shown in Figure 1. The different structure of the distribution of bright and faint galaxies is apparent. Some characteristics of these distributions are given in Table I. α_{min} (α_{max}) is the strength of the measure of the region of the set where the concentration of points is maximum (minimum); α_0 is the most frequent α value. The following inequalities hold:

$$\alpha_{min}^B < \alpha_{min}^F \; ; \; \alpha_0^B < \alpha_0^F \; ; \; \alpha_{max}^B < \alpha_{max}^F \; ; \tag{5}$$

where a $B(F)$ superindex stands for bright (faint) galaxies. These inequalities, when interpreted in terms of equations (1) and (2), imply that bright galaxies are more clustered than faint galaxies.

Similar qualitative results have been found for every sample we have analyzed.

TABLE I

	$Sample 15B$	$Sample 15F$
$Num.\ Glxs.$	269	268
v_{max}	5000	5000
v_{min}	1000	1000
M_{max}	−19.49	−19.00
M_{min}	−25.00	−19.49
D_0	1.72	1.99
D_1	1.56	1.87
D_2	1.46	1.71
α_{min}	1.21	1.45
α_0	1.90	2.12
α_{max}	2.82	3.06

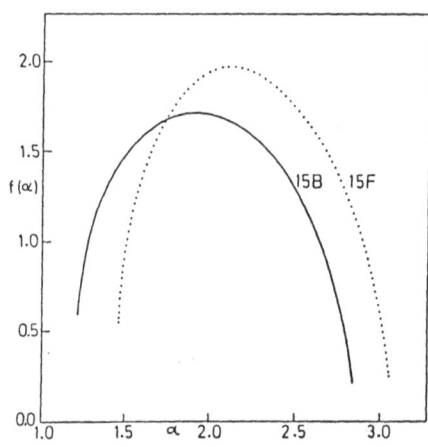

REFERENCES (1) J.N. Fry, Ap.J. Letters 277,L5 (1984); R. Schaeffer Astr. Ap. 134,L15 (1984); S. Maurogordato and M. Lachieze-Rey Ap.J. 320,13 (1987). (2) P.J.E. Peebles, "The Large Scale Structure of the Universe", Princeton University Press, 1980. (3) B.B. Mandelbrot, "The Fractal Geometry of Nature", Freeman, New York, 1982; T.I. Hasley et al., Phys.Rev. A33,1141 (1986). (4) B.J.T. Jones, V.J. Martinez, E. Saar and J. Einasto, Ap.J. Letters (in press). (5) H.G.E. Hentschel and I. Procaccia, Physica 8D,435 (1983). (6) J. Einasto, A. Klypin, E. Saar and S.F. Shandarin, M.N.R.A.S. 206,529 (1984). (7) D. Zugan et al. 1988 preprint; A.J.S. Hamilton 1988 preprint; S. Phillips and T. Shanks, these proceedings.

A BALLOON-BORNE EXPERIMENT DEVOTED TO THE SEARCH FOR EXTRAGALACTIC BACKGROUND ANISOTROPIES IN THE FAR IR

S. Masi , G. Dall' Oglio , P. de Bernardis , M. De Petris ,
M. Gervasi , E. Giovannozzi , G. Guarini , M. Epifani
F. Melchiorri , G. Moreno
Dipartimento di Fisica , Universita' "La Sapienza" , Roma
A. Boscaleri, I. Guidi, V. Natale
I.R.O.E.-C.N.R. , Firenze

Several cosmological scenarios have been proposed , in which a strong extragalactic background in the far infrared region is expected (see Bond , Carr , Hogan (1986) for a review). The Berkeley-Nagoya group claimed to have detected such an emission in a recent rocket-borne experiment carried out in the submillimetre range (Matsumoto et al. (1988)).

Both the dipole anisotropy and higher order fluctuations should be present in such a background (de Bernardis et al. 1987). The detection of these deviations from isotropy should represent a fundamental step in confirming the cosmological origin of the background.

Here we describe a balloon-borne experiment carried out in July , 1987 and devoted to the search for large scale anisotropies in the far infrared diffuse emission.

Many experimental problems must be taken into account in an experiment devoted to this kind of measurements : the main problem is the atmospheric emission which is very high at these wavelengths also at balloon altitude : the only way to overcome it is to use a differential technique , modulating between two fields at the same altitude on the horizon , thus reducing drastically the atmospheric contribution . However , a residual atmospheric offset can be present in the signal if the modulation is not perfectly parallel to the horizon.

The experiment consisted of a 4 channel high throughput differential photometer , with a $7°$ field of view and a $15°$ p-p sinusoidal modulation . The four wavelength bands were peacked at $167\mu m, 230\mu m, 379\mu m, 600\mu m$ with HPBW between $105 - 200\mu m$, $210 - 290\mu m$, $325 - 430\mu m$, $340 - 800\mu m$. The modulation was achieved by means of four aluminum wobbling mirrors. The detectors were Ge bolometers cooled at $1.0°K$. The ac signal from the preamplifier , after a further amplification and prefiltering was demodulated by two lock-in amplifiers driven by two reference signals in phase and in quadrature . An in-flight calibrator allowed us to monitor the detectors responsivity during the flight. The experimental apparatus is shown in fig. 1 . The experiment was flown from the balloon facility of Trapani - Milo (Italy) on July 29 , 1987 , 6^{15} U.T. and the flight was over 20 hours later near Arenosillo (Spain) .

The calibrators worked well during the flight , and the signals from the calibrators were used in order to measure the in-flight responsivity. Also a moon transit was used (fig. 2) for the same purpose , giving consistent results. The in-flight responsivity was lower than during the laboratory tests , due to the formation of ice on the detector windows .

Sky anisotropies at an angular scale of $15°$ have been detected in an high latitude region observed during the flight : the level of these fluctuations is $\lambda I_\lambda \simeq 10^{-12} W\, cm^{-2} sr^{-1}$ in

M. Mezzetti et al. (eds.), Large Scale Structure and Motions in the Universe, 401–402.
© *1989 by Kluwer Academic Publishers.*

the wavelength band between 350 and 800μm. The interpretation of this result is discussed in Francesco Melchiorri's talk .

figure 1. Sketch of the experimental apparatus.

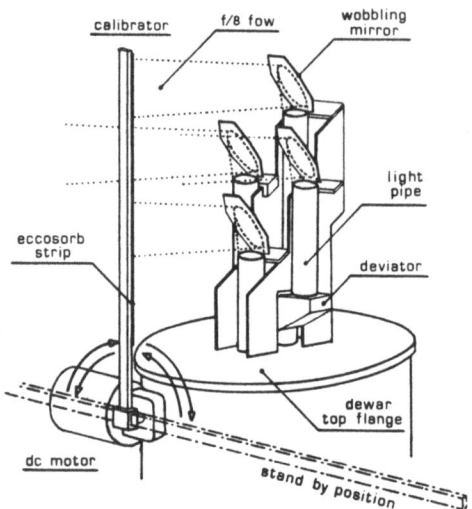

figure 2. The moon signal was used for in-flight calibration of the photometer

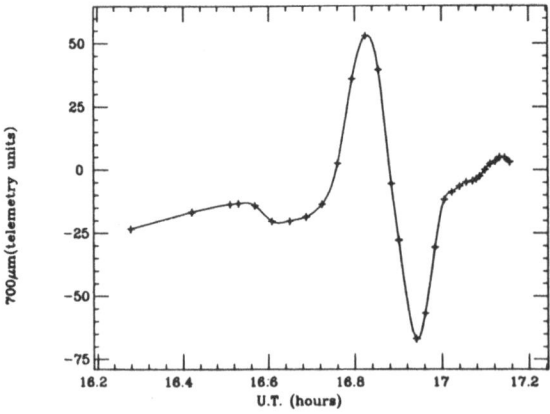

Bond J.R. , Carr B.J. , Hogan C.J. ,*Spectrum and anisotropy of the cosmic infrared background* , 1986 , Ap.J. , **306** , 428

Matsumoto T. , Hayakava S. , Matsuo H. , Murakami H. , Sato S. , Lange A.E. , Richards P.L. ,*The submillimeter spectrum of the cosmic background radiation* , Ap.J. in press , (1988) .

de Bernardis P. , Fabbri R. , Masi S. , Melchiorri F. , Olivo B. , Pecorella W. ,*The anisotropy in the distribution of extragalactic infrared sources and background* , in *The large scale structure of the Universe* , *IAU symposium 130* , Audouze J. , Szalay A. eds. *1987*

SUBCLUSTERING AND EVOLUTION IN THE COMA CLUSTER

A. MAZURE[1], G. MATHEZ[2], Y. MELLIER[2]

[1] CNRS & Université de Montpellier

[2] CNRS & Observatoire de Toulouse

We have recently (Mellier et al, 1988) investigated how the dynamical youth, indicated by the substructure established in the center of the Coma cluster (Mazure & Proust 1985, Gerbal et al. 1986, Fichet & Webster 1987), may be compatible with the advanced evolution stage suggested by the observed luminosity segregation (Capelato et al. 1980, Mellier et al. 1988). We have verified that the latter does not result from a spurious effect such as field contamination.

From high resolution contour density maps of a sample of 1630 presumed Coma members, we show that this cluster exhibits 9 density peaks (Fig 1) well correlated with the locations of bright galaxies. Monte Carlo simulations are used to establish the significance levels of these subsystems.

A composite sample is constructed , consisting of velocities with respect to the brightest galaxy in each subsystem. The histogram is not consistent with the predictions based on a simple Gaussian. Two Gaussians are needed of respective widths 300 and 1400 km s^{-1} (Fig 2). The composite sample shows a strong dependence on magnitude of the velocity dispersion.

All these results can be interpreted in terms of mass segregation instability inside the subclusters, the cluster as a whole being likely by far younger than its subsystems. As a consequence of the small velocity dispersion (300 km s^{-1}) derived , the content of dark matter in the subclusters appears rather small.

REFERENCES.

Capelato, H., Gerbal, D., Mathez, G., Mazure, A., Salvador-Solé, E., Sol, H., 1980, Astrophys. J., 241, 521.
Fitchett, M., Webster, R., 1987, Astrophys. J.,317, 653.
Gerbal, D., Mathez, G., Mazure, A., Salvador-Solé, E., 1986 , "About the Coma Cluster. Proceedings of the UAI Symposium n° 117 /Dark Matter in the Universe, p. 112.
Mazure, A., Proust, D., 1985 (preprint).
Mellier, Y., Mathez, G., Mazure, A., Chauvineau, B., Proust, D., 1988, Astron. Astrophys. (sous presse).

M. Mezzetti et al. (eds.), Large Scale Structure and Motions in the Universe, 403–404.
© *1989 by Kluwer Academic Publishers.*

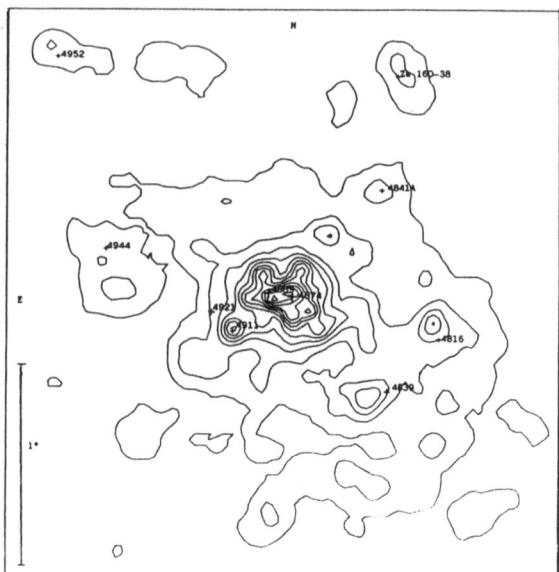

Figure 1.

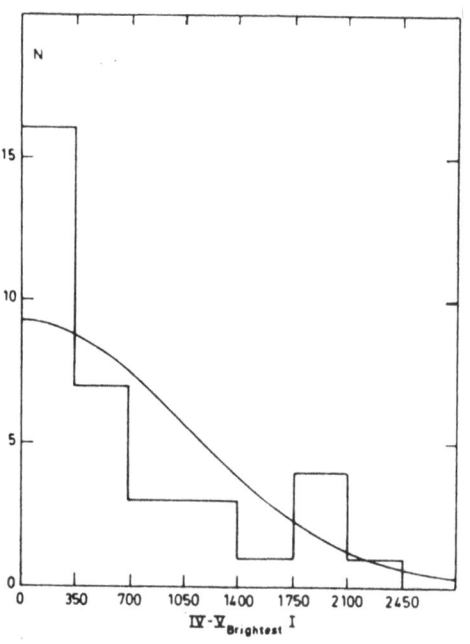

Figure 2.

Angular Structure in the Southern Cluster Catalogue

Colin McGill and H.M.P. Couchman
Canadian Institute for Theoretical Astrophysics,
University of Toronto,
Toronto, ON M5S 1A1,
Canada

Abstract. Measures of angular clustering in the Southern Cluster survey are presented and compared with similar results for the Northern catalogue. We find that the angular two-point correlation function for the two cluster samples are similar, except that the Northern Cluster sample possesses a long tail of positive correlation. In part, this can be accounted for by allowing for galactic absorption.

Introduction

The Southern cluster catalogue [Abell, Corwin and Olowin 1988] was constructed from deep IIIa-J southern sky survey plates. The intention was to mimic the procedure used by Abell in the compilation of the Northern catalogue. Thus the two catalogues taken together provide, nominally, a homogeneous all sky survey of rich clusters of galaxies out to a redshift of ~ 0.2. Unfortunately, few redshifts are presently available in the South. For this reason we compare only angular measures of structure in the two catalogues. In particular we have estimated the two-point angular correlation function and the r.m.s. variation of the cluster number density field when smoothed over various scales. The catalogues were also examined for large-scale density gradients, and the effects of these gradients on the clustering measures.

Comparison of $w_{cc}(\theta)$.

In order to compare the Northern and Southern cluster samples, it was necessary to restrict the clusters to richness class 1 or greater. In addition, regions of the sky were selected which were relatively free from galactic absorption. The regions selected had similar cluster densities which suggested that the samples had roughly similar depths.

The cluster distributions in the selected regions were then smoothed to test for remaining density gradients. We discovered that no matter how small the region selected, significant density gradients always remain. It is important to allow for these gradients when calculating measures of clustering. The gradients are consistent with being caused by absorption in our galaxy.

To compare the clustering more explicitly, the estimated function was fitted to the form

$$w(\theta) = \left(\frac{\theta}{\theta_\circ}\right)^{-\gamma}.$$

The best-fit parameters for various samples are given in Table 1. It can be seen that values for the parameter γ increase when allowance is made for the large-scale density gradients. The Northern

405

M. Mezzetti et al. (eds.), Large Scale Structure and Motions in the Universe, 405–406.
© *1989 by Kluwer Academic Publishers.*

Sample	θ_0	γ
N, $R > 0$	0.71 ± 0.09	0.91 ± 0.07
S, $R > 0$	0.69 ± 0.07	1.36 ± 0.12
S	0.51 ± 0.04	1.33 ± 0.10
N^*, $R > 0$	0.66 ± 0.08	1.05 ± 0.08
S^*, $R > 0$	0.60 ± 0.06	1.47 ± 0.16
S^*	0.46 ± 0.04	1.51 ± 0.14

Table 1. Power-law fits to angular correlation functions. The S (N) refers to the Southern (Northern) cluster catalogues. If $R > 0$, only clusters of richness class greater than zero have been analysed. The * signifies that we have allowed for large-scale density gradients in calculating the correlation function.

sample also has a significantly lower value of γ; here the difference is caused by the long positive tail. For a fuller discussion, see Couchman, McGill and Olowin [1988].

Two tests of the data were also carried out. The scaling relation [Peebles 1980] was found to be obeyed to $\sim$ 20%. We also examined the inter-plate, intra-plate correlation functions; it is possible that the correlation is caused by plate-to-plate variations [see also Sutherland, this volume]. There does seem to be some effect, but it is richness dependent. Richer clusters are significantly more affected than poorer ones; richness class zero clusters are almost unaffected. Indeed, smoothing the number of galaxies in a cluster across the sky shows a definite correlation with cluster density. This suggests that plate-to-plate variations may be cause cluster richnesses to be muddled. Hence, the increase in correlation length with richness may be partly accounted for by this effect, but not the essential correlation function.

We also calculated the rms fluctuation of the cluster number density field when smoothed over various scales. The results from this analysis indicate that clusters occur in groups of $\sim$ 4; the groups themselves are more or less randomly distributed.

Conclusions

Clustering in the Southern Cluster Survey is consistent with that exhibited by the Northern clusters. Large-scale density gradients, probably caused by variable galactic absorption, significantly affect the resulting correlation function and should be properly accounted for. Plate-to-plate variations may be partly responsible for the increase of correlation length with richness, but they do not seem to be fundamentally responsible for the correlation seen.

References

Abell, G.O., Corwin, H.G. and Olowin, R.P. (1988) Preprint.

Couchman, H.M.P., McGill, C. and Olowin, R.P. (1988) Preprint.

Peebles, P.J.E. (1980) *The Large-Scale Structure of the Universe* Princeton University Press.

Sutherland, W. (1988) This volume.

SOME REMARKS ON THE (L_X, σ_v) RELATION FOR CLUSTERS OF GALAXIES

Adam Michalec
Astronomical Observatory of the Jagellonian University
30-244 Cracow
Poland

On the basis of X-ray observation of the Abell clusters from "Einstein" and the published new value of L_X(0.5-3.5 keV) for clusters of galaxies (Jones and Forman, 1984) and a compilation of velocity dispersion for the Abell clusters (Struble and Rood, 1987) I present a list of 50 clusters for study of relation between L_X and σ_v. This relation was suggested on theoretical grounds by Solinger and Tucker (1972) and studied systematically at "Einstein" by Quintana and Melnick (1982) for 31 clusters.

Discussion of results:

1 – The value of the best-fit slope (in the relation $L_X \propto \sigma_v^n$) represented by the straight line (in the log-log scale) is $n = 2.80 \pm 0.56$ for 50 clusters presented at Fig. 1, which is in good agreement with the assumption of thermal bremsstrahlung as a source of X-ray emission of galaxy clusters (see: Quintana and Melnick, 1982).

2 – There is still an observational effect: L_X and σ_v are functions of redshift (Michalec 1984), these clusters of greater "z" have a l s o greater values of L_X and σ_v.
The mediana test applied to the assumed relation $L_X = f(z)$ and $\sigma_v = g(z)$ with $z_{med} = 0.050$ gives $\chi^2 = 5.12$ and $\chi^2 = 9.70$, respectively, while with 1 degree of freedom and 5% confidence level $\chi^2 = 3.84$, which is an indication that the galaxy clusters with high L_X and σ_v (and $z > 0.05$) are considerably different from those with low L_X and σ_v (and $z < 0.05$).

3 – The more precise relation $L_X = F(\sigma_v)$ allows for an estimation of dynamical parameters of a cluster, in particular its mass, from X-ray observations.

4 – To eliminate observational effects it is necessary to have a larger sample of clusters with L_X and σ_v known simultaneously.

M. Mezzetti et al. (eds.), Large Scale Structure and Motions in the Universe, 407–408.
© 1989 by Kluwer Academic Publishers.

408

Acknowledgements. Part of this work was done while I was visiting at Scuola Normale Superiore, Pisa, Italy.

References

Jones, C. and Forman, W. 1984, Astrophys.J. 276, 38.
Michalec, A. 1984, Proceed. of a Course and Workshop on Plasma Astroph.,
    Varenna, ESA SP-207, p.223.
Quintana, H. and Melnick, J. 1982, Astron. J. 87, 972.
Solinger, A.B. and Tucker, W.H. 1972, Astrophys. J. 175, L107.
Struble, M.F. and Rood, H.J. 1987, Astrophys. J. Suppl. 63, 543.

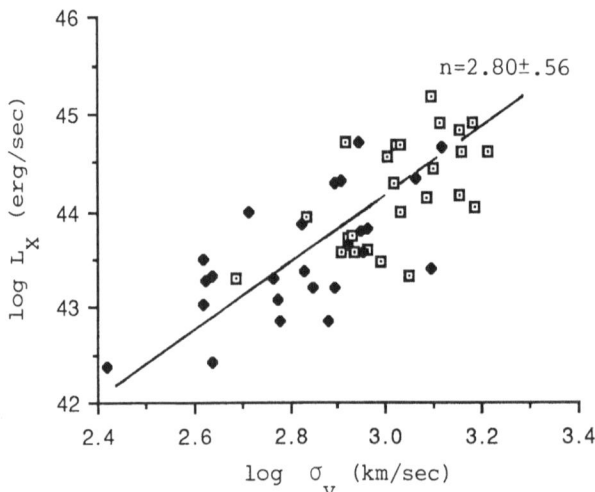

Figure 1. X-ray luminosity vs velocity dispersion correlation based on "Einstein" fluxes.
♦ - cluster with z < 0.050
▣ - cluster with z > 0.050

APPLICATION OF THE METHOD OF STATISTICAL REDUCTION TO THE A1314 CLUSTER OF GALAXIES

Adam Michalec
Astronomical Observatory of the Jagellonian University
30-244 Cracow
Poland

The first results of application of the statistical reduction method (Zieba A. 1975) for the distribution of galaxies, to 12 Abell clusters of galaxies was presented in 1986 by Michalec and Zieba S. (1988).

In this paper I present the results of application of the statistical reduction method (N=2, u=2 variant; where N is the reduction level, and u the number of elementary domains within the fundamental domain; for details see: Garncarek, Kuklewski and Rudnicki 1988) to the cluster of galaxies A1314.

It involved a study at the distribution of radio objects within the cluster (Wilson and Vallée 1982) after converting the (α, δ) coordinates to the rectangular coordinates (x, y) in PSS scale, centered at IC 712, as well as the distribution of optical objects (Michalec, unpublished data) and the superposition of both.

Optical dimensions of A1314: 82×82 arc minutes; $N_G = 50$.
Radio dimensions of A1314: 128×128 arc minutes; $N_R = 56$.
When we restrict ourselves to the optical dimensions of A1314, the number of radio sources within the field is $n_R = 40$.

Brief discussion of results:

1 - The radio sources within the cluster are more randomly distributed than the galaxies, as in the case of the radio sources from the 4C catalogue and galaxies from the Uppsala Catalogue (Zieba S. 1977).

2 - The superposition of (Opt. + Rad.) map of the A1314 cluster results in a sudden drop of the value of structure index S:
$S_{Opt} = 14.6 \pm 0.10$, $S_{Rad} = 10.9 \pm 0.71$, $S_{O+R} = 1.9 \pm 0.13$.
For the random distribution: $S = 1$, so one can conclude that the galaxies and the radio sources situated in the direction of A 1314 practically avoid each other. The conclusion is in accordance with the fact that there are few radio sources connected with optical objects in the A1314 (3 probable and 1 possible cluster members; see: Wilson and Vallée 1982).

M. Mezzetti et al. (eds.), Large Scale Structure and Motions in the Universe, 409–410.
© 1989 by Kluwer Academic Publishers.

3 - The result obtained stimulated us to apply the statistical reduction method to other Abell clusters as well, which will be presented in a separate work.

Acknowledgements. Part of this work was done while I was visiting at Scuola Normale Superiore, Pisa, Italy.

References

Garncarek, Z., Kuklewski, J. and Rudnicki, K. 1988, Acta Cosm. 15, 63.
Michalec, A. and Zieba, S. 1988, Acta Cosm. 15, 167.
Wilson, A.S. and Vallée, J.P. 1982, Astron. Astrophys. Suppl. 47, 601.
Zieba, A. 1975, Acta Cosm. 3, 75.
Zieba, S. 1977, Acta Cosm. 6, 75.

EXACT SOLUTION OF THE DIRAC EQUATION

M. Missana
Osservatorio Astronomico di Brera
via Brera 28
20121 Milano
Italy

ABSTRACT.   A solution is given of the Dirac equation for an electron
in the field of an electromagnetic wave and of a constant potential;
its interest is noted in the study of the galactic spectra.

From the study of the photons scattered by the bound electrons of the
hydrogen atoms, in a rarefied gas, it follows that there is a change of
wavelength due to the generalized Compton effect, beyond the wellknown
absorption and emission effects; the scattering cross-section largely
increases when the wavelength of the incoming photons becomes near to
the wavelengths of the spectral lines of the scattering atoms. Part of
the effects obtained with this first approximation theory are confirmed
by the measurements made in the spectra of the solar chromosphere, of
some O and B-type stars and of some diffuse nebula as shown in the article Missana (1987). For the galaxies of cosmological interest however
few spectra can be considered, for now. In the spectra of NGC6052 , in
which we have the wavelengths λ' of 14 lines (λ are the corresponding laboratory wavelengths) given by Du Puy et al. (1969), if we as-
sume that $\lambda' - \lambda = \Delta + \beta \lambda$ with Δ and β constants, by means of the
least squares method it can be deduced $\Delta = 4 \pm 3$ A° (1A°=0.1nm) which is
the generalized Compton effect and $\beta = 0.015 \pm 0.0006$, which takes into account the effects of the resonance scattering and of Doppler. Also
in the article of Schmidt et al. (1986) the spectra are given of 15
galaxies with wavelengths λ' of 8 corresponding emission lines.
From the differences of these λ' by means of the quoted formulae it
can be deduced that the spectra of most galaxies show a generalized Compton effect with respect to the other galaxies. In particular we have
$\Delta = 16 \pm 5$ A° for PC 1302+4721 and $\Delta = 13 \pm 5$ A° for PC 1701+4713,
in the average with respect to the remainig 13 galaxies. In the astrophysical articles it is usually assumed that the large positive values
observed for β are due to the expansion of the Universe and that the
contribution due to the scattering of light can be neglected. Aim of
the present communication is to notice that such an assumption should
be rejected from a logical standpoint, at least untill an average scattering cross-section will be available for the diffusion of the photons

411

M. Mezzetti et al. (eds.), Large Scale Structure and Motions in the Universe, 411–412.
© *1989 by Kluwer Academic Publishers.*

by the atoms of the intergalactic gas. Indeed the dispersive theory of the light propagation in a low energy scattering medium gives a lost of energy of the photons which grows with the distance source-observer (Missana,1986). As a first step in this calculation, the exact solution ψ is given of the Dirac equation for an electron in the field of an electromagnetic wave of frequency $\omega = 2\pi c/\lambda$ and a costant potential V ; c is the light velocity in vacuo. Let we assume a 4-potential $A_\mu \equiv (0, a_0 \cos\varphi', 0, V)$ with $\varphi' = \omega(t-z/c)$, where t is the time, an orthogonal coordinate system $\overline{x} = (x, y, z)$; we have:

$$\psi \cong \begin{vmatrix} U_1 -B\left[U_1(P^1+iP^2)D +U_2(1-P^3D)\right]\cos\varphi' \\ U_2 -B\left[U_1(1-P^3D) -U_2(P^1-iP^2)D\right]\cos\varphi' \\ \left[U_1 P^3+U_2(P^1-iP^2)\right]D -B\left[U_1(P^1+iP^2)D+U_2(1-P^3D)\right]\cos\varphi' \\ \left[U_1(P^1+iP^2)-U_2 P^3\right]D +B\left[U_1(1-P^3D)-U_2(P^1-iP^2)D\right]\cos\varphi' \end{vmatrix} \exp\left[-iS(t,z)/\hbar\right]$$

with $\overline{P}\equiv(P^1,P^2,P^3)$ constant vector, $B=ia_0 e/(2E'')$, $D=c/(mc^2+E')$, $E'=c(\overline{P}^2+m^2c^2)^{\frac{1}{2}}$, $E =E'+e$ V , $E''=E'-cP^3$, m and e mass and charge of the electron , U_1 and U_2 arbitrary constants and at last

$$S(t,z)= E t-\overline{P}\cdot\overline{x}+\left\{8c e a_0 \overline{P}^2\sin\varphi' +e^2 a_0^2\left[2\varphi'+\sin(2\varphi')\right]\right\}/(8\omega E'') \quad.$$

The representation adopted for the Dirac matrices is that of Dirac(1959) and the proof of that result can be obtained with the method of Volkow (1935). The computations are shown in the paper (Missana,1987b).

REFERENCES

Dirac P.A.M.:1959,I principi della meccanica quantistica,Torino,p.357.
Du Puy D.L and Veny J.B.:1969,Contribution from the Kitt Peak National Observatory,n_0 470.
Schmidt Marteen,Schneider D.P.,Gunn J.E.:1986,Ap.J.310,518.
Missana M.:1986,in Structure and Evolution of Active Galactic Nuclei,
     (eds.G.Giuricin et al.,D.Reidel,Dordrecht),p.645.
Missana M.:1987,Astrophys.Space Sci.136,167.
Missana M.:1987b,'Dispersion of the light in the sideral spaces',Osservatorio di Brera,Milano.
Volkow D.M.:1935,Zeits.fur Phys.94,25.

A DISTANT X-RAY SELECTED CLUSTER WITH A COOLING FLOW: 1E0839.9+2938

R. Nesci[1], G.C. Perola[1], I. M. Gioia[2,3], T. Maccacaro[2,3], R. E. Schild[2], and A. Wolter[2].

1) Istituto Astronomico dell'Universita' di Roma, Italy
2) Harvard-Smithsonian Center for Astrophysics, Cambridge, MA
3) Istituto di Radioastronomia del CNR, Bologna, Italy

During the systematic identification of the Extended Medium Sensitivity Survey sources detected with the Einstein Observatory, the cluster of galaxies 1E 0839.9+2938 was discovered.

The optical spectrum of the dominant galaxy (G1), obtained at the MMT with the FOGS, shows the presence of a number of emission lines ([OII], H β, [OIII], [OI], H α, [NII], [SII]) as well as the typical absorption lines found in elliptical galaxies, at a redshift of z=0.193 (see Fig. 1).

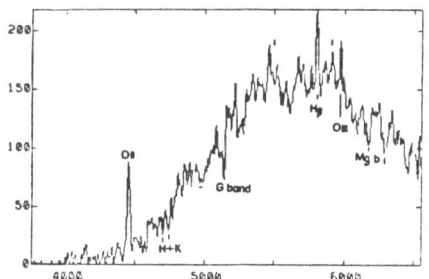 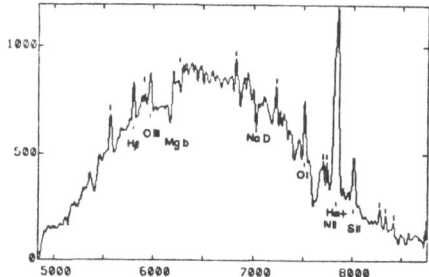

Fig. 1: *Optical spectra of galaxy G1 with FOGS. Left: blue grating; right: red grating.*

The equivalent widths of the emission lines, corrected for underlying stellar absorptions, are reported in Table 1. Relative line intensities are similar to those found in cD galaxies at the center of cooling flow clusters (Hu et al., 1985, Ap. J. Suppl. 59, 447, Johnstone et al., 1987, MNRAS, 224, 75.).

Table 1
Equivalent widths, line ratios relative to H α and typical values for cooling flows

	E.W.(Å)	Flux ratio	Average for cooling flows
[OII]3727/29	100.	0.74	1.5±1.0
H β	9.	0.19	0.31±0.12
[OIII]5007	9.	0.19	0.26±0.12
[OI]6300	15.	0.33	0.33±0.18
H α	46.	1.00	1.00
[NII]6583	71.	1.54	1.43±0.37
[SII]6717/30	24.	0.52	1.02±0.37

M. Mezzetti et al. (eds.), Large Scale Structure and Motions in the Universe, 413–414.
© 1989 by Kluwer Academic Publishers.

A comparison of the spatial profile of the H α + [NII] line with a stellar profile shows that the emitting region is probably spatially resolved by the present observation, with an intrinsic FWHM of about 1.6". This implies an extension of about 7 kpc (H_0=50 km s^{-1} Mpc^{-1} and q_0 = 0), comparable to the core emitting region of A1795 (12 kpc; Cowie et al., 1983, Ap. J., 272, 29.).

Spectra of five other cluster galaxies were obtained. Redshifts derived from absorption lines are listed in Table 2. The resulting r.m.s. radial velocity dispersion of the cluster is about 1300 km s^{-1}, typical of a rich cluster (Dressler and Shectman, 1988, A. J., 95, 284.).

V and R photometry for 13 galaxies in the cluster (within 80" from the central galaxy) was performed with the CfA CCD camera at the FLWO 24" telescope and is reported in Table 2. The color of the dominant galaxy is ⁻0.2 mag bluer than the expected value (0.97) for an elliptical at the redshift of the cluster.

The dominant galaxy was also detected at 6 cm with the VLA at a flux level of 5 mJy. Table 3 collects all the relevant data in the X-ray, optical and radio wavebands.

Table 2 Photometric data and redshifts for some galaxies.				Table 3 Relevant data for galaxy G1		
Galaxy	V	V-R	z			
G1	17.53	0.82	0.193	V (within 10" diameter)	17.53	
G2	18.83	0.89		V-R	0.82	
G3	18.87	0.80		z	0.193	
G4	19.31	0.84		M_v (k corrected)	-23.36	
G6	18.50	0.85		X-ray Lum. (0.3-3.5 keV)	3.9×10^{44}	erg s^{-1}
G7	19.09	0.97		Radio Lum. (4.8 GHz)	1.1×10^{24}	W Hz^{-1}
G9	19.24	0.90	0.194	Morphological Hubble type	E2	
G11	18.56	0.78	0.198			
G17	18.33	0.91	0.200			
G20	19.04	0.86				
G25	19.78	0.56				
G27	18.45	0.42	0.186			
G30	19.00	0.92				
G31			0.197			

All the observational evidence available suggests that we have detected one of the most distant cooling flows known to date. It looks like that optical spectroscopy of X-ray selected sources can be used as an efficient technique to reveal the cooling flow phenomenon at moderate/high redshifts with ground based telescopes. At these redshifts, only the Hubble Space Telescope will have a resolving power high enough to reveal the existence of extended structures by narrow-band imaging at suitable emission line wavelengths.

I. M. Gioia acknowledges financial support from the Smithsonian Institution under the "Research Opportunities Fund" program. Partial financial support for this work has come from the Smithsonian Scholarly Studies Grant SS88-3-87 and from NASA contract NAS-30751.

RADIO GALAXIES AS LARGE–SCALE COSMOLOGICAL PROBES

D. NICHOLSON
Department of Astronomy, University of Edinburgh

ABSTRACT. We are working on an all-sky sample of $\simeq$ 400 radio-selected elliptical galaxies to provide a powerful probe of clustering and streaming velocities in the universe. An optical programme is currently underway to obtain B & I CCD frames for all objects and accurate spectroscopic redshifts. This is now about 40% complete. Radio galaxies are optically very homogenous, and we have been able to estimate distances photometrically via optical profile fitting, as the galaxies obey a luminosity-size relation (Kormendy 1977). Our accuracy of redshift prediction (26%) is nearly as good as can be achieved using velocity-dispersion data (Fig.1). As yet we have only performed the simple test of looking for a dipole term in the velocities. Fig.2 shows a plot of velocity (in the microwave background frame) against angle from the Rubin-Ford (RF) apex; no correlation is evident, in contrast to the reported RF effect $v \simeq 800\cos\theta \mathrm{kms}^{-1}$, which is in conflict with our present data at about the $2 \cdot 5\sigma$ level (the number of points on Fig.2 should increase by a factor 4 in the final sample). It is clear that our method can provide adequate sensitivity to test for streaming motions at $\langle z \rangle = 0 \cdot 02$, and we hope to push the search out further; streaming velocities of $1000 \mathrm{kms}^{-1}$ would be detectable at $\langle z \rangle = 0 \cdot 05$.

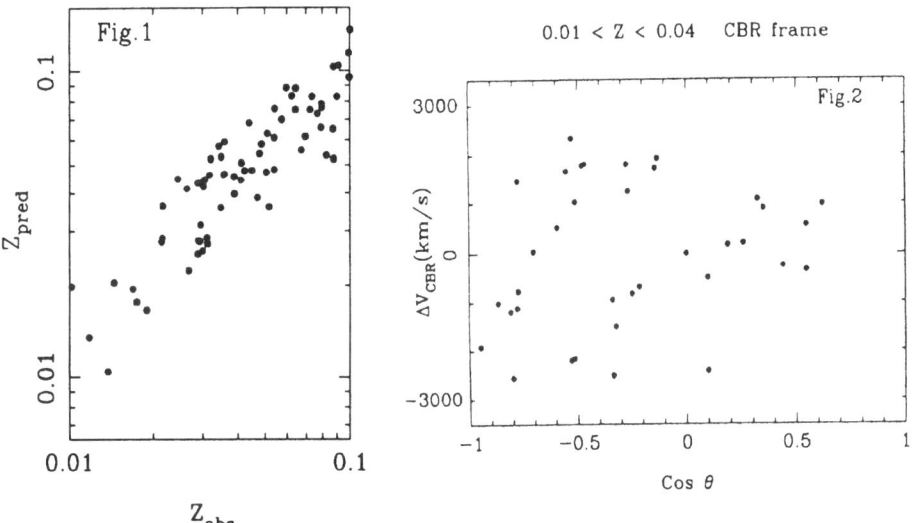

Figure 1. The results of the photometric distance indicator.

Figure 2. Peculiar velocities deduced from Fig.1, plotted against cosine of the angle away from the RF apex, in the CMB frame.

415

M. Mezzetti et al. (eds.), Large Scale Structure and Motions in the Universe, 415–416.
© *1989 by Kluwer Academic Publishers.*

416

There is some controversy about the degree of clustering of Abell clusters and its dependence on richness (Sutherland 1988). We can provide an independent check on this since radio galaxies occupy a range of environments from sub-Abell groups to the richest of clusters (Longair & Seldner 1979; Prestage & Peacock 1988). Using our sample, we have made the first determination of the 3D two-point correlation function for radio galaxies. We find a strong signal (with a correlation length $\sim 15h^{-1}$Mpc)–but only for those galaxies in the range of radio power expected *a priori* to lie in systems with average Abell richness $R \simeq 0$ (Fig.3). It will be important to confirm this result by roughly doubling the present statistics.

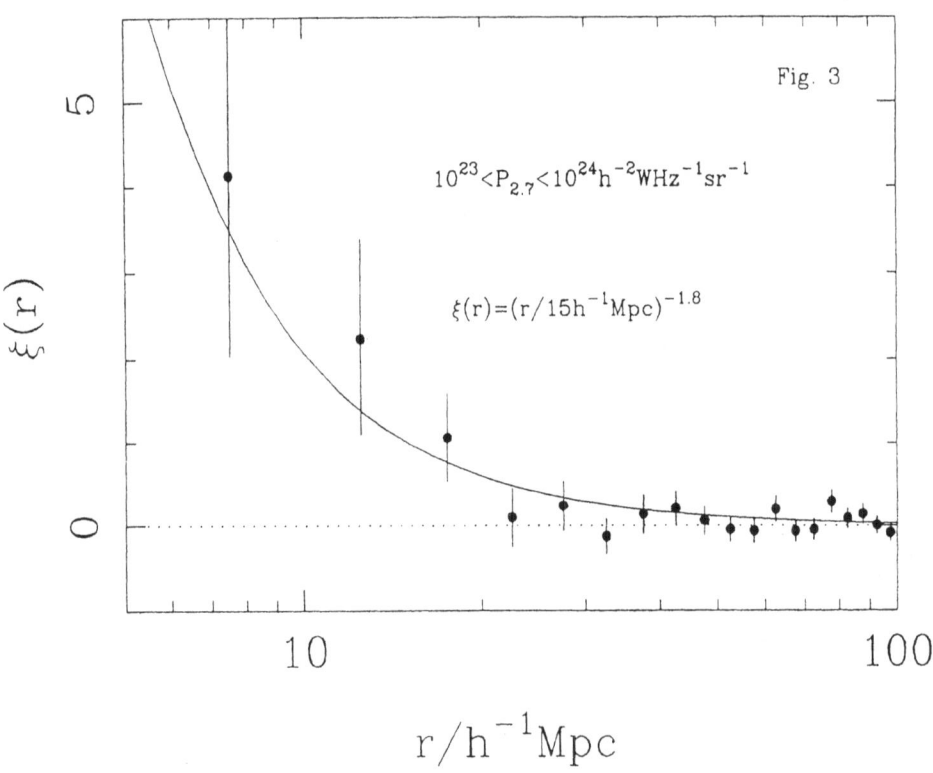

Figure 3. The two-point correlation function estimated from 120 galaxies with spectroscopic redshifts and radio powers in the range $10^{23} \lesssim P_{2.7} \lesssim 10^{24}h^{-2}$WHz^{-1}sr^{-1}. No significant clustering is seen in other power ranges.

References

Kormendy, J., 1977. Ap. J. **218** 333
Longair, M.S., & Seldner, M., 1979. MNRAS **189**, 433
Prestage, R.M., & Peacock, J.A., 1988. MNRAS **230**, 131
Sutherland, W.J., 1988. MNRAS in press

The 3-D distribution for a magnitude limited galaxy sample to Bj=16.5
using techniques of hierarchical clustering

Q.A.Parker and H.T.MacGillivray
Royal Observatory
Edinburgh
EH9 3HJ

ABSTRACT. Techniques of hierarchical clustering were applied to a
magnitude limited, near complete sample of 107 galaxies with slit
spectra redshifts to Bj=16.5 in ESO/SERC survey field 349 near the SGP.
This multivariate statistical analysis enabled a quantitative and
unbiased description of the galaxies in this sample into well defined
groups, associations and field galaxies.

RESULTS

The sample for analysis consists of 107 galaxy redshifts accurate to
better than 75 Km/s in one UKST field, 85% complete to Bj=16.5. Details
of the observations and reductions are given by Parker et al. (1987). A
cone diagram for the entire sample is given in Fig.1 which illustrates
the clumpy nature of the galaxy distribution to 180 h^{-1}Mpc with tight
knots and looser aggregates separated by large empty regions. A redshift
histogram highlights an obvious dichotomy with two galaxy distributions
separated by a 20 Mpc void which has a projected extent of 5X10 Mpc.
    A hierarchical clustering analysis was applied to each of the two
major redshift distributions (after defining 3-D co-ordinates where each
co-ordinate is expressed in Mpc relative to a mid redshift reference
plane and the field centre). Such methods are a powerful means of
unravelling and describing inherent structure in data in a quantitative
and objective manner (Murtagh & Heck, 1987). Galaxies can be classified
simultaneously in 3-D into groups and clusters. No model assumptions for
the clusters themselves are required. Fig.2 gives the tree diagram or
'dendrogram' from classifying the first redshift grouping (30 galaxies)
according to the 'Ward' criterion which seeks tight, minimum variance
spherical clusters (Ward 1969). All levels of substructure are revealed
from single objects through to binaries, groups and clusters.
Other clustering methods applied to both samples gave much the same
results indicating a robust classification. A detailed account of the
analyses and results are given by Parker et al. (in preparation). Galaxy
groups from both samples had velocity dispersions <200 Km/s with a mean
separation of 7 Mpc, whilst the (X,Y,Z) co-ordinate dispersions were all
generally < 2.0 Mpc.   417

M. Mezzetti et al. (eds.), Large Scale Structure and Motions in the Universe, 417–418.
© *1989 by Kluwer Academic Publishers.*

418

Comparison of the clustering results from more traditional analyses (e.g. de Vaucouleurs, 1974), are favourable. The techniques adopted here have the ability to identify not only well defined groups but also field galaxies in an independent and objective manner giving much material for further study. The good redshift accuracy, adopted co-ordinate system and lack of rich clusters in the field mean that velocity uncertainties and the 'finger of God effect' do not seriously impair the clustering results.

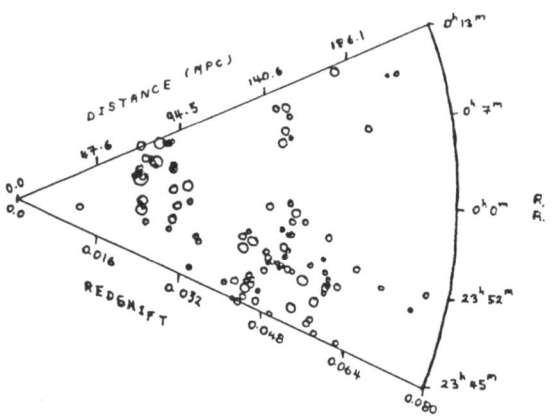

Fig.1. Cone diagram of R.A. versus redshift for the whole galaxy sample.

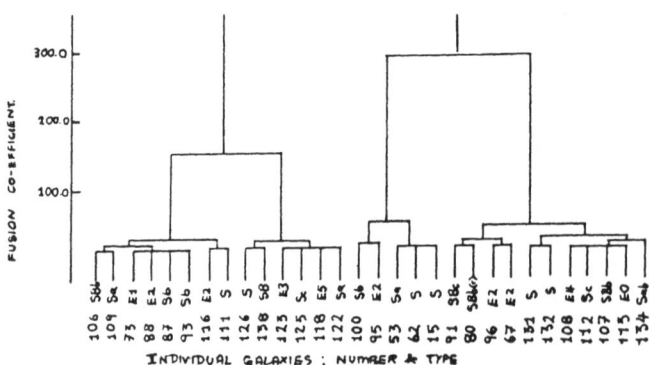

Fig.2. Dendrogram for the first major redshift distribution in the field. The abscissa is the "fusion co-efficient" and indicates the level where neighbouring objects fuse according to the chosen clustering criterion.

REFERENCES

Murtagh,F.,& Heck,A.,1987. "Multivariate data Analysis", D.Reidel, Dordrecht,Holland.
Parker,Q.A.,MacGillivray,H.T,Hill,P.W.,& Dodd,R.,1986. M.N.R.A.S.,220,901.
Ward,J.H.,1963. J.Amer.Stat.Ass.,58,236.
de Vaucouleurs,G.,1974. "Stars & stellar systems", vol IX,557,Eds. Sandage,A.,sandage.M., & Kristian,J., Univ.Chicago press.

GALAXY CORRELATIONS AS A FUNCTION OF LUMINOSITY

S.Phillipps[1] and T. Shanks[2]
[1]Department of Applied Mathematics and Astronomy.
University College Cardiff, Wales, U.K.

[2]Department of Physics,
University of Durham, Durham, U.K.

It is difficult to determine the variation of galaxy clustering as a function of luminosity from normal 3-D redshift surveys because such surveys of galaxy distances contain few intrinsically faint objects (the majority of galaxies in a magnitude limited sample being bright "M^* galaxies"). On the other hand it is clear that apparently faint galaxies seen projected near bright galaxies known to be at a particular distance are likely, in a statistical sense, to be at this same distance. This then enables us to obtain the correlation function of galaxies of different intrinsic luminosities by counting the excess (over the average background) number of galaxies around redshift survey (i.e. known distance) galaxies, as a function of apparent magnitude. This allows us to utilise very large numbers of faint companions to bright galaxies and hence gives good statistics on the faint galaxy correlation function.

From counts around 260 galaxies in Durham redshift survey fields (containing in total some 100000 galaxies down to B $\simeq$ 21 on UKSTU plates), we find that there is no significant difference between the correlation function slope for bright and faint galaxies (i.e. the ratio of bright to faint galaxies does not change significantly with distance from the redshift survey galaxy).

Calculation of the relative amplitude of the clustering of the bright and faint galaxies unfortunately requires knowledge of the galaxy luminosity function (since the existence of many faint galaxies around our redshift survey galaxies, for instance, could mean either that faint objects are very clustered or that faint objects are generally very common). Nevertheless, we can say that even if we allow fairly wide bounds on the possible shape of the overall luminosity function, then bright and faint galaxies must be equally clustered to within a factor 2.

References
Phillipps, S. & Shanks, T., 1987a, *Mon.Not.R.astr.Soc.*, 212, 657
Phillipps, S. & Shanks, T., 1987b, *Mon.Not.R.astr.Soc.*, 229,621.

M. Mezzetti et al. (eds.), Large Scale Structure and Motions in the Universe, 419.

VOID STRUCTURE IN THE LYMAN ALPHA FOREST

M. PIERRE, P.A. SHAVER, A. IOVINO
European Southern Observatory
Karl-Schwarzschild Str. 2
D-8046 Garching bei München, W. Germany

ABSTRACT. using a model of the present structure of the Universe, it is shown that voids are less pronounced in the ditribution of Lyα absorbers at high redshift than they are in the distribution of galaxies today.

1. MODEL

The model, inspired by the bubble like structures revealed by the CfA survey (Lapparent at al. 1986), consists of closed cells completely filling the space (Voronoi tesselation). The problem is scaled to a mean void size of 25 Mpc and the cell walls are given a constant thickness of 4 Mpc (*cf* CfA survey). The walls are then populated at random without any additional clustering (Fig. 1), in order to produce the lowest possible correlation for a cellular topology (Pierre et al. 1988).

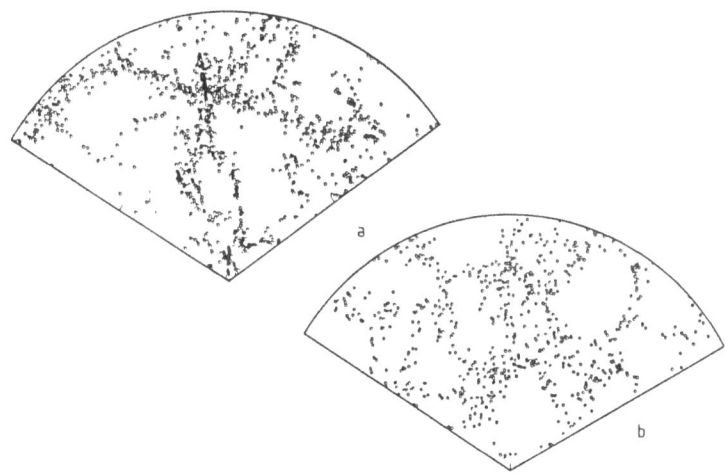

Fig. 1. Comparison of the model universe (b) with the CfA survey (a). The slice is $12°$ thick and $10\,000\ kms^{-1}$ deep.

M. Mezzetti et al. (eds.), Large Scale Structure and Motions in the Universe, 421–422.
© *1989 by Kluwer Academic Publishers.*

Then, assuming the Lyα absorbers follow this structure, the model is shifted to z=2.5 and used for creating random lines of sight. Taking into account peculiar velocities, line widths and line blending this yields a synthetic Lyα forest.

2. RESULTS

Finally, we derive from the line list so obtained the one-dimensional two-point correlation function. Fig. 2 shows our theoretical result compared with the observations of Webb (1987).

The correlation given by the model is much higher. There are two possible explainations: (i) The Lyα absorbers occupy the voids. (ii) The structure has evolved. In this latter case, agreement with the observations can be reached if the "filling factor of voids" is about 25% instead of 64% (as given by 25 Mpc voids with a 4 Mpc wall thickness). This would correspond to a void size of 7 Mpc with a wall thickness of 4 Mpc, or alternatively 19 Mpc and 10 Mpc respectively.

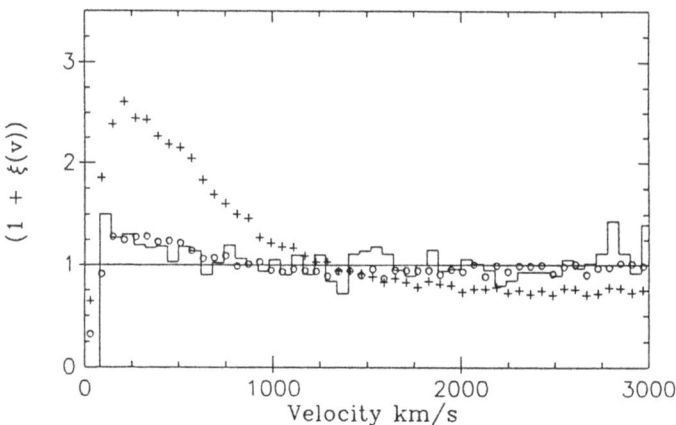

Fig. 2. One-dimensional two-point correlation function.
▢ : Observed data for the Lyα absorbers (Webb 1987)
+ : Model representing today's Universe.(void filling factor 64%)
○ : Model with smaller voids.(void filling factor 28%)

REFERENCES

de Lapparent,V.,Geller,M.J.,Huchra,J.P. 1896. *Astrophys. J.*302, L1

Pierre,M.,Shaver,P.A.,Iovino,A. 1988, *Astron. Astrophys. Lett.* (in press)

Webb,J.K. 1987. in *Observational Cosmology* (ed. A. Hewitt, G. Burbidge, L.Z. Fang; Reidel,Dordrecht). p. 803

MEMBERSHIP AND M/L OF NEARBY GROUPS OF GALAXIES

D. Puche and C. Carignan
Département de Physique and
Observatoire astronomique du Mont Mégantic
Université de Montréal
Montréal, Québec, H3C 3J7
Canada

ABSTRACT. We present a method for determining membership and dynamical mass-to-light ratios for nearby groups of galaxies. By computing an approximate centroid for a group and adding galaxies radially from there on, we are able to determine a dynamical mass from four virial-type mass estimators. It is shown that by adding non-member galaxies a break in the slope of mass versus radius is produced. This makes it possible to eliminate interlopers and to obtain accurate mass-to-light ratios for nearby groups. The technique is applied to the Sculptor group, where we find a mass-to-light ratio consistent with the mean M/L estimated for other groups, and much less than was previously estimated. These results suggest that high M/L values for groups are probably caused by the presence of interlopers or the fact that they are simply unbound systems. If the M/L ratios prove to be, as in the Sculptor group, almost half an order of magnitude lower than was previously estimated, this would have important consequences on the formation of galaxies and on clustering at larger scales.

1. Introduction

The determination of group membership has always been based on subjective criteria (De Vaucouleurs 1959). Even so called automated techniques (Gott and Turner 1977, Huchra and Geller 1982) were using cuttoff parameters most often determined by what should be "reasonnable" for small groups of galaxies. This problem is very important in the derivation of dynamical mass-to-light (M/L) ratios. By including non-member galaxies ("interlopers") with larger spatial and/or kinematical separations, the dynamical mass estimators like the virial theorem will tend to produce overestimates of the masses. This, in turn, will be reflected in an overestimate of the M/L ratios and lead one to believe that small groups contain much more "dark matter" than there actually is.

M. Mezzetti et al. (eds.), Large Scale Structure and Motions in the Universe, 423–424.
© 1989 by Kluwer Academic Publishers.

2. Application of the technique

We use the radial velocities, distances and projected tangential distances to derive the mass of the Sculptor group using the average of the Heisler, Tremaine and Bahcall (1985) (HTB) mass estimators. We extend the number of galaxies as a function of radial distance from the centroid formed by NGC 247, 253 and 7793. Accurate distances are needed for each of the members. This of course will make the technique more difficult to apply to distant groups.

A break in the slope determined by the first few models means that the galaxy added is probably an interloper. This allows us to eliminate NGC 24 and NGC 45 (Puche and Carignan 1988). From this result we can compute a total mass for the Sculptor group of 2.0×10^{12} $M_\odot$. This gives us a M/L for the group of $(M/L_B)_{dyn} = 83 \pm 20$ $M_\odot / L_\odot$, which disfavors previous determinations as high as $(M/L_B)_{dyn} \simeq 500$.

3. Conclusions

We show that the HTB estimators allow to establish dynamical membership of galaxies in nearby small groups. This technique was applied to the Sculptor group and the M/L ratio for this group was found. The technique could be applied to other nearby groups for which sufficient three dimensionnal data exist. It appears that dynamical M/L were greatly overestimated, and that a more accurate selection of group members would reduce the average value of this parameter. This determination of the amount of dark matter on intermediate scales suggests that the total mass versus total luminous mass (M_{tot}/M_{lum}) is closer to a constant on all scales, and would then be in agreement with cold dark matter scenarios for the formation of galaxies and large scale structures.

4. References

De Vaucouleurs, G., 1959, *Astrophys. J.*, **130**, 718.
Gott, J.R., and Turner, E.L., 1977, *Astrophys. J.*, **213**, 309.
Heisler, J., Tremaine, S., and Bahcall, J.N., 1985, *Astrophys. J.*, **298**, 8. (HTB)
Huchra, J.P., and Geller, M.J., 1982, *Astrophys. J.*, **257**, 423.
Puche, D., and Carignan, C., 1988, *Astron. J.*, **95**, 1025.

The Tidal Torque on the Local Group of Galaxies

Somak Raychaudhury
Institute of Astronomy,
Cambridge, U.K.

ABSTRACT. The quadrupolar component of the gravitational field at the barycentre of the Local Group of galaxies due to neighbouring galaxies is estimated. It is seen that the Local Group does not gain an appreciable amount of angular momentum through tidal interaction with other galaxies or groups.

If the net gravitational potential at the barycentre of the Local Group of galaxies (LG) due to all other neighbouring galaxies is denoted by ψ, then the torque on the Local Group due to this field is $\Gamma_i = \epsilon_{ijk}\, q_{jl}\, \mathcal{Q}_{lk}$ where $q_{\alpha\beta}$ is the quadrupole moment of the LG, and

$$\mathcal{Q}_{kl} = \left.\frac{\partial^2 \psi}{\partial r_k\, \partial r_l}\right|_{\substack{barycentre \\ of\ LG}} = \sum_{\alpha \notin LG} \frac{G\, M_\alpha}{|\mathbf{r} - \mathbf{r}_\alpha|^3} \left[\frac{3\,(\mathbf{r} - \mathbf{r}_\alpha)_k\,(\mathbf{r} - \mathbf{r}_\alpha)_l}{|\mathbf{r} - \mathbf{r}_\alpha|^2} - \delta_{kl} \right].$$

More than 90% of the luminous matter in the LG rests in M 31 and the Galaxy, and so does most of the mass, unless there are other equally massive nearby galaxies hidden behind the plane of the Galaxy. The barycentre thus lies on the line joining the Galaxy and M 31. To choose its actual position, we adopt $M_A/M_G = 2$.

The sample of galaxies used to evaluate $\mathcal{Q}_{kl}$ has been drawn from Kraan-Korteweg's catalogue of nearby galaxies (KK86, 1986). Since $\mathcal{Q}_{kl} \propto r^{-3}$, only closer and brighter galaxies are expected to be important, and so we have used only the galaxies with velocities $\leq 1000\ kms^{-1}$ with respect to the LG and having a distance less than 12 Mpc. Excluded from the sample were galaxies marked in the catalogue as members of the LG or of the Virgo cluster, and all galaxies within the region $|b| \leq 10°$, because for these galaxies the values quoted for galactic extinction in the literature are very uncertain. The remaining 212 galaxies were used in the actual evaluation of $\mathcal{Q}_{kl}$.

Distances to these galaxies were obtained from their redshifts (corrected for the motion of the sun, taken to be $300\ kms^{-1}$ towards $l = 90°$, $b = 0°$) in terms of the distance to the Virgo cluster using a Virgocentric flow model as described in KK86, with an infall velocity of $220\ kms^{-1}$. The adopted distance of the Virgo cluster in this paper is 20 Mpc. The masses M_α of the galaxies can be replaced by a function of their apparent blue magnitudes $m_B^{0,i}$ corrected for galactic extinction as prescribed by Burstein and Heiles (BH, 1976), if as a starting point one assumes that mass follows light. A constant $M/L_B = 10$ was adopted for all galaxies in the sample.

The estimates show that though individual galaxies like IC 342, M 81, Cen A, NGC 6946 and the South Polar Group (NGC 55, 247, 253, 300) contribute significantly towards the magnitude of the eigenvalues of $\mathcal{Q}_{kl}$, the direction of its axes depend on the large-scale distribution of nearby smaller galaxies. It is also seen that the contribution of galaxies beyond ~ 6 Mpc is small and that the final values obtained are remarkably stable against the inclusion or exclusion of individual objects, within the anticipated errors.

M. Mezzetti et al. (eds.), Large Scale Structure and Motions in the Universe, 425–426.
© 1989 by Kluwer Academic Publishers.

The largest source of error is expected to arise due to incompleteness in the sample, since in measuring the quadrupole component of a gravitational field, voids in the distribution can be as important as clusters. KK86 claims to be complete all over the sky to $13^m.4$. The sample within 12 Mpc is expected to be reasonably complete to this brightness limit, except close to the plane of the galaxy. To account for the contribution of unknown galaxies in this region, the following process of extrapolation has been adopted. One assumes that one is looking at the sky through a medium of varying absorption together with an opaque strip in the region $|b| \leq 10°$, and then, constructing a model of the absorbing medium (BH), attempts to infer the actual distribution of galaxies on the sky from the observed distribution.

The results are as follows: [format: eigenvalue (direction of eigenvectors in l,b)]

      from sample: 0.039 (139, 48);   −0.006 (308, 42);   −0.033 (43, 5)

      after correction: 0.037 (136, 25);   −0.005 (298, 64);   −0.032 (42, 7)

[units: G=1, velocity, distance and mass 100 kms^{-1}, 1 Mpc and $2.3 \times 10^{12} M_\odot$ respectively].

The 3×3 matrix Q_{kl} is traceless and symmetric. If the sample consisted of only one object, then the eigenvalues would have been of the form $(-a/2, -a/2, a)$ and the eigenvector corresponding to the positive eigenvalue would have been in the direction of the object from the barycentre. These eigenvalues show a curious form, corresponding to an almost equal tidal extension and compression along two axes and virtually no effect along the third. The same pattern can be seen in the evaluation of the quadrupolar component of the velocity field in the Local Supercluster (Lilje et al 1986).

The dependence of Q_{kl} on the parameters of the Virgocentric flow model used is not significant. The most sensitive parameter is the value of the infall velocity towards Virgo, and as KK86 demonstrates, changing that to 440 kms^{-1} does not change the distances appreciably, and since apparent magnitudes are used, Q_{kl} varies only as $\sim r^{-1}$. The distances are evaluated as a fraction of the distance to the Virgo cluster, and so Q_{kl} varies as D_{Virgo}, changing which would change the eigenvalues, but not the eigenvectors.

The dynamics of the Galaxy and M31 can then be represented as a two-body system moving under their mutual gravitational attraction in the presence of a tidal field due to external galaxies, $\ddot{r}(t) + GMr(t)/|r(t)|^3 + F(t) = 0$, where M is the combined mass of the system and $F_k(t) = - \sum_l r_l(t) Q_{kl}(t)$ to first order. However, t_0 and M are not known, but from timing arguments, in the Kepler limit, specifying one determines the other. For $\Omega_0 = 1$, $Q_{kl} \propto t^{-2}$, and $t_0 \approx 1.3 \times 10^{10}$ yr if $H_0 = 50$, in which case $M = 4.2 \times 10^{12} M_\odot$. On integrating backwards in time, it was found that the binary orbit of M 31 and the Galaxy remains largely unaffected, at least in the last six billion years, upto which the tidal approximation holds good. On the other hand, three-body orbits involving outlying dwarf members of the LG e.g. WLM can gain considerable angular momentum in the same period. The actual details of the orbits are, however, strongly dependent on the relative masses of the galaxies in question. A detailed description of these results would be published elsewhere (Raychaudhury 1988).

The author is grateful to Professor D. Lynden-Bell who suggested this problem, to Dr R. Kraan-Korteweg for supplying a tape version of her catalogue, and to the Cambridge Philosophical Society for a travel grant. He is supported by the Isaac Newton Studentship.

References: Burstein D. and Heiles C. (1976), *Astrophys. J.* **225**, 40.

Kraan-Korteweg R. (1986), *Astron. Astrophys. Suppl. Ser.* **66**, 255 and catalogue privately circulated.

Lilje P. B., Yahil A. and Jones B. J. T. (1986), *Astrophys. J.* **307**, 91.

Raychaudhury S. (1988), to be submitted to *Mon. Not. R. astron. Soc.*

RICH CLUSTERS & LARGE STRUCTURES: THE SEARCH FOR ALIGNMENTS

G. RHEE, M. van HAARLEM, P. KATGERT and H. LATOUR
Sterrewacht Leiden
P.O.Box 9513
2300 RA Leiden
the Netherlands

We present results of a survey of Abell clusters. Areas including 2 Mpc around each cluster have been digitized using the Leiden plate scanning machine. We discuss the alignment of Abell clusters with position angles based on galaxy distributions and on the X-ray morphology of the clusters in our sample which have been observed with Einstein.

The Leiden astroscan machine was used to digitize POSS plate areas around Abell clusters. The sample consists of 108 rich cluster with; $z \leq 0.1$, $9^h \leq \alpha \leq 18^h$, $b \leq 30°$, $R \geq 1$. We use the sample to search for two effects as test of galaxy formation theories. The first test is the so called Binggeli effect. Binggeli found that Abell clusters with pair distances less than $30 \sim 40$ Mpc $H_o = 50$ tend to point to one another. The second test is alignment of the first-ranked galaxy with its cluster. To define the cluster position angle we have used 1)the galaxy distribution dereived from the digitised plates 2)the EINSTEIN observatory IPC x-ray data if available for our cluster sample.

To search for cluster-cluster alignments we plot D (the distance to the nearest neighbour in Mpc) versus $\Delta\Theta$ (the difference between the cluster p.a. and the direction to the nearest neighbour). An (factor of ~ 2) excess of clusters with $\Delta\Theta < 45$ and $D < 20$Mpc is present.

We find convincing first ranked galaxy cluster alignments. The effect is stronger for clusters with BM class 1 (clusters which are dominated by a single central cD galaxy). The effect is also stronger if we exclude weakly elongated clusters which suggests that the scatter in the histogram is due to the uncertainties in the cluster position angles.

We find a good correlation of the x-ray position angle with the first ranked galaxy position angle. Using x-rays to define the cluster position angle we find a weak alignment for clusters with neighbours < 30Mpc away and an antialignment for clusters with neighbours > 30 Mpc away. The distribution of hot gas thus has some knowledge of the direction to the nearest neighbour.

M. Mezzetti et al. (eds.), Large Scale Structure and Motions in the Universe, 427.
© *1989 by Kluwer Academic Publishers.*

VLA HI OBSERVATIONS OF THE BRIGHTEST SPIRALS IN HYDRA I

O.-G. Richter[1], H.C. Ferguson[2] and J.H. van Gorkom[3,4]

1)   Space Telescope Science Institute
3700 San Martin Drive, Baltimore, MD 21218, U.S.A.
Also affiliated to the Astrophysics Division of the
Space Science Department of E.S.A.
2)   Center for Astrophysical Sciences, Johns Hopkins University
Homewood Campus, Baltimore, MD 21218, U.S.A.
3)   National Radio Astronomy Observatory[1], Very Large Array
P.O. Box 0, Socorro, NM 87801, U.S.A.
4)   Department of Astronomy, Columbia University
538 W. 120th St., New York, NY 10027, U.S.A.

ABSTRACT   In VLA HI images of the Hydra I cluster we have detected the three largest spirals and ten other galaxies. The ratio of HI diameter to optical diameter tends to increase with projected distance from the cluster center, similar to the trend seen in the Virgo cluster. The relatively small HI disk of NGC 3312 supports the hypothesis (Gallagher, 1978) that ram pressure sweeping is occurring in this galaxy.

1. INTRODUCTION

The Hydra I (A 1060) cluster of galaxies is the nearest large cluster beyond the Virgo cluster. It therefore represents an ideal target for comparison with the abundant data on Virgo. Both clusters are of roughly equal size and global structure and have similar total masses, X-ray luminosities, velocity dispersions, and morphological type distributions. However, while the Virgo cluster still suffers from infall of galaxies, the Hydra I cluster appears rather isolated in redshift space and only loosely coupled to the rest of the Hydra-Centaurus supercluster. It is thus probably more relaxed than the Virgo cluster. In VLA observations carried out in February 1987 the three brightest spirals and a number of fainter cluster galaxies were detected. Here a few preliminary results are presented. A more thorough analysis of the radio, optical and X-ray data for the cluster, including upper limits for the undetected galaxies and multicolor CCD photometry for the brighter spirals, is currently in progress.

2. NEW RESULTS

The brightest galaxies were clearly detected. For a few of the galaxies we made the first determination of the redshift. Non-detection of a galaxy does not necessarily

M. Mezzetti et al. (eds.), Large Scale Structure and Motions in the Universe, 429–430.

mean that the galaxy has a low HI content since the observed velocity range is quite limited. Synthesis observations of galaxies in the Virgo cluster (van Gorkom *et al.*, 1984; Warmels, 1985) have shown that the HI diameter of cluster spirals tends to increase with distance from the central galaxy. The results for the Hydra I galaxies are shown in Fig. 1. The trend appears to be similar for the Hydra I galaxies, but there appears to be an offset in the sense that the HI diameters of the Hydra I galaxies are larger than the Virgo galaxies at the same distance from the cluster center. In spite of the relatively larger HI disks of the Hydra I galaxies, there is still evidence that the gas is being removed from a number of galaxies. NGC 3312 is red and "anemic" on its west side and blue and rich in dust features and H II regions on its east side. With a velocity relative to the cluster of 840 km s^{-1}, it appears certain that it moves supersonically and must be strongly influenced by the intracluster gas. Indeed, already Gallagher (1978) suggested that it is undergoing ram pressure stripping. The HI barely extends beyond the disk of the galaxy and is distributed somewhat asymmetrically with respect to the optical image. This suggests that the tangential component of the galaxy's velocity is significant and is pointed away from the cluster center. If the galaxy is on a roughly radial orbit, then the direction of its tangential velocity implies that it has already passed through the cluster core and is on our side of the core. In NGC 3314a we found an extended HI tail associated with some very faint optical emission. It is not currently clear if this represents a collection of dwarf galaxies close to NGC 3314a or truly a tail in the HI (and light) distribution of NGC 3314a itself.

REFERENCES

Gallagher,J.S.: 1978, *Astrophys. J.*, **223**, 386.

van Gorkom,J.H., Balkowski,C. and Kotanyi,C.: 1984, *Clusters and Groups of Galaxies*, eds. F. Mardirossian, G. Giuricin, and M. Mezzetti, Reidel, Dordrecht, p. 261.

Warmels,R.H.: 1985, *The Virgo Cluster of Galaxies*, Proceedings of an ESO workshop, eds. O.-G. Richter and B. Binggeli, ESO, Garching, p. 51.

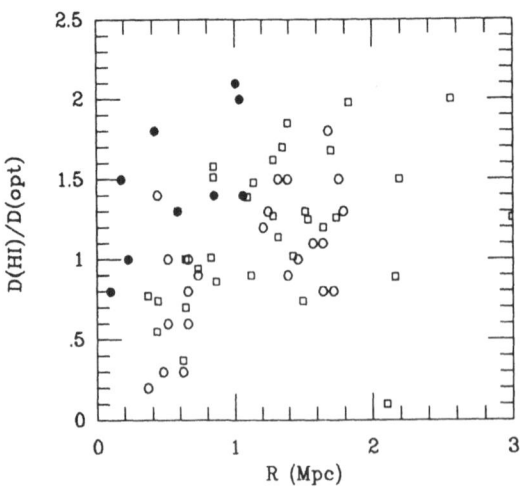

Fig. 1: Ratio of HI diameter to optical diameter versus distance from the cluster center. The filled circles are the Hydra I data. The HI diameter is taken at a contour of $N_{HI} = 5 \cdot 10^{19}$ cm^{-2}. The optical diameter is at an isophote of approximately 25^m per square arcsec. The open circles are VLA observations of Virgo cluster galaxies from van Gorkom *et al.*, (1984). The open squares are Virgo cluster observations from Warmels (1985).

DARK MATTER IN ELLIPTICAL GALAXIES

R.P. Saglia, G. Bertin, M. Stiavelli
Scuola Normale Superiore
I-56100 PISA
Italy

Photometric and kinematical data (stellar velocity dispersion) for a set of bright elliptical galaxies (NGC 3379, 4374, 4472, 4486, 4636, 7562, 7619, 7626) have been analyzed in terms of self-consistent anisotropic models (f_∞) under the assumption of constant mass-to-light ratio [1]. The results are consistent with the possibility that dark matter in these ellipticals is absent or, at least, approximately distributed like the luminous matter. However, the above mentioned kinematical data do not pose a stringent constraint, since they extend out to a fraction of R_e only. In fact, a violation of the constant M/L assumption might be required when reliable velocity dispersion data will become available on a larger radial extent or for the small number of ellipticals where kinematical information in the outer regions already exists in the form of gas rotation curves or X-ray data.

Here we present a study of NGC 4278, an El hydrogen-rich galaxy, that has been mapped in the 21 cm line radiation [2]. Under the assumption of constant M/L, we have fitted the anisotropic f_∞ models to the R band CCD photometry of [3]; for simplicity we have excluded from the fit the innermost ($R \leq 6"$) regions that are affected by seeing. In this way we determine the scale length and the dimensionless central potential Ψ of the model. The kinematical fit to the velocity dispersion data along the major axis of the galaxy (taken from [4]) provides the velocity scale. The mass and the M/L$_B$ ratio of NGC 4278 derived from this analysis are $3.3 \times 10^{11} M_\odot$ and $22 M_\odot/L_\odot$ (distance D=16.4 Mpc). The rotation curve of the best fit model is then completely determined and can be compared with the HI observations.

In the Figure we show the results of the various fits for NGC 4278. The photometric and velocity dispersion fits prove to be very good. The agreement between the theoretical rotation curve and the HI data is less satisfactory. In the inner region ($R \leq 30"$) the predicted rotational velocity is too high, while in the outer parts ($R \geq 3'$) it shows a Keplerian decrease not totally compatible with the data. Thus we think that for this galaxy we have sufficient evidence for the presence of dark matter not distributed as the luminous matter.

431

M. Mezzetti et al. (eds.), Large Scale Structure and Motions in the Universe, 431–432.
© *1989 by Kluwer Academic Publishers.*

432

Dark matter could be in the form of a diffuse "corona"; a preliminary study indicates that the inclusion of such a dark component would not affect significantly the luminosity profile associated with the best fit one-component model and would imply a flatter rotation curve consistent with the available HI data.

[1] Bertin, G., Saglia, R.P., Stiavelli, M. 1988, Ap.J. <u>330</u> (July 1).
[2] Raimond, E., Faber, S.M., Gallagher III, S.J., Knapp, G.R. 1981
      Ap.J. <u>246</u>, 708.
[3] Peletier, R.F., Davies, R.L., Illingworth, G., Davis, L., Cawson,
      M. 1988 in preparation.
[4] Schechter, P.L., Gunn, J.E. 1979 Ap.J. <u>229</u>, 472.

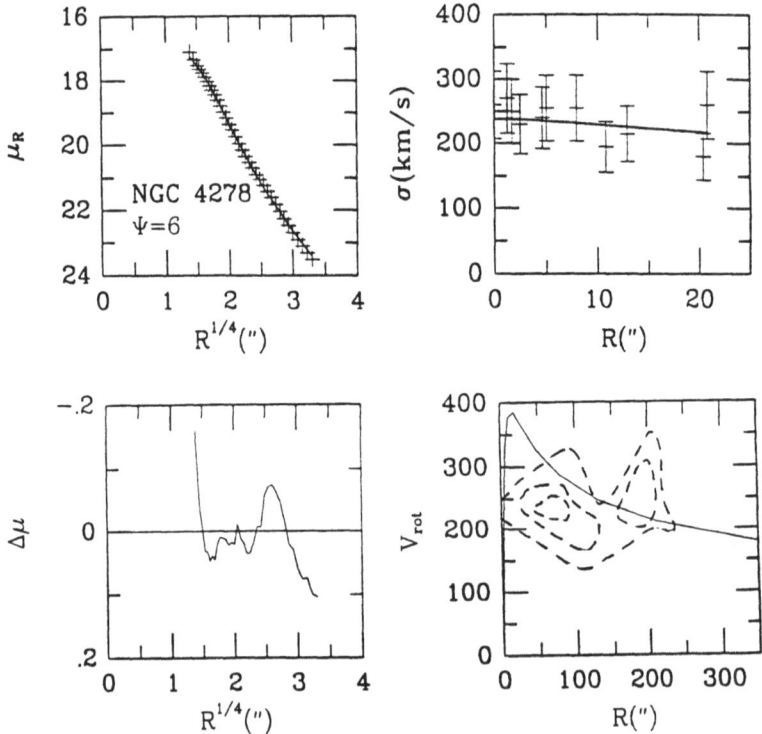

Top left: the R magnitudes for NGC 4278 (crosses) measured by [3] and the magnitudes μ_∞ of the $\Psi = 6$ model (solid line).
Bottom left: the residuals $\Delta\mu = \mu - \mu_\infty$ between the data and the model.
Top right: the velocity dispersions measured by [4] and the projected velocity dispersions of the $\Psi = 6$ model (solid line). Bottom right: the rotation curve of the $\Psi = 6$ model (solid line) and the distribution of HI velocities (dashed contours) along the kinematical SW semimajor axis (data from [2] ; adopted inclination i = 45°).

TOY NUMERICAL MODELS: VARIABILITY OF CLUSTER CLASSICAL PARAMETERS WITH DISTANCE

R. Scaramella[1], F. Mardirossian[2], M. Mezzetti[2]
G. Giuricin[2]
[1]International School for Advanced Studies
Trieste, Italy.
[2]Dept. of Astronomy, University of Trieste
Trieste, Italy.

1. INTRODUCTION

The aim of this work is to analyze the variation with distance of the values of the classical parameters of galaxy clusters (i.e.: total luminosity L, velocity dispersion V, virial radius R, virial mass M, and mass-to-light ratio M/L), to investigate on possible systematic effects.

To perform this, we artificially move away nearby clusters, taking into account the dilution of information by adopting two extreme models: i) the maximum of information available is used at any distance (Case A); ii) only information above completeness in magnitude is used (Case B).

2. SELECTION OF CLUSTERS

For our purpose, we need clusters which contain at list more than 50 galaxies with known magnitude and redshift. We have taken from the literature data for Virgo, Coma, Perseus and A1367.

The first step is to define members of clusters, rejecting probable interlopers. This has been done dropping all objects at a projected distance from the cluster center larger than 1 Mpc (H_0=100 km.s^{-1}.Mpc^{-1}), and those with a difference between their radial velocity and the mean cluster velocity exceeding three standard deviations.

3. MODELING THE INCREASE IN THE DISTANCE

For each cluster, the sample is complete in magnitude up to m_{compl}, and galaxies are observed up to m_{cut} (we discarded objects fainter than 10^8 L$_\odot$, if present).

433

M. Mezzetti et al. (eds.), Large Scale Structure and Motions in the Universe, 433–434.
© 1989 by Kluwer Academic Publishers.

3.1 CASE A

At each step we drop galaxies having $L < L_{cut}$ (corresponding to m_{cut}); we assume a Schechter-type luminosity function with $a = -1.25$ and $L^* = 3.10^{10} L_\odot$; we estimate $\emptyset^*$ from the number N_g of galaxies with $L > L_{compl}$ (corresponding to m_{compl}), excluding the two brightest members (possible cD galaxies).

We than estimate the classical parameters: total luminosity L, velocity dispersion V, luminosity-weighted virial radius R, virial mass M, and mass-to-light ratio M/L.

The cluster is moved away until we have less than 5 galaxies to be used in the estimate of $\emptyset^*$.

3.2 CASE B

The procedure is similar to Case A; the only difference is that we consider only galaxies brighter than L_{compl} to estimate the parameters L, V, R, M, and M/L.

This case corresponds to the maximum dilution of dynamical information on the system.

4. PRELIMINARY RESULTS

The "true" behavior of a real cluster would depend on the details of the observational technique used, but should be reasonably bounded by Case A and Case B.

An important quantity to be used as an indicator of the reliability of the derived cluster parameters is a measure of the real amount of information available on the luminosity. Indeed, the L determination is the most difficult. We have defined the parameter $\beta = L_{est}/L_{above}$, where L_{above} is the known luminosity of the galaxies above completeness, and L_{est} is the estimated luminosity below completeness; β gives an estimate of the extrapolated amount of information.

We feel that there is the possibility that classical parameters, especially M/L, derived from cases in which $\beta > 1$, could not give a faithful representation of the physical status of clusters (if this is possible within the classical virial approach and data selection procedure).

ON THE LARGE SCALE ANISOTROPY OF THE COSMIC MICROWAVE BACKGROUND

Roberto Scaramella
International School for Advanced Studies, Trieste, Italy

Nicola Vittorio
Dipartimento di Fisica, Università dell'Aquila, Italy

1. METHOD

In the framework of the gravitational instability theories, the primordial density fluctuations are commonly described as a 3-D random, isotropic gaussian field. Under this assumption, the CMB temperature distribution is itself a 2-D random gaussian field. It is convenient to expand the temperature fluctuations in spherical harmonics, whose coefficients a_l^m's are stochastic variables, with zero average value and variance given by: $a_l^2 \equiv \langle |a_l^m|^2 \rangle = \{2^{n-1} A \, \Gamma(3-n) \, \Gamma\left(\frac{2l+n-1}{2}\right)\}/\{r_0^{(n+3)} \left[\Gamma\left(\frac{4-n}{2}\right)\right]^2 \Gamma\left(\frac{2l+5-n}{2}\right)\}$ (1). This expression has been obtained (Peebles, 1982a) by assuming in a flat universe for the density fluctuations a scale free power spectrum , $|\delta_k^{(in)}|^2 = A \cdot k^n$, where A is the normalization constant, n is the spectral index, $r_0 \equiv 2cH_0^{-1}$, and H_0 is the present value of Hubble parameter. As usual, in Eq. (1) we did not consider either the monopole component, which is of course unobservable by difference measurements, or the dipole component, dominated by our peculiar motion relative to the CMB. The statistics of the CMB temperature field is completely described by the two point angular a.c.f. $C(\alpha) \equiv \langle \Delta T/T(\hat{n}_1) \cdot \Delta T/T(\hat{n}_2) \rangle$, where $\alpha = \cos^{-1}(\hat{n}_1 \cdot \hat{n}_2)$ and the average is taken over the whole sky and over the ensemble constituted by all the possible realizations of last scattering surfaces. We take into account the finite resolution of the antenna, by modeling the beam with a gaussian of dispersion σ. Then, the smoothed a.c.f. is given by (see, e.g., Wilson and Silk, 1981): $C(\alpha, \sigma) = \frac{1}{4\pi} \sum_{l=2}^{\infty} a_l^2 \, (2l + 1) \, P_l(\cos \alpha) \, \exp\left\{-\left[(l + \frac{1}{2})\sigma\right]^2\right\}$ (2). The effect of the beam, as expected, acts as a low band pass filter, which severely attenuates harmonics of order $l \gg 1/\sigma$. It is convenient to consider the smoothed a.c.f. normalized at zero angular separation: $R(\alpha, \sigma) \equiv [C(\alpha, \sigma)/C(0, \sigma)]$. We considered quite a wide range of values $(-3 \lesssim n \leq 3)$ to study the trend of the considered quantities. Eq. (2) allows one to derive from the observations the relevant information on the CMB rms temperature fluctuation, i.e. $C^{1/2}(0, \sigma)$, and, then, to relate directly results obtained in experiments which differ, e.g., in the subtraction technique and/or for the values of α and/or σ. In fact, for a single subtraction experiment, we have $C(0, \sigma) = \frac{1}{2} \left(\frac{\Delta T}{T}\big|_{obs}\right)^2 \cdot [1 - R(\alpha, \sigma)]^{-1}$ (3). or, for a double subtraction experiment, $C(0, \sigma) = \left(\frac{\Delta T}{T}\big|_{obs}\right)^2 \cdot [\frac{3}{2} - 2R(\alpha, \sigma) + \frac{1}{2} R(2\alpha, \sigma)]^{-1}$ (4).

M. Mezzetti et al. (eds.), Large Scale Structure and Motions in the Universe, 435–436.
© 1989 by Kluwer Academic Publishers.

2. CONFRONTATION WITH THE OBSERVATIONS

Melchiorri *et al.* (1981; hereafter M) reported a positive detection of CMB temperature fluctuation, commonly considered as an upper limit, because of the uncertainties in possible galactic contamination. The experimental configuration involved a single beam subtraction with $\alpha = 6°$ and $\sigma = 2°.2$. The deduced upper limit is, at the 90% confidence level, $\Delta T/T < 4 \cdot 10^{-5}$. More recently, Davies *et al.* (1987; hereafter D) reported also a positive detection of CMB temperature fluctuations. The experimental configuration involved a triple beam subtraction, with $\alpha = 8°.2$ and $\sigma = 3°.5$. After applying the Likelihood Ratio method, a value of $C(0,\sigma)^{1/2} = 0.10 \, mK/T_b$ was obtained. According to the a.c.f. assumed by D, we get the value $\Delta T/T(8°.2, 3°.5) \cong 3.0 \cdot 10^{-5}$ (cfr. Eq. (4)). The Princeton Group (Fixsen et al., 1981; hereafter F) placed an upper limit on the a.c.f. : $C(\alpha, 2°.9) < 1.4 \cdot 10^{-9}$ for $10° < \alpha < 180°$. More recently, the RELIC satellite borne experiment (as quoted by Lukash and Novikov, 1987; hereafter R) set $C(20°, 2°.4) < 5.5 \cdot 10^{-10}$ at the 95 % confidence level. In order to have a comparison among these different experiments, using Eq. (3), for a given n, we convert the M upper limit on $\Delta T/T$ to an upper limit on $C(0, 2°.2)$. Then, using Eq. (2), we evaluate, again for a given n, $C(0, 3°.5)$, the result that M would have had with a larger beam. Then we take the F upper limit on $C(10°, 2°.9)$ and, again through Eq. (2), we evaluate $C(0, 3°.5)$. The same procedure is then applied to the R limit on the a.c.f. Similarly, we obtain the same quantity from $\Delta T/T$, obtained above for the D experiment and Eq. (4). We have then for each n four values of $C(0, 3°.5)$: three upper limits (M, F, and R) and one 'detection' (D). Within our assumptions the comparison between these different experiments implies that the D 'detection' is consistent with the F and R upper limits only if $n > 1$. The M upper limit implies a rms temperature value smaller than the one implied by the D experiment. Also, it is more stringent than the F and R upper limits for $n > 1$ (for $n < 1$ the opposite holds). The above result is fairly independent on the density parameter Ω_0. In fact, also for $0.2 \lesssim \Omega_0 < 1$, all the values would have the same relative amplitude. In fact the angles involved are smaller than that subtended by the possible space intrinsic curvature (Peebles, 1981, 1982a).

3. CONCLUSIONS

We presented a general discussion of the dependence of the large scale fluctuations of the CMB on the primordial spectral index n. From the theoretical point of view, the value $n = 1$ is highly motivated from the standard inflationary scenario. However, the possibility of having values different from unity for the primordial spectral index should be kept in mind and may arise, for example, in power law inflationary scenarios. If the D experiment really provides a measure of the primordial CMB temperature fluctuations and the RELIC result sets a realistic upper limit on the quadrupole anisotropy we have indications that n should be greater than 1. This conclusion is hinted at from a comparison of the $C(0,\sigma)$'s derived from F and D experiments and is then fairly independent of the assumed value of the density parameter Ω_0. All the discussion is independent of the 'chemistry' of the universe (i.e., presence/type of dark matter). In fact, the intermediate and large scale CMB temperature fluctuations are determined by scales which experienced uninterrupted growth, preserving the primordial spectrum. The assumptions we made are that of a gaussian, adiabatic, scale-free initial density fluctuation spectrum and that we live in a typical region of the Universe which is well described by ensemble averages.

DISTRIBUTION OF GALAXIES IN A REGION OF THE COMA SUPERCLUSTER

E. Slezak, G. Mars, A., Bijaoui, A.
Nice Observatory - B.P. 139
F-06003 Nice Cedex - FRANCE
C. Balkowski, P. Fontanelli
Paris Observatory, DAEC, UA 173, Paris VII University
F92195 Meudon Principal Cedex - FRANCE

The existence of a filament of galaxies was shown by Fontanelli (1984) in the Coma Cluster region. We have processed a IIIaJ Palomar Schmidt plate on the eastern end of this filament in order to confirm that feature on the distribution of the faint galaxies. Two processing techniques were used : real time analysis with a PDS microdensitometer and an off-line analysis using image segmentation and radial profiles.

We have applied a real time reduction technique performing in real time in image segmentation. The galaxy selection was made using a classical Bayesian classification based on the diagram of the integrated density versus the area. All the non stellar objects where visually inspected using the PDS and a morphological type was attributed to the galaxies. A catalogue of 7582 galaxies giving their coordinates, their magnitude, their area and the morphological type is available (Slezak et al., 1988a).

An histogram of all the galaxy magnitudes shows that the catalogue is quite complete up to the 19th magnitude (figure 1, from Slezak et al., 1988b). The asymmetry in the distribution of the bright objects between the north-west and the south-east part of the plate found by Fontanelli (1984) is also present for the fainter objects in the catalogue. Figure 2 (a-d) show the distribution of the galaxies from 13.5^{th} to 17th magnitude. 56 galaxy condensations were identified using an image segmentation on the galaxy density map.

We have also developed an automated off-line procedure to achieve accurate detection and measurement of galaxies on the same plate. Ellipse fitting and radial profile determination are its main characteristics. Star/galaxy discrimination were performed using a set of bayesian classifications. On a field of $1°5 \times 1°5$ the off-line procedure has been compared to the real time one. The rate of detected galaxies increases by a factor 2, but that does not appreciably modify the characteristics of the density maps.

M. Mezzetti et al. (eds.), Large Scale Structure and Motions in the Universe, 437–438.
© *1989 by Kluwer Academic Publishers.*

438

REFERENCES

Fontanelli, P. : 1984, Astron. Astrophys. 138, 85.
Slezak, E., Mars, G., Bijaoui, A., Balkowski, C., Fontanelli, P. : 1988a,
    Astron. Astrophys Suppl. Ser., in press.
Slezak, E., Bijaoui, A., Mars, G. : 1988b, Astron. Astrophys., in press.

CAPTIONS FOR FIGURES

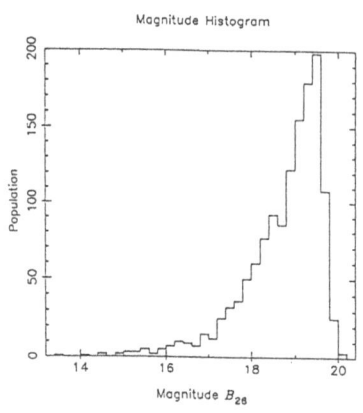

Figure 1 :
Histogram of the galaxy
magnitudes from Slezak
et al. (1988b).

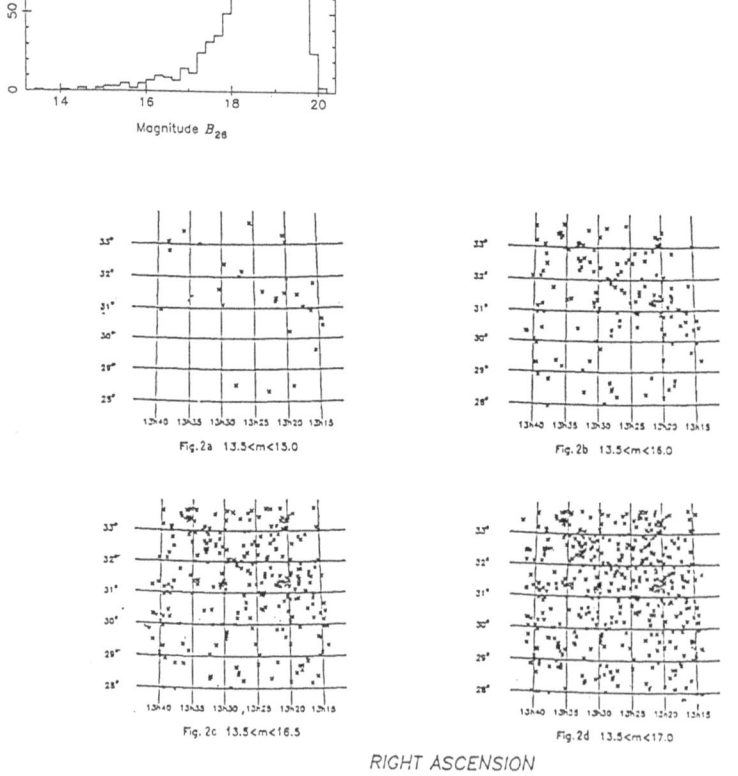

Figure 2 :   Plot of the brightest galaxies
            a) 13.5 < m < 15,   b) 13.5 < m < 16,
            c) 13.5 < m < 16.5, d) 13.5 < m < 17.

PECULIAR VELOCITIES OF NEARBY SPIRAL GALAXIES

L. Staveley-Smith[1] and R.D. Davies[2]

1. *Anglo-Australian Observatory, P.O. Box 296, Epping, NSW 2121, Australia.*
2. *NRAL, Jodrell Bank, Macclesfield, Cheshire SK11 9DL, U.K.*

We report measurements of peculiar velocities from a sample of over 300 intermediate-type spiral galaxies. The galaxies have been observed at λ21cm at Jodrell Bank and Parkes (Staveley-Smith & Davies 1987,1988; Davies, Staveley-Smith & Murray 1988) and include, as a subset, an updated version of the *Sbc* data utilized by Hart & Davies (1982). Relative distances are obtained using the blue Tully-Fisher relation ($\sigma \sim 0.5^m$), the diameter-linewidth relation ($\sigma \sim 0.6^m$) and the HI mass-linewidth relation ($\sigma \sim 0.7^m$). Peculiar velocities are derived by letting the weighted mean distance (in velocity units) equal the weighted mean recession velocity. This sets up a fiducial Hubble flow which is a good approximation to the near field Hubble flow because of the isotropic nature of the sample.

Galaxy peculiar velocities are examined in the foregrounds of two superclusters: Perseus-Pisces ($\ell = 135°$, b=$-25°$) and Hydra-Centaurus ($\sim \ell = 286°$, b=24°). Fig.1 shows the peculiar velocities in the CMB frame: (a) contains galaxies within 22.5° of the Supergalactic plane; (b) is an orthogonal view containing galaxies within 22.5° of a plane whose normal lies in the Supergalactic plane at longitude 245°. Galaxies in the direction of the Perseus-Pisces supercluster appear to be at rest with respect to the CMB with an unweighted mean peculiar velocity of $\bar{v} = -45 \pm 165$ km s^{-1} for the 33 galaxies within 45°. Galaxies within and in the foreground of the Hydra-Centaurus region have a component of streaming directed away from us. The 55 galaxies within 45° give $\bar{v} = 372 \pm 99$ km s^{-1}. Although significant, it is much smaller than the mean *flow* of 913 ± 181 km s^{-1} reported by Lynden-Bell *et al.* (1988) for elliptical galaxies within 60° of the 'Great Attractor' which lies $\sim 25°$ away (we obtain a mean peculiar velocity of 397 ± 91 km s^{-1} for spirals in the identical cone). Our results for Hydra-Centaurus using the HI mass-linewidth relation, which does not depend on Galactic absorption, are $\bar{v} = 474 \pm 109$ km s^{-1}; the more accurate blue Tully-Fisher relation gives $\bar{v} = 529 \pm 84$ km s^{-1} for the 30 galaxies with suitable photometry. These results suggest that Lynden-Bell *et al.* may have overestimated streaming velocities in this region, although differences in survey depth and distribution may partly account for this. Inspection of Fig.1 reveals that the streaming direction is misaligned with the Great Attractor for our sample (it is closer to the CMB apex at the position of the Hydra Cluster than to the

M. Mezzetti et al. (eds.), Large Scale Structure and Motions in the Universe, 439–440.
© 1989 by Kluwer Academic Publishers.

Great Attractor). Peculiar velocities in the Hydra-Centaurus region are however in qualitative agreement with the *IRAS* predictions (Strauss & Davis 1988; Yahil 1988). Results in Perseus-Pisces appear to differ in that more flow towards Perseus-Pisces is predicted using the *IRAS* counts than is actually observed.

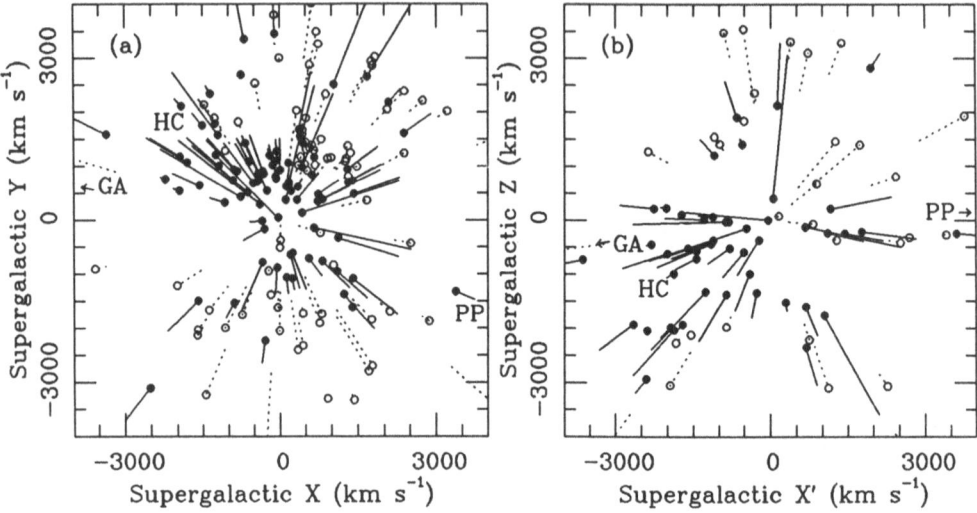

Figure 1: Peculiar velocities (from the diameter-linewidth relation) of spirals within 22.5° of two orthogonal planes intersecting at the Hydra-Centaurus (HC) and Perseus-Pisces (PP) superclusters. The normals to the planes are (a) $\ell = 47°$, b=6° and (b) $\ell = 331°$, b=−64° (Galactic coordinates). Points are plotted at their predicted distances. Filled points and solid lines represent outward motion; empty points and dotted lines represent inward motion. (a) should be compared with fig.12c of Yahil (1988) and fig.8 of Faber & Burstein (1988). X' is in the Supergalactic XY plane at $\ell = 134°$, b=−25°.

References

Davies, R.D., Staveley-Smith, L. & Murray, J.D., 1988. *Submitted.*

Faber, S.M. & Burstein, D., 1988. In *Large Scale Motions in the Universe,* Vatican Study Week.

Hart, L. & Davies, R.D., 1982. *Nature,* **297,** 191.

Lynden-Bell, D., Faber, S.M., Burstein, D., Davies, R.L., Dressler, A., Terlevich, R.J. & Wegner, G., 1988. *Astrophys. J.,* **326,** 19.

Staveley-Smith, L. & Davies, R.D., 1987. *Mon. Not. R. astr. Soc.,* **224,** 953.

Staveley-Smith, L. & Davies, R.D., 1988. *Mon. Not. R. astr. Soc.,* **231,** 833.

Strauss, M.A. & Davis, M., 1988. In *Large Scale Structure in the Universe,* IAU Symposium 130.

Yahil, A., 1988. In *Large Scale Motions in the Universe,* Vatican Study Week.

BRIGHT QUASARS AND THE LOCAL SUPERCLUSTER

Jack W. Sulentic
Department of Physics and Astronomy
University of Alabama, Tuscaloosa USA

Introduction

The hypothesis that some or all quasars originate in ejection events from galaxies (Arp 1987) implies that many of the brightest quasars will be associated with the brightest nearby galaxies. This is particularly true if the dispersion in intrinsic luminosities among, assumed local, quasars is small. This is a difficult prediction to test for at least two reasons: 1) quasar samples are incomplete even at very bright apparent magnitude levels (e.g. m$\leq$17.5) and 2) the correlation scale is unknown but can be expected to be quite large (many degrees). The latter conclusion is forced upon us by the observation that bright quasars are found in regions of the sky where there are few bright galaxies. Recent tests (Sulentic 1988) involving quasars (m$\leq$17.5) and low redshift galaxies (m$\leq$11.0) imply a correlation between these two classes.

A New Test

We report here on a new test involving a subset of the 42 (including LMC and SMC) brightest galaxies on the sky (m$\leq$10.0). This sample is interpreted here as representative of the distribution of galaxies in the nearest part of the Local Supercluster. We compare the distribution of these galaxies with that of the PBQS (Schmidt and Green 1983) confirmed quasars which have been surveyed over 10714 deg^2 of the sky. This sample includes many of the brightest known quasars and is reasonably complete to an average limiting magnitude of B= 16.16. Twenty of the brightest galaxies fall within the PBQS survey area and 13 of these cluster conspicuously within a few degrees of the supergalactic plane near the Virgo cluster ($\alpha \approx 12^h30^m$; $\delta \approx +23°$). The PBQS quasars, particularly those of high redshift, show a similar clustering.

If we calculate the surface density of PBQS quasars in an arbitrary area of $\approx$1250 deg^2 centered on the Virgo Cluster (α=12^{h}30^m; δ= +12°) we obtain σ=0.02 deg^{-2}. The value for the remainder of the PBQS survey area is σ=0.01 deg^{-2}. We have included in this test sample all 115 objects selected in PBQS. Schmidt and Green (1983) define only 93 of these objects as quasars based upon a (redshift dependent) absolute magnitude criterion (M$_B$ $\leq$-23). The overdensity of bright galaxies in this same area is a factor of about 3-4.

M. Mezzet. i et al. (eds.), Large Scale Structure and Motions in the Universe, 441–442.

The High Redshift Excess

Twenty of the PBQS quasars have redshifts higher than $z=1.0$ with only 2 higher than 2.0. These objects show a much higher overdensity in the region of Virgo where $\sigma=0.005$ deg^{-2} which is four times higher than the remainder of the PBQS survey. This overdensity rises as high as a factor of six if we define an arbitrary region based on the positions of the 13 bright galaxies centered near Virgo. This overdensity appears to be an underestimate because the limiting magnitudes for search fields $W(8^h\text{-}10^h30^m)$ and $E(16^h\text{-}22^h)$ of this region are systematically fainter (see Schmidt and Green 1983). Examination of the extinction estimates (A_B) for the PBQS fields derived from Burstein and Heiles (1984) allow us to ascribe very little of the overdensity to the effects of galactic obscuration. Thus the overdensity of brightest galaxies **and high redshift quasars** in the direction of Virgo are comparable.

Implications

We interpret these results as consistent with a picture where quasars (especially high redshift quasars) are local and, in this sample, arise from three sources: 1) the Local Group, 2) the nearest bright galaxies beyond the Local Group which concentrate in the direction of Virgo and 3) possibly the intrinsically brightest quasars associated with the Virgo cluster itself. Other investigators have noted an apparent association between fainter quasars and the brightest galaxies in the Virgo Cluster (He et al. 1984; Arp 1986).

The observed overdensity may be telling us that there are two kinds of quasar: 1) those with $z \leq 0.5$ with redshifts indicative of distance, and other observations consistent with those distances, and 2) those with $z \geq 1.0$ which are local. We do not advocate such an arbitrary division of the quasar class especially since Markarian 205 would fall in the former category. We note that a gravitational lens amplification explanation for this overdensity would not produce a redshift dependent effect.

References

Arp, H. (1986), *J. Astrophys. Astr.*, **7**,77.

Arp, H. (1987), *Quasars, Redshifts and Controversies*, (Interstellar Media: Berkeley).

Burstein, D. and Heiles, C. (1984), *Ap.J. Suppl. Ser.*, **54**, 33.

He, X-T. et al. (1984), *MNRAS*, **211**, 443.

Schmidt, M. and Green, R. (1983), *Ap.J.*, **269**, 352.

Sulentic, J. (1988), *Tests of the Discordant Redshift Hypothesis*, in **New Ideas in Astronomy**, eds. F. Bertola, J. Sulentic and B. Madore, (Cambridge U. Press: England).

THE DYNAMICS AND STRUCTURE OF THE MULTIPLE CLUSTER SC 0627-54

Peter F. Teague
Department of Physics and Astronomy
Northwestern University
Evanston IL 60208   U.S.A.

ABSTRACT.   The mini-supercluster or multiple cluster system, SC 0627-54, was recently studied as part of an extensive program on 10 rich clusters of galaxies undertaken at the Mount Stromlo and Siding Spring Observatories (Australia) with David Carter. The total data-base includes multi-object spectroscopy of almost 1200 galaxies (300 in the field of SC 0627-54), Einstein Observatory X-ray images, and COSMOS photometry. The thorough analysis of the velocity and X-ray data has enabled us to investigate the various dynamical and structural properties of each of the three clusters in SC 0627-54. From this we gain insight to the processes occurring within each cluster as well as the dynamical status and history of the whole system.

1.   DATA AND DISCUSSION

SC 0627-54 is a small supercluster containing three density enhancements spread along a North-South plane. Two rich clusters, 0627-54 S and 0627-54 N, are separated by about one degree on the sky ($\sim$ 3 Mpc, using H_0 = 100 km/s/Mpc) with a compact group of galaxies located mid-way between the two, with significant extended X-ray emission centered on each of the three clusters. The whole system can be considered as a multiple cluster (or perhaps an interacting binary cluster) similar to those reported in Karachentsev and Shcherbanovskii [1] and Ulmer et al. [2], [3].

The large number of good quality spectra required for a rigorous dynamical study of the clusters was obtained using the multi-object fiber-optic coupling device on the Anglo-Australian Telescope (AAT). Spectra, sufficient to yield a velocity with an accuracy of about 50 km/s, for up to 50 galaxies down to a magnitude of B=20 can be obtained in about three hours on the AAT. The velocity results and analysis for the whole cluster sample are given in Teague, Carter and Gray [4].

The individual velocity histograms for the South and North clusters (with 159 and 52 members respectively) are each well-fitted by a Gaussian distribution with parameters $\overline{cz}$ = 15154 ± 89 km/s, σ = 1144 ± 70 km/s, and $\overline{cz}$ = 16334 ± 167 km/s, σ = 1215 ± 130 km/s respectively. The

443

M. Mezzetti et al. (eds.), Large Scale Structure and Motions in the Universe, 443–444.
© *1989 by Kluwer Academic Publishers.*

complete histogram for the whole system is approximately Gaussian, and the distribution of velocities for the 30 or so central group galaxies is flat with a very large dispersion (over 2000 km/s).

Reduction of the Einstein IPC X-ray images at the Harvard-Smithsonian Center for Astrophysics produced X-ray surface brightness contour maps, X-ray radial brightness profiles, and total X-ray luminosities for the clusters. An isothermal-isothermal gas model was fitted to the X-ray radial profiles; values were obtained for parameters such as the core radius, gas temperature, central electron density, central gas pressure, gas mass, and the total cluster mass.

A color-dot velocity map plotted on the sky shows that there is no significant small-scale spatial substructure apparent in the velocity distribution (VD) of either the S or N cluster, and in particular there is no evidence for rotation. The VD for the central group is quite flat and very broad, which is rather different from the VDs of the S and N clusters. This could be caused by a simple line-of-sight overlap of the VDs of the S and N clusters; however, the group is tight and centrally concentrated. If the S and N clusters are bound to each other then one might expect a compact group in between, situated at a Lagrangian point, or perhaps the group is direct evidence that the two systems have interacted in the past (through a tidal or collisional encounter). The VDs of the S and N clusters do not show any significant signs of disruption. The velocity dispersion profile (with projected radius from the cluster center) for the two clusters is rather constant, implying isotropic motions in the cluster and ruling out strongly anisotropic (radial or tangential) dynamical models (see Kent and Gunn [5], Kent and Sargent [6]).

The galaxy surface distribution and the X-ray morphology match each other quite well. The elongation and orientation of the X-ray and optical distributions of the S and N clusters could be partly caused by a previous interaction. The small amount of X-ray emission from the central group could be a remnant from a gravitational encounter. The β parameter (from the X-ray gas models) implies that there is much more energy per unit mass in the gas than in the galaxies.

So, to summarize, it appears that the two relaxed clusters have interacted in the past, and further work will show whether or not the whole system is in the process of collapsing and forming one large cluster.

REFERENCES

1. Karachentsev, I.D., and Shcherbanovskii, A.L., 1978, Soviet Astron., 22, 257.
2. Ulmer, M.P., and Cruddace, R.G., 1981, Ap. J. (Letters), 246, L99.
3. Ulmer, M.P., Cruddace, R.G., and Kowalski, M.P., 1985, Ap. J., 290, 551.
4. Teague, P.F., Carter, D., and Gray, P.M., 1988, Ap. J. Suppl., accepted.
5. Kent, S.M., and Gunn, J.E., 1982, Astron. J., 87, 945.
6. Kent, S.M., and Sargent, W.L.W., 1983, Astron. J., 88, 697.

DISCRETE SOURCE CONTRIBUTIONS TO THE ANISOTROPIES OF THE MICROWAVE BACKGROUND RADIATION

LUIGI TOFFOLATTI, ALBERTO FRANCESCHINI, LUIGI DANESE
Department of Astronomy, University of Padova
vicolo dell'Osservatorio 5, 35122 Padova, Italy
GIANFRANCO DE ZOTTI
Padova Astronomical Observatory
vicolo dell'Osservatorio 5, 35122 Padova, Italy

We have computed the temperature fluctuations of the microwave background radiation (MBR) due to unresolved, randomly distributed sources, for a wide interval of angular scales (from a few arcseconds to several tens of degrees) and for wavelengths ranging from several cm to below 1 mm. For wavelengths $\geq 6\ cm$ and angular scales $\geq 30''$ the present estimates follow almost directly from the observed counts. The large amount of available data on the space distribution of radio sources has allowed a reliable extrapolation of the 5 GHz counts down to $\approx 1\ \mu Jy$, and, hence, of $\Delta T/T$ down to scales of a few arcseconds. In addition, extensive studies of the high frequency spectra of compact radio sources, during the last few years, have substantially narrowed down the uncertainties in the extrapolations of source counts to mm wavelengths, so that we estimate the predicted values of $\Delta T/T$ to be reliable to within $\approx 30\%$ for $\lambda \geq 1\ cm$ and a factor of ~ 2 down to $\lambda \approx$ a few mm.

The predicted temperature fluctuations at 5 GHz reported in Fig. 1 show a broad peak centered at the angular scale $\theta_0 \simeq 10'$ and sink down rapidly on smaller scales ($\theta_0 \leq 10 \div 15\ arcsec$). From the same figure we see, in particular, that the most sensitive searches for anisotropies on angular scales in the range $1' \leq \theta_0 \leq 100'$ have already come close to the source confusion limit.

As shown by Fig. 2, only in a quite narrow spectral window near the peak of the microwave background spectrum, source confusion is not a problem down to $\Delta T/T \leq 10^{-5}$. For frequencies $\nu \geq 100\ GHz$, the galaxy populations whose far-IR spectra are dominated by thermal dust emission give rise to $\Delta T/T$ levels which make it hard to detect truly primordial fluctuations of the MBR.

We have also investigated the effect of clustering of radiosources. If their positions are strongly correlated in the space, as suggested by a recent study of Peacock *et al.* (1988, Proceedings of IAU Symposium N° **130**, *Evolution of Large Scale Structure in the Universe*), albeit only for sources within a relatively narrow range of radio luminosity, non-Poisson contributions to fluctuations would dominate on some angular scales. On the other hand, the most sensitive data on $\Delta T/T$ at $\lambda = 6\ cm$ already provide significant constraints on the amplitude of the correlation function, $\xi(r)$, of radio sources. If we adopt for radio galaxies with radio power $P_5 < 2.6\ 10^{24}(W/Hz/sr)$ the usual representation $\xi(r) = (r_o/r)^\gamma$ with $\gamma \simeq 1.8$, the upper limit of Lasenby and Davies (1983, M.N.R.A.S. **203**, 137) for $\theta_0 = 10'$ implies $r_o \leq 22(\frac{50}{H_0})\ Mpc$ at the 95% confidence level.

M. Mezzetti et al. (eds.), Large Scale Structure and Motions in the Universe, 445–446.
© *1989 by Kluwer Academic Publishers.*

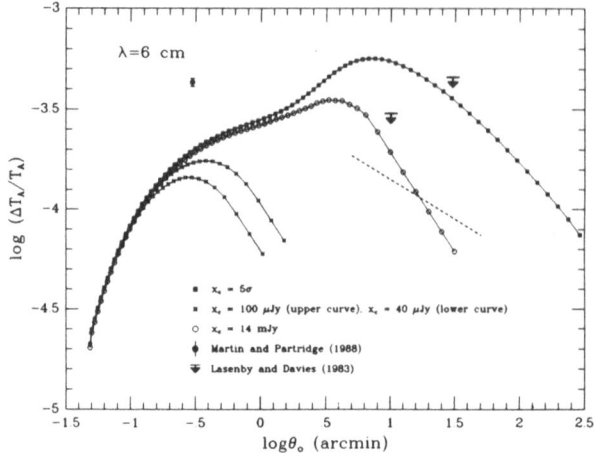

Fig. 1 Predicted antenna temperature fluctuations at $\lambda = 6$ cm for various source detection limits. Also plotted are the sky fluctuations reported by Martin and Partridge (1988, Ap.J. **324**, 794) at $\theta_0 = 18''$ and the upper limits by Lasenby and Davies (1983) at $\theta_0 = 10'$ and $30'$. The lower curves curves refer to the various resolution limits of the two surveys. The dashed line is the estimated contribution due to spatial autocorrelation of the steep-spectrum radio galaxies (computed with $x_c = 14$ mJy): a $\xi(r) = (r_0/r)^{1.8}$ with $r_0 = 30(\frac{50}{H_0})Mpc$ in the luminosity range $2.6 \ 10^{23} < P_5(W/Hz/sr) < 2.6 \ 10^{24}$ has been assumed following Peacock *et al.*, (1988).

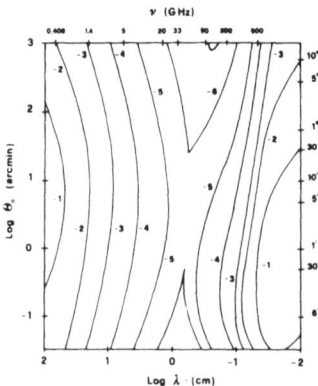

Fig. 2 Curves of constant intensity fluctuations of the MBR in the plane (λ, θ_0) for a random distribution of sources. The plotted values of $\log \Delta T_A/T_A$ refer to a source detection limit $x_c = 5 \cdot \sigma$. The estimate of the temperature fluctuation levels due to far-IR selected sources is based on the model by Franceschini, Danese, De Zotti and Xu (1988b, M.N.R.A.S., in press), fitting the IRAS data at 60 μm. This model has been extrapolated to the longer wavelengths assuming a modified Planck spectrum of the form $\nu^2 B_\nu(T_d)$ (see Chini, Krugel and Kreysa, 1986, A.A. **167**, 315).

VORONOI TESSELLATIONS AND THE
LARGE SCALE STRUCTURE OF THE UNIVERSE

RIEN VAN DE WEYGAERT and VINCENT ICKE
Sterrewacht Leiden, Postbus 9513
2300 RA Leiden, The Netherlands

We present a geometrical description of the large scale structure of the Universe as a *Voronoi foam*, using a kinematical model for its evolution, based on the *Bubble Theorem* (Icke 1984). We give examples of 2–D Voronoi tessellations (see Icke & Van de Weygaert 1987,) and the first results on 3–D Voronoi tessellations are shown, namely examples of cells (see Fig. 1) and some slices through a 3–D Voronoi foam. These have been calculated with a new program which is among the fastest available.

The Voronoi foams closely resemble the appearance of the large scale structure. They account for important features, such as the huge *voids* in the galaxy distribution, the *filaments* and *clusters* (Einasto *et al.* 1980; Kirshner *et al.* 1981, 1987; De Lapparent *et al.* 1986.) The N–body simulations (e.g. Centrella & Melott 1983) also produce a distribution of matter on scales of 20–50 Mpc with a void-and-filament or sponge-like topology.

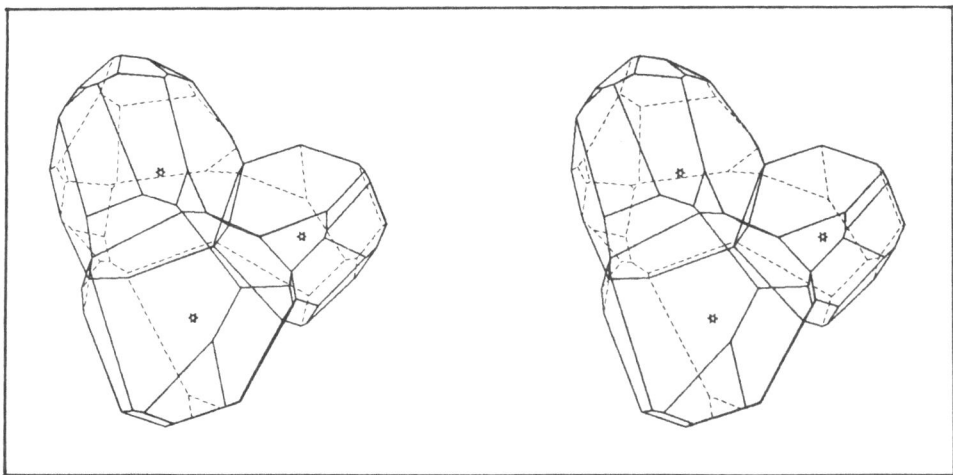

Figure 1. *Stereo plot of three three-dimensional Voronoi cells sharing a common line. In a stereo viewer, the dashed lines will appear at the rear of the picture. When attempting stereo fusion with crossed eyes, the pictures must be reversed to obtain the correct depth.*

M. Mezzetti et al. (eds.), Large Scale Structure and Motions in the Universe, 447–448.

Centrella, J., & Melott, A., 1983 *Nature* **305**, 196.

De Lapparent, V., Geller, M., & Huchra, J., 1986 *Astrophys. J. Letters* **302**, L1.

Einasto, J., Joêveer, M., & Saar, E., 1980 *Monthly Notices Roy. astron. Soc.* **193**, 353.

Icke, V., 1984 *Monthly Notices Roy. astron. Soc.* **206**, 1P.

Icke, V., & Van de Weygaert, R., 1987 *Astron. Astrophys.* **184**, 16.

Kirshner, R.P., Oemler, A., Schechter, P.L., & Shectman, S.A., 1981 *Astrophys. J. Letters* **248**, L57.

Kirshner, R.P., Oemler, A., Schechter, P.L., & Shectman, S.A., 1987 *Astrophys. J.* **314**, 493.

POSITION ANGLE DISTRIBUTIONS OF THE BRIGHTEST GALAXIES IN ABELL CLUSTERS

E. van KAMPEN
Sterrewacht Leiden
P.O.Box 9513
2300 RA Leiden
the Netherlands

An analysis of the distribution of galaxy position angles in clusters should help to discriminate between the various formation scenarios of clusters of galaxies, since these predict different evolution of the angular momentum of the member galaxies. We have applied this test to a complete sample of clusters drawn from Abell's catalogue. The sample is defined as follows:

$$9^h \leq \alpha \leq 18^h, R \geq 1, \delta \geq -25^o, b \geq 30^o, z \leq 0.1$$

We have produced digitized images of each of the 108 clusters using the Leiden Astroscan. These images are reduced by computer to produce lists of galaxies to $M_r \sim 19.5$ for each cluster. We used normalized intensity weighted moments to calculate the position angle and the ellipticity of the 20 brightest cluster galaxies, and their errors.

We searched for anisotropies in the position angle distribution using two basic tests. Using a χ^2-test we found no deviations from a uniform distribution. Using the $\cos 2\theta$ and $\sin 2\theta$ Fourier-components, we found no preferred orientations. The observed distributions follow the theoretical distributions for the random (isotropic) case very closely.

If ellipticals should tend to align perpendicular to spirals, preferred orientations would be obscured. We tested this for the 10 clusters for which morphological types were available, but found no dependence. We are still trying to obtain morphological types by computer for all 108 clusters, to be able to test this on the whole sample.

Since there is no preferred orientation in the position angle distribution of the 15-20 brightest galaxies, they are not aligned with anything. But we again confirm the strong alignment of the first-ranked galaxy with its cluster. It appears to be stronger for clusters with their first-ranked galaxy positioned near the centre. We also found an alignment of the first-ranked galaxy with the X-ray image of its cluster.

449

M. Mezzetti et al. (eds.), Large Scale Structure and Motions in the Universe, 449.
© *1989 by Kluwer Academic Publishers.*

LARGE SCALE RADIO EMISSION IN THE COMA CLUSTER

Venturi T.[1], *Giovannini G.*[1,2], *Feretti L.*[1,2]

1 Dipartimento di Astronomia - Università di Bologna
2 Istituto di Radioastronomia CNR - Bologna

New radio observations at 327 MHz of the Coma cluster of galaxies, obtained with the Westerbork Synthesis Radio Telescope, have added information on the two regions of diffuse radio emission present in the Coma cluster: Coma-C and 1253+275, lying respectively in the center of the cluster and in a peripheral region at about 1 degree from the cluster center, near Coma-A.

Coma-C is the prototype of the halo sources in cluster of galaxies, while the nature of 1253+275 is uncertain (see Giovannini G. et al.: Astron. Astrophys. **150**, 302 and references therein).

Coma-C shows a very broad and approximately regular morphology; its linear dimension corresponds to 650 kpc with H_o=100 km/sec Mpc.

A convolved map obtained after subtraction of the point sources present in the field is shown in fig. 1.

The radio morphology of 1253+275 is elongated in the NW-SE direction, its major and minor axis being respectively 27' and 15' (560 x 315 kpc). These new observations show that also this source has a large scale low brightness emission, suggesting that its nature could be similar to that of Coma-C rather than the relic emission of a single galaxy.

We have derived the equipartition parameters of Coma-C and 1253+275 assuming a filling factor of unity and k=1. An average brightness of 0.0018 and 0.0022 $mJy/arcsec^2$, and a depth of 500 kpc and 300 kpc were assumed for these two sources respectively. The results are shown in table 1. The values we obtained are similar, implying that the two sources are characterized by similar physical conditions.

Since we expect that these extended, relaxed sources are in equilibrium with the external medium, we derive that the external pressure necessary for the confinement must be similar in the cluster center and in the region where 1253+275 lies. This is in contrast with the slope of the thermal gas density derived from the X-ray observations, therefore either the gas density decrease is smoother than expected, or the gas density in the region near Coma A is denser and/or hotter than that predicted by the current models of gas distribution.

451

M. Mezzetti et al. (eds.), Large Scale Structure and Motions in the Universe, 451–452.
© *1989 by Kluwer Academic Publishers.*

Table 1

Equipartition parameters

Source	U_{min} erg/cm^3	P_{min} $dyne/cm^2$	B_{eq} $gauss$
Coma–C	0.29×10^{-13}	0.18×10^{-13}	0.56×10^{-6}
1253+275	0.37×10^{-13}	0.23×10^{-13}	0.63×10^{-6}

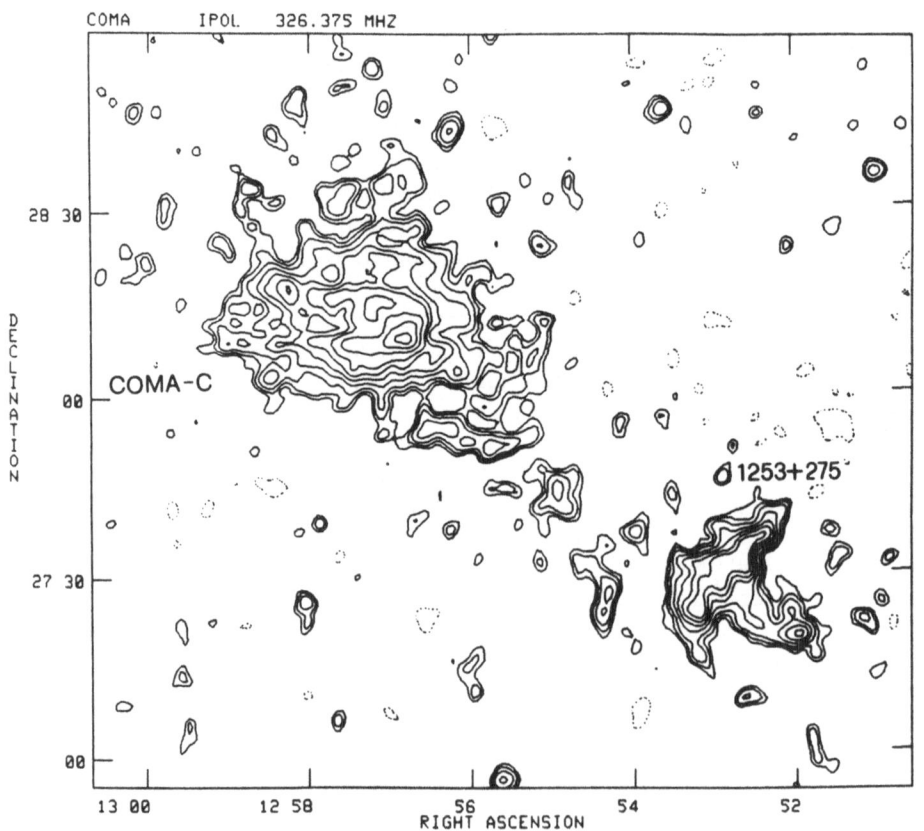

Isocontour map of the extended radio emission from the Coma Cluster at 327 MHz. The restoring HPBW is $140'' \times 175''$ (RA x DEC). Pointlike sources have been subtracted; the map is not corrected for the primary beam attenuation. Contour levels are: -8, 5, 7, 10, 15, 20, 25, 30, 40, 50, 60 mJy/beam.

A GREAT CONCENTRATION OF GALAXY CLUSTERS IN THE SOUTHERN SKY

G. Vettolani [†], G. Baiesi–Pillastrini [b], R. Scaramella [*], G. Zamorani [°], G. Chincarini [◊].

[†] Istituto di Radioastronomia CNR, Bologna, Italy.
[b] Dipartimento di Astronomia, Università di Bologna, Bologna, Italy.
[*] International School for Advanced Studies (S.I.S.S.A.), Trieste, Italy.
[°] Osservatorio Astronomico, Trieste, Italy.
[◊] Osservatorio Astronomico di Brera, Milano, Italy.

ABSTRACT

We present some results from a preliminary analysis of the ACO southern cluster catalogue. We derived a multiparameter relation for distance estimates which enables us to identify, through percolation technique, gross features of the cluster distribution which deserve priority in further studies. Most remarkable among these features is a nearby ($z \simeq 0.04$), unusually high concentration of clusters in the Hydra Centaurus zone. Notably this concentration lies only $\simeq$ 10° away from the reported direction of large–scale motions and $\simeq$ 35° from that of the Cosmic Microwave Background dipole.

The ACO catalogue [1] lists clusters of galaxies found by Abell, Corwin and Olowin according to the Abell selection criteria. It covers all the southern sky with $\delta \leq -17°30'$ and lists 1611 entries, 274 of which are in the overlapping zone with the Abell catalogue and are also listed by the latter. Besides angular coordinates and some morphological classification, ACO lists for each cluster the magnitudes for the first, the third, and the tenth brightest galaxy, together with the number of galaxies (N_g) between $m(3)$ and $m(3) + 2$, corrected for the background counts.

Redshifts for 71 clusters are available and we used them to calibrate a linear, multiparameter relation for $\log(z)$ as a function of the listed number N_g and magnitudes of the 1st, 3rd, and 10th brightest galaxy. We discarded two data which gave discrepancies greater than three sigmas from the average relation and probably had spurious redshifts due to foreground contamination. The relation we use, due to the 69 surviving calibrating clusters, has a dispersion in $\log(z)$ of $\sigma = 0.09$. With the aid of the above relation we find that for rich clusters ($N_g \geq 50$) ACO is fairly complete for $cz < 40 \cdot 10^3 \, km \, s^{-1}$. We also find a significant incompleteness with galactic latitude (b^{II}) which is well fitted by a $csc(|b^{II}|)$ law, but with a steeper decrease than that observed for Abell catalogue [2].

We proceeded to a preliminary identification of structures through percolation techniques on both the "reference" catalogue (i.e. that with $\log(z) = \log(z_{estim})$), and 100 random catalogues in which the redshift for each cluster was drawn according to a gaussian distribution in $\log(z)$ with variance σ^2, centered on $\log(z_{estim})$ [$\Delta z/z \simeq 0.2$]. An analysis of the results obtained in this way allowed us to identify possible structures with various degrees of robustness (defined in terms of observed frequency of the structure in the random

453

M. Mezzetti et al. (eds.), Large Scale Structure and Motions in the Universe, 453–454.

catalogues), together with the frequency of the related cluster membership. Most remarkable among these features is a great concentration of relatively nearby clusters ($N_{cl} \approx 20$, of which ≈ 8 are rich and 5 have measured redshift; percolation radius $R_p = 2500 \, km \, s^{-1}$), dubbed by us "$\alpha$-region", whose baricenter lies around $cz \approx 12 \cdot 10^3 \, km \, s^{-1}$, $l^{II} \approx 310°$, and $b^{II} \approx 30°$.

Because of its richness and its interesting closeness to the direction of MW dipole vector [3] and to the reported direction of bulk flows [4] and tidal distortions of peculiar velocity field [5] –see Figure–, the α-region deserves priority in further study, because of its possible cosmological relevance. We also note that because of its low galactic latitude the cluster richness of the α-region is probably underestimated, and the region itself could also extend in the avoidance zone. Were this structure to be greatly responsible for the peculiar motion of the Local Group, one would need to increase the value of the mass of a recently proposed model [4] by an order of magnitude because of the tripling of the distance from us. On the other hand, according to the spherical infall model (if applicable), the required overdensity would be smaller: $\Omega^{0.6} \, (\delta\rho/\rho) \cong 14\%$ would be sufficient.

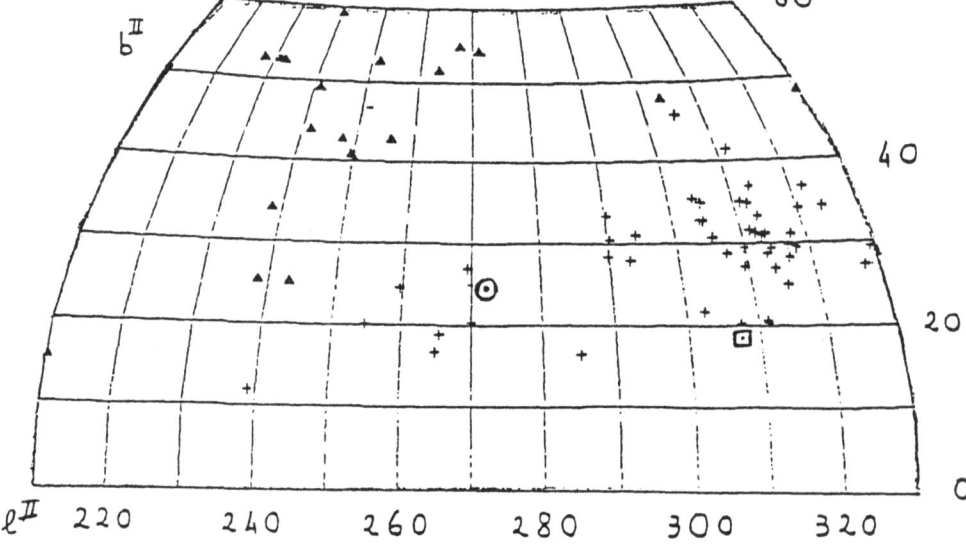

Abell (triangles)and ACO (crosses) clusters within $20 \cdot 10^3 \, km \, s^{-1}$; circle and square denote MW dipole [3] and Great Attractor [4] directions, respectively.

REFERENCES

[1] Abell, G.O., Corwin Jr., H.G., and Olowin, R.P, **ACO** catalogue, to be published.
[2] Bahcall, N.A., and Soneira, R.M., 1983, Ap.J. **270**, 20.
[3] Lubin, P., and Villela, T., 1986, in *Galaxy Distances and Deviations from Universal Expansion*, Madore, B.F., and Tully, R.B., eds. D.Reidel Pub. Co.
[4] D.Burstein, this conference, and references therein.
[5] Lilje, P., Yahil, A., and Jones, B.T.J., 1986, Ap.J. **307**, 91.

MONTE CARLO SIMULATIONS OF A CBR
ANISOTROPY EXPERIMENT

N. Vittorio
Dipartimento di Fisica , Universitá de l'Aquila, Italy
P. de Bernardis, S. Masi
Dipartimento di Fisica , Universitá "La Sapienza" , Roma
R. Scaramella
International School for Advanced Studies, Trieste, Italy

A recent experiment (Davies *et al.*, 1987, D hereafter) detected a temperature fluctuation of the microwave sky. The data were analysed with the likelihood ratio method by assuming a gaussian form for the CBR temperature correlation function. We analysed the published data, using the same method, but considering the temperature correlation functions predicted by gravitational instability scenarios, where the density fluctuation power spectrum is usually assumed to be a power law, with a spectral index n: $|\delta_k|^2 = A k^n$.

If the density perturbations, responsible for the observed large scale structure of the universe, are gaussian distributed and adiabatic, one can predict the functional form of the smoothed correlation function , $C(\alpha, \sigma) = C_0 R(\alpha, \sigma)$, where σ is the antenna beam size and α is the beamthrow. In this case, at variance with the gaussian correlation function, one expects anticorrelation on angular scales around $\pi/2$, for reasonable choices of the spectral index n (Scaramella and Vittorio, 1988) Also, the correlation can be larger than suggested by the gaussian form on the angular scales of several degrees, which are of interest here. The quantity C_0 is the variance of the cosmic background radiation (CBR) temperature fluctuations and is the value to be determined. We used the correlation function expected for primordial spectral indeces $-3 < n < 3$. This is a quite wide range of values, which nevertheless shows the dependence of the result on different assumptions. For each choice of the spectral index, we varied C_0 in order to find the value C_{0M} which maximizes the Likelihood (L). For assessing a non-zero rms temperature fluctuation, we computed the Likelihood Ratio (LR) defined as $L(C_{0M})/L(0)$. This quantity is a measure of the posterior confidence of having $C_0 \neq 0$ (the greater the ratio the better the odds).

The Davies et al. data are consistent ($LR \simeq 10$) with either white noise ($n = 0$) or scale invariant ($n = 1$) primordial density fluctuations and these are slightly p In order to test the significance of the maximum likelihood result, we performed a detailed comparison of this observation with the theoretical predictions, by simulating the microwave sky and by taking into account, for the first time, both the actual sky coverage and the instrumental noise of the specific experiment. We random generated 200 strips of the sky which are realizations of the theoretical microwave sky, expected in a flat universe, where density fluctuations are gaussian distributed, adiabatic, and scale–invariant. Then we generated a data set of 70 points per strip where each data point represents the result of a double subtraction, with the D values: $\alpha = 8°.3$ and $\sigma = 3°.5$. For simulating the receiver noise, we

455

added to the "true" CBR temperature fluctuations at each point an uncorrelated gaussian "noise" of amplitude (rms) σ_{err}. We can use our Monte Carlo simulations in order to state a firm upper limit on $C_0^{1/2}$. In fact, we can test the hypothesis $H_0 : C_0 = 0$, against the hypothesis $H_1 : C_0 = C_0^*$ using as a test statistic the Likelihood Ratio LR , whose distribution has been found via the Monte Carlo simulations. From these, we found that H_1 is preferred (i.e., the LR is greater than the measured value $LR \simeq 10$) in 95% of the cases if the SNR is 2 , i.e. if $C_o^{*1/2} = 0.44\,mK$. On the contrary, if $C_0 = 0$, we have that H_1 is preferred only in 0.5% of the cases. This means that $0.44\,mK$ is an upper limit on $C_o^{1/2}$ at a confidence level of 95%, with a power against the H_0 hypothesis of 99.5% (a very significant upper limit !).

In the D experiment SNR=0.75 ($C_o^{1/2} = 0.16\,mK, \sigma_{err} = 0.22\,mK$) and $LR \simeq 10$. The corresponding distribution we have found from the Monte Carlo simulations is marginally consistent with this result. In fact, 95% of the cases with $LR \gtrsim 10$ have $C_o^{1/2} \gtrsim C_s^{1/2} = 0.17\,mK$) . Increasing the SNR to unity ($C_0^{1/2} = 0.22\,mK$; $\sigma_{err} = 0.22\,mK$) slightly changes $C_s^{1/2}$ and the theoretical predictions are still marginally consistent with the experimental result.

On the contrary, one can try to evaluate the required sensitivity of the experiment in order to have a positive detection with an higher confidence level. For example, if one lowers the experimental noise down to $\sigma_{err} = 0.16\,mK$ (which corresponds to doubling the D integration time) , the D result should be compared with a distribution which has still a $SNR = 1$, but $C_o = 0.16\,mK$. For such a distribution, in the 95% of the cases $C_o^{1/2} \gtrsim 0.13\,mK$ and $LR \gtrsim 10$.

We want to point out that the likelihood analysis provides only the best estimate of the model parameters, but not a confidence interval. Then, it is necessary to check the L results using numerical simulations of the experiment, which seems to be the only way to build up the probability distribution for the parameters.

REFERENCES

Davies R.D., Watson R., Daintree E.J., Hopkins J., Lasenby A.N. Beckman J., Sanchez-Almeida J., Rebolo R. *Nature*, **326**, 6112 (1987).

Scaramella R. and Vittorio N. , *Ap.J.Lett.*, in press (1988).

FORMATION OF CONDENSATIONS IN AN EXPANDING WORLD MODEL

M. I. Wanas
Department of Astronomy
Faculty of Science
Cairo University
Egypt

ABSTRACT. The problem of forming inhomogeneities in an expanding world model is investigated. The world model used has been obtained [1] as a unique solution of the unperturbed field equations of the generalized field theory [2]. This model is free from particle horizons, and has the parameters $(k=-1, \Omega_o =0.75, q_o =0)$. The geometric structure used to construct this model is an absolute parallelism space [3].

The solution of the perturbed equations shows how condensations are formed in the model. The behaviour of matter in the model depends on its distance from the centre of perturbation. It is shown that at a certain distance $S = S_e$ from the centre, matter is stationary. If $S > S_e$ matter recedes away from the centre, while if $S < S_e$ matter falls towards the centre of perturbation. This may throw some light on the way of forming galaxies and clusters of galaxies in the Universe.

REFERENCES
[1] M. I. Wanas (1987) To appear in the proceedings of the 4th Asian-Pacific Regional Meeting of the IAU, held in Beijing, China 4–9 Oct. 1987.
[2] F.I. Mikhail and M. I. Wanas (1977) Proc.Roy. Soc. Lond. A356, 471.
[3] H. P. Robertson (1932) Ann. Math. Princeton (2), 33, 496.
    M. I. Wanas (1986) Astrophys. Space Sc. 127, 21.

M. Mezzetti et al. (eds.), Large Scale Structure and Motions in the Universe, 457.

WHERE DID GALAXIES COME FROM?

M. Zabierowski

Institute of Physics, Wrocław,
Technical University, Wybrzeże
Stanisława Wyspiańskiego No 27,
50-370 Wrocław, Poland

ABSTRACT. In the framework of the concept of vacuum fluctuation the existence of huge matter configuration was explained by Tryon /1973/. Now the idea of quantum like coherence is recognized as a foundation of creational process with consequences at the astrophysical level - stars and galaxies may be under control of quantum mechanism.

The mass of the typical star s
$$m_s \approx m_{Sun} \approx 10^{33} \text{ g} ,$$
the number of stars in the typical galaxy g
$$n_s \approx 10^{11} , \quad \left(m_g \approx 10^{44} \text{ g}\right),$$
and the number of galaxies in the Universe U
$$n_g \approx 10^{11} \approx n_s , \qquad m_U \approx 10^{55} \text{ g}$$
have been explained in the framework of the anthropic principle /Carter, 1973; Wheeler, 1976; Zabierowski, 1988/ which relates astrophysical objects with Man.

One can hardly distinguish between galaxy aggregates and statistical fluctuations of field galaxies. The so-called groups, filaments, superclusters, cells, ... of galaxies are not easily seen in contradistinction to the galaxies and stars. Stars and galaxies have been recognized without specific statistical methods. They are remarkable for their compact structure, shape and other astrophysical parameters. They may serve as the most fundamental quanta of the macro-world. The differences among stars themselves /similarly as among galaxies themselves/ are completely negligible in comparision with the differences among clouds of cosmic dust /Grabińska,1985; 1986; 1988/; galaxy systems are also unlike - there is not the only one of its sort /Grabińska, 1986; Lachieze-Rey and Maurogordato 1986; Lachiéze-Rey 1988/. Stars form a unique class of astronomical objects which are physically extremely uniform, even more than elementary particles /particularly if one accepts the continuous field approach, Grabińska, 1979; Grabińska and Turko, 1976;1979/.

M. Mezzetti et al. (eds.), Large Scale Structure and Motions in the Universe, 459–460.
© 1989 by Kluwer Academic Publishers.

Galaxies and stars form the two classes of extremely homogeneous objects.
 Stars and galaxies - giant configurations of the
"bulk" mass /classical mass, the sharp distinction between
the micro- and macro-levels is usually postulated/ - may
be considered also as fermions. Then the typical galaxy g
state consists of n_s stars and n_s "quantum" stars yield

$$n_s + 1 \cong n_s \qquad \text{/here } 10^{11} + 1 \cong 10^{11}/$$

different states. In this approach, the microphysical laws
are treated as directly responsible for the features of
the astrophysical bodies /thus we have not dealing with
the classical understanding of the observer postulated by
the anthropic cosmological principle in its weaker sense/.
So, it is possible to make only n_s different states of
galaxies putting n_s stars together; and $n_g n_s$ stars are
required to fill up all the possible galaxy g states in U.
Preferring the simplest hypothesis compatible with the
observational evidence, i.e. assuming that the Universe
contains only those stars which are collected in galaxies
/we neglect the new results of Grabińska requiring a
modification of our heuristics/ , the mass of the Universe
may be estimated in the following way

$$m_U = n_g m_g = n_g m_s n_s = (n_s + 1)m_s n_s$$
$$\cong n_s^2 m_s$$
$$\approx \left[(10^{11})^2 + o(10^{11})^2 \right] 10^{33} g \cong 10^{55} g ,$$

where $\approx$ stands for the observational predicate /empirical
identity/, $\cong$ and $=$ mark the exact identities, we
cannot verify the exact relation marked by the equals sign
$\cong$, observational verification gives only $\approx$. Searches of
intergalactic barionic dark matter /Grabińska, 1985, 1988/
may serve as a measure of the principal uncertainties.

References

Carter, B.: 1974, in: Confrontation of Cosmological
 Theories with Observational Data, M.Longair /ed./,291.
Grabińska, T.: 1979, Zeit. Phys. C2, 191.
Grabińska, T.: 1985, Astrophys. Space Science 115, 369.
Grabińska, T.: 1986, Geodezja-AGH-Kraków 1000, z.87, 84.
Grabińska, T.: 1986, in: COSMOS-an Educational Challenge,
 J.J. Hunt /ed./, European Space Agency, Paris,303-308.
Grabińska, T.: 1988, in: From Stars to Quasars, S. Gru-
 dzińska /ed./, N. Copernicus University at Toruń, Toruń;
 in:Large-Scale Structure and Motions in the Universe, G.
 Giuricin et al. /eds./, Kluwer Acad. Publ.;to be publish.
Grabińska, T. and Turko, L.:1978, Zeit.Phys. C1, 377.
Lachiéze-Rey, M.:1988, to be published in Astr. Astrophys.
 'Fractals and Galaxy Distribution', IRF-SAp-CEN-CEDEX.
Wheeler, J.:1976,'Genesis and Observership',prep.Princeton

Percolation and Radiogalaxy distribution

L. Zaninetti
Istituto di Fisica Generale
Via Pietro Giuria 1, 10125
Torino , Italy

ABSTRACT : The spatial distribution of radio-galaxies as given by the percolation theory is examined and the behavior of the correlation coefficient is explored .

1. Introduction

Recently the hierachical distribution of galaxies has been explained through the percolation theory of Miller 1983 , Shulman and Seiden 1986 , Charlton and Schramm 1986 .

This theory explains the observed galaxy-galaxy power law correlation (Groth and Peebles 1977, Soneira and Peebles 1978) and describes the appearance of voids and clusters .

Here, on the basis of the percolation theory, we try to answer the following question:

given a certain probability of galaxies to be radiogalaxies, what is the distribution of the non-thermal sources in a such universe ?

Due to the complexity of the problems involved we will start by reproducing the galaxy distribution as given by a percolating process and then we model a radiogalaxy distribution .

2. Radiogalaxy Distribution

Suppose that galaxies form following a process similar to that adopted in Shulman and Seiden, 1987 . The starting hypotheses are:

(i) in a cube of $51 \times 51 \times 51$ cells we place a galaxy (spontaneously formed in the center) . This will induce galaxy formation in the neighbouring cells with the same probability (pc) and the procedure repeats itself ,

(ii) the number of new galaxies grows exponentially, and in order to prevent anomalous growth we stop the process when a maximum number of new galaxies is reached (new). The procedure continues due to the presence of a certain probability of spontaneous formation (ps) ,

(iii) the procedure stops once all the new galaxies are formed outside the cube ,

(iv) every cell is allowed to have only one galaxy .

We introduce a correlation coefficient :

$$\xi(r) = \frac{n(r) - n_0(r_f)}{n_0(r_f)} \quad , \tag{1}$$

here r is the distance from the center and takes the value r_f on the face and n (r) the density of galaxies in a volume $(2r)^3$. Plotting $\log\xi$ (r) versus log r we obtain a straight line of slope -1 , that becomes steeper due to the presence of the boundaries.

M. Mezzetti et al. (eds.), Large Scale Structure and Motions in the Universe, 461–462.

In order to study the behavior of the curve, we perform a simple test of linear regression with dependent variable Log ξ and independent variable Log r :

$$Log\,\xi = a + b\,Log\,r \quad . \tag{2}$$

The regression coefficient , b , and the regression constant , a , together the Pearson product-moment correlation coefficient rr , are reported in the table .

pc	ps	new	rr	a	b
0.251	0.05	1000	-0.86	1.24	-1.06
0.251	0.01	1000	-0.89	1.59	-1.26
0.251	0.005	1000	-0.94	1.65	-1.41
0.251	0.002	1000	-0.93	1.90	-1.59
0.251	0.001	1000	-0.92	2.30	-1.85

In the absence of clear rules on the formation of radiogalaxies we introduce a certain probability , pr , of the galaxies to be radiosources ; these randomly selected sources could be projected on the face of the cube , see fig. 1 .

References

Charlton,J.,C.,Schramm,D.N. : 1986 , *Astrophys. J.* **310**, 26

Grooth,E.J., Peebles , P.J.E. 1977 , *Astrophys. J.* **217**,385

Miller,R.,H. 1983, *Astrophys. J.* **270** , 390

Schulman,L.S.,Seiden,P.E. : 1986 ,*Astrophys. J.* **311**, 1

Soneira,R.M.,Peebles,P.J.E. 1978 ,*Astron. J.* **83**, 45

Fig. 1 The radiogalaxy distribution arising from the simulation . The three views are projections along the three axes of the cube . Every cell is represented by a square proportional to the number of objects on the line of sight . Given a coordinate system x , y, z the plane considered is x=0 .

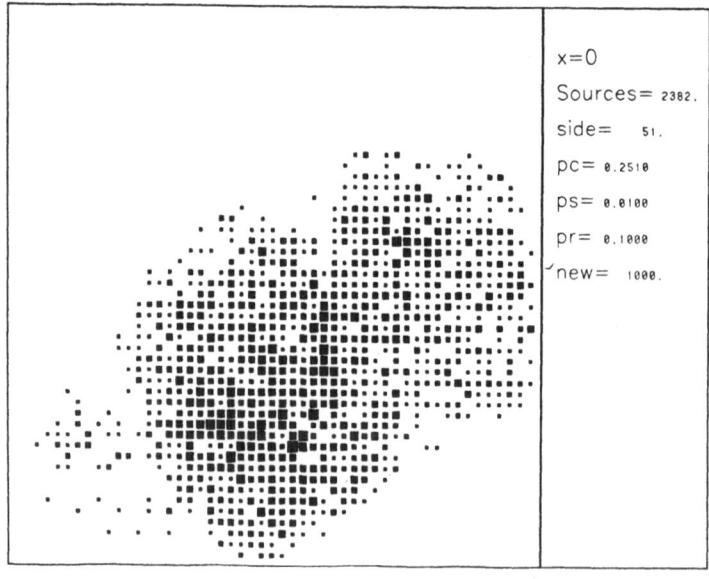

x=0
Sources= 2382.
side= 51.
pc= 0.2510
ps= 0.0100
pr= 0.1000
new= 1000.

INDEX OF NAMES